Practical Electronics
for Inventors

Practical Electronics for Inventors

Paul Scherz

McGraw-Hill

New York   San Francisco   Washington, D.C.   Auckland   Bogotá
Caracas   Lisbon   London   Madrid   Mexico City   Milan
Montreal   New Delhi   San Juan   Singapore
Sydney   Tokyo   Toronto

Cataloging-in-Publication Data is on file with the Library of Congress

McGraw-Hill

A Division of The McGraw·Hill Companies

6 7 8 9 10 11   QWK   0 9 8 7 6 5 4 3 2

ISBN 0-07-058078-2

The sponsoring editor for this book was Scott Grillo, the editing supervisor was Steven Melvin,
and the production supervisor was Sherri Souffrance.
It was set in Palatino by North Market Street Graphics.
Phoenix Color was printer and binder.

 This book is printed on recycled, acid-free paper containing a minimum of
50% recycled, de-inked fiber.

CONTENTS

CHAPTER 8   Filters   **247**

CHAPTER 9   Oscillators and Timers   **267**

PREFACE

Inventors in the field of electronics are individuals who possess the knowledge, intuition, creativity, and technical know-how to turn their ideas into real-life electrical gadgets. It is my hope that this book will provide you with an intuitive understanding of the theoretical and practical aspects of electronics in a way that fuels your creativity.

What Makes This Book Unique

Balancing the Theory with the Practical (Chapter Format)

A number of electronics books seem to throw a lot of technical formulas and theory at the reader from the start before ever giving the reader an idea of what a particular electrical device does, what the device actually looks like, how it compares with other devices similar to it, and how it is used in applications. If practical information is present, it is often toward the end of the chapter, and by this time, the reader may have totally lost interest in the subject or may have missed the "big picture," confused by details and formulas. *Practical Electronics for Inventors* does not have this effect on the reader. Each chapter is broken up into sections with the essential practical information listed first. A typical chapterone on junction field-effect transistors (JFETs)is outlined below.

- Basic Introduction and Typical Applications (three-lead device; voltage applied to one lead controls current flow through the other two leads. Control lead draws practically no current. Used in switching and amplifier applications.)

- Favors (*n*-channel and *p*-channel; *n*-channel JFET's resistance between its conducting lead increases with a negative voltage applied to control lead; *p*-channel uses a positive voltage instead.)

- How the JFETs Work (describes the semiconductor physics with simple drawings and captions)

- JFET Water Analogies (uses pipe/plunger/according contraption that responds to water pressure)

- Technical Stuff (graphs and formulas showing how the three leads of a JFET respond to applied voltages and currents. Important terms are defined.)

- Example Problems (a few example problems that show how to use the theory)

- Basic Circuits (current driver and amplifier circuits used to demonstrate how the two flavors of JFETs are used.)

- Practical Consideration (types of JFETs: small-signal, high-frequency, dual JFETs; voltage, current, and other important ratings and specifications, along with a sample specification table)

- Applications (complete circuits: relay driver, audio mixer, and electric field-strength meter)

By receiving the practical information at the beginning of a chapter, readers can quickly discover whether the device they are reading about is what the doctor ordered. If not, no great amount of time will have been spent and no brain cells will have been burned in the process.

Clearing up Misconceptions

Practical Electronics for Inventors aims at answering many of the often misconceived or rarely mentioned concepts in electronics such as displacement currents through capacitors, how to approach op amps, how photons are created, what impedance matching is all about, and so on. Much of the current electronics literature tends to miss many of these subtle points that are essential for a better understanding of electrical phenomena.

Worked-out Example Problems

Many electronics books list a number of circuit problems that tend to be overly simplistic or impractical. Some books provide interesting problems, but often they do not explain how to solve them. Such problems tend to be like exam problems or homework problems, and unfortunately, you have to learn the hard waysolving them yourself. Even when you finish solving such problems, you may not be able to check to see if you are correct because no answers are provided. Frustration! *Practical Electronics for Inventors* will not leave you guessing. It provides the answers, along with a detailed description showing how the problem was solved.

Water Analogies

Analogies can provide insight into unfamiliar territory. When good analogies are used to get a point across, learning can be fun, and an individual can build a unique form of intuition. *Practical Electronics for Inventors* provides the reader with numerous mechanical water analogies for electrical devices. These analogies incorporate springs, trapdoors, balloons, et cetera, all of which are fun to look at and easy to understand. Some of the notable water analogies in this book include a capacitor water analogy, various transistor water analogies, and an operational amplifier water analogy.

Practical Information

Practical Electronics for Inventors attempts to show the reader the subtle tricks not taught in many conventional electronics books. For example, you will learn the difference between the various kinds of batteries, capacitors, transistors, and logic families. You will also learn how to use test equipment such as an oscilloscope and multimeter and logic probes. Other practical things covered in this book include deciphering transistor and integrated circuit (IC) labels, figuring out where to buy electrical components, how to avoid getting shocked, and places to go for more in-depth information about each subject.

Built Circuits

A reader's enthusiasm for electronics often dies out when he or she reads a book that lacks practical real-life circuits. To keep your motivation going, *Practical Electronics for Inventors* provides a number of built circuits, along with detailed explanations of how they work. A few of the circuits that are presented in this book include power supplies, radio transmitter and receiver circuits, audio amplifiers, microphone preamp circuits, infrared sensing circuits, dc motors/RC servo/stepper motor driver circuits, and light-emitting diode (LED) display driver circuits. By supplying already-built circuits, this book allows readers to build, experiment, and begin thinking up new ways to improve these circuits and ways to use them in their inventions.

How to Build Circuits

Practical Electronics for Inventors provides hands-on instruction for designing and construction circuits. There are tips on drawing schematics, using circuit simulator programs, soldering, rules on safety, using breadboards, making printed circuit boards, heat sinking, enclosure design, and what useful tools to keep handy. This book also discusses in detail how to use oscilloscopes, multimeters, and logic probes to test your circuits. Troubleshooting tips are also provided.

Notes on Safety

Practical Electronics for Inventors provides insight into how and why electricity can cause bodily harm. The book shows readers what to avoid and how to avoid it. The book also discusses sensitive components that are subject to destruction form electrostatic discharge and suggests ways to avoid harming these devices.

Interesting Side Topics

In this book I have included a few side topics within the text and within the Appendix. These side topics were created to give you a more in-depth understanding of the physics, history, or some practical aspect of electronics that rarely is presented in a conventional electronics book. For example, you will find a section on power distribution and home wiring, a section on the physics of semiconductors, and a section on the physics of photons. Other side topics include computer simulation programs, where to order electronics components, patents, injection molding, and a historical timeline of inventions and discoveries in electronics.

Who Would Find This Book Useful

This book is designed to help beginning inventors invent. It assumes little or no prior knowledge of electronics. Therefore, educators, students, and aspiring hobbyists will find this book a good initial text. At the same time, technicians and more advanced hobbyists may find this book a useful reference source.

CHAPTER I

Introduction to Electronics

Perhaps the most common predicament a newcomer faces when learning electronics is figuring out exactly what it is he or she must learn. What topics are worth covering, and in which general order should they be covered? A good starting point to get a sense of what is important to learn and in what general order is presented in the flowchart in Fig. 1.1. This chart provides an overview of the basic elements that go into designing practical electrical gadgets and represents the information you will find in this book. The following paragraphs describe these basic elements in detail.

At the top of the chart comes the theory. This involves learning about voltage, current, resistance, capacitance, inductance, and various laws and theorems that help predict the size and direction of voltages and currents within circuits. As you learn the basic theory, you will be introduced to basic passive components such as resistors, capacitors, inductors, and transformers.

Next down the line comes discrete passive circuits. Discrete passive circuits include current-limiting networks, voltage dividers, filter circuits, attenuators, and so on. These simple circuits, by themselves, are not very interesting, but they are vital ingredients in more complex circuits.

After you have learned about passive components and circuits, you move on to discrete active devices, which are built from semiconductor materials. These devices consist mainly of diodes (one-way current-flow gates), transistors (electrically controlled switches/amplifiers), and thyristors (electrically controlled switches only).

Once you have covered the discrete active devices, you move on to discrete active/passive circuits. Some of these circuits include rectifiers (ac-to-dc converters), amplifiers, oscillators, modulators, mixers, and voltage regulators. This is where things start getting interesting.

To make things easier on the circuit designer, manufacturers have created integrated circuits (ICs) that contain discrete circuits—like the ones mentioned in the last paragraph—that are crammed onto a tiny chip of silicon. The chip usually is housed within a plastic package, where tiny internal wires link the chip to external metal terminals. Integrated circuits such as amplifiers and voltage regulators are referred to as *analog devices,* which means that they respond to and produce signals

1

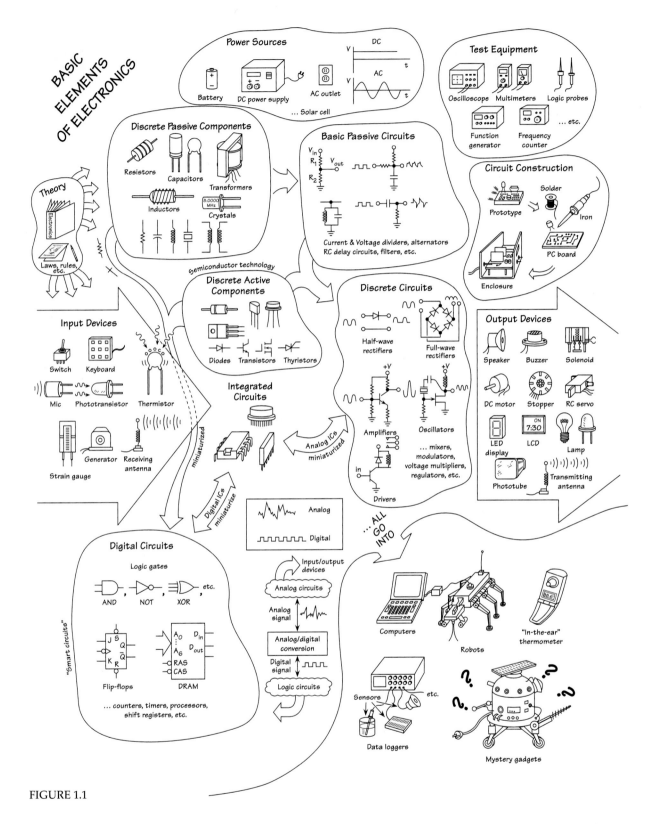

FIGURE 1.1

of varying degrees of voltage. (This is unlike *digital* ICs, which work with only two voltage levels.) Becoming familiar with integrated circuits is a necessity for any practical circuit designer.

Digital electronics comes next. Digital circuits work with only two voltage states, *high* (e.g., 5 V) or *low* (e.g., 0 V). The reason for having only two voltage states has to do with the ease of data (numbers, symbols, control information) processing and storage. The process of encoding information into signals that digital circuits can use involves combining bits (1's and 0's, equivalent to *high* and *low* voltages) into discrete-meaning "words." The designer dictates what these words will mean to a specific circuit. Unlike analog electronics, digital electronics uses a whole new set of components, which at the heart are all integrated in form. A huge number of specialized ICs are used in digital electronics. Some of these ICs are designed to perform logical operations on input information, others are designed to count, while still others are designed to store information that can be retrieved later on. Digital ICs include logic gates, flip-flops, shift registers, counters, memories, processors, and the like. Digital circuits are what give electrical gadgets "brains." In order for digital circuits to interact with analog circuits, special analog-to-digital (A/D) conversion circuits are needed to convert analog signals into special strings of 1's and 0's. Likewise, digital-to-analog conversion circuits are used to convert strings of 1's and 0's into analog signals.

Throughout your study of electronics, you will learn about various input-output (I/O) devices (transducers). Input devices convert physical signals, such as sound, light, and pressure, into electrical signals that circuits can use. These devices include microphones, phototransistors, switches, keyboards, thermistors, strain gauges, generators, and antennas. Output devices convert electrical signals into physical signals. Output devices include lamps, LED and LCD displays, speakers, buzzers, motors (dc, servo, stepper), solenoids, and antennas. It is these I/O devices that allow humans and circuits to communicate with one another.

And finally comes the construction/testing phase. This involves learning to read schematic diagrams, constructing circuit prototypes using breadboards, testing prototypes (using multimeters, oscilloscopes, and logic probes), revising prototypes (if needed), and constructing final circuits using various tools and special circuit boards.

CHAPTER 2

Theory

This chapter covers the basic concepts of electronics, such as current, voltage, resistance, electrical power, capacitance, and inductance. After going through these concepts, this chapter illustrates how to mathematically model currents and voltages through and across basic electrical elements such as resistors, capacitors, and inductors. By using some fundamental laws and theorems, such as Ohm's law, Kirchoff's laws, and Thevenin's theorem, the chapter presents methods for analyzing complex networks containing resistors, capacitors, and inductors that are driven by a power source. The kinds of power sources used to drive these networks, as we will see, include direct current (dc) sources, alternating current (ac) sources (including sinusoidal and nonsinusoidal periodic sources), and nonsinusoidal, nonperiodic sources. At the end of the chapter, the approach needed to analyze circuits that contain nonlinear elements (e.g., diodes, transistors, integrated circuits, etc.) is discussed.

As a note, if the math in a particular section of this chapter starts looking scary, don't worry. As it turns out, most of the nasty math in this chapter is used to prove, say, a theorem or law or is used to give you an idea of how hard things can get if you do not use some mathematical tricks. The actual amount of math you will need to know to design most circuits is surprisingly small; in fact, algebra may be all you need to know. Therefore, when the math in a particular section in this chapter starts looking ugly, skim through the section until you locate the useful, nonugly formulas, rules, etc. that do not have weird mathematical expressions in them.

2.1   Current

Current (symbolized with an I) represents the amount of electrical charge ΔQ (or dQ) crossing a cross-sectional area per unit time, which is given by

$$I = \frac{\Delta Q}{\Delta t} = \frac{dQ}{dt}$$

The unit of current is called the *ampere* (abbreviated amp or A) and is equal to one coulomb per second:

$$1\,A = 1\,C/s$$

Electric currents typically are carried by electrons. Each electron carries a charge of $-e$, which equals

$$-e = -1.6 \times 10^{-19} C$$

Benjamin Franklin's Positive Charges

Now, there is a tricky, if not crude, subtlety with regard to the direction of current flow that can cause headaches and confusion later on if you do not realize a historical convention initiated by Benjamin Franklin (often considered the father of electronics). Anytime someone says "current I flows from point A to point B," you would undoubtedly assume, from what I just told you about current, that electrons would flow from point A to point B, since they are the things moving. It seems obvious. Unfortunately, the conventional use of the term *current*, along with the symbol I used in the equations, assumes that positive charges are flowing from point A to B! This means that the electron flow is, in fact, pointing in the opposite direction as the current flow. What's going on? Why do we do this?

The answer is *convention*, or more specifically, Benjamin Franklin's convention of assigning positive charge signs to the mysterious things (at that time) that were moving and doing work. Sometime later a physicist by the name of Joseph Thomson performed an experiment that isolated the mysterious moving charges. However, to measure and record his experiments, as well as to do his calculations, Thomson had to stick with using the only laws available to him—those formulated using Franklin's positive currents. However, these moving charges that Thomson found (which he called *electrons*) were moving in the opposite direction of the conventional current I used in the equations, or moving against convention.

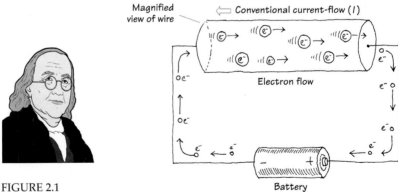

FIGURE 2.1

What does this mean to us, to those of us not so interested in the detailed physics and such? Well, not too much. I mean that we could pretend that there were positive charges moving in the wires, and electrical devices and things would work out fine. In fact, all the formulas used in electronics, such as Ohm's law ($V = IR$), "pretend" that the current I is made up of positive charge carriers. We will always be stuck with this convention. In a nutshell, whenever you see the term *current* or the symbol I, pretend that positive charges are moving. However, when you see the term *electron flow*, make sure you realize that the conventional current flow I is moving in the opposite direction.

2.2 Voltage

When two charge distributions are separated by a distance, there exists an electrical force between the two. If the distributions are similar in charge (both positive or both negative), the force is opposing. If the charge distributions are of opposite charge (one positive and the other negative), the force is attractive. If the two charge distributions are fixed in place and a small positive unit of charge is placed within the system, the positive unit of charge will be influenced by both charge distributions. The unit of charge will move toward the negatively charged distribution ("pulled" by the negatively charged object and "pushed" by the positively charged object). An *electrical field* is used to describe the magnitude and direction of the force placed on the positive unit of charge due to the charge distributions. When the positive unit of charge moves from one point to another within this configuration, it will change in potential energy. This change in potential energy is equivalent to the work done by the positive unit of charge over a distance. Now, if we divide the potential energy by the positive unit of charge, we get what is called a *voltage* (or *electrical potential*—not to be confused with electrical potential energy). Often the terms *potential* and *electromotive force* (emf) are used instead of *voltage*.

Voltage (symbolized V) is defined as the amount of energy required to move a unit of electrical charge from one place to another (potential energy/unit of charge). The unit for voltage is the *volt* (abbreviated with a V, which is the same as the symbol, so watch out). One volt is equal to one joule per coulomb:

$$1\ V = 1\ J/C$$

In terms of electronics, it is often helpful to treat voltage as a kind of "electrical pressure" similar to that of water pressure. An analogy for this (shown in Fig. 2.2) can be made between a tank filled with water and two sets of charged parallel plates.

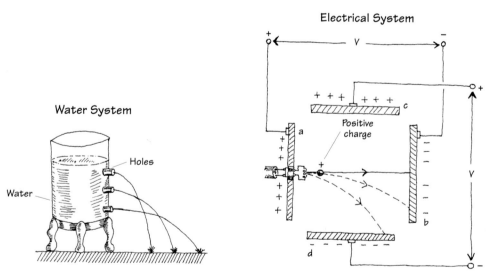

FIGURE 2.2

In the tank system, water pressure is greatest toward the bottom of the tank because of the weight from the water above. If a number of holes are drilled in the side of the tank, water will shoot out to escape the higher pressure inside. The further down the hole is drilled in the tank, the further out the water will shoot from it. The exiting beam of water will bend toward the ground due to gravity.

Now, if we take the water to be analogous to a supply of positively charged particles and take the water pressure to be analogous to the voltage across the plates in the electrical system, the positively charged particles will be drawn away from the positive plate (a) and move toward the negatively charged plate (b). The charges will be "escaping from the higher voltage to the lower voltage (analogous to the water escaping from the tank). As the charges move toward plate (b), the voltage across plates (c) and (d) will bend the beam of positive charges toward plate (d)—positive charges again are moving to a lower voltage. (This is analogous to the water beam bending as a result of the force of gravity as it escapes the tank.) The higher the voltage between plates (a) and (b), the less the beam of charge will be bent toward plate (d).

Understanding voltages becomes a relativity game. For example, to say a point in a circuit has a voltage of 10 V is meaningless unless you have another point in the circuit with which to compare it. Typically, the earth, with its infinite charge-absorbing ability and net zero charge, acts as a good point for comparison. It is considered the 0-V reference point or *ground* point. The symbol used for the ground is shown here:

FIGURE 2.3

There are times when voltages are specified in circuits without reference to ground. For example, in Fig. 2.4, the first two battery systems to the left simply specify one battery terminal voltage with respect to another, while the third system to the right uses ground as a reference point.

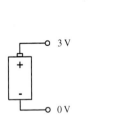 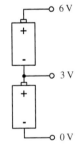 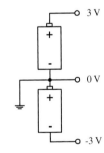

FIGURE 2.4

2.3 Resistance

Resistance is the term used to describe a reduction in current flow. All conductors intrinsically have some resistance built in. (The actual cause for the resistance can be a number of things: electron-conducting nature of the material, external heating, impurities in the conducting medium, etc.). In electronics, devices called *resistors* are specifically designed to resist current. The symbol of a resistor used in electronics is shown next:

FIGURE 2.5

If a voltage is placed between the two ends of a resistor, a current will flow through the resistor that is proportional to the magnitude of the voltage applied across it. A man by the name of Ohm came up with the following relation (called *Ohm's law*) to describe this behavior:

$$V = IR$$

R is called the *resistance* and is given in units of volts per ampere, or ohms (abbreviated Ω):

$$1\,\Omega = 1\,V/A$$

Electrical Power

Some of the kinetic energy within the electron current that runs through a resistor is converted into thermal energy (vibration of lattice atoms/ions in the resistor). The power lost to these collisions is equal to the current times the voltage. By substituting Ohm's law into the power expression, the power lost to heating can be expressed in two additional forms. All three forms are expressed in the following way:

$$P = IV = I^2R = V^2/R$$

2.4   DC Power Sources

Power sources provide the voltage and current needed to run circuits. Theoretically, power sources can be classified as *ideal voltage sources* or *ideal current sources*.

An ideal voltage source is a two-terminal device that maintains a fixed voltage drop across its terminals. If a variable resistive load is connected to an ideal voltage source, the source will maintain its voltage even if the resistance of the load changes. This means that the current will change according to the change in resistance, but the voltage will stay the same (in $I = V/R$, I changes with R, but V is fixed).

Now a fishy thing with an ideal voltage source is that if the resistance goes to zero, the current must go to infinity. Well, in the real world, there is no device that can supply an infinite amount of current. Instead, we define a *real voltage source* (e.g., a battery) that can only supply a maximum finite amount of current. It resembles a perfect voltage source with a small resistor in series.

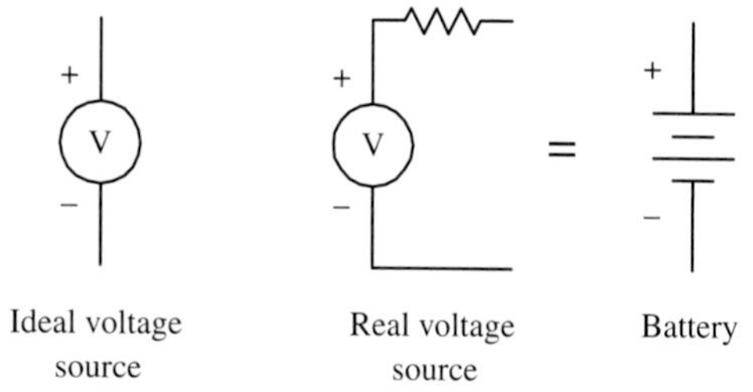

Ideal voltage   Real voltage   Battery
source   source

FIGURE 2.6

An *ideal current source* is a two-terminal idealization of a device that maintains a constant current through an external circuit regardless of the load resistance or applied voltage. It must be able to supply any necessary voltage across its terminals. Real current sources have a limit to the voltage they can provide, and they do not provide constant output current. There is no simple device that can be associated with an ideal current source.

2.5   Two Simple Battery Sources

The two battery networks shown in Fig. 2.7 will provide the same power to a load connected to its terminals. However, the network to the left will provide three times the voltage of a single battery across the load, whereas the network to the right will provide only one times the voltage of a single battery but is capable of providing three times the current to the load.

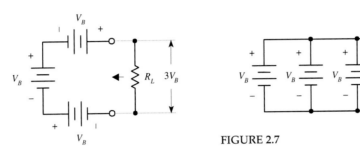

FIGURE 2.7

2.6   Electric Circuits

An electric circuit is any arrangement of resistors, wires, or other electrical components that permits an electric current to flow. Typically, a circuit consists of a voltage source and a number of components connected together by means of wires or other conductive means. Electric circuits can be categorized as *series circuits, parallel circuits,* or series and parallel combination circuits.

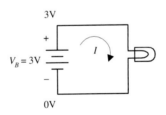

Basic Circuit

A simple light bulb acts as a load (the part of the circuit on which work must be done to move current through it). Attaching the bulb to the battery's terminals as shown to the right, will initiate current flow from the positive terminal to the negative terminal. In the process, the current will power the filament of the bulb, and light will be emitted. (Note that the term *current* here refers to conventional positive current—electrons are actually flowing in the opposite direction.)

Series Circuit

Connecting load elements (light bulbs) one after the other forms a series circuit. The current through all loads in a series circuit will be the same. In this series circuit, the voltage drops by a third each time current passes through one of the bulbs. With the same battery used in the basic circuit, each light will be one-third as bright as the bulb in the basic circuit. The effective resistance of this combination will be three times that of a single resistive element (one bulb).

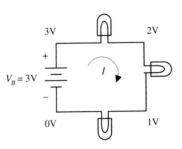

FIGURE 2.8

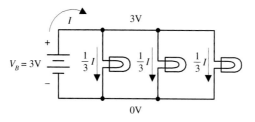

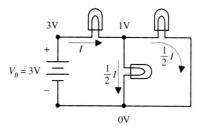

FIGURE 2.8   (*Continued*)

Parallel Circuit

A parallel circuit contains load elements that have their leads attached in such a way that the voltage across each element is the same. If all three bulbs have the same resistance values, current from the battery will be divided equally into each of the three branches. In this arrangement, light bulbs will not have the dimming effect as was seen in the series circuit, but three times the amount of current will flow from the battery, hence draining it three times as fast. The effective resistance of this combination will be one-third that of a single resistive element (one bulb).

Combination of Series and Parallel

A circuit with load elements placed both in series and parallel will have the effects of both lowering the voltage and dividing the current. The effective resistance of this combination will be three-halves that of a single resistive element (one bulb).

Circuit Analysis

Following are some important laws, theorems, and techniques used to help predict what the voltages and currents will be within a purely resistive circuit powered by a direct current (dc) source such as a battery.

2.7   Ohm's Law

Ohm's law says that a voltage difference V across a resistor will cause a current $I = V/R$ to flow through it. For example, if you know R and V, you plug these into Ohm's law to find I. Likewise, if you know R and I, you can rearrange the Ohm's law equation to find V. If you know V and I, you can again rearrange the equation to find R.

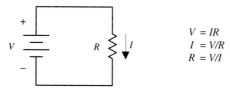

$$V = IR$$
$$I = V/R$$
$$R = V/I$$

FIGURE 2.9

2.8   Circuit Reduction

Circuits with a number of resistors usually can be broken down into a number of series and parallel combinations. By recognizing which portions of the circuit have resistors in series and which portions have resistors in parallel, these portions can be reduced to a single equivalent resistor. Here's how the reduction works.

Resistors in Series

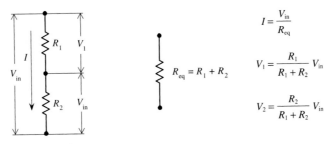

FIGURE 2.10

When two resistors R_1 and R_2 are connected in series, the sum of the voltage drops across each one (V_1 and V_2) will equal the applied voltage across the combination (V_{in}).

$$V_{in} = V_1 + V_2$$

Since the same current I flows through both resistors, we can substitute IR_1 for V_1 and IR_2 for V_2 (using Ohm's law). The result is

$$V_{in} = IR_1 + IR_2 = I(R_1 + R_2) = IR_{eq}$$

The sum $R_1 + R_2$ is called the *equivalent resistance* for two resistors in series. This means that series resistors can be simplified or reduced to a single resistor with an equivalent resistance R_{eq} equal to $R_1 + R_2$.

To find the current I, we simply rearrange the preceding equation or, in other words, apply Ohm's law, taking the voltage to be V_{in} and the resistance to be R_{eq}:

$$I = \frac{V_{in}}{R_{eq}}$$

To figure out the individual voltage drops across each resistor in series, Ohm's law is applied:

$$V_1 = IR_1 = \frac{V_{in}}{R_{eq}} R_1 = \frac{R_1}{R_1 + R_2} V_{in}$$

$$V_2 = IR_2 = \frac{V_{in}}{R_{eq}} R_2 = \frac{R_2}{R_1 + R_2} V_{in}$$

These two equations are called the *voltage divider relations*—incredibly useful formulas to know. You'll encounter them frequently.

For a number of resistors in series, the equivalent resistance is the sum of the individual resistances:

$$R = R_1 + R_2 + \cdots + R_n$$

Resistors in Parallel

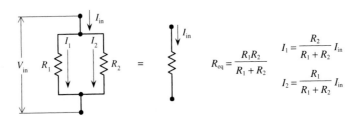

FIGURE 2.11

When two resistors R_1 and R_2 are connected in parallel, the current I_{in} divides between the two resistors in such a way that

$$I_{in} = I_1 + I_2$$

Using Ohm's law, and realizing that voltage across each resistor is the same (both V_{in}), we can substitute V_{in}/R_1 for I_1, and V_{in}/R_2 for I_2 into the preceding equation to get

$$I = \frac{V_{in}}{R_1} + \frac{V_{in}}{R_2} = V_{in}\left(\frac{1}{R_1} + \frac{1}{R_2}\right)$$

The equivalent resistance for these two resistors in series becomes

$$\frac{1}{R_{eq}} = \frac{1}{R_1} + \frac{1}{R_2} \qquad \text{or} \qquad R_{eq} = \frac{R_1 R_2}{R_1 + R_2}$$

To figure out the current through each resistor in parallel, we apply Ohm's law again:

$$I_1 = \frac{V_{in}}{R_1} = \frac{I_{in} R_{eq}}{R_1} = \frac{R_2}{R_1 + R_2} I_{in}$$

$$I_2 = \frac{V_{in}}{R} = \frac{I_{in} R_{eq}}{R_2} = \frac{R_1}{R_1 + R_2} I_{in}$$

These two equations represent what are called *current divider relations*. Like the voltage divider relations, they are incredibly useful formulas to know.

To find the equivalent resistance for a larger number of resistors in parallel, the following expression is used:

$$\frac{1}{R_{eq}} = \frac{1}{R_1} + \frac{1}{R_2} + \cdots + \frac{1}{R_n}$$

Reducing a Complex Resistor Network

To find the equivalent resistance for a complex network of resistors, the network is broken down into series and parallel combinations. A single equivalent resistance for these combinations is then found, and a new and simpler network is formed. This new network is then broken down and simplified. The process continues over and over again until a single equivalent resistance is found. Here's an example of how reduction works:

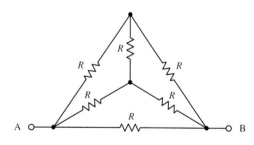

By applying circuit reduction techniques, the equivalent resistance between points A and B in this complex circuit can be found.

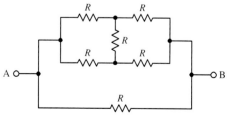

First, we redraw the circuit so that it looks a bit more familiar.

FIGURE 2.12

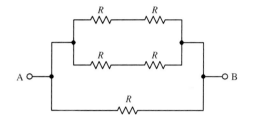

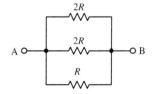

$$R/2$$
A o——/\\/\\——o B

FIGURE 2.12   *(Continued)*

Notice that the vertical resistor can be eliminated by symmetry (there is no voltage difference across the resistor in the particular case). If the resistances were not equal, we could not apply the symmetry argument, and in that case, we would be stuck—at least with the knowledge built up to now.

Next, we can reduce the two upper branches by adding resistors in series. The equivalent resistance for both these branches of resistors in series is 2R.

Finally, to find the equivalent resistance for the entire network, we use the formula for resistor in parallel:

$$\frac{1}{R_{eq}} = \frac{1}{2R} + \frac{1}{2R} + \frac{1}{R} = \frac{2}{R}$$

The final equivalent resistance becomes

$$R_{eq} = \frac{R}{2}$$

2.9   Kirchhoff's Laws

Often, you will run across a circuit that cannot be analyzed with the previous circuit reduction techniques alone. Even if you could find the equivalent resistance by using circuit reduction, it might not be possible for you to find the individual currents and voltages through and across the components within the network. For example, when circuits get complex, using the current and voltage divider equations does not work. For this reason, other laws or theorems are needed. One set of important laws is *Kirchhoff's laws.*

Kirchhoff's laws provide the most general method for analyzing circuits. These laws work for either linear or nonlinear elements, no matter how complex the circuit gets. Kirchhoff's two laws are stated as follows:

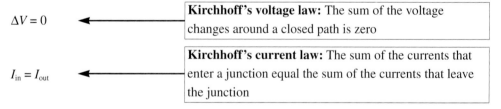

FIGURE 2.13

In essence, *Kirchhoff's voltage law* is a statement about the conservation of energy. If an electric charge starts anywhere in a circuit and is made to go through any loop in a circuit and back to its starting point, the net change in its potential is zero. *Kirchhoff's current law,* on the other hand, is a statement about the conservation of charge flow through a circuit. Here is a simple example of how these laws work.

Say you have the following circuit:

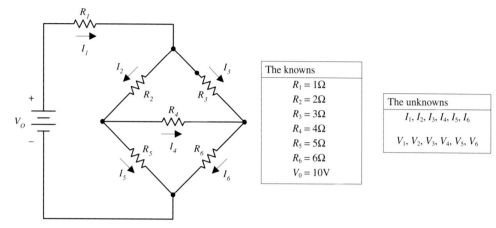

FIGURE 2.14

By applying Kirchhoff's laws to this circuit, you can find all the unknown currents I_1, I_2, I_3, I_4, I_5, and I_6, assuming that R_1, R_2, R_3, R_4, R_5, R_6, and V_0 are known. After that, the voltage drops across the resistors, and V_1, V_2, V_3, V_4, V_5, and V_6 can be found using $V_n = I_n R_n$.

To solve this problem, you apply Kirchhoff's voltage law to enough closed loops and apply Kirchhoff's current law to enough junctions so that you end up with enough equations to counterbalance the unknowns. After that, it is simply a matter of doing some algebra. Here is how to apply the laws in order to set up the final equations:

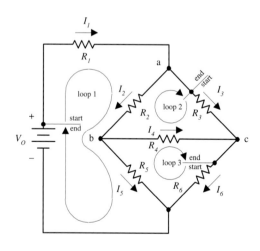

Equations resulting after applying Kirchhoff's current law:

$$I_1 = I_2 + I_3 \quad \text{(at junction a)}$$
$$I_2 = I_5 + I_4 \quad \text{(at junction b)}$$
$$I_6 = I_4 + I_3 \quad \text{(at junction c)}$$

Equations resulting after applying Kirchhoff's voltage law:

$$V_0 - I_1 R_1 - I_2 R_2 - I_5 R_5 = 0 \quad \text{(around loop 1)}$$
$$-I_3 R_3 + I_4 R_4 + I_2 R_2 = 0 \quad \text{(around loop 2)}$$
$$-I_6 R_6 + I_5 R_5 - I_4 R_4 = 0 \quad \text{(around loop 3)}$$

To determine the sign of the voltage drop across the resistors and battery used in setting up Kirchhoff's voltage equations, the convention to the left was used.

loop trace

a ——⌇⌇——• b
→ R
I

$$\Delta V_{ab} = -IR$$

loop trace

a •——⌇⌇——• b
→ R
I

$$\Delta V_{ba} = +IR$$

loop trace

a •——┤|⊢——• b
V

$$\Delta V_{ab} = V_a - V_b = +V_0$$

loop trace

a •——┤|⊢——• b
V

$$\Delta V_{ba} = V_b - V_a = -V_0$$

FIGURE 2.15

Here, there are six equations and six unknowns. According to the rules of algebra, as long as you have an equal number of equations and unknowns, you can usually figure out what the unknowns will be. There are three ways I can think of to solve for the unknowns in this case. First, you could apply the old "plug and chug" method, better known as the *substitution method,* where you combine all the equations together and try to find a single unknown and then substitute it back into another equation, and so forth. A second method, which is a lot cleaner and perhaps easier, involves using matrices. A book on linear algebra will tell you all you need to know about using matrices to solve for unknowns.

A third method that I think is useful—practically speaking—involves using a trick with determinants and Cramer's rule. The neat thing about this trick is that you do not have to know any math—that is, if you have a mathematical computer program or calculator that can do determinants. The only requirement is that you are able to plug numbers into a grid (determinant) and press "equals." I do not want to spend too much time on this technique, so I will simply provide you with the equations and use the equations to find one of the solutions to the resistor circuit problem.

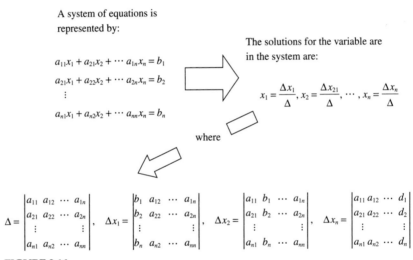

A system of equations is represented by:

$$a_{11}x_1 + a_{21}x_2 + \cdots a_{1n}x_n = b_1$$
$$a_{21}x_1 + a_{22}x_2 + \cdots a_{2n}x_n = b_2$$
$$\vdots$$
$$a_{n1}x_1 + a_{n2}x_2 + \cdots a_{nn}x_n = b_n$$

The solutions for the variable are in the system are:

$$x_1 = \frac{\Delta x_1}{\Delta}, x_2 = \frac{\Delta x_{21}}{\Delta}, \cdots, x_n = \frac{\Delta x_n}{\Delta}$$

where

$$\Delta = \begin{vmatrix} a_{11} & a_{12} & \cdots & a_{1n} \\ a_{21} & a_{22} & \cdots & a_{2n} \\ \vdots & & & \vdots \\ a_{n1} & a_{n2} & \cdots & a_{nn} \end{vmatrix}, \quad \Delta x_1 = \begin{vmatrix} b_1 & a_{12} & \cdots & a_{1n} \\ b_2 & a_{22} & \cdots & a_{2n} \\ \vdots & & & \vdots \\ b_n & a_{n2} & \cdots & a_{nn} \end{vmatrix}, \quad \Delta x_2 = \begin{vmatrix} a_{11} & b_1 & \cdots & a_{1n} \\ a_{21} & b_2 & \cdots & a_{2n} \\ \vdots & & & \vdots \\ a_{n1} & b_n & \cdots & a_{nn} \end{vmatrix}, \quad \Delta x_n = \begin{vmatrix} a_{11} & a_{12} & \cdots & d_1 \\ a_{21} & a_{22} & \cdots & d_2 \\ \vdots & & & \vdots \\ a_{n1} & a_{n2} & \cdots & d_n \end{vmatrix}$$

FIGURE 2.16

For example, you can find Δ for the system of equations from the resistor problem by plugging all the coefficients into the determinant and pressing the "evaluate" button on the calculator or computer:

$$\begin{aligned} I_1 - I_2 - I_3 &= 0 \\ I_2 - I_5 - I_4 &= 0 \\ I_6 - I_4 - I_3 &= 0 \\ I_1 + 2I_2 + 5I_5 &= 10 \\ -3I_3 + 4I_4 + 2I_2 &= 0 \\ -6I_6 + 5I_5 - 4I_4 &= 0 \end{aligned} \qquad \Delta = \begin{vmatrix} 1 & -1 & -1 & 0 & 0 & 0 \\ 0 & 1 & 0 & -1 & -1 & 0 \\ 0 & 0 & -1 & -1 & 0 & 1 \\ 1 & 2 & 0 & 0 & 5 & 0 \\ 0 & 2 & -3 & 4 & 0 & 0 \\ 0 & 0 & 0 & -4 & 5 & -6 \end{vmatrix} = -587$$

Now, to find, say, the current through R_5 and the voltage across it, you find ΔI_5 and then use $I_5 = \Delta I_5/\Delta$ to find the current. Then you use Ohm's law to find the voltage. Here is how it is done:

$$\Delta I_5 = \begin{vmatrix} 1 & -1 & -1 & 0 & 0 & 0 \\ 0 & 1 & 0 & -1 & 0 & 0 \\ 0 & 0 & -1 & -1 & 0 & 1 \\ 1 & 2 & 0 & 0 & 10 & 0 \\ 0 & 2 & -3 & 4 & 0 & 0 \\ 0 & 0 & 0 & -4 & 0 & -6 \end{vmatrix} = -660$$

$$I_5 = \frac{\Delta I_5}{\Delta} = \frac{-660}{-587} = 1.12 \text{ A}$$

$$V_5 = I_5 R_5 = (1.12 \text{ A})(5 \text{ }\Omega) = 5.6 \text{ V}$$

To solve for the other currents, simply find the other ΔI's and divide by Δ.

However, before you get too gung-ho about playing around with systems of equations, you should look at a special theorem known as *Thevenin's theorem*. Thevenin's theorem uses some very interesting tricks to analyze circuits, and it may help you avoid dealing with systems of equations.

2.10   Thevenin's Theorem

Say that you are given a complex circuit such as that shown in Fig. 2.17. Pretend that you are only interested in figuring out what the voltage will be across terminals A and F (or any other set of terminals, for that matter) and what amount of current will flow through a load resistor attached between these terminals. If you were to apply Kirchhoff's laws to this problem, you would be in trouble—the amount of work required to set up the equations would be a nightmare, and then after that, you would be left with a nasty system of equations to solve.

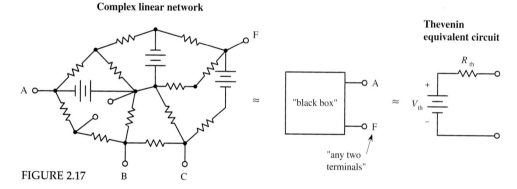

FIGURE 2.17

Luckily, a man by the name of Thevenin came up with a theorem, or trick, to simplify the problem and produce an answer—one that does not involve "hairy" mathematics. Using what Thevenin discovered, if only two terminals are of interest, these two terminals can be extracted from the complex circuit, and the rest of the circuit can be considered a "black box." Now the only things left to work with are these two terminals. By applying Thevenin's tricks (which you will see in a second), you will discover that this black box, or any linear two-terminal dc network, can be represented by a voltage source in series with a resistor. (This statement is referred to as *Thevenin's theorem*.) The voltage source is called the *Thevenin voltage* (V_{th}), and the resistance of the resistor is called the *Thevenin resistance* (R_{th}); the two together form what is called the *Thevenin equivalent circuit*. From this simple equivalent circuit you can easily calculate the current flow through a load placed across its terminals by using $I = V_{th}/(R_{th} + R_{load})$.

Now it should be noted that circuit terminals (black box terminals) actually may not be present in a circuit. For example, instead, you may want to find the current and voltage across a resistor (R_{load}) that is already within a complex network. In this case, you must remove the resistor and create two terminals (making a black box) and then find the Thevenin equivalent circuit. After the Thevenin equivalent circuit is found, you simply replace the resistor (or place it across the terminals of the Thevenin equivalent circuit), calculate the voltage across it, and calculate the current through it by applying Ohm's law [$I = V_{th}/(R_{th} + R_{load})$] again.

However, two important questions remain: What are the tricks? and What are V_{th} and R_{th}? V_{th} is simply the voltage across the terminals of the black box, which can be either measured or calculated. R_{th} is the resistance across the terminals of the black box when all the batteries within it are shorted (the sources are removed and replaced with wires), and it too can be measured or calculated.

Perhaps the best way to illustrate how Thevenin's theorem works is to go through a couple of examples.

EXAMPLE I

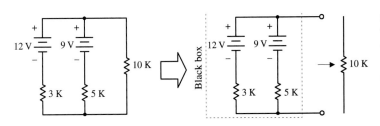

Say you are interested in the voltage across the 10kΩ resistor and the current that will run through this circuit.

First, you remove the resistor to free up a set of terminals, making the black box.

Finding V_{th},

CASE 1:   Calculate the voltage across the terminals. This is accomplished by finding the loop current I, taking the sources to be opposing:

$$I = \frac{12\,\text{V} - 9\,\text{V}}{3\,\text{k}\Omega + 5\,\text{k}\Omega} = 0.375\ \text{mA}$$

Then, to find V_{th}, you can used either of these expressions:

$12\,\text{V} - (0.375\ \text{mA})(3\ \text{k}\Omega) = 10.875\,\text{V}$
or
$9\,\text{V} - (0.375\ \text{mA})(5\ \text{k}\Omega) = 10.875\,\text{V}$

CASE 2:   Simply measure the V_{th} across terminals using a voltmeter.

Finding R_{th},

CASE 1:   Short the internal voltage supplies (batteries) with a wire, and calculate the equivalent resistance of the shorted circuit. R_{th} becomes equivalent to two resistors in parallel:

$$\frac{1}{R_{th}} = \frac{1}{3\ \text{k}\Omega} + \frac{1}{5\ \text{k}\Omega}?\qquad R_{th} = 1875\ \Omega$$

CASE 2:   Simply measure R_{th} across the terminals using an ohmmeter. (*Note:* Do not short a real battery—this will kill it. Remove the battery, and replace it with a wire instead.)

Finding V_{th}

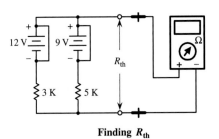

Finding R_{th}

FIGURE 2.18

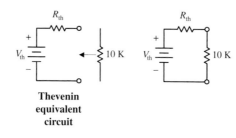

Now that V_{th} and R_{th} are known, the equivalent Thevenin circuit can be made. Using the Thevenin equivalent circuit, reattach the 10-kΩ resistor and find the voltage across it (V_R). To find the current through the resistor (I_R), use Ohm's law:

$$V_R = \frac{R}{R + R_{th}} V_{th}$$

$$V_R = \frac{10\ k\Omega}{10\ k\Omega + 1.875\ k\Omega} (12.875\ V) = 10.842\ V$$

$$I_R = \frac{V_R}{R} = \frac{10.842\ V}{10\ k\Omega} = 1.084\ mA$$

FIGURE 2.18 (*Continued*)

EXAMPLE 2

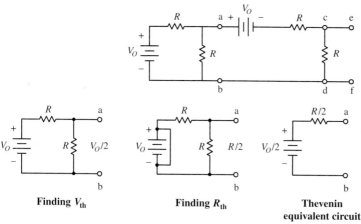

First, find the Thevenin equivalent circuit for everything to the left of points (a) and (b). Remove the "load," or everything to the right of points (a) and (b), and then find V_{th} and R_{th} across points (a) and (b). Here you use the voltage divider to find V_{th}:

$$V_{th} = \frac{R}{R + R} V_0 = \frac{V_0}{2}$$

To find R_{th}, short the battery, and find the equivalent resistance for two resistors in parallel:

$$R_{th} = \frac{R \times R}{R + R} = \frac{R^2}{2R} = \frac{R}{2}$$

Now let's find the Thevenin equivalent circuit of everything to the left of (c) and (d). Here, you can use the Thevenin equivalent circuit for everything to the left of (a) and (b) and insert it into the circuit. To calculate the new V_{th} and R_{th}, you apply Kirchhoff's law:

$$V_{th} = \frac{V_0}{2} - V_0 = -\frac{V_0}{2}$$

(Note that since no current will flow through a resistor in this situation, there will be no voltage drop across the resistor, so in essence, you do not have to worry about the resistors.)

To find R_{th}, you short the batteries and calculate the equivalent resistance, which in this case is simply two resistors in series:

$$R_{th} = \frac{R}{2} + R = \frac{3}{2} R$$

Now let's find the voltage across points (*e*) and (*f*). Simply take the preceding Thevenin equivalent circuit for everything to the left of (*c*) and (*d*) and insert it into the remaining circuit. To find the voltage across (*e*) and (*f*), you use the voltage divider:

$$V_{ef} = \left(\frac{R}{R + \frac{3}{2}R}\right)\left(\frac{V_0}{2}\right) = -\frac{V_0}{5}$$

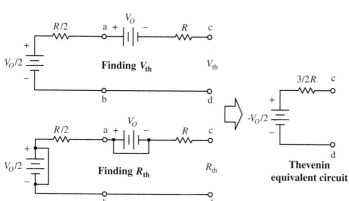

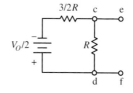

FIGURE 2.19

2.11   Sinusoidal Power Sources

A *sinusoidal power source* is a device that provides a voltage across its terminals that alternates sinusoidally with time. If a resistive load is connected between the terminals of a sinusoidal power source, a sinusoidal current will flow through the load with the same frequency as that of the source voltage (the current and voltage through and across the resistor will be in phase). Currents that alternate sinusoidally with time are called *alternating currents* or *ac currents*.

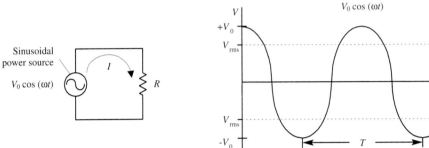

FIGURE 2.20

The voltage produced by the sinusoidal source can be expressed as $V_0 \cos(\omega t)$, where V_0 is the peak voltage [voltage when $\cos(\omega t) = 1$], and ω is the angular frequency (the rate at which the waveform progresses with time, given in radians or degrees per second). You could use $V_0 \sin(\omega t)$ to express the source voltage too—there is no practical difference between the two, except for where $t = 0$ is defined. For now, stick to using $V_0 \cos(\omega t)$—it happens to work out better in the calculations that follow.

The current through the resistor can be found by substituting the source voltage $V_0 \cos(\omega t)$ into V in Ohm's law:

$$I = \frac{V}{R} = \frac{V_0 \cos(\omega t)}{R}$$

The period (time it takes for the wave pattern to repeat itself) is given by $T = 2\pi/\omega = 1/f$, where f is the cycling frequency [number of cycles (360 degrees or 2π radians) per second].

By plugging $V_0 \cos(\omega t)$ into the power expression $P = IV = I^2 R$, the instantaneous power at any time can be found. Practically speaking, knowing the exact power at an instant in time is not very useful. It is more useful to figure out what the average power will be over one period. You can figure this out by summing up the instantaneous powers over one period (the summing is done by integrating):

$$\overline{P} = \frac{1}{T}\int_0^T I^2 R \, dt = \frac{1}{T}\int_0^T \frac{V_0^2 \cos^2(\omega t)}{R} \, dt = \frac{V_0^2}{2R}$$

2.12   Root Mean Square (rms) Voltage

In electronics, ac voltages typically are specified with a value equal to a dc voltage that is capable of doing the same amount of work. For sinusoidal voltages, this value is $1/\sqrt{2}$ times the peak voltage (V_0) and is called the *root mean square* or *rms voltage* (V_{rms}), given by

$$V_{rms} = \frac{V_0}{\sqrt{2}} = (0.707)V_0$$

Household line voltages are specified according to rms values. This means that a 120-V ac line would actually have a peak voltage that is $\sqrt{2}$ (or 1.414) times greater than the rms voltage. The true expression for the voltage would be $120\sqrt{2}\cos(\omega t)$, or $170\cos(\omega t)$.

Using the power law ($P = IV = V^2/R$), you can express the average power dissipated by a resistor connected to a sinusoidal source in terms of rms voltage:

$$\overline{P} = \frac{V_{rms}^2}{R}$$

2.13 Capacitors

If you take two oppositely charged parallel plates (one set to $+Q$ and the other set to $-Q$) that are fixed some distance apart, a potential forms between the two. If the two plates are electrically joined by means of a wire, current will flow from the positive plate through the wire to the negative plate until the two plates reach equilibrium (both become neutral in charge). The amount of charge separation that accumulates on the plates is referred to as the *capacitance*. Devices especially designed to separate charges are called *capacitors*. The symbol for a capacitor is shown below.

FIGURE 2.21

By convention, a capacitor is said to be charged when a separation of charge exists between the two plates, and it is said to be charged to Q—the charge on the positive plate. (Note that, in reality, a capacitor will always have a net zero charge overall—the positive charges on one plate will cancel the negative charges on the other.)

The charge Q on a capacitor is proportional to the potential difference or voltage V that exists between the two plates. The proportionality constant used to relate Q and V is called the *capacitance* (symbolized C) and is determined by the following relation:

$$Q = CV \qquad C = \frac{Q}{V}$$

C is always taken to be positive. The unit of capacitance is the *farad* (abbreviated F), and one farad is equal to one coulomb per volt:

$$1\,F = 1\,C/V$$

Typical capacitance values range from about 1 pF (10^{-12} F) to about 1000 μF (10^{-3} F).

If a capacitor is attached across a battery, one plate will go to $-Q$ and the other to $+Q$, and no current will flow through the capacitor (assuming that the battery has been attached to the capacitor for some time). This seems to make sense, since there is a physical separation between the two plates. However, if an accelerating or alternating voltage is applied across the capacitor's leads, something called a *displacement current* will flow. This displacement current is not conventional current, so to speak, but is created when one plate is charging or discharging with time. As charges move into place on one of the capacitor's plates, a changing magnetic field is produced, which, in turn, induces an electric current to flow out the other plate.

To find the displacement current, you plug $Q = CV$ into the expression for the definition of current ($I = dQ/dt$):

$$I = \frac{dQ}{dt} = \frac{d(CV)}{dt} = C\frac{dV}{dt}$$

By rearranging this equation, you can calculate the voltage across the capacitor as a function of the displacement current:

$$V = \frac{1}{C}\int I\,dt$$

Unlike a resistor, a capacitor does not dissipate energy. Instead, a capacitor stores energy in the form of electrical fields between its plates, which can be recovered later when it discharges. To find the energy W stored in a capacitor, you take both the expression for electrical power $P = IV$ and the expression for the definition of power $P = dW/dt$ and combine them; then you do some math to get W:

$$W = \int VI\,dt = \int VC\frac{dV}{dt} = \int CV\,dV = \frac{1}{2}CV^2$$

Capacitors in Parallel

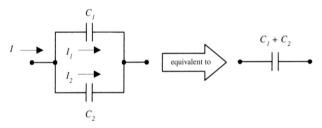

FIGURE 2.22

To find the equivalent capacitance for two capacitors in parallel, you apply Kirchhoff's current law to the junction to the left in the circuit in this figure, which gives $I = I_1 + I_2$. Making use of the fact that the voltage V is the same across C_1 and C_2, you can substitute the displacement currents for each capacitor into Kirchhoff's current expression as follows:

$$I = C_1\frac{dV}{dt} + C_2\frac{dV}{dt} = (C_1 + C_2)\frac{dV}{dt}$$

$C_1 + C_2$ is called the *equivalent capacitance* for two capacitors in parallel and is given by

$$C_{eq} = C_1 + C_2$$

When a larger number of capacitors are placed in parallel, the equivalent capacitance is found by the following expression:

$$C_{eq} = C_1 + C_2 + \cdots + C_n$$

Capacitors in Series

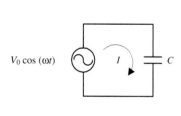

FIGURE 2.23

To find the equivalent capacitance of two capacitors in series, you apply Kirchhoff's voltage law. Since the current *I* must flow through each capacitor, Kirchhoff's voltage expression becomes

$$V = \frac{1}{C_1} \int I \, dt + \frac{1}{C_2} \int I \, dt = \left(\frac{1}{C_1} + \frac{1}{C_2} \right) \int I \, dt$$

$1/C_1 + 1/C_2$ is called the *equivalent capacitance* for two capacitors in series and is given by

$$\frac{1}{C_{eq}} = \frac{1}{C_1} + \frac{1}{C_2} \qquad \text{or} \qquad C_{eq} = \frac{C_1 C_2}{C_1 + C_2}$$

The equivalent capacitance for a larger number of capacitors in series is

$$\frac{1}{C_{eq}} = \frac{1}{C_1} + \frac{1}{C_2} + \cdots + \frac{1}{C_n}$$

2.14 Reactance of a Capacitor

A capacitor connected to a sinusoidal voltage source will allow displacement current to flow through it because the voltage across it is changing (recall that $I = C \, dV/dt$ for a capacitor). For example, if the voltage of the source is given by $V_0 \cos(\omega t)$, you can plug this voltage into *V* in the expression for the displacement current for a capacitor, which gives

$$I = C \frac{dV}{dt} = -\omega C V_0 \sin(\omega t)$$

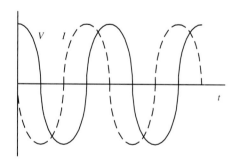

FIGURE 2.24

Maximum current or peak current I_0 occurs when $\sin(\omega t) = -1$, at which point $I_0 = \omega C V_0$. The ratio of peak voltage V_0 to peak current resembles a resistance and is given in ohms. However, because the physical phenomenon for "resisting" is different (a balance between physical plate separation and induced displacement currents) from that of a traditional resistor (heating), the effect is given the name *capacitive reactance* (abbreviated X_C) and is expressed as

$$X_C = \frac{V_0}{I_0} = \frac{V_0}{\omega C V_0} = \frac{1}{\omega C}$$

As ω goes to infinity, X_C goes to 0, and the capacitor acts like a short (wire) at high frequencies (capacitors like to pass current at high frequencies). As ω goes to 0, X_C goes to infinity, and the capacitor acts like an open circuit (capacitors do not like to pass low frequencies).

By applying a trigonometric identity to the current expression, you can express the current in terms of cosines for the purpose of comparing its behavior with that of the voltage source:

$$I = -I_0 \sin(\omega t) = I_0 \cos(\omega t + 90°)$$

When compared with the voltage, the current is out of phase by +90°, or +π/2 radians. The displacement current through a capacitor therefore leads the voltage across it by 90° (see Fig. 2.24).

2.15   Inductors

An *inductor* is an electrical component whose behavior is opposite to that of a capacitor (it "resists" changes in current while passing steady-state or dc current). An inductor resembles a springlike coil or solenoid, shown below.

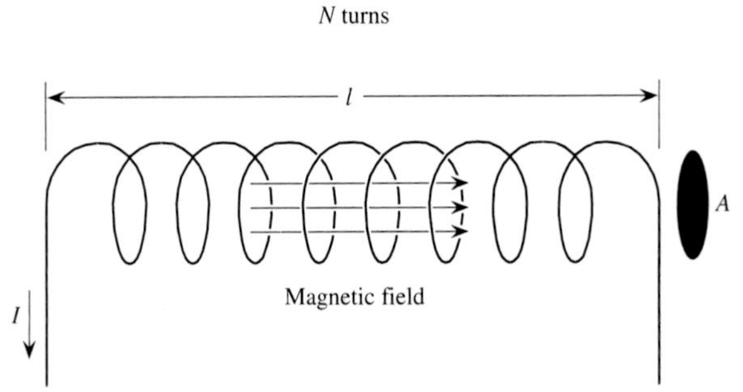

FIGURE 2.25

The inductor coil has a loop area A and a constant number of turns per unit length (N/l). If a current I flows through the inductor, by Ampere's law, a magnetic flux $\Phi = BA = \mu NIA/l$ will form, and it will point in the opposite direction as the current flow. Here μ is the permeability of the material on which the coil is wound (typically an iron or air core). In free space, $\mu = \mu_0 = 4\pi \times 10^{-7}\,\text{N}/\text{A}^2$. According to Faraday's law, the voltage induced across the terminals of the coil as a result of the magnetic fields is

$$V = N\frac{d\Phi}{dt} = \frac{\mu N^2 A}{l}\frac{dI}{dt} = L\frac{dI}{dt}$$

The constant $\mu N^2 A/l$ is called the *inductance*, which is simplified into the single constant L.

Inductance has units of *henrys* (abbreviated H) and is equal to one weber per ampere. Typical values for inductors range from 1 μH (10^{-6} H) to about 1 H.

To find the expression for the current through an inductor, you can rearrange the preceding equation to get the following:

$$I = \frac{1}{L}\int V\,dt$$

Similar to capacitors, inductors do not dissipate energy like resistors; instead, they store electrical energy in the form of magnetic fields about their coils. The energy W

stored in an inductor can be determined by using the electrical power expression $P = IV$ and the expression for the definition of power $P = dW/dt$ and then doing some math:

$$W = \int IV\, dt = \int IL\frac{dI}{dt}\, dt = \int LI\, dt = \frac{1}{2}LI^2$$

Inductors in Series

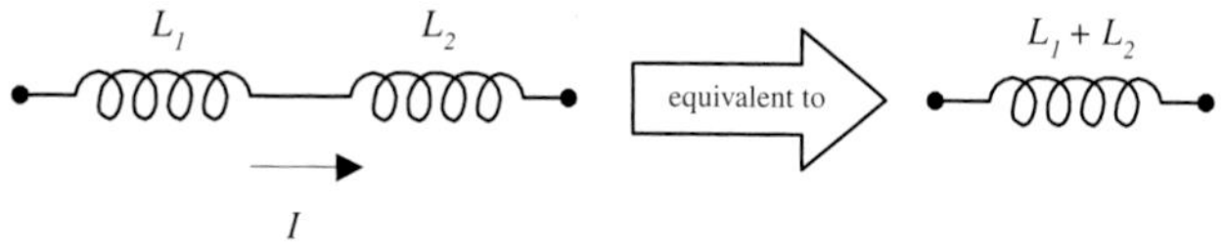

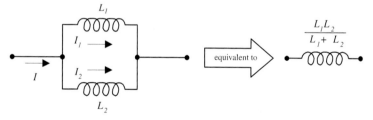

FIGURE 2.26

The equivalent inductance of two inductors in series can be found by applying Kirchhoff's voltage law. Taking the voltage drop across L_1 to be to be $L_1\, dI/dt$ and the voltage drop across L_2 to be $L_2 dI/dt$, you get the following expression:

$$V = L_1\frac{dI}{dt} + L_2\frac{dI}{dt} = (L_1 + L_2)\frac{dI}{dt}$$

$L_1 + L_2$ is called the *equivalent inductance* for two inductors in series and is given by

$$L_{eq} = L_1 + L_2$$

To find the equivalent inductance for a larger number of inductors in series, the following equation is used:

$$L_{eq} = L_1 + L_2 + \cdots + L_n$$

Inductors in Parallel

FIGURE 2.27

To find the equivalent inductance for two capacitors in parallel, you apply Kirchhoff's current law to the junction to the left in the circuit in the figure, which gives $I = I_1 + I_2$. Making use of the fact that the voltage V is the same across L_1 and L_2, current I_1 becomes $1/L_1 \int V\, dt$, and current I_2 becomes $1/L_2 \int V\, dt$. The final expression for I becomes

$$I = \frac{1}{L_1}\int V\, dt + \frac{1}{L_2}\int V\, dt = \left(\frac{1}{L_1} + \frac{1}{L_2}\right)\int V\, dt$$

$1/L_1 + 1/L_2$ is called the *equivalent inductance* for two inductors in parallel and is given by

$$\frac{1}{L_{eq}} = \frac{1}{L_1} + \frac{1}{L_2} \qquad \text{or} \qquad L_{eq} = \frac{L_1 L_2}{L_1 + L_2}$$

To find the equivalent inductance for a larger number of inductors in parallel, the following equation is used:

$$\frac{1}{L_{eq}} = \frac{1}{L_1} + \frac{1}{L_2} + \cdots + \frac{1}{L_n}$$

2.16   Reactance of an Inductor

An inductor connected to a sinusoidal voltage source will allow a current $I = 1/L \int V\, dt$ to flow through it.

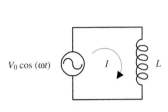
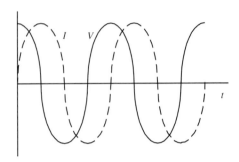

$V_0 \cos(\omega t)$ I L

FIGURE 2.28

For example, if the source voltage is given by $V_0 \cos(\omega t)$, the current through the inductor becomes

$$I = \frac{1}{L} \int V\, dt = \frac{V_0}{\omega L} \sin(\omega t)$$

The maximum current or peak current I through an inductor occurs when $\sin(\omega t) = 1$, at which point it is equal to

$$I_0 = \frac{V_0}{\omega L}$$

The ratio of peak voltage to peak current resembles a resistance and has units of ohms. However, because the physical phenomenon doing the "resisting" (self-induced backward current working against the forward current) is different from that of a traditional resistor (heating), the effect is given a new name, the *inductive reactance* (symbolized X_L), and is stated as

$$X_L = \frac{V_0}{I_0} = \frac{V_0}{V_0/\omega L} = \omega L$$

As ω goes to infinity, X_L goes to infinity, and the inductor acts like an open circuit (inductors do not like to pass high-frequency signals). However, if ω goes to 0, X_L goes to zero, and the inductor acts like a short circuit (inductors like to pass low frequencies, especially dc).

By applying a trigonometric identity, you can express the current in terms of cosines for the purpose of comparing its behavior with that of the voltage source:

$$I = I_0 \sin(\omega t) = I_0 \cos(\omega t - 90°)$$

When compared with the voltage of the source, the current is out of phase by −90°, or −π/2 radians. The current through an inductor therefore lags behind the voltage across it by 90° (see Fig. 2.28)

2.17   Fundamental Potentials and Circuits

In electronics, the four basic kinds of voltage sources include dc, sinusoidal, periodic nonsinusoidal, and nonperiodic nonsinusoidal.

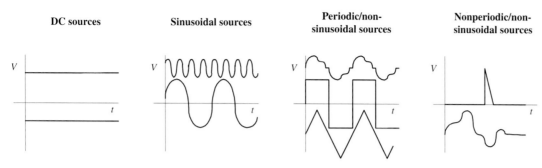

FIGURE 2.29

And the fundamental components in electronics are the resistor, capacitor, and inductor, which are characterized by the voltage and current relations shown below:

$$I = \frac{V}{R}, V = IR$$

$$I = C\frac{dV}{dt}, V = \frac{1}{C}\int I dt$$

FIGURE 2.29 *(Continued)*

$$I = \frac{1}{L}\int V dt, V = L\frac{dI}{dt}$$

Now, in theory, any complex circuit containing these components, and which is powered by any one of these potentials, can be analyzed by using Kirchhoff's laws; the voltage and current expression above and the mathematical expression for the source are plugged into Kirchhoff's voltage and current equations. However, in reality, taking this approach when modeling a circuit can become a nightmare—mathematically speaking. When the circuits get complex and the voltage and current sources start looking weird (e.g., a square wave), setting up the Kirchhoff equations and solving them can require fairly sophisticated mathematics. There are a number of tricks used in electronic analysis that prevent the math from getting out of hand, but there are situations where avoiding the nasty math is impossible. Table 2.1 should give you a feeling for the difficulties ahead.

TABLE 2.1 Math Needed to Analyze Circuits

TYPE OF CIRCUIT	DIFFICULTY	WHAT IT ENTAILS	STATUS
Dc source + resistor network	Easy to analyze	Requires simple algebra and knowing a few laws and theorems.	Already covered
Dc source + *RLC* network	Not so easy to analyze	Requires a familiarity with calculus and differential equations to truly understand what is going on.	Will be covered
Sinusoidal source + *RLC* network	Fairly easy to analyze	Is surprisingly easy and requires essentially no knowledge of calculus or differential equations; however, does require working with complex numbers.	Will be covered
Nonsinusoidal periodic source + *RLC* circuit	Hard to analyze	Requires advanced math such as calculus and Fourier series.	Will be discussed only briefly
Nonsinusoidal nonperiodic source + *RLC* circuit	Hard to analyze	Requires advanced math such as calculus, Fourier analysis, and Laplace transformations.	Will be discussed only briefly

2.18   DC Sources and *RC/RL/RLC* Circuits

Since dc sources connected to purely resistive networks have already been covered, let's now take a look at dc sources connected to circuits that also contain capacitors and inductors.

RC *Circuits*

In the following *RC* circuit, assume that the switch is initially open until time $t = 0$, at which time it is closed and left closed thereafter.

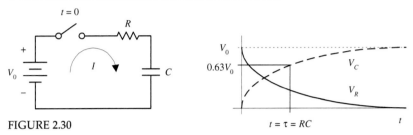

FIGURE 2.30

By applying Kirchhoff's voltage law to the circuit, you can set up the following expression to figure out how the current will behave after the switch is closed:

$$V_0 = IR + \frac{1}{C} \int I \, dt$$

To get rid of the integral, differentiate each term:

$$0 = R\frac{dI}{dt} + \frac{1}{C}I \quad \text{or} \quad \frac{dI}{dt} + \frac{1}{RC}I = 0$$

This expression is a linear, first-order, homogeneous differential equation that has the following solution:

$$I = I_0 e^{-t/RC}$$

The *RC* term in the equation is called the *RC time constant* ($\tau = RC$).

    With the solution for the current now known, the voltage across both the resistor and the capacitor can be found by substituting it into Ohm's law to find V_R and likewise substituting it into the general voltage expression of a capacitor to find V_C:

$$V_R = IR = V_0 e^{-t/RC}$$

$$V_C = \frac{1}{C} \int_0^t I \, dt = V_0 (1 - e^{-t/RC})$$

The graph of V_R and V_C versus time is shown above. Note that when $t = \tau = RC$, V_C reaches $0.632\, V_o$, or 63.2% the maximum voltage.

RL *Circuit*

In the following *RL* circuit, assume that the switch is open until time $t = 0$, at which time it is closed and is left closed thereafter.

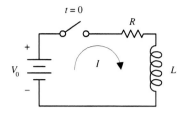
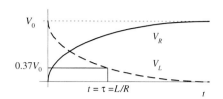

FIGURE 2.31

At the moment the switch is closed and thereafter you can use Kirchhoff's voltage law to come up with an expression for how the current will behave:

$$V_0 = IR + L\frac{dI}{dt} \quad \text{or} \quad \frac{dI}{dt} + \frac{R}{L}I = \frac{V_0}{L}$$

This expression is a linear, first-order, nonhomogeneous differential equation that has the following solution:

$$I = \frac{V_0}{R}(1 - e^{-Rt/L})$$

The L/R term is called the *RL time constant* ($\tau = L/R$).

With the solution for the current now known, the voltage across both the resistor and inductor can be found by substituting it into Ohm's law to find V_R and likewise substituting it into the general voltage expression of an inductor to find V_L:

$$V_R = IR = V_0(1 - e^{-Rt/L})$$

$$V_L = L\frac{dI}{dt} = V_0 e^{-Rt/L}$$

The graph of V_R and V_L versus time is shown above. Note that when $t = \tau = L/R$, V_L reaches $0.37\,V_0$ or 37% maximum voltage.

RLC Circuit

In the following *RLC* circuit, assume that the capacitor is initially charged and that the switch is initially open. Then, at time $t = 0$, the switch is closed and left closed thereafter.

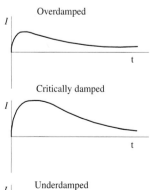
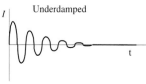

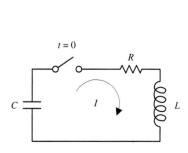

FIGURE 2.32

At the moment the switch is closed and thereafter you can apply Kirchhoff's voltage law to come up with an expression for how the current will behave:

$$\frac{1}{C}I\,dt + IR + L\frac{dI}{dt} = 0 \quad \text{or} \quad \frac{d^2I}{dt^2} + \frac{R}{L}\frac{dI}{dt} + \frac{1}{LC}I = 0$$

This expression is a linear, second-order, homogeneous differential equation that has the following solution:

$$I = \frac{V_0}{(C_1 - C_2)L}\,(e^{C_1 t} - e^{C_2 t})$$

$$C_1 = -\frac{R}{2L} + \sqrt{\frac{R^2}{4L^2} - \frac{1}{LC}} \quad C_2 = -\frac{R}{2L} - \sqrt{\frac{R^2}{4L^2} - \frac{1}{LC}}$$

Unlike with RC and RL circuits, understanding what the solution means or "visualizing the solution" is a bit more difficult. This circuit has a unique set of characteristics that are categorized as follows:

When $R^2 > 4L/C$, we get a condition within the circuit known as *overdamping*. Taking R^2 to be much greater than $4L/C$, the approximate solution for the current is

$$I \approx \frac{V_0}{R}(e^{-t/RC} - e^{-Rt/L})$$

However, when $R^2 = 4L/C$, you get a condition within the circuit called *critical damping*, which has the solution

$$I = \frac{V_0 t}{L}e^{-Rt/2L}$$

And finally, if $R^2 < 4L/C$, you get what is called *underdamping*, which has the solution

$$I = \frac{V_0}{\omega L}e^{-Rt/2L}\sin(\omega t)$$

Note that the solution for the underdamped case is oscillatory, where the angular frequency ω of oscillation can be related to the cycling frequency f by $\omega = 2\pi f$. The period of oscillation can be found using $T = 1/f = 2\pi/\omega$. In the underdamped case, electrical energy is switching between being stored in the electrical fields within the capacitor and being stored in the magnetic fields about the inductor. However, because of the resistor, energy is gradually lost to heating, so there is an exponential decay. The underdamped RLC circuit is called a *resonant circuit* because of its oscillatory behavior. A simple LC circuit, without the resistor, is also a resonant circuit. Figure 2.32 shows the overdamping, critical damping, and underdamping conditions.

2.19 Complex Numbers

Before I touch on the techniques used to analyze sinusoidally driven circuits, a quick review of complex numbers is helpful. As you will see in a moment, a sinusoidal circuit shares a unique trait with a complex number. By applying some tricks, you will be able to model and solve sinusoidal circuit problems using complex numbers and complex number arithmetic, and—this is the important point—you will be able to avoid differential equations in the process.

A *complex number* consists of two parts: a real part and an imaginary part.

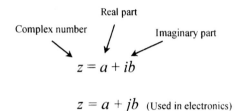

FIGURE 2.33 $z = a + jb$ (Used in electronics)

Both a and b are real numbers, whereas $i = \sqrt{-1}$ is an imaginary unit, thereby making ib an imaginary number or the imaginary part of a complex number. In practice, to avoid confusing the i (imaginary unit) with the symbol i (current). The imaginary unit i is replaced with j.

A complex number can be expressed graphically on what is called an *Argand diagram* or *Gaussian plane,* with the horizontal axis representing the real axis and the vertical axis representing the imaginary axis.

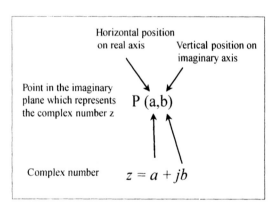

FIGURE 2.34

In terms of the drawing, a complex number can be interpreted as the vector from 0 to P having a magnitude or length of

$$r = \sqrt{a^2 + b^2}$$

and that makes an angle relative to the positive real axis of

$$\theta = \tan^{-1} \frac{b}{a}$$

Now let's go a bit further—for the complex number to be useful in circuit analysis, it must be altered slightly. If you replace a with $r\cos\theta$ and replace b with $r\sin\theta$, the complex number takes on what is called the *trigonometric* (or *polar*) *form* of a complex number.

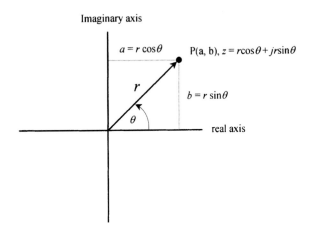

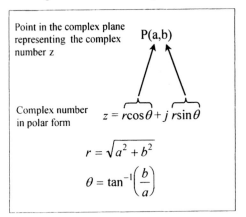

Trigonometric form of complex number

FIGURE 2.35

OK, now you are getting there, just one more thing to cover. Long ago, a man by the name of Euler noticed that the $\cos \theta + j \sin \theta$ part of the trigonometric form of the complex number was related to $e^{j\theta}$ by the following expression:

$$e^{j\theta} = \cos \theta + j \sin \theta$$

(You can prove this by taking the individual power series for $e^{j\theta}$, $\cos \theta$, and $j \sin \theta$. When the power series for $\cos \theta$ and $j \sin \theta$ are added, the result equals the power series for $e^{j\theta}$.) This means that the complex number can be expressed as follows:

$$Z = re^{j\theta}$$

Now there are three ways to express a complex number: $z = a + jb$, $z = r(\cos \theta + j \sin \theta)$, and $z = re^{j\theta}$. Each form is useful in its own way; that is, sometimes it is easier to use $z = a + jb$, and sometimes it is easier to use either $z = r(\cos \theta + j \sin \theta)$ or $z = re^{j\theta}$—it all depends on the situation.

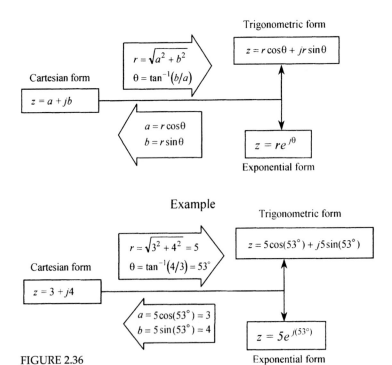

FIGURE 2.36

The model shown in Fig. 2.36 is designed to help you get a feeling for how the various forms of a complex number are related. For what follows, you will need to know the arithmetic rules for complex numbers too, which are summarized in Table 2.2.

TABLE 2.2   Arithmetic with Complex Numbers

FORM OF COMPLEX NUMBER	ADDITION/SUBTRACTION	MULTIPLICATION	DIVISION
	$z_1 \pm z_2 = (a \pm c) + j(b \pm d)$	$z_1 z_2 = (ac - bd) + j(ad - bc)$	$\dfrac{z_1}{z_2} = \dfrac{ac + bd}{c^2 + d^2} + j\left(\dfrac{bc - ad}{c^2 + d^2}\right)$
$z_1 = a + jb$ $z_2 = c + jd$	Example: $z_1 = 3 + j4, z_2 = 5 - j7$	Example: $z_1 = 5 + j2, z_2 = -4 + j3$	Example: $z_1 = 1 + j, z_2 = 3 + 2j$
Cartesian form	$z_1 + z_2 = (3 + 5) + j(4 - 7)$ $= 8 - j3$	$z_1 z_2 = [5(-4) - 2(3)] +$ $j[5(3) + 2(-4)] = -26 + j7$	$\dfrac{z_1}{z_2} = \dfrac{1(3) + 1(2)}{3^2 + 2^2} + j\left(\dfrac{1(3) - 1(2)}{3^2 + 2^2}\right)$ $= \dfrac{5}{13} + j\dfrac{1}{15}$
$z_1 = \cos\theta_1 + j\sin\theta_1$ $z_2 = \cos\theta_2 + j\sin\theta_2$ Polar form (trigonometric)	Can be done but involves using trigonometric identities; it is easier to convert this form into the Cartesian form and then add or subtract.	Can be done but involves using trigonometric identities; it is easier to convert this form into the exponential form and then multiply.	Can be done but involves using trigonometric identities; it is easier to convert this form into the exponential form and then divide.
$z_1 = r_1 e^{\theta_1}$ $z_2 = r_2 e^{\theta_2}$ Polar form (exponential)	Does not make much sense to add or subtract complex numbers in this form because the result will not be in a simplified form, except if r_1 and r_2 are equal; it is better to first convert to the Cartesian form and then add or subtract.	$z_1 z_2 = r_1 r_2 e^{j(\theta_1 + \theta_2)}$ Example: $z_1 = 5e^{j\pi}, z_2 = 2e^{j\pi/2}$ $z_1 z_2 = 5(2)e^{j(\pi + \pi/2)}$ $= 10e^{j(3\pi/2)}$	$\dfrac{z_1}{z_2} = \dfrac{r_1}{r_2} e^{j(\theta_1 - \theta_2)}$ Example: $z_1 = 8e^{j\pi}, z_2 = 2e^{j\pi/3}$ $\dfrac{z_1}{z_2} = \dfrac{8}{2} e^{j(\pi - \pi/3)} = 4e^{j(2\pi/3)}$

Here are some useful relationships to know:

$$j = \sqrt{-1}, \qquad j^2 = -1, \qquad \frac{1}{j} = -j$$
$$e^{j\pi/2} = j, \qquad e^{j\pi} = -1, \qquad e^{j3\pi/2} = -j, \qquad e^{j2\pi} = 1$$
$$\frac{1}{a + jb} = \frac{a - jb}{a^2 + b^2}$$

2.20   Circuits with Sinusoidal Sources

Suppose that you are given the following two circuits that contain linear elements (resistors, capacitors, inductors) driven by a sinusoidal voltage source.

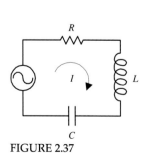

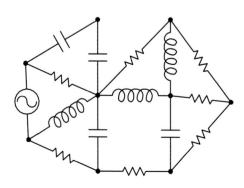

FIGURE 2.37

To analyze the simpler of the two circuits, you could apply Kirchhoff's voltage law and get the following:

$$V_0 \cos(\omega t) = IR + L\frac{dI}{dt} + \frac{1}{C}\int I\,dt$$

which reduces to

$$L\frac{d^2I}{dt^2} + R\frac{dI}{dt} + \frac{1}{C}I = -\omega V_0 \sin(\omega t)$$

This expression is a linear second-order nonhomogeneous differential equation. To find the solution to this equation, you could apply, say, the technique of variation of parameters or the method of underdetermined coefficients. After the solution for the current is found, finding the voltages across the resistor, capacitor, and inductor is a simple matter of plugging the current into the characteristic voltage/current equation for that particular component. However, coming up with the solution for the current in this case is not easy because it requires advanced math.

Now, if things were not bad enough, let's consider the more complex circuit in Fig. 2.39. To analyze this mess, you could again apply Kirchhoff's voltage and current laws to a number of loops and junctions within the circuit and then come up with a system of differential equations. The math becomes even more advanced, and finding a solution becomes ridiculously difficult.

Before I scare you too much with these differential equations, let me tell you about an alternative approach, one that does away with differential equations completely. This alternative approach makes use of what are called *complex impedances*.

2.21   Analyzing Sinusoidal Circuits with Complex Impedances

To make solving sinusoidal circuits easier, it is possible to use a technique that enables you to treat capacitors and inductors like special kinds of resistors. After that, you can analyze any circuit containing resistors, capacitors, and inductors as a "resistor" circuit. By doing so, you can apply to them all the dc circuit laws and theorems that I presented earlier. The theory behind how the technique works is a bit technical, even though the act of applying it is not hard at all. For this reason, if you do not have the time to learn the theory, I suggest simply breezing through this section and pulling out the important results. Here's a look at the theory behind complex impedances.

In a complex, linear, sinusoidally driven circuit, all currents and voltages within the circuit will be sinusoidal in nature. These currents and voltages will be changing with the same frequency as the source voltage (the physics makes this so), and their magnitudes will be proportional to the magnitude of the source voltage at any particular moment in time. The phase of the current and voltage patterns throughout the circuit, however, most likely will be shifted relative to the source voltage pattern. This behavior is a result of the capacitive and inductive effects brought on by the capacitors and inductors.

As you can see, there is a pattern within the circuit. By using the fact that the voltages and currents will be sinusoidal everywhere, and considering that the frequencies of these voltages and current will all be the same, you can come up with a mathematical trick to analyze the circuit—one that avoids differential equations. The trick involves using what is called the *superposition theorem.* The superposition theorem says that the current that exists in a branch of a linear circuit that contains several sinusoidal sources is equal to the sum of the currents produced by each source independently. The proof of the superposition theorem follows directly from the fact that Kirchhoff's laws applied to linear circuits always result in a set of linear equations that can be reduced to a single linear equation with a single unknown. The unknown branch current thus can be written as a linear superposition of each of the source terms with an appropriate coefficient.

What this all means is that you do not have to go to the trouble of calculating the time dependence of the unknown current or voltage within the circuit because you know that it will always be of the form $\cos(\omega t + \phi)$. Instead, all you need to do is calculate the peak value and phase and apply the superposition theorem. To represent currents and voltages and apply the superposition theorem, it would seem obvious to use sine or cosine functions to account for magnitude, phase, and frequency. However, in the mathematical process of superimposing (adding, multiplying, etc.), you would get messy sinusoidal expressions in terms of sines and cosines that would require difficult trigonometric rules and identities to convert the answers into something you could understand. Instead, what you can do to represent the magnitudes and phases of voltages and currents in a circuit is to use complex numbers.

Recall from the section on complex numbers that a complex number exhibits sinusoidal behavior—at least in the complex plane. For example, the trigonometry form of a complex number $z = r \cos\theta + jr \sin\theta$ will trace out a circular path in the complex plane when θ runs from 0 to 360°, or from 0 to 2π radians. If you graph the real part of z versus θ, you get a sinusoidal wave pattern. To change the amplitude of the wave pattern, you simply change the value of r. To set the frequency, you simply multiply θ by some number. To induce a phase shift relative to another wave pattern of the same frequency, you simply add some number (in degrees or radians) to θ. Now, if you replace θ with ωt, replace the r with V_0, and leave a place for a term to be added to θ (place for phase shifts), you come up with an expression for the voltage source in terms of complex numbers. You could do the same sort of thing for currents, too.

Now the nice thing about complex numbers, as compared with sinusoidal functions, is that you can represent a complex number three different ways, either in Cartesian, trigonometric, or exponential form. Having three different options makes the mathematics involved in the superimposing process easier. For example, by converting the number, say, into Cartesian form, you can easily add or subtract terms. By converting the number into exponential form, you can easily multiply and divide terms (terms in the exponent will simply add or subtract).

It should be noted that, in reality, currents and voltages are always real; there is no such thing as an imaginary current or voltage. But then, why are there imaginary parts? The answer is that when you start expressing currents and voltages with real and imaginary parts, you are simply introducing a mechanism for keeping track of the phase. (The complex part is like a hidden part within a machine; its function does not show up externally but does indeed affect the external output—the "real" or important, part as it were.) What this means is that the final answer (the result of the superimposing) always must be converted back into a real quantity. This means that after all the calculations are done, you must convert the complex result into either trigonometric or Cartesian form and then delete the imaginary part.

You may be scratching your head now and saying, "How do I *really* do the superimposing and such? This all seems too abstract or wishy-washy. How do I actually account for the resistors, capacitors, and inductors in the grand scheme of things?" Perhaps the best way to avoid this wishy-washiness is to begin by taking a sinusoidal voltage and converting it into a complex number representation. After that, you can apply it individually across a resistor, capacitor, and then an inductor and see what you get. Important new ideas and concrete analysis techniques will surface in the process.

Let's start by taking the following expression for a sinusoidal voltage

$$V_0 \cos (\omega t)$$

and converting it into a complex expression:

$$V_0 \cos (\omega t) + j V_0 \sin (\omega t)$$

What about the $jV_0\sin(\omega t)$ term? It is imaginary and does not have any physical meaning, so it does not affect the real voltage expression (you need it, however, for the superimposing process). To help with the calculations that follow, the trigonometric form is converted into the exponential form using Euler's relation $r \cos \theta + jr \sin \theta = re^{j\theta}$:

$$V = V_0 e^{j\omega t}$$

Graphically, you can represent this voltage as a vector rotating counterclockwise with angular frequency ω in the complex plane (recall that $\omega = d\theta/dt$, where $\omega = 2\pi f$). The length of the vector represents the maximum value of V, or V_0, while the projection of the vector onto the real axis represents the real part, or the instantaneous value of V, and the projection of the vector onto the imaginary axis represents the imaginary part of V.

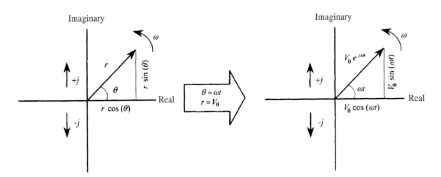

FIGURE 2.38

Now that you have an expression for the voltage in complex form, you can place, say, a resistor, a capacitor, or an inductor across the source and come up with a complex expression for the current through each device. To find the current through a resistor in complex form, you simply plug $V = V_0 e^{j\omega t}$ into V in $I = V/R$. To find the current through a capacitor, you plug $V = V_0 e^{j\omega t}$ into $I = C\,dV/dt$. Finally, to find the current through an inductor, you plug $V = V_0 e^{j\omega t}$ into $I = 1/L\int V\,dt$. The results are shown below:

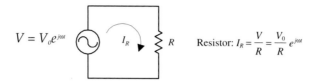

$V = V_0 e^{j\omega t}$
I_R
R
Resistor: $I_R = \frac{V}{R} = \frac{V_0}{R} e^{j\omega t}$
$V = V_0 e^{j\omega t}$
I_C
C
Capacitor: $I_C = C \frac{dV}{dt} = j\omega C V_0 e^{j\omega t}$
$V = V_0 e^{j\omega t}$
I_L
L
Inductor: $I_L = \frac{1}{L} \int V dt = \frac{V_0}{j\omega L} e^{j\omega t}$

FIGURE 2.39

Comparing the phase difference between the current and voltage through and across each component, notice the following: For a resistor, there is no phase difference; for a capacitor, there is a +90° phase difference (because of the $+j$ term); and for an inductor, there is a −90° phase difference (because $1/j = -j$). This agrees with what you learned earlier about reactances. The relationship between the phase differences of the currents and voltages can be shown graphically by using what is called a *phasor diagram*. The phasor diagram is simply a graph in the complex plane that represents a "snapshot" of the current and voltage vectors in the circuit when the time is held fixed. For example, if you set $t = 0$ in the current expression, you get the following phasor diagrams for the three circuits:

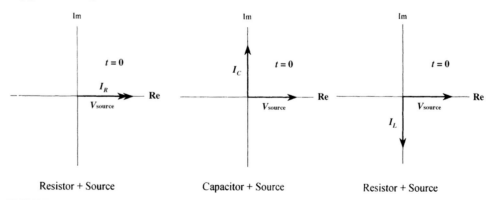

Im
t = 0
I_R
Re
V_{source}
Resistor + Source
Im
I_C
t = 0
Re
V_{source}
Capacitor + Source
Im
t = 0
Re
V_{source}
I_L
Resistor + Source

FIGURE 2.40

The phase difference between the current and voltage is represented by the angle between the I and V vectors.

If you now divide the voltage across each component by the current through it, the $e^{j\omega t}$ terms will cancel, and you get three expressions. The expression you get after dividing the complex current through a resistor by the complex source voltage is simply equal to the resistance R. The expression you get by dividing the current through a capacitor by the source voltage resembles capacitive reactance, but it is imaginary in form. The expression you get by dividing the current through an inductor by the source voltage resembles inductive reactance, but it is also imaginary in form. Since you are using complex numbers to express things, resistances and impedances are collectively referred to as *complex impedance* (symbolized with a Z).

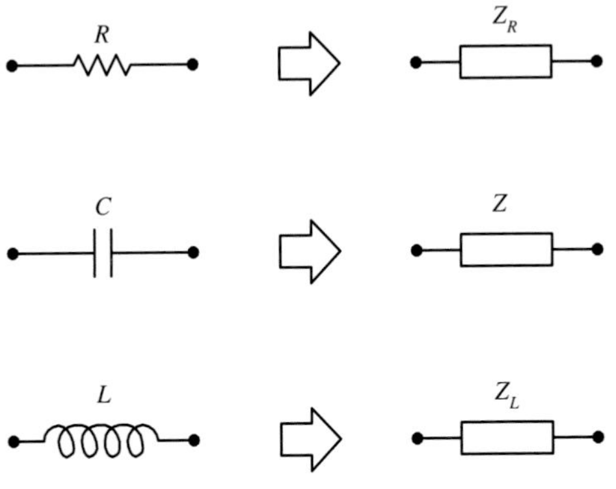

$$Z_R = \frac{V_0 e^{j\omega t}}{\frac{V_0}{R} e^{j\omega t}} = R$$

$$Z_C = \frac{V}{I_C} = \frac{V_0 e^{j\omega t}}{j\omega C V_0 e^{j\omega t}} = \frac{1}{j\omega C}$$

$$Z_L = \frac{V}{I_L} = \frac{V_0 e^{j\omega t}}{\frac{V_0}{j\omega L} e^{j\omega t}} = j\omega L$$

FIGURE 2.41

(An important point to note with impedances is that during the voltage/current division, the $e^{j\omega t}$ terms cancel. The resulting expression is simply a complex number, and it will be a function of only frequency, but not time. This is part of the trick to avoiding the differential equations.)

The nifty thing, now that you have a way to express I, V, and Z in complex form, is that you can plug these terms into all the old laws and theorems that were presented earlier. You do not have to worry about setting up time-dependent equations—the complex numbers will retain all the necessary magnitude, frequency, and phase information needed. The "looks" of the old laws, relations, and theorems will be different, however, because you will have to substitute all R's with Z's and replace dc sources with sinusoidal sources expressed in complex form.

For example, the revised form of Ohm's law (now called the *ac Ohm's law*) looks like this:

$$V(t) = I(t)Z$$

This type of revising can be applied to the other dc laws and theorems, too. Here's a look at the other laws and theorems in revised ac form.

2.22 Impedances in Series and the Voltage Divider

$$I(t) = \frac{V_{in}(t)}{Z_{eq}}$$

$$V_1(t) = \left(\frac{Z_1}{Z_1 + Z_2}\right) V_{in}(t)$$

$$V_2(t) = \left(\frac{Z_2}{Z_1 + Z_2}\right) V_{in}(t)$$

Voltage dividers

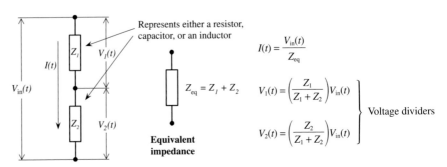

FIGURE 2.42

To find the equivalent impedance for a number of components in series, use

$$Z_{eq} = Z_1 + Z_2 + Z_3 + \cdots + Z_n \qquad \text{(series)}$$

2.23 Impedances in Parallel and the Current Divider

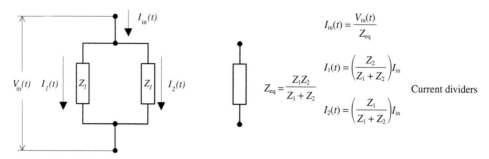

$$I_{in}(t) = \frac{V_{in}(t)}{Z_{eq}}$$

$$I_1(t) = \left(\frac{Z_2}{Z_1 + Z_2}\right) I_{in}$$

$$Z_{eq} = \frac{Z_1 Z_2}{Z_1 + Z_2}$$

Current dividers

$$I_2(t) = \left(\frac{Z_1}{Z_1 + Z_2}\right) I_{in}$$

FIGURE 2.43

To find the equivalent impedance for a larger number of components in parallel, use

$$1/Z_{eq} = 1/Z_1 + 1/Z_2 + 1/Z_3 + \cdots + 1/Z_n \qquad \text{(parallel)}$$

2.24 Applying Kirchhoff's Laws in AC Form

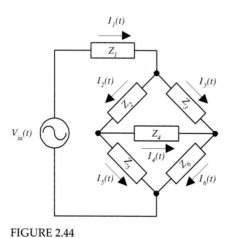

Applying Kirchhoff's current law, you get the following set of equations:

$$I_1(t) = I_2(t) + I_3(t)$$
$$I_2(t) = I_5(t) + I_4(t)$$
$$I_6(t) = I_4(t) + I_3(t)$$

Applying Kirchhoff's voltage law, you get the following set of equations:

$$V_{in}(t) - I_1(t)Z_1 - I_2(t)Z_2 - I_5(t)Z_5 = 0$$
$$-I_3(t)Z_3 + I_4(t)Z_4 + I_2(t)Z_2 = 0$$
$$-I_6(t)Z_6 + I_5(t)Z_5 - I_4(t)Z_4 = 0$$

FIGURE 2.44

Sample Problems

Modeling sinusoidal circuits that contain resistors, capacitors, and inductors is easy. However, solving the resulting equations may be a bit confusing if you are not familiar with the technical tricks needed to do arithmetic with complex numbers. The following examples are designed to show you the subtleties involved in doing the math.

EXAMPLE I (RESISTOR AND CAPACITOR IN SERIES)

Here you will find the equivalent impedance (Z_{eq}) for a resistor and capacitor in series and express the result in Cartesian, trigonometric, and exponential forms.

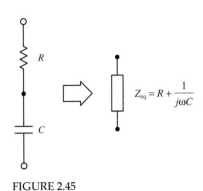

FIGURE 2.45

The equivalent impedance for a resistor and capacitor in series is

$$Z_{eq} = Z_R + Z_C$$

$$Z_{eq} = R + \frac{1}{j\omega C} = R - \frac{j}{\omega C} = R - j\left(\frac{1}{\omega C}\right)$$

To find the exponential form ($re^{j\theta}$) of Z_{eq}, you find the magnitude r and the phase θ.

The magnitude is

$$r = (a^2 + b^2)^{1/2} = \left[R^2 + \left(\frac{1}{\omega C}\right)^2\right]^{1/2}$$

The phase is

$$\theta = \tan^{-1}\left(\frac{b}{a}\right) = \tan^{-1}\left(\frac{1/\omega C}{R}\right) = -\tan^{-1}\left(\frac{1}{\omega RC}\right)$$

Hence the exponential form becomes

$$Z_{eq} = \left[R^2 + \left(\frac{1}{\omega C}\right)^2\right]^{1/2} e^{-j\tan^{-1}\left(\frac{1}{\omega RC}\right)}$$

If $R = 150\ \Omega$, $C = 0.1\ \mu F$, and $\omega = 10^4$ rad/s (the angular frequency of a source connected across the pair), the numerical value of Z_{eq} becomes

$$Z_{eq} = (150 - j2500)\ \Omega \quad \text{(Cartesian form)}$$

$$Z_{eq} = (2504\ \Omega)e^{j(-86.5°)} \quad \text{(polar exponential form)}$$

$$Z_{eq} = (2504\ \Omega)\cos(-86.5°) + j(2504\ \Omega)\sin(-86.5°) \quad \text{(trigonometric form)}$$

If a voltage source $V_0 e^{j\omega t}$ is attached across an RC series network, the current flow through the combination can be found using ac Ohm's law, substituting Z_{eq} for Z:

$$I(t) = \frac{V_0 e^{j\omega t}}{Z_{eq}} = \frac{V_0 e^{j\omega t}}{(120\ \Omega)e^{j(36.9°)}}$$

To find the individual voltages across R and C, the voltage divider can be used:

$$V_R(t) = \left(\frac{Z_R}{Z_R + Z_C}\right)V_0 e^{j\omega t} = \left(\frac{R}{Z_{eq}}\right)V_0 e^{j\omega t}$$

$$V_C(t) = \left(\frac{Z_C}{Z_R + Z_C}\right)V_0 e^{j\omega t} = \left(\frac{1/j\omega C}{Z_{eq}}\right)V_0 e^{j\omega t}$$

After finding $V_R(t)$ and $V_C(t)$, the imaginary part of the complex result must be removed simply by deleting it from the final expressions.

EXAMPLE 2 (RESISTOR AND INDUCTOR IN PARALLEL)

Here, find the equivalent impedance for a resistor and inductor in parallel and express the result in Cartesian, trigonometric, and exponential forms.

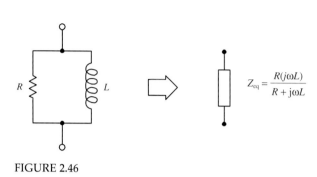

FIGURE 2.46

The equivalent impedances for a resistor and an inductor in parallel are

$$\frac{1}{Z_{eq}} = \frac{1}{Z_R} + \frac{1}{Z_L}$$

$$Z_{eq} = \frac{Z_R Z_C}{Z_R + Z_C} = \frac{R(j\omega L)}{R + j\omega L} = \frac{R(\omega L)^2 + jR^2\omega L}{R^2 + (\omega L)^2}$$

$$Z_{eq} = \frac{R(\omega L)^2}{R^2 + (\omega L)^2} + j\frac{R^2\omega L}{R^2 + (\omega L)^2}$$

To find the exponential form ($re^{j\theta}$) of Z_{eq}, we find the magnitude (r) and the phase (θ):

$$r = [a^2 + b^2]^{1/2} = \left\{ \frac{R^2(\omega L)^4 + R^4(\omega L)^2}{[R^2 + (\omega L)^2]^2} \right\}^{1/2} = \frac{R\omega L}{[R^2 + (\omega L)^2]^{1/2}}$$

$$\theta = \tan^{-1}\left(\frac{R^2\omega L}{R(\omega L)^2} \right) = \tan^{-1}\left(\frac{R}{\omega L} \right)$$

$$Z_{eq} = \frac{R\omega L}{[R^2 + (\omega L)^2]^{1/2}} e^{j\tan^{-1}(R/\omega L)}$$

If $R = 150\Omega$, and $L = 50$ mH, and $\omega = 4 \times 10^3$/s (angular frequency of a source connected across the pair), Z_{eq} becomes:

$Z_{eq} = (96 + j72)\Omega$ (Cartesian form)

$Z_{eq} = (120\Omega)e^{j(36.9°)}$ (polar exponential form)

$Z_{eq} = (120\Omega)\cos(36.9°) + j\sin(36.9°)$ (trigonometric form)

If a voltage source $V_0 e^{j\omega t}$ is connected across the RL parallel network, the individual current $I_R(t)$ and $I_L(t)$ can be found by using the current divider relation:

$$I_R(t) = \left(\frac{Z_L}{Z_R + Z_L} \right) I_{in}(t)$$

$$I_Z(t) = \left(\frac{Z_R}{Z_R + Z_L} \right) I_{in}(t)$$

where

$$I_{in}(t) = \frac{V_0 e^{j\omega t}}{Z_{eq}}$$

Again, as before, the imaginary part of the complex result must be deleted for the current to make sense in the real world.

EXAMPLE 3

The aim of this example is to find the current $I(t)$ and the individual voltages $V_R(t)$, $V_C(t)$, and $V_L(t)$ in an RLC sinusoidally driven circuit. To do this, you find the equivalent impedance for the circuit and then find $I(t)$ using Ohm's law [$I = V(t)/Z_{eq}$]. After finding $I(t)$, you can plug it into Ohm's law again to find the voltage across each device. When going through this example, note the mathematical techniques used—there are a few tricky parts that require adding two complex numbers of different forms.

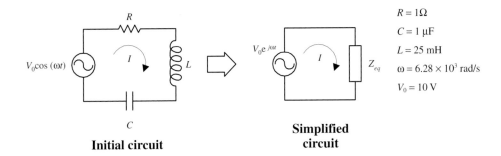

$$R = 1\,\Omega$$
$$C = 1\,\mu F$$
$$L = 25\,mH$$
$$\omega = 6.28 \times 10^3\,\text{rad/s}$$
$$V_0 = 10\,V$$

Initial circuit   **Simplified circuit**

To find the equivalent impedance for this circuit, you add impedances in series and subtitue in the known values:

$$Z_{eq} = Z_R + Z_C + Z_L = R + \frac{1}{j\omega C} + j\omega L = R + j(\omega L - 1/\omega C) = 1\,\Omega - j2.08\,\Omega$$

It will be helpful in the calculations that follow to convert Z_{eq} into exponential form ($re^{j\theta}$):

$$r = \sqrt{a^2 + b^2} = \sqrt{1^2 + (-2.08)^2} = 2.30\,\Omega$$

$$\theta = \tan^{-1}\left(\frac{b}{a}\right) = \tan^{-1}\left(\frac{2.08}{1}\right) = -64.3°$$

$$Z_{eq} = re^{j\theta} = (2.30\,\Omega)e^{j(-64.3°)}$$

To find $I(t)$, you divide the source voltage $V_0 e^{j\omega t}$ by Z_{eq}:

$$I(t) = \frac{V_0 e^{j\omega t}}{Z_{eq}} = \frac{V_0 e^{j\omega t}}{re^{j\theta}} = \frac{V_0}{r}e^{j(\omega t + \theta)} = \frac{(10\,V)}{(2.30\,\Omega)}e^{j(\omega t + 64.3°)} = (4.34\,A)e^{j(\omega t + 64.3°)}$$

To find $V_R(t)$, $V_L(t)$, and $V_C(t)$, you use $V(t) = ZI(t)$:

$$V_R(t) = RI(t) = (1\,\Omega)(4.34\,A)e^{j(\omega t + 64.3°)} = (4.34\,V)e^{j(\omega t + 64.3°)}$$

$$V_L(t) = (j\omega L)I(t) = j(4.34\,A)(157.1\,\Omega)e^{j(\omega t + 64.3°)} = (682\,V)e^{j(\omega t + 64.3° + 90°)}$$

$$= (628\,V)e^{j(\omega t + 154.3°)}$$

> Here we needed to convert $j(4.34\,A)$ into exponential form to multiply
>
> $$r = \sqrt{0^2 + (4.34\,A)^2} = 4.34\,A$$
>
> $$\theta = \tan^{-1}\left(\frac{4.34}{0}\right) = \tan^{-1}(\infty) = 90°$$
>
> $$j(4.34\,A) = 4.34e^{j(90°)}$$

$$V_C(t) = \left(\frac{1}{j\omega C}\right)I(t) = \frac{1}{j}(159.2\,\Omega)(4.34\,A)e^{j(\omega t + 64.3°)} = (691\,V)e^{j(\omega t + 64.3° - 90°)}$$

$$= (691\,V)e^{j(\omega t - 25.7°)}$$

> Here we needed to convert $1/j(159.2\,\Omega)$ into exponential form to multiply. We use $1/j = -j$:
>
> $$r = \sqrt{0^2 + (-159.2\,\Omega)^2} = 159.2\,\Omega$$
>
> $$\theta = \tan^{-1}\left(\frac{-159.2}{0}\right) = \tan^{-1}(-\infty) = -90°$$
>
> $$\frac{1}{j}(159.2\,\Omega) = (159.2\,\Omega)e^{j(-90°)}$$

FIGURE 2.47

Again, like before, $I(t)$, $V_R(t)$, $V_C(t)$, and $V_L(t)$ must be real to make sense in the real world. To make the results real, delete the imaginary parts. The end results become

$$I(t) = (4.34\,A)e^{j(\omega t + 64.3°)} = (4.34\,A)\cos(\omega t + 64.3°) = (4.34\,A)\cos(6.28 \times 10^3 t + 64.3°)$$

$$V_R(t) = (4.34\,V)\cos(\omega t + 64.3°) = (4.34\,V)\cos(6.28 \times 10^3 t + 64.3°)$$

$$V_L(t) = (628\,V)\sin(\omega t + 154.3°) = (628\,V)\sin(6.28 \times 10^3 t + 154.3°)$$

$$V_C(t) = (691\,V)\sin(\omega t + 25.7°) = (691\,V)\sin(6.28 \times 10^3 t - 25.7°)$$

To visualize the phase and magnitudes of $V_R(t)$, $V_C(t)$, $V_L(t)$, and $I(t)$ relative to the source $V(t) = V_0 e^{j\omega t}$, you can draw things out in a phasor diagram using the cartesian form of the results to plot the vectors.

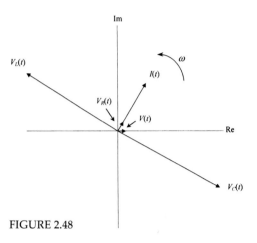

FIGURE 2.48

In terms of the phasor diagram, the real values are the vertical projections of the vectors onto the real axis. All vectors will rotate in a counterclockwise direction at an angular frequency $\omega = 6.28 \times 10^3$ rad/s. All angles between the vectors indicate the phase relationships between the voltages and currents within the circuit.

2.25   Thevenin's Theorem in AC Form

Thevenin's theorem, like the other dc theorems, can be modified so that it can be used in ac linear circuit analysis. The revised ac form of Thevenin's theorem reads: Any complex network of resistors, capacitors, and inductors can be represented by a single sinusoidal power source connected to a single equivalent impedance. For example, if you want to find the voltage across two points in a complex, linear, sinusoidal circuit or find the current and voltage through and across a particular element within, you remove the element, find $V_{th}(t)$, short the sinusoidal sources, find the Z_{th} (Thevenin impedance), and then make the Thevenin equivalent circuit. Figure 2.49 shows the Thevenin equivalent circuit for a complex circuit containing resistors, capacitors, and inductors.

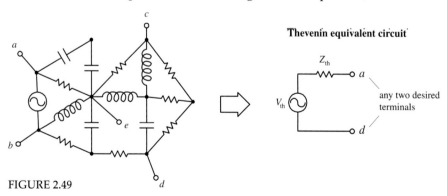

FIGURE 2.49

EXAMPLE

Here you will apply the ac Thevenin's theorem to find the current through a resistor attached to reactive network.

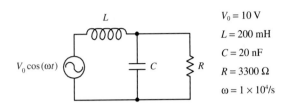

$V_0 = 10$ V

$L = 200$ mH

$C = 20$ nF

$R = 3300$ Ω

$\omega = 1 \times 10^4$/s

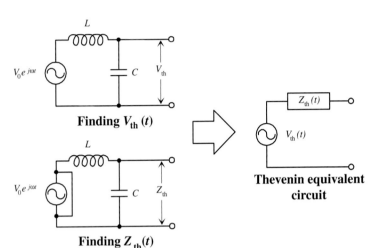

Finding V_{th} (t)

Finding Z_{th} (t)

Thevenin equivalent circuit

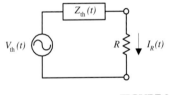

FIGURE 2.50

Suppose that you are interested in finding the current through the resistor in the circuit to the left. You can apply Thevenin's theorem to make solving this problem a breeze.

First, you remove the resistor in order to free up two terminals to make the "black box." Next, you calculate the open circuit, or Thevenin, voltage $V_{th}(t)$ by using the ac voltage divider equation:

$$V_{th}(t) = V_C(t) = \left(\frac{Z_C}{Z_C + Z_L} \right) V_0 e^{j\omega t}$$

$$= \left(\frac{1/j\omega C}{1/j\omega C + j\omega L} \right) V_0 e^{j\omega t} = \left(\frac{1}{1 - \omega^2 LC} \right) V_0 e^{j\omega t}$$

Here

$$\omega^2 LC = (10^4/\text{s})(0.200 \text{ H})(2 \times 10^{-8} \text{ F}) = 0.400$$

Hence,

$$V_{th}(t) = (16.67 \text{ V}) e^{j\omega t}$$

To find Z_{th}, you short the load with a wire and take the impedance of the inductor and capacitor in parallel:

$$Z_{th} = \frac{Z_C \times Z_L}{Z_C + Z_L} = \frac{1/j\omega C \times j\omega L}{1/j\omega C + j\omega L} = \frac{j\omega L}{1 - \omega^2 LC}$$

$$= \frac{j(10^4/\text{s})(0.200 \text{ H})}{1 - 0.4} = j(3333 \ \Omega)$$

Next, you reattach the load resistor to the Thevenin equivalent circuit and find the current by combining Z_{th} and Z_R (or R) of the resistor in series:

$$Z = R + Z_{th} = (3300 + j3333) \ \Omega = (4690 \ \Omega) e^{j(45.3°)}$$

Using Ohm's law we can find the current:

$$I_R(t) = \frac{V_{th}}{Z} = \frac{(16.67 \text{ V}) e^{j\omega t}}{(4690 \Omega) e^{j(45.3°)}} = (3.55 \text{mA}) e^{j(\omega t - 45.3°)}$$

$$I_R(t) = (3.55 \text{mA}) \cos(\omega t - 45.3°)$$

The last expression is the trigonometric form for $I_R(t)$ with the imaginary part removed.

2.26   Power in AC Circuits

In a linear sinusoidally driven circuit, how do you figure out what the average power loses will be over one cycle? You might think that you could simply use $P_{ave} = I_{rms}^2 R$ and substitute the R with Z (equivalent impedance for the entire circuit). It seems like you have applied this substitution to all the other dc laws and theorems, why not here too? Well, in this case things do not work out so nicely. The reason this substitution fails has to do with the physical fact that only resistors in a complex circuit will contribute to power loses—capacitors and inductors only temporarily store and release energy without consuming power in the process. Therefore it is necessary to figure out which part of the complex power (or apparent power) is real and which part is reactive. A simple way to find the real average power is to use the following:

$$P_{ave} = \text{Re}(VI^*) = |VI^*| \cos \phi$$

where VI^* is the complex power, and V and I are the complex rms amplitudes. The $\cos \phi$ in the last term is called the power factor, and using trigonometry, it is the ratio of the real resistance (heating) to the total impedance of the circuit. It is given by the cosine of the phase angle between the voltage applied across the equivalent impedance of the circuit and the current through the impedance. More simply, it can be considered the ratio between the real part of the impedance and the magnitude of the total impedance, as shown below:

$$\text{Power factor} = \cos \phi = \frac{\text{Re } Z}{|Z|} = \frac{\text{Re } Z}{\sqrt{(\text{Re } Z)^2 + (\text{Im } Z)^2}}$$

(Here, $\text{Re } Z = a$, and $\text{Im } Z = b$ in $Z = a + jb$.) For a purely reactive circuit (no resistors), the power factor is zero, but for a purely resistive circuit, the power factor is one.

In terms of applications, circuits that have reactive components will "steal" some of the useful current needed to power a load. To counterbalance this effect, capacitors frequently are thrown into a circuit to limit the reactive behavior and free up useful current needed to power a resistive load.

2.27 Decibels

In electronics, often you will encounter situations where you will need to compare the relative amplitudes or the relative powers between two signals. For example, if an amplifier has an output voltage that is 10 times the input voltage, you can set up a ratio $V_{out}/V_{in} = 10/1 = 10$ and give the ratio a special name—*gain*. If you have a device whose output voltage is smaller than the input voltage, the gain ratio will be less than one. In this case, you call the ratio the *attenuation*. Using ratios to make a comparison between two signals is done all the time—not only in electronics. However, there are times when the range over which the ratio of amplitudes between two signals or the ratio of powers between two signals becomes inconveniently large. For example, if you consider the range over which the human ear can perceive different levels of sound intensity, you would find that this range is very large, from about 10^{-12} to 1 W/m^2. In this case, the ratio of intensities would cover a range from 10^{-12} to 1. This large range could pose a problem if you were to attempt to make a "ratio versus something else" graph, especially if you are plotting a number of points at different ends of the scale—the resolution gets nasty. To avoid such problems, you can use a logarithmic measure. For this, the *decibel* (dB) is used.

By definition, the ratio between two signals with amplitudes A_1 and A_2 in decibels is

$$dB = 20 \log_{10} \frac{A_2}{A_1}$$

This means that $A_2/A_1 = 2$ corresponds to $20 \log_{10}(2) = 6$ dB, $A_2/A_1 = 10$ corresponds to 20 dB, $A_2/A_1 = 1000$ corresponds to 60 dB, $A_2/A_1 = 10^6$ corresponds to 120 dB, etc. If the ratio is less than 1, you get negative decibels. For example, $A_2/A_1 = 0.5$ corresponds to $20 \log_{10}(0.5) = -6$ dB, $A_2/A_1 = 0.001$ corresponds to -60 dB, $A_2/A_1 = 10^{-6}$ corresponds to -120 dB, etc.

If you are using decibels to express the ratio of powers between two signals, use

$$dB = 10 \log_{10} \frac{P_2}{P_1}$$

where P_1 and P_2 are the corresponding powers of the two signals. As long as the two signals have similar waveforms, the power expression will yield the same result (in decibels) as the amplitude expression. (P is proportional to A^2, and $\log_{10} A^2 = 2 \log_{10} A$—this provides the link between the amplitude and power expressions.)

There are instances when decibels are used to specify an absolute value of a signal's amplitude or power. For this to make sense in terms of the decibel equations, a reference amplitude or power must be provided for comparison. For example, to describe the magnitude of a voltage relative to a 1-V reference, you indicate the level in decibels by placing a V at the end of dB, giving units of dBV. In terms of the first equation above, A_1 would be the reference amplitude (1 V), while the amplitude being measured (voltage being measured) would be the A_2 term. Another common unit used is dBm. It is used when defining voltages where the reference voltage corresponds to a 1-mW power input into a load. In acoustics, dB,SPL is used to describe the pressure of one signal in terms of a reference pressure of 20 μPa.

2.28 Resonance in *LC* Circuits

When an *LC* network is driven by a sinusoidal voltage at a special frequency called the *resonant frequency*, an interesting phenomenon occurs. For example, if you drive a series *LC* circuit (shown below) at its resonant angular frequency—which happens when $\omega_0 = 1/\sqrt{LC}$—the effective impedance across the *LC* network goes to zero (acts like a short). This means that current flow between the source and ground will be at a maximum. On the other hand, if you drive a parallel *LC* circuit at its resonant angular frequency—which also happens when $\omega_0 = 1/\sqrt{LC}$—the effective impedance across the network goes to infinity (acts like an open circuit). This means that current flow between the source and ground will be zero. [In nonangular form, the resonant frequency is given by $f_0 = 1/(2\pi\sqrt{LC})$. This we get from $\omega = 2\pi f$.]

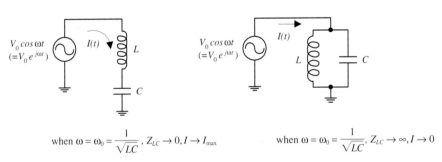

Series *LC* circuit

Parallel *LC* circuit

when $\omega = \omega_0 = \dfrac{1}{\sqrt{LC}}$, $Z_{LC} \to 0, I \to I_{\max}$ when $\omega = \omega_0 = \dfrac{1}{\sqrt{LC}}$, $Z_{LC} \to \infty, I \to 0$

FIGURE 2.51

To get an idea of how the *LC* series resonant circuit works, find the equivalent impedance of the circuit—taking L and C in series:

$$Z_{eq} = Z_L + Z_C = j\omega L + \frac{1}{j\omega C} = j\left(\omega L - \frac{1}{\omega C}\right)$$

Now ask yourself what value of ω when substituted into the preceding equation will drive Z_{eq} to zero? The answer is $1/\sqrt{LC}$—the angular resonant frequency. If we apply the ac Ohm's law, the current goes to infinity at the resonant frequency:

$$I(t) = \frac{V(t)}{Z_{eq}} = \frac{V_0 e^{j\omega t}}{0} = \infty$$

Of course, in real life, the current does not reach infinity—internal resistance and limited output source current take care of that. Now, if you do not like equations, just think that at the resonant frequency the voltages across the capacitor and inductor are exactly equal and opposite in phase. This means that the effective voltage drop across the series pair is zero; therefore, the impedance across the pair also must be zero.

To figure out how the LC parallel resonant circuit works, start by finding the equivalent impedance of the circuit—taking L and C in parallel:

$$\frac{1}{Z_{eq}} = \frac{1}{Z_L} + \frac{1}{Z_C} = \frac{1}{j\omega L} + \frac{1}{(1/j\omega C)} = j\left(\omega C - \frac{1}{\omega L}\right)$$

Rearranging this equation, you get

$$Z_{eq} = j\,\frac{1}{(1/\omega L - \omega C)}$$

If you let $\omega_0 = 1/\sqrt{LC}$, then Z_{eq} goes to infinity. This means that the current must be zero (Ohm's law):

$$I(t) = \frac{V(t)}{Z_{eq}} = \frac{V_0 e^{j\omega t}}{\infty} = 0$$

Again, if you do not like equations, consider the following: At resonance, you know that the impedance and voltage across L are equal in magnitude but opposite in phase (direction) with respect to C. From this you can infer that an equal but opposite current will flow through L and C. In other words, at one moment a current is flowing upward through L and downward through C. The current through L runs into the top of C, while the current from C runs into the bottom of the inductor. At another moment the directions of the currents reverse (energy is "bounced" back in the other direction; L and C act as an oscillator pair that passes the same amount of energy back and forth, and the amount of energy is determined by the sizes of L and C). This internal current flow around the LC loop is referred to as a *circulating current*. Now, as this is going on, no more current will be supplied through the network by the power supply. Why? The power source cannot "feel" a potential difference across the network. Another way to put this would be to say that if an external current were to be supplied through the LC network, it would mean that one of the elements (L or C) would have to be passing more current than the other. However, at resonance, this does not happen because the L and C currents are equal and pointing in opposite directions.

2.29   Resonance in *RLC* Circuits

RLC circuits, like *LC* circuits, also experience resonance effects at $\omega_0 = 1/\sqrt{LC}$. However, because you now have a resistor, a zero impedance (as you saw with the series LC circuit) or an infinite impedance (as you saw with the parallel LC circuit) is brought up or down in size. For example, in the RLC series circuit shown next, when the driving frequency reaches the resonant frequency, the LC combination acts like a short. However, because there is a resistor now, the total current flow will not be infinite. Instead, it will be equal to $V(t)/R$ (Ohm's law). Within the parallel RLC circuit, the LC section effectively will go to infinity at the resonant frequency. However, because of the resistor, a current equal to $V(t)/R$ will sneak through the resistor.

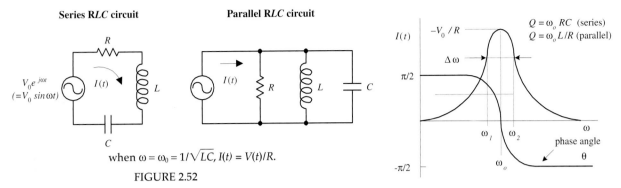

Series *RLC* circuit   Parallel *RLC* circuit

when $\omega = \omega_0 = 1/\sqrt{LC}$, $I(t) = V(t)/R$.

FIGURE 2.52

The magnitude and phase of the current (relative to the applied voltage) for both *RLC* circuits are shown in the graph in Fig. 2.52. For both circuits, when *R* gets smaller, the current "mountain peak" becomes higher and narrower in width. In electronics, the width of the peak is defined as the distance between the two *half-power points* ω_2 and ω_1. [*Half-power* means $\frac{1}{2}P_{max}$ dissipation by the resistor. It corresponds to a voltage equal to $(1/\sqrt{2})V_0$, or a current equal to $(1/\sqrt{2})I_0$.] The distance between half-power points is referred to as the *bandwidth*:

Bandwidth $= \Delta\omega = (\omega_2 - \omega_1)$

The *quality factor* is a term used to describe the sharpness of the peak on the graph in Fig. 2.53. It is defined by

$$Q = \frac{\omega_0}{\Delta\omega} = \frac{\omega_0}{\omega_2 - \omega_1}$$

For a series *RLC* circuit, $Q = \omega_0 L/R$. For a parallel *RLC* circuit, $Q = \omega_0 RC$.

2.30   Filters

By combining resistors, capacitors, and inductors in special ways, you can design circuits that are capable of passing certain frequency signals while rejecting others. This section examines four basic kinds of filters, namely, low-pass, high-pass, bandpass, and notch filters.

Low-Pass Filters

The simple *RL* and *RC* filter circuits shown in Fig. 2.53 act as low-pass filters—they pass low frequencies but reject high frequencies.

RC LOW-PASS FILTER

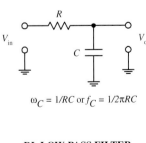

$\omega_C = 1/RC$ or $f_C = 1/2\pi RC$

RL LOW-PASS FILTER

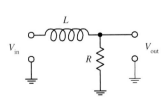

$\omega_C = R/L$ or $f_C = R/2\pi L$

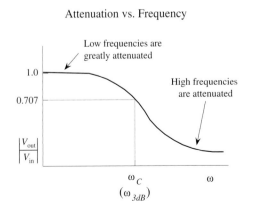

Attenuation vs. Frequency

Low frequencies are greatly attenuated

High frequencies are attenuated

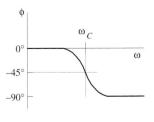

Phase Shift vs. Frequency

FIGURE 2.53

To get an idea of how these two filters work, treat both as frequency-sensitive voltage dividers—the frequency-sensitive part coming from the reactive elements (C and L). For the RC filter, when a high-frequency signal is applied to the input, the capacitor's impedance decreases. This reduction in impedance results in a leakage of input current through the capacitor to ground—the output voltage approaches the input voltage. At low frequencies, the capacitor's impedance is large; only a small amount of current passes through the capacitor to ground—the output voltage drop is small. For the RL filter, when a high-frequency signal is applied to the input, the inductor's impedance is large; very little current passes to the output—the output voltage is small. At low frequencies, the inductor's impedance decreases; most of the input current passes through the inductor—the output voltage approaches the input voltage.

In terms of the equations, you can express the relationship between the input and output voltages by setting up voltage divider relations. For the RC low-pass filter, you get the following voltage divider:

$$V_{out} = \frac{Z_C}{Z_C + R} V_{in}$$

As you will see in a moment, this expression is more useful (in terms of convention) if you rearrange it into the following form:

$$\frac{V_{out}}{V_{in}} = \frac{Z_C}{Z_L + Z_R} = \frac{1/j\omega C}{1/j\omega C + R} = \frac{1}{1 + j\omega RC}$$

To give this expression a more intuitive look, convert it into exponential form by using the following trick: Let $1 + j\omega RC = Ae^{j\alpha}$, where the magnitude is $A = \sqrt{1^2 + (\omega RC)^2}$ and the phase is $\alpha = \tan^{-1}(\omega RC)$:

$$\frac{V_{out}}{V_{in}} = \frac{1}{Ae^{j\alpha}} = \frac{1}{A} e^{-j\alpha} = \frac{1}{\sqrt{1^2 + (\omega RC)^2}} e^{-j\tan^{-1}(\omega RC)}$$

Now, in electronics, what is useful is the *magnitude of the ratio of voltages*. To find this, you simply treat the V_{out}/V_{in} term as a complex number (e.g., $Z = re^{j\theta}$) and remove the imaginary exponential term ($e^{j\theta}$) to get the magnitude r. Instead of using r to designate the magnitude of the ratio, you use straight brackets:

$$\left| \frac{V_{out}}{V_{in}} \right| = \frac{1}{\sqrt{1 + (\omega RC)^2}}$$

$|V_{out}/V_{in}|$ is called the *attenuation*. The attenuation provides a measure of how much of the input voltage gets passed to the output. For an RC low-pass filter, when ω goes to zero, the ωRC term goes to zero, and the attenuation approaches 1. This means that low frequencies are more easily passed through to the output. When ω gets large, the attenuation approaches 0. This means that high frequencies are prevented from reaching the output. When the frequency of the input signal to an RC low-pass filter reaches what is called the *cutoff frequency*, given by

$\omega_C = 1/RC$ (angular form)

$f_C = 1/2\pi RC$ (conventional form)

the output voltage drops to $1/\sqrt{2}$ the input voltage (equivalent to the output power dropping to half the input power).

There is one last thing you must consider when describing the output signal from a low-pass filter. You must determine how much the output signal is shifted in phase relative to the input signal. For an RC low-pass filter, the phase shift is $\phi = -\tan^{-1}(\omega RC/1)$, which is just the exponential term in the V_{out}/V_{in} equation presented earlier. As ω goes to 0, the phase shift goes to 0. However, as ω goes to infinity, the phase shift goes to $-90°$. The phase shift also can be written in terms of the cutoff frequency ($\omega_C = 1/RC$):

$$\phi = -\tan^{-1}(\omega RC) = -\tan^{-1}(\omega/\omega_C)$$

Notice that when $\omega = \omega_C$, the phase shift is equal to $-45°$.

To determine the attenuation and phase shift for an RL low-pass filter, you would apply the same tricks you used on the RC filter. The end result gives us

$$\left|\frac{V_{out}}{V_{in}}\right| = \frac{1}{\sqrt{1 + (\omega L/R)^2}}$$

The cutoff frequency and phase shift follow:

$$\omega_C = R/L$$
$$f_C = R/2\pi L$$

and

$$\phi = -\tan^{-1}(\omega L/R) = -\tan^{-1}(\omega/\omega_C)$$

Attenuation Expressed in Decibels

Because the range of the attenuation of a filter can take on many orders of magnitude, it is desirable to express attenuation in terms of decibels. Attenuation in decibels is given by

$$A_{dB} = 20 \log \left|\frac{V_{out}}{V_{in}}\right|$$

To get a feeling for how the decibel expression for attenuation correlates with the nondecibel expression, take a look at Fig. 2.54.

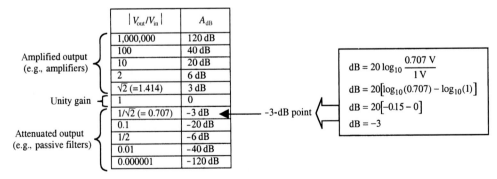

$\|V_{out}/V_{in}\|$	A_{dB}
1,000,000	120 dB
100	40 dB
10	20 dB
2	6 dB
$\sqrt{2}$ (=1.414)	3 dB
1	0
$1/\sqrt{2}$ (= 0.707)	-3 dB
0.1	-20 dB
1/2	-6 dB
0.01	-40 dB
0.000001	-120 dB

Amplified output (e.g., amplifiers)

Unity gain

Attenuated output (e.g., passive filters)

-3-dB point

$$dB = 20 \log_{10} \frac{0.707 \text{ V}}{1 \text{ V}}$$
$$dB = 20[\log_{10}(0.707) - \log_{10}(1)]$$
$$dB = 20[-0.15 - 0]$$
$$dB = -3$$

FIGURE 2.54

Notice that when $|V_{out}/V_{in}| = 1\sqrt{2}$ (the half-power condition), $A_{dB} = -3$ dB. This is called the *-3-dB point*. This point corresponds to the half-power point or cutoff frequency for a filter. Also notice that both positive and negative A_{dB} values are possible. If A_{dB} is positive, the attenuation is not actually called *attenuation*, but instead, it is referred to as the *gain*. When dealing with amplifiers (devices that have a larger output voltage and power), positive decibels are used. Negative decibels represent *true attenuation*—output voltage and power are smaller than input voltage and power. When describing passive filters, stick with negative decibels.

To describe the rate of rise or fall in attenuation as the frequency changes, descriptions such as "6 dB/octave" or "20 dB/decade" are used. Here, 6 dB/octave simply means that if the frequency changes by a factor of 2 (an octave), the attenuation changes by 6 dB. Twenty dB/decade means that if the frequency changes by a factor of 10 (a decade), as the attenuation increases or decreases by 20 dB.

High-Pass Filters

The *RC* and *RL* circuits shown below act as high-pass filters—they pass high frequencies but reject low frequencies.

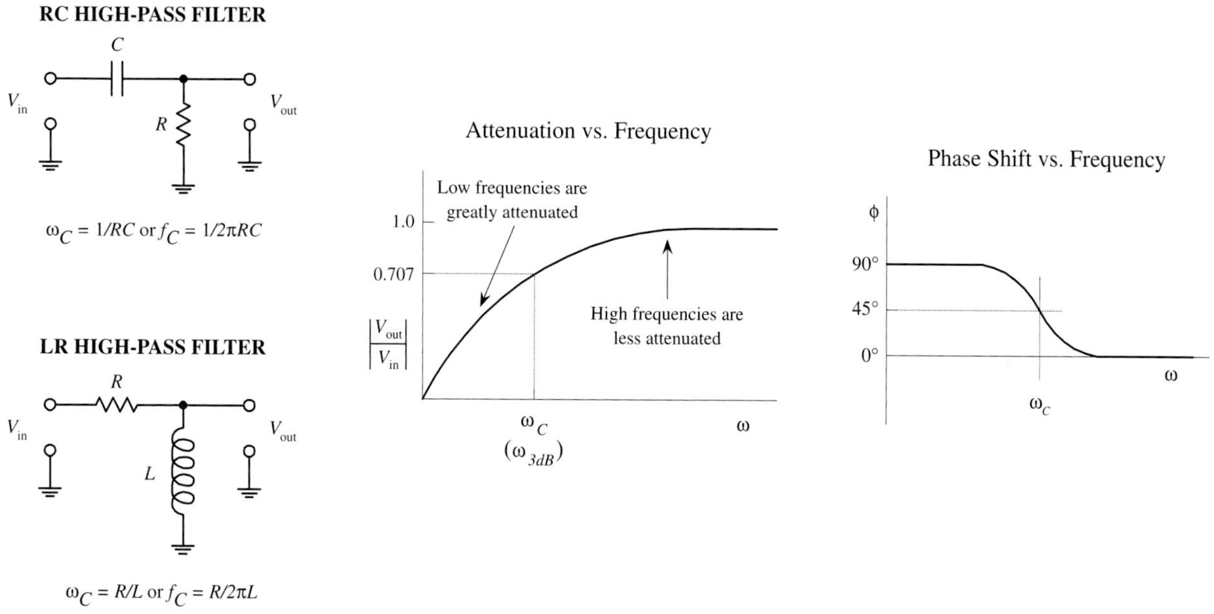

RC HIGH-PASS FILTER

$\omega_C = 1/RC$ or $f_C = 1/2\pi RC$

LR HIGH-PASS FILTER

$\omega_C = R/L$ or $f_C = R/2\pi L$

Attenuation vs. Frequency

Low frequencies are greatly attenuated

High frequencies are less attenuated

Phase Shift vs. Frequency

FIGURE 2.55

As with the low-pass filters, you can get an idea of how these two filters work by treating them as frequency-sensitive voltage dividers. For an *RC* high-pass filter, when the input frequency is high, the capacitor's impedance is small; most of the input current passes through to the output—the output voltage approaches the input voltage. At low frequencies, the capacitor's impedance is high; very little current passes to the output—the output voltage is small. For an *RL* filter, high frequencies pass to the output because the inductor's impedance is high—very little current leaks through the inductor to ground. At low frequencies, the inductor's impedance is small; most of the input current is sunk through the inductor to ground—the output voltage is small.

If you were to do the math, you would find that the attenuation for an *RC* high-pass filter would be

$$\left|\frac{V_{out}}{V_{in}}\right| = \frac{R}{\sqrt{R^2 + (1/\omega C)^2}}$$

When the ω goes to 0, the attenuation goes to 0, but if ω goes to infinity, the attenuation approaches 1. The cutoff frequency for this filter is given by

$\omega_C = 1/RC$
$f_C = 1/2\pi RC$

For an RL high-pass filter, the attenuation is given by

$\omega_C = R/L$
$f_C = R/2\pi L$

The output phase shift for both filters, expressed in terms of the cutoff frequency, is

$\phi = \tan^{-1}(\omega_C/\omega)$

As ω goes 0, ϕ goes to +90°; as ω goes to infinity, ϕ goes to 0°. At the cutoff frequency, $\omega = 45°$. See the graph in Fig. 2.55.

Bandpass Filter

By using a parallel LC resonant network as a voltage divider element (see Fig. 2.56), you can create a filter that is capable of passing a narrow band of frequencies that are near the LC network's resonant frequency (ω_0). Such a filter is referred to as a *bandpass filter*.

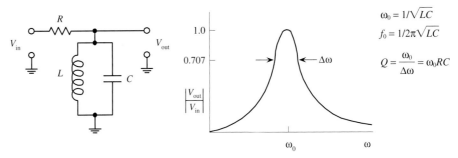

FIGURE 2.56

To figure out how this filter works, start by finding the equivalent impedance of the LC section—you take L and C in parallel:

$$\frac{1}{Z_{LC}} = \frac{1}{Z_L} + \frac{1}{Z_C} = \frac{1}{j\omega L} - \frac{\omega C}{j} = j\left(\omega C - \frac{1}{\omega L}\right)$$

Solving for Z_{LC}, you get

$$Z_{LC} = j\,\frac{1}{(1/\omega L) - \omega C}$$

As ω approaches $\omega_0 = 1/\sqrt{LC}$, the impedance of the parallel LC network approaches infinity; only input signals that are near the resonant frequency will be passed to the output (these signals will not be diverted to ground). The quality factor for this filter is the same as that of a parallel LC resonant circuit:

$$Q = \frac{\omega_0}{\Delta\omega} = \omega_0\,RC$$

Notch Filter

By using a series LC resonant network as a voltage divider element (see Fig. 2.57), you can design a filter that is capable of rejecting all frequencies except those near the resonant frequency (ω_0) of the LC series network.

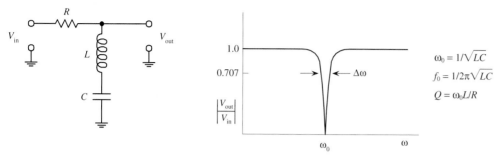

FIGURE 2.57

To figure out how this filter works, start by finding the equivalent impedance of the LC section—we take L and C in series:

$$Z_{LC} = Z_L + Z_C = j\omega L - j\frac{1}{\omega C} = j\left(\omega L - \frac{1}{\omega C}\right)$$

As ω approaches $\omega_0 = 1/\sqrt{LC}$, the impedance of the series LC network approaches zero; current is diverted through the LC network to ground—the output voltage drops. This explains the "notch" in the graph. The quality factor for this filter is the same as that of a series LC resonant circuit:

$$Q = \frac{\omega_0}{\Delta\omega} = \omega_0 L / R$$

2.31 Circuits with Periodic Nonsinusoidal Sources

Suppose that you are given a periodic nonsinusoidal source voltage (e.g., a square wave, triangle wave, ramp, etc.) that is used to drive a circuit containing resistors, capacitors, and inductors. How do you analyze the circuit? The circuit is not dc, so you cannot use dc theorems on it. The circuit is not sinusoidal, so you cannot directly apply complex impedances on it. What do you do?

If all else fails, you might assume that the only way out would be to apply Kirchhoff's laws on it. Well, before going any further, how do you mathematically represent the source voltage in the first place? That is, even if you could set up Kirchhoff's equations and such, you still have to plug in the source voltage term. For example, how do you mathematically represent a square wave? In reality, coming up with an expression for a periodic nonsinusoidal source is not so easy. However, for the sake of argument, let's pretend that you can come up with a mathematical representation of the waveform. If you plug this term into Kirchhoff's laws, you would again get differential equations (you could not use complex impedance then, because things are not sinusoidal).

To solve this dilemma most efficiently, it would be good to avoid differential equations entirely and at the same time be able to use the simplistic approach of complex impedances. The only way to satisfy both these conditions is to express a nonsinusoidal wave as a superposition of sine waves. In fact, a man by the name of Fourier discovered just such a trick. He figured out that a number of sinusoidal waves of different frequencies and amplitudes could be added together in a special manner to produce a superimposed pattern of any nonsinusoidal periodic wave pattern. More technically stated, a periodic nonsinusoidal waveform can be represented as a Fourier series of sines and cosines, where the waveform is a summation over a set of discrete, harmonically related frequencies.

For example, a square voltage can be represented mathematically by the following expression:

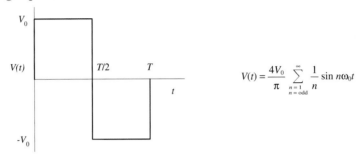

$$V(t) = \frac{4V_0}{\pi} \sum_{\substack{n=1 \\ n=\text{odd}}}^{\infty} \frac{1}{n} \sin n\omega_0 t$$

FIGURE 2.58

Taking the first three terms of the series ($n = 3$), you get the superimposed wave pattern shown in Fig. 2.59. The appearance of the superimposed wave pattern is not exactly sinusoidal at $n = 3$, but if you keep adding terms in the series, the result will appear more squarelike in shape. If you add all the terms in the series (let n go to infinity), you would indeed get a square wave pattern.

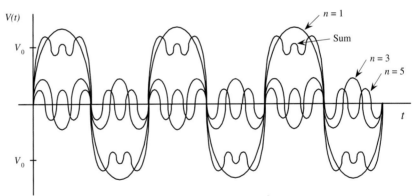

FIGURE 2.59

Other popular wave patterns such as the triangle wave and the ramp-voltage wave pattern also can be expressed in terms of a Fourier series:

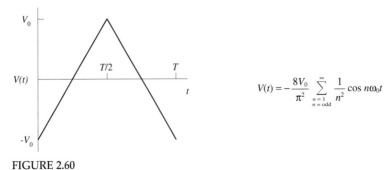

$$V(t) = -\frac{8V_0}{\pi^2} \sum_{\substack{n=1 \\ n=\text{odd}}}^{\infty} \frac{1}{n^2} \cos n\omega_0 t$$

FIGURE 2.60

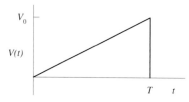

$$V(t) = \frac{V_0}{2} - \frac{V_0}{\pi} \sum_{n=1}^{\infty} \frac{1}{n} \sin (n\omega_0 t)$$

FIGURE 2.60 (*Continued*)

Now, suppose that you attach one of these nonsinusoidal voltages to an *RLC* circuit. To figure out what will happen inside the circuit (currents, voltages, phases shifts, etc.), you could substitute the series term for the voltage source into the ac theorems and laws. (You can use complex impedances to model the behavior of resistors, capacitors, and inductors in the process.) The hard part comes when you consider the series itself. How do you handle the infinite number of terms in the source voltage expression? Do you have to add an infinite number of terms? This would be impractical. Instead, you can use good old approximation techniques. For example, instead of summing to infinity, you could approximate the voltage term by only taking the first four terms in the series. These terms could then be plugged into the equations.

The actual mathematical techniques needed to create a Fourier series for a nonsinusoidal periodic waveform, as well as the math required to set up and solve the equations that contain the series, are beyond the scope of this book. There are a number of more advanced books that discuss the techniques for applying Fourier analysis to nonsinusoidal periodic circuit analysis.

2.32   Circuits with Nonperiodic Sources

A source that is nonsinusoidal as well as nonperiodic also can be represented as a superposition of sine waves like the Fourier series. But now, instead of summing over a set of discrete, harmonically related frequencies, the waves have a continuous spectrum of frequencies. A nonperiodic function can be thought of as a periodic function with an infinite period. Two common methods used to analyze waveforms that start or stop abruptly but continue to infinity in either the positive or negative direction make use of Fourier transforms and Laplace transforms. Again, if you are interested in analyzing such circuits, I recommend consulting a more advanced book on the subject.

2.33   Nonlinear Circuits and Analyzing Circuits by Intuition

So far this chapter has only covered the techniques used for analyzing circuits that contain linear elements (e.g., resistors, capacitors, inductors). Now, the term *linear* refers to a device that has a response that is proportional to the applied signal. For example, doubling the voltage applied across a resistor doubles the current flow through it. For a capacitor, doubling the frequency of the applied voltage across it doubles the current through it. For an inductor, doubling the frequency of the voltage across it halves the current flow through it. As you saw previously in this chapter, the techniques used for analyzing linear circuits are fairly straightforward—there is a law or theorem that tells you what will happen.

However, as it turns out, most of the exciting and useful devices in electronics, such as diodes, transistors, operational amplifiers, integrated circuits, and the like, do not act in a linear manner. For example, a diode has a nonlinear relationship between the applied voltage and the resulting current flow through it, which is described by the approximation relation $I = I_0(e^{eV/kT} - 1)$. If you were to attempt to plug this expression into Kirchhoff's laws, etc., you would be in trouble—the math would get very ugly.

To analyze circuits containing nonlinear elements, you need to find a new approach. Now, you may be thinking, "Oh no, not more math!" Well, I'll give you some good news, you do not really have to learn any more math. In fact, what you will be doing is using intuition. It may sound strange, perhaps, but practically speaking, intuition is the best hope you have for analyzing nonlinear circuits. (As it turns out, the intuition approach will work for linear circuits, too.) This approach involves gaining a basic understanding of how each nonlinear device (or linear device) behaves in a general way—say, getting a general idea of what the current and voltage relationships look like through and across its various leads. After that, you start examining simple circuits containing the device—looking specifically at the circuit's overall behavior (current-voltage responses, etc.) at its input and output leads. In the process of examining these simple circuits, you begin looking for patterns that will, in turn, give more insight into how the device works and how it can be used in applications. For example, you may notice that a voltage divider (two resistors in series) keeps popping up in simple transistor circuits, such as an amplifier circuit. (By selecting particular values for resistors in a voltage divider, you can set the biasing voltage at the transistor's base/gate—the base/gate is attached at the junction between the two resistors.) By recognizing such patterns, you can begin to come up with new ideas as to how these individual devices can be incorporated into other kinds of circuits. Once you have a sense of how the device works and how it is used, you can move on to larger and more complex circuits.

To attack more complex circuits, you again look for patterns within. However, these new patterns become groupings of simple circuits, or *functional groups*—not groupings of individual circuit elements. For example, to figure out how a simple ac-to-dc power supply works, you break the power supply up into individual simple circuits, or functional groups, and then you start figuring out how each individual group functions as a whole. After that, you figure out how all the functional groups interact with each other. For example, after breaking apart the ac-to-dc power supply, you find the following functional groups: a transformer (used to step down the ac voltage to a lower ac voltage), a bridge rectifier circuit (used to convert the ac signal from the transformer into a pulsing dc signal), and a filter circuit (attached to the output of the bridge rectifier, which is used to eliminate the ripple or the pulsing part of the dc signal). In this example, figuring out how the supply works is not a matter of knowing exactly how, say, an individual diode or resistor behaves within one of the functional groups but becomes a matter of figuring out how each functional group behaves and how these functional groups behave collectively. You do not actually have to model each and every part and then come up with a mathematical expression for the final circuit. Doing so would be ridiculous.

The rest of this book will concentrate on the *intuitive approach* to understanding electronic circuits. Each chapter that follows is designed in such a way as to be useful to an *inventor.* This means that the information provided will allow an individual to gain the insight and practical know-how to start building his or her own circuits. The format for each section throughout the remainder of this book will begin with a basic explanation of what a particular device does (or what a particular circuit does) and how it is typically used in applications. After that, I will discuss how the device works (the physics) and then discuss the real kinds of devices that exist (devices you would find in a store or catalog). I will also discuss device specifications, such as power ratings, biasing voltages, and so on. At the end of each section, I will finally take a look at some simple circuits to complete the learning process.

CHAPTER 3

Basic Electronic Circuit Components

3.1 Wires, Cables, and Connectors

Wires and cables provide low-resistance pathways for electric currents. Most electrical wires are made from copper or silver and typically are protected by an insulating coating of plastic, rubber, or lacquer. Cables consist of a number of individually insulated wires bound together to form a multiconductor transmission line. Connectors, such as plugs, jacks, and adapters, are used as mating fasteners to join wires and cable with other electrical devices.

3.1.1 Wires

A wire's diameter is expressed in terms of a *gauge number*. The gauge system, as it turns out, goes against common sense. In the gauge system, as a wire's diameter increases, the gauge number decreases. At the same time, the resistance of the wire decreases. When currents are expected to be large, smaller-gauge wires (large-diameter wires) should be used. If too much current is sent through a large-gauge wire (small-diameter wire), the wire could become hot enough to melt. Table 3.1 shows various characteristics for B&S-gauged insulated copper wire at 20°C. For rubber-insulated wire, the allowable current should be reduced by 30 percent.

TABLE 3.1   Copper Wire Characteristics According to Gauge Number

GAUGE	DIAMETER (IN)	ALLOWABLE CURRENT (A)	FT/LB	FT/Ω
8	0.128	50	20.01	0.628
10	0.102	30	31.82	0.999
12	0.081	25	50.59	1.588
14	0.064	20	80.44	2.525
16	0.051	10	127.9	4.016

(Continues)

TABLE 3.1   Copper Wire Characteristics According to Gauge Number (*Continued*)

GAUGE	DIAMETER (IN)	ALLOWABLE CURRENT (A)	FT/LB	FT/Ω
18	0.040	5	203.4	6.385
20	0.032	3.2	323.4	10.15
22	0.025	2.0	514.2	16.14
24	0.020	1.25	817.7	25.67
26	0.016	0.80	1300.0	40.81
28	0.013	0.53	2067.0	64.90
30	0.010	0.31	3287.0	103.2

Wire comes in solid core, stranded, or braided forms.

Solid Core

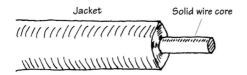

This wire is useful for wiring breadboards; the solid-core ends slip easily into breadboard sockets and will not fray in the process. These wires have the tendency to snap after a number of flexes.

Stranded Wire

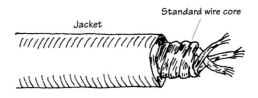

The main conductor is comprised of a number of individual strands of copper. Stranded wire tends to be a better conductor than solid-core wire because the individual wires together comprise a greater surface area. Stranded wire will not break easily when flexed.

Braided Wire

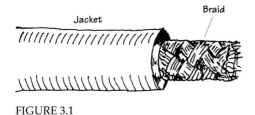

A braided wire is made up of a number of individual strands of wire braided together. Like stranded wires, these wires are better conductors than solid-core wires, and they will not break easily when flexed. Braided wires are frequently used as an electromagnetic shield in noise-reduction cables and also may act as a wire conductor within the cable (e.g., coaxial cable).

FIGURE 3.1

Kinds of Wires

Pretinned Solid Bus Wire

This wire is often referred to as *hookup wire.* It includes a tin-lead alloy to enhance soderability and is usually insulated with polyvinyl chloride (PVC), polyethylene, or Teflon. Used for hobby projects, preparing printed circuit boards, and other applications where small bare-ended wires are needed.

FIGURE 3.2

Speaker Wire

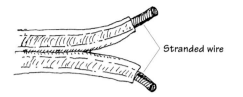

Stranded wire

This wire is stranded to increase surface area for current flow. It has a high copper content for better conduction.

Magnetic Wire

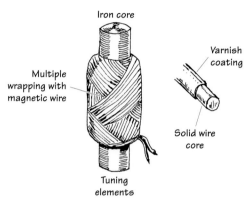

Iron core

Varnish coating

Multiple wrapping with magnetic wire

Solid wire core

Tuning elements

This wire is used for building coils and electromagnets or anything that requires a large number of loops, say, a tuning element in a radio receiver. It is built of solid-core wire and insulated by a varnish coating. Typical wire sizes run from 22 to 30 gauge.

FIGURE 3.2   (*Continued*)

3.1.2   Cables

A cable consists of a multiple number of independent conductive wires. The wires within cables may be solid core, stranded, braided, or some combination in between. Typical wire configurations within cables include the following:

FIGURE 3.3

Twin Lead

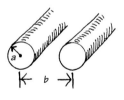

Coaxial

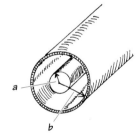

Ribbon and Plane

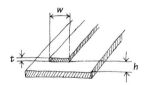

Twisted Pair

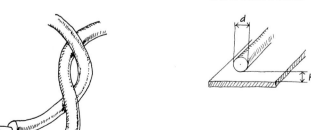

Wire and Plane

Strip Line

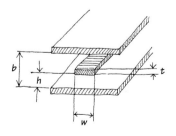

Kinds of Cables

Paired Cable

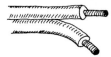

This cable made from two individually insulated conductors. Often it is used in dc or low-frequency ac applications.

Twisted Pair

This cable is composed of two interwound insulated wires. It is similar to a paired cable, but the wires are held together by a twist.

Twin Lead

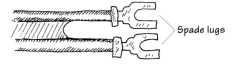

Spade lugs

This cable is a flat two-wire line, often referred to as 300-Ω line. The line maintains an impedance of 300 Ω. It is used primarily as a transmission line between an antenna and a receiver (e.g., TV, radio). Each wire within the cable is stranded to reduce skin effects.

Shielded Twin Lead

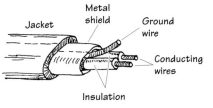

Metal shield
Jacket
Ground wire
Conducting wires
Insulation

This cable is similar to paired cable, but the inner wires are surrounded by a metal-foil wrapping that's connected to a ground wire. The metal foil is designed to shield the inner wires from external magnetic fields—potential forces that can create noisy signals within the inner wires.

Unbalanced Coaxial

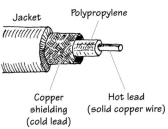

Jacket
Polypropylene
Copper shielding (cold lead)
Hot lead (solid copper wire)

This cable typically is used to transport high-frequency signals (e.g., radio frequencies). The cable's geometry limits inductive and capacitive effects and also limits external magnetic interference. The center wire is made of solid-core copper wire and acts as the hot lead. An insulative material, such as polyethylene, surrounds the center wire and acts to separate the center wire from a surrounding braided wire. The braided wire, or copper shielding, acts as the cold lead or ground lead. Coaxial cables are perhaps the most reliable and popular cables for transmitting information. Characteristic impedances range from about 50 to 100 Ω.

Dual Coaxial

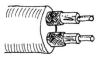

This cable consists of two unbalanced coaxial cables in one. It is used when two signals must be transferred independently.

Balanced Coaxial

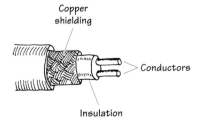

Copper shielding
Conductors
Insulation

This cable consists of two solid wires insulated from one another by a plastic insulator. Like unbalanced coaxial cable, it too has a copper shielding to prevent noise pickup. Unlike unbalanced coaxial cable, the shielding does not act as one of the conductive paths; it only acts as a shield against external magnetic interference.

Ribbon

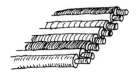

This type of cable is used in applications where many wires are needed. It tends to flex easily. It is designed to handle low-level voltages and often is found in digital systems, such as computers, to transmit parallel bits of information from one digital device to another.

FIGURE 3.4

Multiple Conductor

This type of cable consists of a number of individually wrapped, color-coded wires. It is used when a number of signals must be sent through one cable.

Fiberoptic

Copper-braided shielding Optical fibers

Jacket

Fiberoptic cable is used in the transport of electromagnetic signals, such as light. The conducting-core medium is made from a glass material surrounded by a fiberoptic cladding (a glass material with a higher index of refraction than the core). An electromagnetic signal propagates down the cable by multiple total internal reflections. It is used in direct transmission of images and illumination and as waveguides for modulated signals used in telecommunications. One cable typically consists of a number of individual fibers.

FIGURE 3.4 (Continued)

3.1.3 Connectors

FIGURE 3.5

The following is a list of common plug and jack combinations used to fasten wires and cables to electrical devices. Connectors consist of plugs (male-ended) and jacks (female-ended). To join dissimilar connectors together, an adapter can be used.

117-Volt

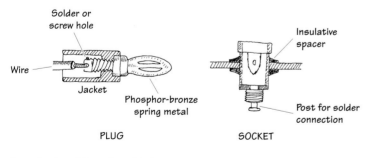

Ground

PLUG   SOCKET

This is a typical home appliance connector. It comes in unpolarized and polarized forms. Both forms may come with or without a ground wire.

Banana

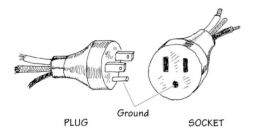

Solder or screw hole

Wire

Jacket

Phosphor-bronze spring metal

Insulative spacer

Post for solder connection

PLUG   SOCKET

This is used for connecting single wires to electrical equipment. It is frequently used with testing equipment. The plug is made from a four-leafed spring tip that snaps into the jack.

Spade Lug/Barrier Strip

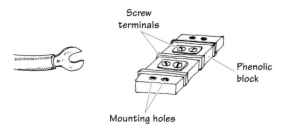

Screw terminals

Phenolic block

Mounting holes

This is a simple connector that uses a screw to fasten a metal spade to a terminal. A barrier strip often acts as the receiver of the spade lugs.

Crimp

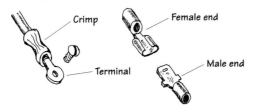

Crimp connectors are color-coded according to the wire size they can accommodate. They are useful as quick, friction-type connections in dc applications where connections are broken repeatedly. A crimping tool is used to fasten the wire to the connector.

Alligator

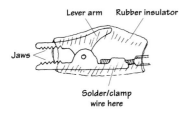

Alligator connectors are used primarily as temporary test leads.

Phone

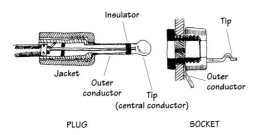

These connectors accept shielded braid, but they are larger in size. They come in two- or three-element types and have a barrel that is 1¼ in (31.8 mm) long. They are used as connectors in microphone cables and for other low-voltage, low-current applications.

Phono

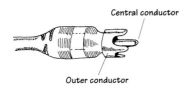

Phono connectors are often referred to as RCA plugs or pin plugs. They are used primarily in audio connections.

F-Type

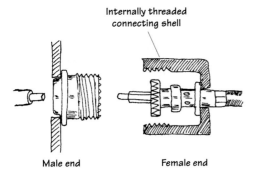

F-type connectors are used with a variety of unbalanced coaxial cables. They are commonly used to interconnect video components. F-type connectors are either threaded or friction-fit together.

Tip

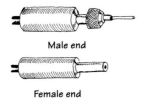

Here, the plug consists of a solid metallic tip that slides into the jack section. The two are held together by friction. Wires are either soldered or screwed into place under the plastic collar.

FIGURE 3.5   *(Continued)*

Mini

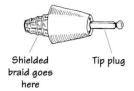

Shielded braid goes here   Tip plug

These connectors are used to join wires with shielded braid cables. The tip of the plug makes contact with the center conductor wire, whereas the cylindrical metal extension (or barrel) makes contact with the braid. Such connectors are identified by diameter and number of threads per inch.

PL-259

These are often referred to as UHF plugs. They are used with RG-59/U coaxial cable. Such connectors may be threaded or friction-fit together.

BNC

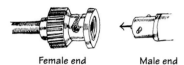

Female end   Male end

BNC connectors are used with coaxial cables. Unlike the F-type plug, BNC connectors use a twist-on bayonet-like locking mechanism. This feature allows for quick connections

T-Connector

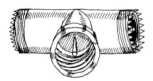

T-connectors consist of two plug ends and one central jack end. They are used when a connection must be make somewhere along a coaxial cable.

DIN Connector

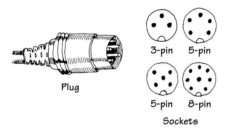

Plug

3-pin   5-pin

5-pin   8-pin

Sockets

These connectors are used with multiple conductor wires. They are often used for interconnecting audio and computer accessories.

Meat Hook

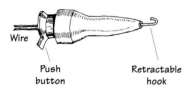

Wire

Push button   Retractable hook

These connectors are used as test probes. The spring-loaded hook opens and closes with the push of a button. The hook can be clamped onto wires and component leads.

D-Connector

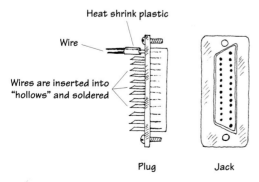

Heat shrink plastic

Wire

Wires are inserted into "hollows" and soldered

Plug   Jack

D-connectors are used with ribbon cable. Each connector may have as many as 50 contacts. The connection of each individual wire to each individual plug pin or jack socket is made by sliding the wire in a hollow metal collar at the backside of each connector. The wire is then soldered into place.

FIGURE 3.5   (*Continued*)

3.1.4 Wiring and Connector Symbols

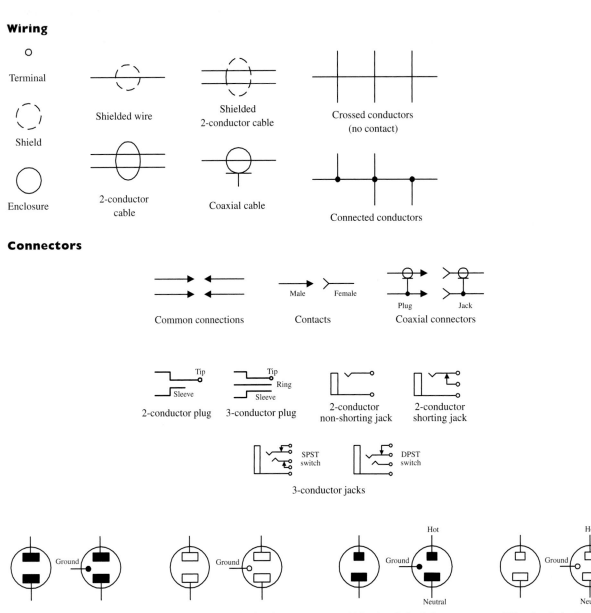

FIGURE 3.6

3.1.5 High-Frequency Effects within Wires and Cables

Weird Behavior in Wires (Skin Effect)

When dealing with simple dc hobby projects, wires and cables are straightforward—they act as simple conductors with essentially zero resistance. However, when you replace dc currents with very high-frequency ac currents, weird things begin to take place within wires. As you will see, these "weird things" will not allow you to treat wires as perfect conductors.

First, let's take a look at what is going on in a wire when a dc current is flowing through it.

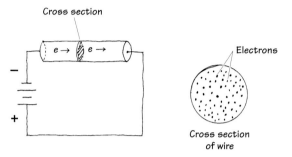

FIGURE 3.7

A wire that is connected to a dc source will cause electrons to flow through the wire in a manner similar to the way water flows through a pipe. This means that the path of any one electron essentially can be anywhere within the volume of the wire (e.g., center, middle radius, surface).

Now, let's take a look at what happens when a high-frequency ac current is sent through a wire.

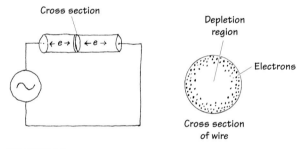

FIGURE 3.8

An ac voltage applied across a wire will cause electrons to vibrate back and forth. In the vibrating process, the electrons will generate magnetic fields. By applying some physical principles (finding the forces on every electron that result from summing up the individual magnetic forces produced by each electron), you find that electrons are pushed toward the surface of the wire. As the frequency of the applied signal increases, the electrons are pushed further away from the center and toward the surface. In the process, the center region of the wire becomes devoid of conducting electrons.

The movement of electrons toward the surface of a wire under high-frequency conditions is called the *skin effect*. At low frequencies, the skin effect does not have a large effect on the conductivity (or resistance) of the wire. However, as the frequency increases, the resistance of the wire may become an influential factor. Table 3.2 shows just how influential skin effect can be as the frequency of the signal increases (the table uses the ratio of ac resistance to dc resistance as a function of frequency).

TABLE 3.2 AC/DC Resistance Ratio as a Function of Frequency

WIRE GAUGE	R_{AC}/R_{DC}			
	10^6 Hz	10^7 Hz	10^8 Hz	10^9 Hz
22	6.9	21.7	68.6	217
18	10.9	34.5	109	345
14	17.6	55.7	176	557
10	27.6	87.3	276	873

One thing that can be done to reduce the resistance caused by skin effects is to use stranded wire—the combined surface area of all the individual wires within the conductor is greater than the surface area for a solid-core wire of the same diameter.

Weird Behavior in Cables (Lecture on Transmission Lines)

Like wires, cables also exhibit skin effects. In addition, cables exhibit inductive and capacitive effects that result from the existence of magnetic and electrical fields within the cable. A magnetic field produced by the current through one wire will induce a current in another. Likewise, if two wires within a cable have a net difference in charge between them, an electrical field will exist, thus giving rise to a capacitive effect.

Coaxial Cable   **Paired Cable**

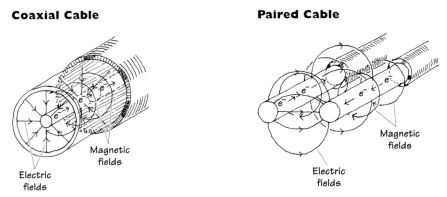

FIGURE 3.9   Illustration of the electrical and magnetic fields within a coaxial and paired cable.

Taking note of both inductive and capacitive effects, it is possible to treat a cable as if it were made from a number of small inductors and capacitors connected together. An equivalent inductor-capacitor network used to model a cable is shown in Fig. 3.10.

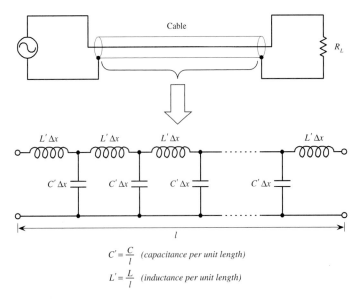

$$C' = \frac{C}{l} \quad \text{(capacitance per unit length)}$$

$$L' = \frac{L}{l} \quad \text{(inductance per unit length)}$$

FIGURE 3.10   The impedance of a cable can be modeled by treating it as a network of inductors and capacitors.

To simplify this circuit, we apply a reduction trick; we treat the line as an infinite ladder and then assume that adding one "run" to the ladder (one inductor-capacitor section to the system) will not change the overall impedance Z of the cable. What this means—mathematically speaking—is we can set up an equation such that $Z = Z +$ (*LC* section). This equation can then be solved for Z. After that, we find the limit as Δx goes to zero. The mathematical trick and the simplified circuit are shown below.

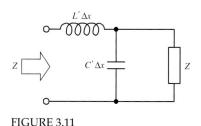

$$Z = j\omega L'\Delta x = \frac{Z/j\omega C'\Delta x}{Z + 1/j\omega C'\Delta x} = j\omega L'\Delta x + \frac{Z}{1 + j\omega C'Z\Delta x}$$

When $\Delta x \rightarrow$ small,

$$Z = \sqrt{L'/C'} = \sqrt{\frac{L/l}{C/l}} = \sqrt{L/C}$$

FIGURE 3.11

By convention, the impedance of a cable is called the *characteristic impedance* (symbolized Z_0). Notice that the characteristic impedance Z_0 is a real number. This means that the line behaves like a resistor despite the fact that we assumed the cable had only inductance and capacitance built in.

The question remains, however, what are L and C? Well, figuring out what L and C should be depends on the particular geometry of the wires within a cable and on the type of dielectrics used to insulate the wires. You could find L and C by applying some physics principles, but instead, let's cheat and look at the answers. The following are the expressions for L and C and Z_0 for both a coaxial and parallel-wire cable:

Coaxial

	L (H/m)	C (F/m)	$Z_0 = \sqrt{L/C}$ (Ω)
	$\dfrac{\mu_0 \ln(b/a)}{2\pi}$	$\dfrac{2\pi\varepsilon_0 k}{\ln(b/a)}$	$\dfrac{138}{\sqrt{k}}\log\dfrac{b}{a}$

Parallel Wire

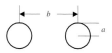

	$\dfrac{\mu_0 \ln(D/a)}{\pi}$	$\dfrac{\pi\varepsilon_0 k}{\ln(D/a)}$	$\dfrac{276}{\sqrt{k}}\log\dfrac{D}{a}$

FIGURE 3.12   Inductance, capacitance, and characteristic impedance formulas for coaxial and parallel wires.

Here, k is the dielectric constant of the insulator, $\mu_0 = 1.256 \times 10^{-6}$ H/m is the permeability of free space, and $\varepsilon_0 = 8.85 \times 10^{-12}$ F/m is the permitivity of free space. Table 3.3 provides some common dielectric materials with their corresponding constants.

Often, cable manufacturers supply capacitance per foot and inductance per foot values for their cables. In this case, you can simply plug the given manufacturer's values into $Z = \sqrt{L/C}$ to find the characteristic impedance of the cable. Table 3.4 shows capacitance per foot and inductance per foot values for some common cable types.

TABLE 3.3 **Common Dielectrics and Their Constants**

MATERIAL	DIELECTRIC CONSTANT (k)
Air	1.0
Bakelite	4.4–5.4
Cellulose acetate	3.3–3.9
Pyrex glass	4.8
Mica	5.4
Paper	3.0
Polyethylene	2.3
Polystyrene	5.1–5.9
Quartz	3.8
Teflon	2.1

TABLE 3.4 **Capacitance and Inductance per Foot for Some Common Transmission-Line Types**

CABLE TYPE	CAPACITANCE/FT (pF)	INDUCTANCE/FT (μH)
RG-8A/U	29.5	0.083
RG-11A/U	20.5	0.115
RG-59A/U	21.0	0.112
214-023	20.0	0.107
214-076	3.9	0.351

Sample Problems (Finding the Characteristic Impedance of a Cable)

EXAMPLE 1

An RG-11AU cable has a capacitance of 21.0 pF/ft and an inductance of 0.112 μH/ft. What is the characteristic impedance of the cable?

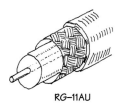

RG–11AU

FIGURE 3.13

You are given the capacitance and inductance per unit length: $C' = C/\text{ft}$, $L' = L/\text{ft}$. Using $Z_0 = \sqrt{L/C}$ and substituting C and L into it, you get

$$Z_0 = \sqrt{L/C}$$

$$Z_0 = \sqrt{\frac{0.112 \times 10^{-6}}{21.0 \times 10^{-12}}} = 73\,\Omega$$

EXAMPLE 2

What is the characteristic impedance of the RG-58/U coaxial cable with polyethylene dielectric ($k = 2.3$) shown below?

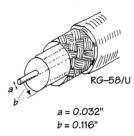

RG–58/U

$a = 0.032''$
$b = 0.116''$

FIGURE 3.14

$$Z_0 = \frac{138}{\sqrt{k}} \log \frac{b}{a}$$

$$Z_0 = \frac{138}{\sqrt{2.3}} \log \left(\frac{0.116}{0.032} \right) = 91 \times 0.056 = 51\,\Omega$$

EXAMPLE 3

Find the characteristic impedance of the parallel-wire cable insulated with polyethylene ($k = 2.3$) shown below.

$$Z_0 = \frac{276}{\sqrt{k}} \log \frac{D}{a}$$

$$Z_0 = \frac{276}{\sqrt{2.3}} \log \frac{0.270}{0.0127} = 242\Omega$$

FIGURE 3.15

Impedance Matching

Since a transmission line has impedance built in, the natural question to ask is, How does the impedance affect signals that are relayed through a transmission line from one device to another? The answer to this question ultimately depends on the impedences of the devices to which the transmission line is attached. If the impedance of the transmission line is not the same as the impedance of, say, a load connected to it, the signals propagating through the line will only be partially absorbed by the load. The rest of the signal will be reflected back in the direction it came. Reflected signals are generally bad things in electronics. They represent an inefficient power transfer between two electrical devices. How do you get rid of the reflections? You apply a technique called *impedance matching*. The goal of impedance matching is to make the impedances of two devices—that are to be joined—equal. The impedance-matching techniques make use of special matching networks that are inserted between the devices.

Before looking at the specific methods used to match impedances, let's first take a look at an analogy that should shed some light on why unmatched impedances result in reflected signals and inefficient power transfers. In this analogy, pretend that the transmission line is a rope that has a density that is analogous to the transmission line's characteristic impedance Z_0. Pretend also that the load is a rope that has a density that is analogous to the load's impedance Z_L. The rest of the analogy is carried out below.

Unmatched Impedances ($Z_0 < Z_L$)

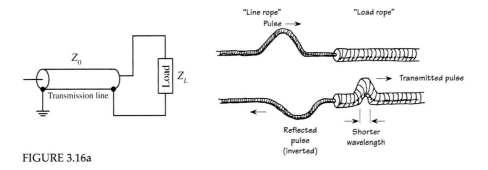

FIGURE 3.16a

A low-impedance transmission line that is connected to a high-impedance load is analogous to a low-density rope connected to a high-density rope. In the rope analogy, if you impart a pulse at the left end of the low-density rope (analogous to sending an electrical signal through a line to a load), the pulse will travel along without problems until it reaches the high-density rope (load). According to the laws of physics, when the wave reaches the high-density rope, it will do two things. First, it will induce a smaller-wavelength pulse within the high-density rope, and second, it will induce a similar-wavelength but inverted and diminished pulse that rebounds back toward the left end of the low-density rope. From the analogy, notice that only part of the signal energy from the low-density rope is transmitted to the high-density rope. From this analogy, you can infer that in the electrical case similar effects will occur—only now you are dealing with voltage and currents and transmission lines and loads.

Unmatched Impedances ($Z_0 > Z_L$)

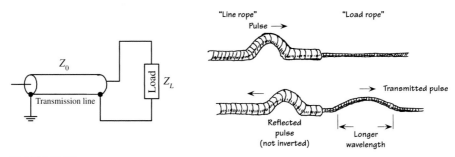

FIGURE 3.16b

A high-impedance transmission line that is connected to a low-impedance load is analogous to a high-density rope connected to a low-density rope. If you impart a pulse at the left end of the high-density rope (analogous to sending an electrical signal through a line to a load), the pulse will travel along the rope without problems until it reaches the low-density rope (load). At that time, the pulse will induce a longer-wavelength pulse within the low-density rope and will induce a similar-wavelength but inverted and diminished pulse that rebounds back toward the left end of the high-density rope. From this analogy, again you can see that only part of the signal energy from the high-density rope is transmitted to the low-density rope.

Matched Impedances ($Z_0 = Z_L$)

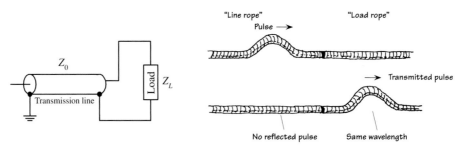

FIGURE 3.16c

Connecting a transmission line and load of equal impedances together is analogous to connecting two ropes of similar densities together. When you impart a pulse in the "transmission line" rope, the pulse will travel along without problems. However, unlike the first two analogies, when the pulse meets the load rope, it will continue on through the load rope. In the process, there will be no reflection, wavelength change, or amplitude change. From this analogy, you can infer that if the impedance of a transmission line matches the impedance of the load, power transfer will be smooth and efficient.

Standing Waves

Let's now consider what happens to an improperly matched line and load when the signal source is producing a continuous series of sine waves. You can, of course, expect reflections as before, but you also will notice that a superimposed standing-wave pattern is created within the line. The standing-wave pattern results from the interaction of forward-going and reflected signals. Figure 3.17 shows a typical resulting standing-wave pattern for an improperly matched transmission line attached between a sinusoidal transmitter and a load. The standing-wave pattern is graphed in terms amplitude (expressed in terms of V_{rms}) versus position along the transmission line.

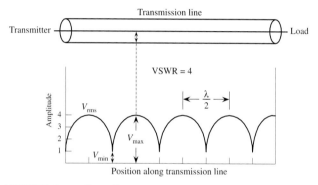

FIGURE 3.17 Standing waves on an improperly terminated transmission line. The VSWR is equal to V_{max}/V_{min}.

A term used to describe the standing-wave pattern is the *voltage standing-wave ratio* (VSWR). The VSWR is the ratio between the maximum and minimum rms voltages along a transmission line and is expressed as

$$\text{VSWR} = \frac{V_{rms,max}}{V_{rms,min}}$$

The standing-wave pattern shown in Fig. 3.17 has VSWR of 4/1, or 4.

Assuming that the standing waves are due entirely to a mismatch between load impedance and characteristic impedance of the line, the VSWR is simply given by either

$$\text{VSWR} = \frac{Z_0}{R_L} \quad \text{or} \quad \text{VSWR} = \frac{R_L}{Z_0}$$

whichever produces a result that is greater than 1.

A VSWR equal to 1 means that the line is properly terminated, and there will be no reflected waves. However, if the VSWR is large, this means that the line is not properly terminated (e.g., a line with little or no impedance attached to either a short or open circuit), and hence there will be major reflections.

The VSWR also can be expressed in terms of forward and reflected waves by the following formula:

$$\text{VSWR} = \frac{V_F + V_R}{V_F - V_R}$$

To make this expression meaningful, you can convert it into an expression in terms of forward and reflected power. In the conversion, you use $P = IV = V^2/R$. Taking P to be proportional to V^2, you can rewrite the VSWR in terms of forward and reflected power as follows:

$$\text{VSWR} = \frac{\sqrt{P_F} + \sqrt{P_R}}{\sqrt{P_F} - \sqrt{P_R}}$$

Rearranging this equation, you get the percentage of reflected power and percentage of absorbed power in terms of VSWR:

$$\% \text{ reflected power} = \frac{P_R}{P_F} = \left[\frac{\text{VSWR} - 1}{\text{VSWR} + 1}\right]^2 \times 100\%$$

$$\% \text{ absorbed power} = 100\% - \% \text{ reflected power}$$

EXAMPLE (VSWR)

Find the standing-wave ratio (VSWR) of a 50-Ω line used to feed a 200-Ω load. Also find the percentage of power that is reflected at the load and the percentage of power absorbed by the load.

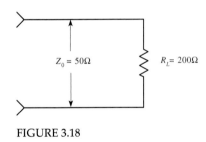

FIGURE 3.18

$$\text{VSWR} = \frac{Z_0}{R_L} = \frac{200}{50} = 4$$

VSWR is 4:1

$$\% \text{ reflected power} = \frac{\text{VSWR} - 1^2}{\text{VSWR} + 1} \times 100\%$$

$$= \frac{4 - 1^2}{4 + 1} \times 100\% = 36\%$$

$$\% \text{ absorbed power} = 100\% - \% \text{ reflected power} = 64\%$$

Techniques for Matching Impedances

This section looks at a few impedance-matching techniques. As a rule of thumb, with most low-frequency applications where the signal's wavelength is much larger than the cable length, there is no need to match line impedances. Matching impedances is usually reserved for high-frequency applications. Moreover, most electrical equipment, such as oscilloscopes, video equipment, etc., has input and output impedances that match the characteristic impedances of coaxial cables (typically 50 Ω). Other devices, such as television antenna inputs, have characteristic input impedances that match the characteristic impedance of twin-lead cables (300 Ω). In such cases, the impedance matching is already taken care of.

IMPEDANCE-MATCHING NETWORK

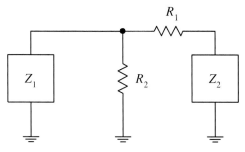

FIGURE 3.19

A general method used to match impedance makes use of the impedance-matching network shown here. To match impedances, choose

$$R_1 = \sqrt{Z_2/(Z_2 - Z_1)}$$

$$R_2 = Z_1 \sqrt{Z_2/(Z_2 - Z_1)}$$

The attenuation seen from the Z_1 end will be $A_1 = R_1/Z_2 + 1$. The attenuation seen from the Z_2 end will be $A_2 = R_1/R_2 + R_1/Z_1 + 1$.

For example, if $Z_1 = 125\ \Omega$, and $Z_2 = 50\ \Omega$, then R_1, R_2, A_1, and A_2 are

$$R_1 = Z_2\sqrt{(Z_2 - Z_1)} = \sqrt{125(125 - 50)} = 97\Omega$$

$$R_2 = Z_1 \sqrt{\frac{Z_2}{Z_2 - Z_1}} = 50 \sqrt{\frac{125}{125 - 50}} = 65\Omega$$

$$A_1 = \frac{R_1}{Z_2} + 1 = \frac{96.8}{125} + 1 = 1.77$$

$$A_2 = \frac{R_1}{R_2} + \frac{R_1}{Z_1} + 1 = \frac{96.8}{64.6} + \frac{96.8}{50} + 1 = 4.43$$

IMPEDANCE TRANSFORMER

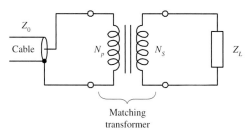

Matching
transformer

FIGURE 3.20

Here, a transformer is used to make the characteristic impedance of a cable with the impedance of a load. By using the formula

$$N_P/N_S = \sqrt{Z_0/Z_L}$$

you can match impedances by choosing appropriate values for N_P and N_S so that the ratio N_P/N_S is equal to $\sqrt{Z_0/Z_L}$.

For example, if you wish to match an 800-Ω impedance line with an 8-Ω load, you first calculate

$$\sqrt{Z_0/Z_L} = \sqrt{800/8} = 10$$

To match impedances, you select N_P (number of coils in the primary) and N_S (number of coils in the secondary) in such a way that $N_P/N_S = 10$. One way of doing this would be to set N_P equal to 10 and N_S equal to 1. You also could choose N_P equal to 20 and N_S equal to 2 and you would get the same result.

BROADBAND TRANSMISSION-LINE TRANSFORMER

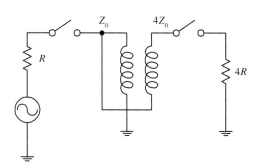

FIGURE 3.21

A broadband transmission-line transformer is a simple device that consists of a few turns of miniature coaxial cable or twisted-pair cable wound about a ferrite core. Unlike conventional transformers, this device can more readily handle high-frequency matching (its geometry eliminates capacitive and inductive resonance behavior). These devices can handle various impedance transformations and can do so with incredibly good broadband performance (less than 1 dB loss from 0.1 to 500 MHz).

QUARTER-WAVE SECTION

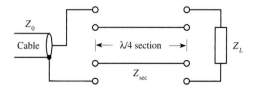

FIGURE 3.22

A transmission line with characteristic impedance Z_0 can be matched with a load with impedance Z_L by inserting a wire segment that has a length equal to one quarter of the transmitted signal's wavelength ($\lambda/4$) and which has an impedance equal to

$$Z_{sec} = \sqrt{Z_0 Z_L}$$

To calculate the segment's length, you must use the formula $\lambda = v/f$, where v is the velocity of propagation of a signal along the cable and f is the frequency of the signal. To find v, use

$$v = c/\sqrt{k}$$

where $c = 3.0 \times 10^8$ m/s, and k is the dielectric constant of the cable's insulation.

For example, say you wish to match a 50-Ω cable that has a dielectric constant of 1 with a 200-Ω load. If you assume the signal's frequency is 100 MHz, the wavelength then becomes

$$\lambda = \frac{v}{f} = \frac{c/\sqrt{k}}{f} = \frac{3 \times 10^8/1}{100 \times 10^6} = 3 \text{ m}$$

To find the segment length, you plug λ into $\lambda/4$. Hence the segment should be 0.75 m long. The wire segment also must have an impedance equal to

$$Z_{sec} = \sqrt{(50)(200)} = 100 \ \Omega$$

STUBS

Short-circuited stub

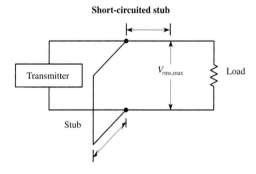

Pair of open-ended matching stubs

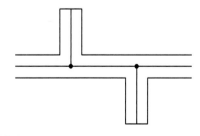

FIGURE 3.23

A short length of transmission line that is open ended or short-circuit terminated possesses the property of having an impedance that is reactive. By properly choosing a segment of open-circuit or short-circuit line and placing it in shunt with the original transmission line at an appropriate position along the line, standing waves can be eliminated. The short segment of wire is referred to as a *stub*. Stubs are made from the same type of cable found in the transmission line. Figuring out the length of a stub and where it should be placed is fairly tricky. In practice, graphs and a few formulas are required. A detailed handbook on electronics is the best place to learn more about using stubs.

3.2 Batteries

A battery is made up of a number of *cells*. Each cell contains a positive terminal, or *cathode*, and a negative terminal, or *anode*. (Note that most other devices treat *anodes* as positive terminals and *cathodes* as negative terminals.)

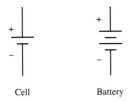

Cell   Battery

FIGURE 3.24

When a load is placed between a cell's terminals, a conductive bridge is formed that initiates chemical reactions within the cell. These reactions produce electrons in the anode material and remove electrons from the cathode material. As a result, a potential is created across the terminals of the cell, and electrons from the anode flow through the load (doing work in the process) and into the cathode.

A typical cell maintains about 1.5 V across its terminals and is capable of delivering a specific amount of current that depends on the size and chemical makeup of the cell. If more voltage or power is needed, a number of cells can be added together in either series or parallel configurations. By adding cells in series, a larger-voltage battery can be made, whereas adding cells in parallel results in a battery with a higher current-output capacity. Figure 3.25 shows a few cell arrangements.

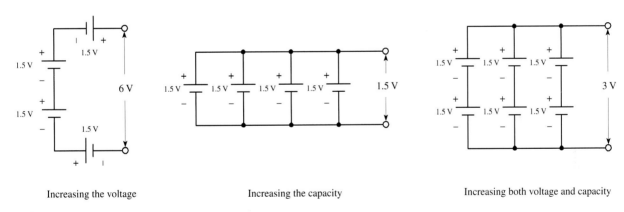

Increasing the voltage   Increasing the capacity   Increasing both voltage and capacity

FIGURE 3.25

Battery cells are made from a number of different chemical ingredients. The use of a particular set of ingredients has practical consequences on the battery's overall performance. For example, some cells are designed to provide high open-circuit voltages, whereas others are designed to provide large current capacities. Certain kinds of cells are designed for light-current, intermittent applications, whereas others are designed for heavy-current, continuous-use applications. Some cells are designed for pulsing applications, where a large burst of current is needed for a short period of time. Some cells have good shelf lives; other have poor shelf lives. Batteries that are designed for one-time use, such as carbon-zinc and alkaline batteries, are called *primary batteries*. Batteries that can be recharged a number of times, such as nickel-cadmium and lead-acid batteries, are referred to as *secondary batteries*.

3.2.1 How a Cell Works

A cell converts chemical energy into electrical energy by going through what are called *oxidation-reduction reactions* (reactions that involve the exchange of electrons).

The three fundamental ingredients of a cell used to initiate these reactions include two chemically dissimilar metals (positive and negative electrodes) and an electrolyte (typically a liquid or pastelike material that contains freely floating ions). The following is a little lecture on how a simple lead-acid battery works.

For a lead-acid cell, one of the electrodes is made from pure lead (Pb); the other electrode is made from lead oxide (PbO_2); and the electrolyte is made from an sulfuric acid solution ($H_2O + H_2SO_4 \rightarrow 3H^+ + SO_4^{2-} + OH^-$).

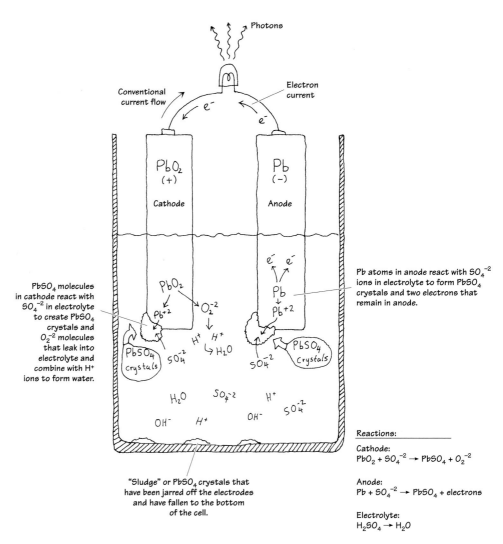

FIGURE 3.26

When the two chemically dissimilar electrodes are placed in the sulfuric acid solution, the electrodes react with the acid (SO_4^{-2}, H^+ ions), causing the pure lead electrode to slowly transform into $PbSO_4$ crystals. During this transformation reaction, two electrons are liberated within the lead electrode. Now, if you examine the lead oxide electrode, you also will see that it too is converted into $PbSO_4$ crystals. However, instead of releasing electrons during its transformation, it releases O_2^{2-} ions. These ions leak out into the electrolyte solution and combine with the hydrogen ions to form H_2O (water). By placing a load element, say, a light bulb, across the electrodes, electrons will flow from the electron-abundant lead electrode, through the bulb's filament, and into the electron-deficient lead oxide electrode.

As time passes, the ingredients for the chemical reactions run out (the battery is drained). To get energy back into the cell, a reverse voltage can be applied across the cell's terminals, thus forcing the reactions backward. In theory, a lead-acid battery can be drained and recharged indefinitely. However, over time, chunks of crystals will break off from the electrodes and fall to the bottom of the container, where they are not recoverable. Other problems arise from loss of electrolyte due to gasing during electrolysis (a result of overcharging) and due to evaporation.

3.2.2 Primary Batteries

Primary batteries are one-shot deals—once they are drained, it is all over. Common primary batteries include carbon-zinc batteries, alkaline batteries, mercury batteries, silver oxide batteries, zinc-air batteries, and silver-zinc batteries. Here are some common battery packages:

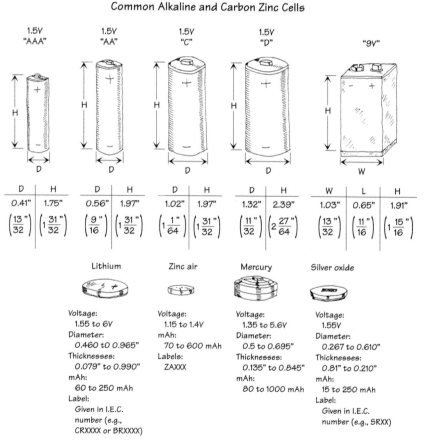

FIGURE 3.27

Table 3.5 lists some common primary battery types, along with their corresponding characteristics.

TABLE 3.5   Common Primary Batteries and Their Characteristics

CELL TYPE	ANODE (−)	CATHODE (+)	MAXIMUM VOLTAGE (THEORETICAL) (V)	MAXIMUM CAPACITY (THEORETICAL) (Ah/kG)	WORKING VOLTAGE (PRACTICAL) (V)	ENERGY DENSITY (Wh/kG)	SHELF LIFE AT 25°C (80% CAPACITY) (MONTHS)
Carbon-zinc	Zn	MnO_2	1.6	230	1.2	65	18
Alkaline-MnO_2	Zn	MnO_2	1.5	230	1.15	65	30
Mercury	Zn	HgO	1.34	185	1.2	80	36
Silver oxide	Zn	AgO	1.85	285	1.5	130	30
Zinc-air	Zn	O_2	1.6	815	1.1	200	18
Lithium	Li	$(CF)_n$	3.6	2200	3.0	650	120
Lithium	Li	CrO_2	3.8	750	3.0	350	108
Magnesium	Mg	MnO_2	2.0	270	1.5	100	40

3.2.3   Comparing Primary Batteries

Carbon-Zinc Batteries

Carbon-zinc batteries are general-purpose, nonrechargeable batteries made from cells that have open-circuit voltages of 1.6 V. They are used for low to moderate current drains. The voltage-discharge curve over time for a carbon-zinc battery is nonlinear, whereas the current output efficiency decreases at high current drains. Carbon-zinc batteries have poor low-temperature performance but good shelf lives. Carbon-zinc batteries are used to power such devices as power toys, consumer electronic products, flashlights, cameras, watches, and remote control transmitters.

Zinc Chloride Batteries

A zinc chloride battery is a heavy-duty variation of a zinc-carbon battery. It is used in applications that require moderate to heavy current drains. Zinc chloride batteries have better voltage discharge per time characteristics and better low-temperature performance than carbon-zinc batteries. Zinc chloride batteries are used in radios, flashlights, lanterns, fluorescent lanterns, motor-driven devices, portable audio equipment, communications equipment, electronic games, calculators, and remote control transmitters.

Alkaline Batteries

Alkaline batteries are general-purpose batteries that are highly efficient under moderate continuous drain and are used in heavy-current or continuous-drain applications. Their open-circuit voltage is about 0.1 V less than that of carbon-zinc cells, but compared with carbon-zinc cells, they have longer shelf lives, higher power capacities, better cold temperature performance, and weigh about 50 percent less. Alkaline batteries are interchangeable with carbon-zinc and zinc chloride batteries. Alkaline batteries are used to power such things as video cameras, motorized toys, photoflashes, electric shavers, motor-driven devices, portable audio equipment, communications equipment, smoke detectors, and calculators. As it turns out, alkaline batteries come in both nonrechargeable and rechargeable forms.

Mercury Batteries

Mercury batteries are very small, nonrechargeable batteries that have open-circuit voltages of around 1.4 V per cell. Unlike carbon-zinc and alkaline batteries, mercury batteries maintain their voltage up to a point just before the die. They have greater capacities, better shelf lives, and better low-temperature performance than carbon-zinc, zinc chloride, and alkaline batteries. Mercury batteries are designed to be used in small devices such as hearing aids, calculators, pagers, and watches.

Lithium Batteries

Lithium batteries are nonrechargeable batteries that use a lithium anode, one of a number of cathodes, and an organic electrolyte. Lithium batteries come with open-circuit voltages of 1.5 or 3.0 V per cell. They have high energy densities, outstanding shelf lives (8 to 10 years), and can operate in a wide range of temperatures, but they have limited high current-drain capabilities. Lithium batteries are used in such devices as cameras, meters, cardiac pacemakers, CMOS memory storage devices, and liquid-crystal displays (LCDs) for watches and calculators.

Silver Oxide Batteries

Silver oxide batteries come with open-circuit voltages of 1.85 V per cell. They are used in applications that require high current pulsing. Silver oxide batteries have flat voltage discharge characteristics up until death but also have poor shelf lives and are expensive. These batteries are used in such devices as alarms, backup lighting, and analog devices. As it turns out, like alkaline batteries, they too come in nonrechargeable and rechargeable forms.

Zinc-Air Batteries

Zinc-air batteries are small, nonrechargeable batteries with open-circuit voltages of 1.15 to 1.4 V per cell. They use surrounding air (O_2) as the cathode ingredient and contain air vents that are taped over during storage. Zinc-air batteries are long-lasting, high-performance batteries with excellent shelf lives and have reasonable temperature performance (about 0 to 50°C, or 32 to 122°F). These batteries typically are used in small devices such as hearing aids and pagers.

3.2.4 Secondary Batteries

Secondary batteries, unlike primary batteries, are rechargeable by nature. The actual discharge characteristics for secondary batteries are similar to those of primary batteries, but in terms of design, secondary batteries are made for long-term, high-power-level discharges, whereas primary batteries are designed for short discharges at low power levels. Most secondary batteries come in packages similar to those of primary batteries, with the exception of, say, lead-acid batteries and special-purpose batteries. Secondary batteries are used to power such devices as laptop computers, portable power tools, electric vehicles, emergency lighting systems, and engine starting systems.

Here are some common packages for secondary batteries:

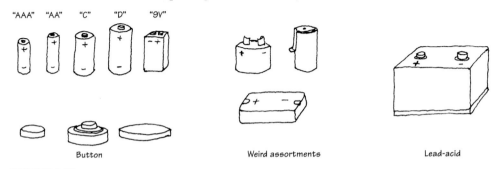

"AAA"   "AA"   "C"   "D"   "9V"

Button   Weird assortments   Lead-acid

FIGURE 3.28

Table 3.6 lists some common secondary battery types and their characteristics.

TABLE 3.6   Secondary Batteries and Their Characteristics

BATTERY TYPE	ANODE (−)	CATHODE (+)	MAXIMUM VOLTAGE (V)	MAXIMUM CAPACITY (THEORETICAL) (Ah/kG)	WORKING VOLTAGE (PRACTICAL) (V)	ENERGY DENSITY (Wh/kG)
Lead-acid	Pb	PbO_2	2.1	55	2.0	37
Edison (Ni-Fe)	Fe	NiO_x	1.5	195	1.2	29
NiCad	Cd	NiO_x	1.35	165	1.2	33
Silver-cadmium	Cd	AgO	1.4	230	1.05	55
Cadmium-air	Cd	Air (O_2)	1.2	475	0.8	90
Silver-zinc	Zn	AgO	1.85	285	1.5	100
Zinc-air	Zn	Air (O_2)	1.6	815	1.1	150

Lead-Acid and Nickel-Cadmium Batteries

LEAD-ACID BATTERIES

Lead-acid batteries are rechargeable batteries with open-circuit voltages of 2.15 V per cell while maintaining a voltage range under a load from 1.75 to 1.9 V per cell. The cycling life (number of times the battery can recharged) for lead-acid batteries is around 1000 cycles. They come in rapid, quick, standard, and trickle-charging rate types. Lead-acid batteries have a charge retention time (time until the battery reaches 80 percent of maximum) of about 18 months. They contain a liquid electrolyte that requires servicing (replacement). Six lead-acid cells make up a car battery.

NICKEL-CADMIUM BATTERIES

Nickel-cadmium batteries contain rechargeable cells that have open-circuit voltages of about 1.2 V. They are often interchangeable with carbon-zinc and alkaline batteries. For the first two-thirds of its life, a nickel-cadmium battery's discharge curve is relatively flat, but after that, its curve begins to drop. Nickel-cadmium batteries weigh about a third as much as carbon-zinc batteries. Placing these batteries in parallel is not recommended (series is OK). These batteries are used in such devices as toys, consumer electronic products, flashlights, cameras, photographic equipment, power tools, and appliances.

Charging Secondary Batteries

To charge a secondary battery, a voltage greater than the open-circuit battery voltage must be applied across its terminals. Current then enters the battery at a rate determined by the difference between the applied voltage and the open-circuit voltage. The rate of charge (or discharge) is determined by the battery's capacity divided by the charge/discharge time. In general, decreasing the charge rate increases the charge efficiency.

When recharging a battery, it is important not to overcharge it. Overcharging a battery may lead to side reactions (reactions other than those which act to convert the electrodes back to their original states). These reactions may result in electrolysis, whereby some of the ingredients of the electrolyte may be converted into a gas or fine mist that escapes from the battery. The net effect of overcharging a battery reduces its efficiency.

3.2.5 Battery Capacity

Batteries are given a *capacity rating* that indicates how much electrical energy they are capable of delivering over a period of time. The capacity rating is equal to the initial current drawn multiplied by the time until the battery dies. The units for battery capacities are ampere-hours (Ah) and millampere-hours (mAh).

EXAMPLE

Suppose that you have a 1.5-V D-sized cell with a capacity rating of 1000 mAh. How long will the battery last if it is to be used to supply current to a 100-Ω resistive load?

The first thing to do is to calculate the initial current drawn:

$$I = \frac{V}{R} = \frac{1.5 \text{ V}}{1000 \text{ }\Omega} = 1.5 \text{ mA}$$

To find the discharge time, take the capacity-rating expression and substitute into it the calculated current:

$$\frac{\text{Capacity rating}}{\text{Initially drawn current}} = \frac{1000 \text{ mAh}}{1.5 \text{ mA}} = 666 \text{ h}$$

If the resistance is reduced to 10 Ω, the discharge time dramatically shrinks to 6.6 hours.

Notably, as the current demand increases, the current-delivering capacity of a battery decreases (the preceding calculations thus are not incredibly reliable). Excessive heat generated inside a battery and efficiency losses due to higher temperatures prevent large output current for extended periods of time.

3.2.6 Note on Internal Voltage Drop of a Battery

Batteries have an internal resistance that is a result of the imperfect conducting elements that make up the battery (resistance in electrodes and electrolyte). When a battery is fresh, the internal resistance is low (around a few tenths of a volt), but as the battery is drained, the resistance increases (conducting electrons become fewer in number as the chemical reactants run out). The following circuit shows a more realistic representation of a battery that has a load attached to it.

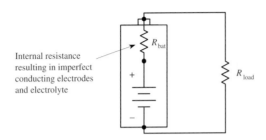

Internal resistance resulting in imperfect conducting electrodes and electrolyte

FIGURE 3.29

Using this new representation of the battery, notice that when a load resistor is attached to it, you get a circuit containing an ideal battery connected to two resistors in series (internal-resistance resistor and load resistor). In effect, the internal-resistance resistor will provide an additional voltage drop within the circuit. The actual supplied voltage "felt" by the load resistor will then be smaller than what is labeled on the battery's case. In reality, it is not important to know the exact value of the internal resistance for a battery. Instead, a battery is placed in a circuit, and the voltage across the battery is measured with a voltmeter. (You cannot remove the battery from the circuit and then measure it; the reading will be tainted by the internal resistance of the meter.) Measuring the voltage across a battery when a load is attached to it gives the true voltage that will be "felt" by the load.

Take note, internal resistance may limit a battery's ability to deliver high currents needed for pulse applications, as in photoflashes and radio signaling. A reliable battery designed for pulsing applications is a silver oxide battery.

3.3   Switches

A *switch* is a mechanical device that interrupts or diverts electric current flow within a circuit.

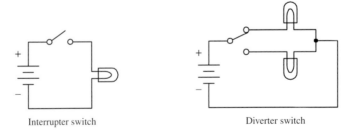

Interrupter switch   Diverter switch

FIGURE 3.30

3.3.1   How a Switch Works

Two slider-type switches are shown in Fig. 3.31. The switch in Fig. 3.31*a* acts as an interrupter, whereas the switch in Fig. 3.31*b* acts as a diverter.

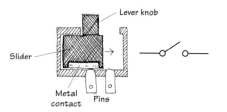
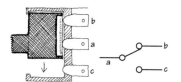

FIGURE 3.31

(a) When the lever is pushed to the right, the metal strip bridges the gap between the two contacts of the switch, thus allowing current to flow. When the lever is pushed to the left, the bridge is broken, and current will not flow.

(b) When the lever is pushed upward, a conductive bridge is made between contacts a and b. When the lever is pushed downward, the conductive bridge is relocated to a position where current can flow between contact a and c.

Other kinds of switches, such as push-button switches, rocker switches, magnetic-reed switches, etc., work a bit differently than slider switches. For example, a magnetic-reed switch uses two thin pieces of leaflike metal contacts that can be forced together by a magnetic field. This switch, as well as a number of other unique switches, will be discussed later on in this section.

3.3.2 Describing a Switch

A switch is characterized by its number of *poles* and by its number of *throws*. A pole represents, say, contact a in Fig. 3.31b. A throw, on the other hand, represents the particular contact-to-contact connection, say, the connection between contacts a and contact b or the connection between contact a and contact c in Fig. 3.31b. In terms of describing a switch, the following format is used: (number of poles) "P" and (number of throws) "T." The letter P symbolizes "pole," and the letter T symbolizes "throw." When specifying the number of poles and the number of throws, a convention must be followed: When the number of poles or number of throws equals 1, the letter S, which stands for "single," is used. When the number of poles or number of throws equals 2, the letter D, which stands for "double," is used. When the number of poles or number of throws exceeds 2, integers such as 3, 4, or 5 are used. Here are a few examples: SPST, SPDT, DPST, DPDT, DP3T, and 3P6T. The switch shown in Fig. 3.31a represents a single-pole single-throw switch (SPST), whereas the switch in Fig. 3.31b represents a single-pole double-throw switch (SPDT).

Two important features to note about switches include whether a switch has momentary contact action and whether the switch has a center-off position. Momentary-contact switches, which include mainly pushbutton switches, are used when it is necessary to only briefly open or close a connection. Momentary-contact switches come in either normally closed (NC) or normally open (NO) forms. A normally closed pushbutton switch acts as a closed circuit (passes current) when left untouched. A normally open pushbutton switch acts as an open circuit (broken circuit) when left untouched. Center-off position switches, which are seen in diverter switches, have an additional "off" position located between the two "on" positions. It is important to note that not all switches have center-off or momentary-contact features—these features must be specified.

Symbols for Switches

SPST SWITCHES
Throw switch
Normally open push-button
Normally closed push-button
SPDT SWITCHES
Throw switch
Normally open/closed push-button
DPST SWITCHES
Throw switch
Normally open push-button
DPDT SWITCHES
Throw switch
Normally open/closed push-button
SP(n)T SWITCHES
Multiple contact slider switch
Multiple contact rotary switch (SP8T)
(n)P(m)T SWITCHES
3P4T
2-deck rotary (DP8T)

FIGURE 3.32

3.3.3 *Kinds of Switches*

FIGURE 3.33

Toggle Switch

ON
1 2
1
2
SPST
1
2
3
1
2
3
SPDT
1 2
3 4
5 6
3
4
1
5
2
6
DPDT

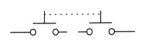

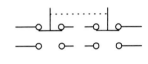

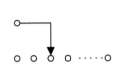

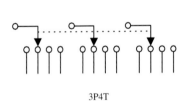

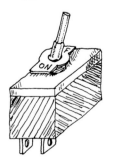

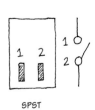

Pushbutton Switch

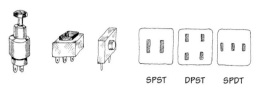

SPST   DPST   SPDT

Snap Switch

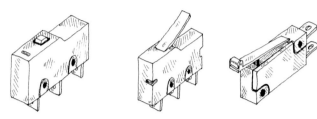

Rotary Switch

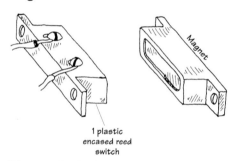

Magnetic Reed Switch

1 plastic
encased reed
switch

A reed switch consists of two closely spaced leaflike contacts that are enclosed in an air-tight container. When a magnetic field is brought nearby, the two contacts will come together (if it is a normally open reed switch) or will push apart (if it is a normally closed reed switch).

Binary-Coded Switches

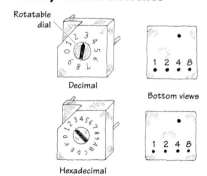

Rotatable dial

Decimal

Hexadecimal

Bottom views

1 2 4 8

These switches are used to encode digital information. A mechanism inside the switch will "make" or "break" connections between the switch pairs according to the position of the dial on the face of the switch. These switches come in either true binary/hexadecimal and complementary binary/hexadecimal forms. The charts below show how these switches work:

True Binary/Hexadecimal

Type		Position	Code			
			1	2	4	8
Hexadecimal	Decimal	0				
		1	•			
		2		•		
		3	•	•		
		4			•	
		5	•		•	
		6		•	•	
		7	•	•	•	
		8				•
		9	•			•
		A		•		•
		B	•	•		•
		C			•	•
		D	•		•	•
		E		•	•	•
		F	•	•	•	•

Complimentary-Binary/Hexadecimal

Type		Position	Code			
			1	2	4	8
Hexadecimal	Decimal	0	•	•	•	•
		1		•	•	•
		2	•		•	•
		3			•	•
		4	•	•		•
		5		•		•
		6	•			•
		7				•
		8	•	•	•	
		9		•	•	
		A	•		•	
		B			•	
		C	•	•		
		D		•		
		E	•			
		F				

FIGURE 3.33   (*Continued*)

DIP Switch

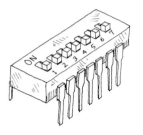

DIP stands for "dual-inline package." The geometry of this switch's pin-outs allows the switch to be placed in IC sockets that can be wired directly into a circuit board.

Mercury Tilt-Over

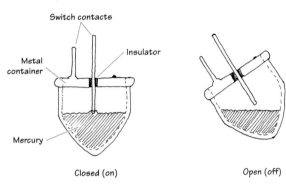

Switch contacts

Metal container

Insulator

Mercury

Closed (on)   Open (off)

This type of switch is used as a level-sensing switch. In a normally closed mercury tilt-over switch, the switch is "on" when oriented vertically (the liquid mercury will make contact with both switch contacts). However, when the switch is tilted, the mercury will be displaced, hence breaking the conductive path.

FIGURE 3.33   (Continued)

3.3.4   Simple Switch Applications

Simple Security Alarm

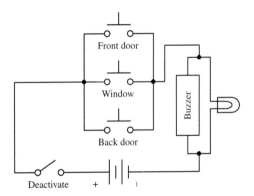

Front door

Window

Buzzer

Back door

Deactivate   +

Here's a simple home security alarm that's triggered into action (buzzer and light go on) when one of the normally open switches is closed. Magnetic reed switches work particularly well in such applications.

FIGURE 3.34

Dual-Location On/Off Switching Network

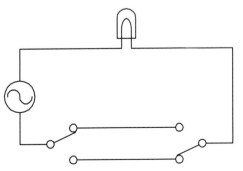

Here's a switch network that allows an individual to turn a light on or off from either of two locations. This setup is frequently used in household wiring applications.

FIGURE 3.35

Current-Flow Reversal

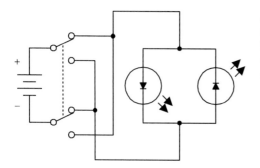

A DPDT switch, shown here, can be used to reverse the direction of current flow. When the switch is thrown up, current will flow throw the left light-emitting diode (LED). When the switch is thrown down, current will flow throw the right LED. (LEDs only allow current to flow in one direction.)

FIGURE 3.36

Multiple Selection Control of a Voltage-Sensitive Device via a Two-Wire Line

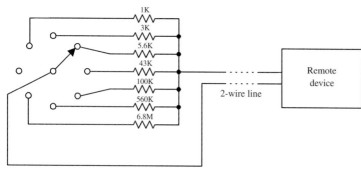

FIGURE 3.37

Say you want to control a remote device by means of a two-wire line. Let's also assume that the remote device has seven different operational settings. One way of controlling the device would be to design the device in such a way that if an individual resistor within the device circuit were to be altered, a new function would be enacted. The resistor may be part of a voltage divider, may be attached in some way to a series of window comparators (see op amps), or may have an analog-to-digital converter interface. After figuring out what valued resistor enacts each new function, choose the appropriate valued resistors and place them together with a rotary switch. Controlling the remote device becomes a simple matter of turning the rotary switch to select the appropriate resistor.

3.4   Relays

Relays are electrically actuated switches. The three basic kinds of relays include mechanical relays, reed relays, and solid-state relays. For a typical mechanical relay, a current sent through a coil magnet acts to pull a flexible, spring-loaded conductive plate from one switch contact to another. Reed relays consist of a pair of reeds (thin, flexible metal strips) that collapse whenever a current is sent through an encapsulating wire coil. A solid-state relay is a device that can be made to switch states by applying external voltages across *n*-type and *p*-type semiconductive junctions (see Chap. 4). In general, mechanical relays are designed for high currents (typically 2 to 15 A) and relatively slow switching (typically 10 to 100 ms). Reed relays are designed for moderate currents (typically 500 mA to 1 A) and moderately fast switching (0.2 to 2 ms). Solid-state relays, on the other hand, come with a wide range of current ratings (a few microamps for low-powered packages up to 100 A for high-power packages) and have extremely fast switching speeds (typically 1 to 100 ns). Some limitations of both reed relays and solid-state relays include limited switching arrangements (type of switch section) and a tendency to become damaged by surges in power.

Mechanical Relay   Reed Relay

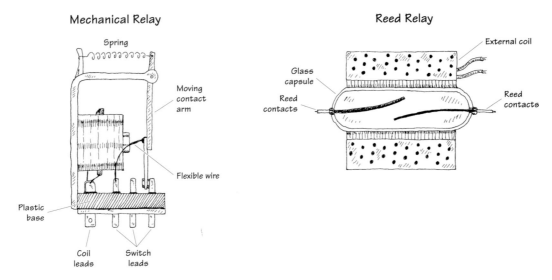

FIGURE 3.38

A mechanical relay's switch section comes in many of the standard manual switch arrangements (e.g., SPST, SPDT, DPDT, etc.). Reed relays and solid-state relays, unlike mechanical relays, typically are limited to SPST switching. Some of the common symbols used to represent relays are shown below.

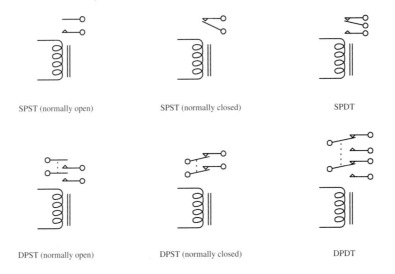

SPST (normally open)   SPST (normally closed)   SPDT

DPST (normally open)   DPST (normally closed)   DPDT

FIGURE 3.39

The voltage used to activate a given relay may be either dc or ac. For, example, when an ac current is fed through a mechanical relay with an ac coil, the flexible-metal conductive plate is pulled toward one switch contact and is held in place as long as the current is applied, regardless of the alternating current. If a dc coil is supplied by an alternating current, its metal plate will flip back and forth as the polarity of the applied current changes.

Mechanical relays also come with a latching feature that gives them a kind of memory. When one control pulse is applied to a *latching relay*, its switch closes. Even when the control pulse is removed, the switch remains in the closed state. To open the switch, a separate control pulse must be applied.

3.4.1 Specific Kinds of Relays

Subminiature Relays

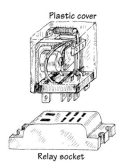

Plastic cover

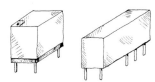

Relay socket

Typical mechanical relays are designed for switching relatively large currents. They come with either dc or ac coils. Dc-actuated relays typically come with excitation-voltage ratings of 6, 12, and 24 V dc, with coil resistances (coil ohms) of about 40, 160, and 650 Ω, respectively. Ac-actuated relays typically come with excitation-voltage ratings of 110 and 240 V ac, with coil resistances of about 3400 and 13600 Ω, respectively. Switching speeds range from about 10 to 100 ms, and current ratings range from about 2 to 15 A.

Miniature Relays

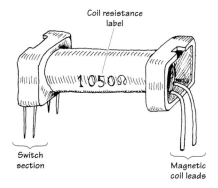

Miniature relays are similar to subminiature relays, but they are designed for greater sensitivity and lower-level currents. They are almost exclusively actuated by dc voltages but may be designed to switch ac currents. They come with excitation voltages of 5, 6, 12, and 24 V dc, with coil resistances from 50 to 3000 Ω.

Reed Relays

Coil resistance label

1050Ω

Switch section

Magnetic coil leads

Two thin metal strips, or reeds, act as movable contacts. The reeds are placed in a glass-encapsulated container that is surrounded by a coil magnet. When current is sent through the outer coil, the reeds are forced together, thus closing the switch. The low mass of the reeds allows for quick switching, typically around 0.2 to 2 ms. These relays come with dry or mercury-wetted contacts. They are dc-actuated and are designed to switch lower-level currents, and come with excitation voltages of 5, 6, 12, and 24 V dc, with coil resistances around 250 to 2000 Ω. Leads are made for PCB mounting.

Solid-State Relays

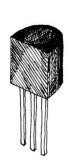
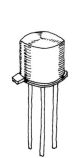

These relays are made from semiconductor materials. Solid-state relays include transistors (FETs, BJTs) and thyristors (SCRs, triacs, diacs, etc.). Solid-state relays do not have a problem with contact wear and have phenomenal switching speeds. However, these devices typically have high "on" resistances, require a bit more fine tuning, and are much less resistant to overloads when compared with electromechanical relays. Solid-state devices will be covered later in this book.

FIGURE 3.40

3.4.2   A Few Notes about Relays

To make a relay change states, the voltage across the leads of its magnetic coil should be at least within ±25 percent of the relay's specified control-voltage rating. Too much voltage may damage or destroy the magnetic coil, whereas too little voltage may not be enough to "trip" the relay or may cause the relay to act erratically (flip back and forth).

The coil of a relay acts as an inductor. Now, inductors do not like sudden changes in current. If the flow of current through a coil is suddenly interrupted, say, a switch is opened, the coil will respond by producing a sudden, very large voltage across its leads, causing a large surge of current through it. Physically speaking, this phenomenon is a result of a collapsing magnetic field within the coil as the current is terminated abruptly. [Mathematically, this can be understood by noticing how a large change in current (dI/dt) affects the voltage across a coil ($V = L\,dI/dt$).] Surges in current that result from inductive behavior can create menacing voltage spikes (as high as 1000 V) that can have some nasty effects on neighboring devices within the circuit (e.g., switches may get zapped, transistors may get zapped, individuals touching switches may get zapped, etc.). Not only are these spike damaging to neighboring devices, they are also damaging to the relay's switch contacts (contacts will suffer a "hard hit" from the flexible-metal conductive plate when a spike occurs in the coil).

The trick to getting rid of spikes is to use what are called *transient suppressors*. You can buy these devices in prepackaged form, or you can make them yourself. The following are a few simple, homemade transient suppressors that can be used with relay coils or any other kind of coil (e.g., transformer coils). Notably, the switch incorporated within the networks below is only one of a number of devices that may interrupt the current flow through a coil. In fact, a circuit may not contain a switch at all but may contain other device (e.g., transistors, thyristors, etc.) that may have the same current-interrupting effect.

FIGURE 3.41

DC-Driven Coil

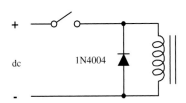

Placing a diode in reverse bias across a relay's coil eliminates voltage spikes by going into conduction before a large voltage can form across the coil. The diode must be rated to handle currents equivalent to the maximum current that would have been flowing through the coil before the current supply was interrupted. A good general-purpose diode that works well for just such applications is the 1N4004 diode.

AC-Driven Coil

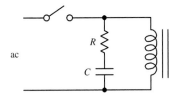

When dealing with ac-actuated relays, using a diode to eliminate voltage spikes will not work—the diode will conduct on alternate half-cycles. Using two diodes in reverse parallel will not work either—the current will never make it to the coil. Instead, an *RC* series network placed across the coil can be used. The capacitor absorbs excessive charge, and the resistor helps control the discharge. For small loads driven from the power line, setting $R = 100\ \Omega$ and $C = 0.05\ \mu\text{F}$ works fine for most cases.

3.4.3   Some Simple Relay Circuits

DC-Actuated Switch

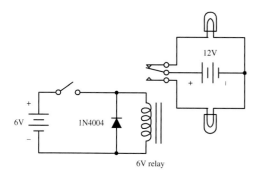

Here, a dc-powered SPDT relay is used to relay current to one of two light bulbs. When the switch in the control circuit is opened, the relay coil receives no current; hence the relay is relaxed, and current is routed to the upper bulb. When the switch in the control circuit is closed, the relay coil receives current and pulls the flexible-metal conductive plate downward, thus routing current to the lower bulb. The diode acts as a transient suppressor. Note that all components must be selected according to current and voltage ratings.

AC-Actuated Switch

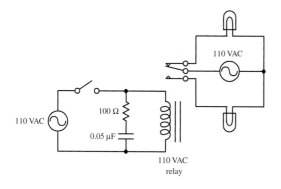

Here, an ac-actuated relay is used to switch ac current to one of two ac-rated light bulbs. The behavior in this circuit is essentially the same as in the preceding circuit. However, currents and voltages are all ac, and an *RC* network is used as a transient suppressor. Make sure that resistor and capacitor are rated for a potential transient current that is as large as the typical coil current.

Relay Driver

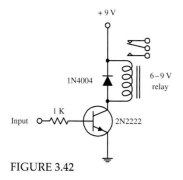

FIGURE 3.42

If a relay is to be driven by an arbitrary control voltage, this circuit can be used. The *npn* bipolar transistor acts as a current-flow control valve. With no voltage or input current applied to the transistor's base lead, the transistor's collector-to-emitter channel is closed, hence blocking current flow through the relay's coil. However, if a sufficiently large voltage and input current are applied to transistor's base lead, the transistor's collector-to-emitter channel opens, allowing current to flow through the relay's coil.

3.5   Resistors

Resistors are electrical devices that act to reduce current flow and at the same time act to lower voltage levels within circuits. The relationship between the voltage applied across a resistor and the current through it is given by $V = IR$.

There are numerous applications for resistors. Resistors are used to set operating current and signal levels, provide voltage reduction, set precise gain values in precision circuits, act as shunts in ammeters and voltage meters, behave as damping agents in oscillators, act as bus and line terminators in digital circuits, and provide feedback networks for amplifiers.

Resistors may have fixed resistances, or they may be designed to have variable resistances. They also may have resistances that change with light or heat exposure (e.g., photoresistors, thermistors).

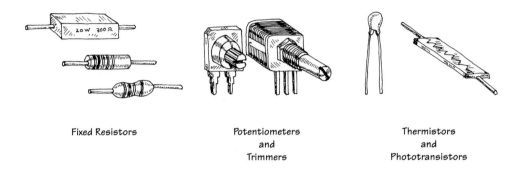

Fixed Resistors   Potentiometers   Thermistors
                              and   and
                            Trimmers   Phototransistors

FIGURE 3.43

3.5.1   How a Resistor Works

An electric current is full of electrons teaming with kinetic energy. The job of a resistor is to absorb some of this energy and send it elsewhere. When electrons enter one end of a resistor, some of them will "collide" with the atoms within, transferring some of their energy into atomic vibrations. These vibrations, in turn, act as a "drum" and transfer their energy to neighboring air molecules or perhaps a metal heat-sink device. The overall effect of the collisions within the resistor results in a smaller, less energetic current.

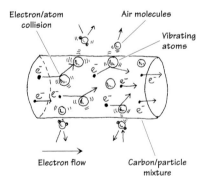

FIGURE 3.44

A real-life example of a resistor is a simple *wire-wound resistor*. This device consists of a resistive wire whose overall resistance increases with length. Coiling the wire around an insulative cylinder makes the device compact. A *carbon-composition resistor,* on the other hand, uses a mixture of carbon power and a gluelike binder for its resistive element. When more carbon is added to the mixture, the device becomes less resistive. Other resistors use metal films, metal particle mixes, or some other kinds of materials as their resistive element.

3.5.2 Basic Resistor Operation

The following are some very useful resistor networks that are employed frequently within complex circuits.

Current Limiter

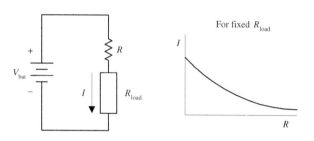

$$I = \frac{V_{bat}}{R + R_{load}}$$

A resistor (R) placed in series with a load resistance (R_{load}) acts to reduce current (I) through the load. Increasing R decreases I. At the same time, the voltage across the load decreases (see next example for voltage-drop characteristics).

Voltage Divider

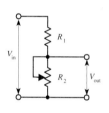 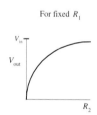 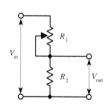 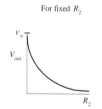

This is the voltage divider relation that is used to describe the whole process:

$$V_{out} = \frac{R_2}{R_2 + R_1} V_{in}$$

Current Divider

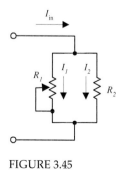 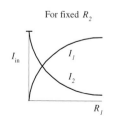

$$I_1 = \frac{R_2}{R_1 + R_2} I_{in}$$

$$I_1 = \frac{R_1}{R_1 + R_2} I_{in}$$

When current I_{in} enters the top junction, part of it goes through R_1, while the other part goes through R_2. If the resistance of R_1 increases, more current is diverted through R_2. The graph and equations in the figure describe how the current through R_1 and R_2 is influenced by changing R_1.

FIGURE 3.45

3.5.3   Types of Fixed Resistors

Carbon Film

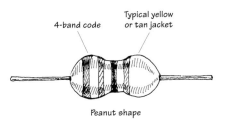

This is one of the most popular resistors used today. It is made by depositing a carbon film onto a small ceramic cylinder. A small spiral groove cut into the film controls the amount of carbon between the leads, hence setting the resistance. Such resistors show excellent reliability, excellent solderability, noise stability, moisture stability, and heat stability. Typical power ratings range from ¼ to 2 W. Resistances range from about 10 Ω to 1 MΩ, with tolerances around 5 percent.

Carbon Composition

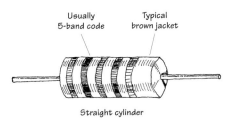

This is another very popular resistor made from a mixture of carbon powder and gluelike binder. To increase the resistance, less carbon is added. These resistors show predictable performance, low inductance, and low capacitance. Power ratings range from about ⅛ to 2 W. Resistances range from 1 Ω to about 100 MΩ, with tolerances around ±5 percent.

Metal Oxide Film

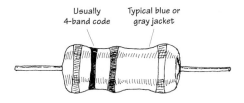

This is a general-purpose resistor that uses a ceramic core coated with a metal oxide film. These resistors are mechanically and electrically stable and readable during high-temperature operation. They contain a special paint on their outer surfaces making them resistant to flames, solvents, heat, and humidity. Typical resistances range from 1 Ω to 200 kΩ, with typical tolerances of ±5 percent.

Precision Metal Film

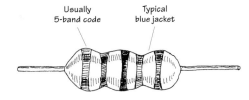

This is a very accurate, ultra-low-noise resistor. It uses a ceramic substrate coated with a metal film, all encased in an epoxy shell. These resistors are used in precision devices, such as test instruments, digital and analog devices, and audio and video devices. Resistances range from about 10 Ω to 2 MΩ, with power rating from ⅛ to about ½ W and tolerances of ±1 percent.

High-Power Wire Wound

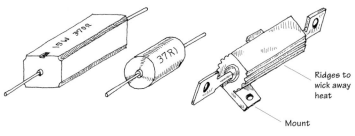

These resistors are used for high-power applications. Types include vitreous enamel coated, cement, and aluminum housed wire-wound resistors. Resistive elements are made from a resistive wire that is coiled around a ceramic cylinder. These are the most durable of the resistors, with high heat dissipation and high temperature stability. Resistances range from 0.1 Ω to about 150 kΩ, with power ratings from around 2 W to as high as 500 W, or more.

FIGURE 3.46

Photoresistors and Thermistors

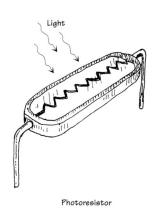

Photoresistor

Thermistor

These are special types of resistors that change resistance when heat or light is applied. Photoresistors are made from semiconductive materials, such as cadmium sulfide. Increasing the light level decreases the resistance. Photoresistors are covered in greater detail in Chap. 5. Thermistors are temperature-sensitive resistors. Increasing the temperature decreases the resistance (in most cases).

FIGURE 3.46   (*Continued*)

3.5.4   *Understanding Resistor Labels*

Resistors use either a series of painted bands or written labels to specify resistance values. Other things that may be on the label include tolerances (the percentage uncertainty between the labeled resistance and the actual resistance), a temperature coefficient rating (not common), and a reliability level rating (reliability that the resistor will maintain its tolerance over a 1000-hour cycle). Here are some common labeling schemes used today.

FIGURE 3.47

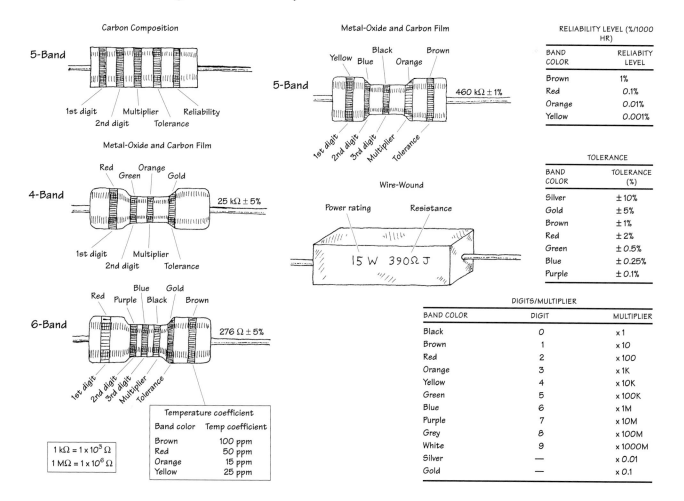

3.5.5 Power Ratings for Resistors

It is important to note that two resistors may have the same resistance values but different power ratings (wattage rating). Resistors with higher power ratings can dissipate heat generated by a current more effectively. Replacing a high-power resistor with a lower-powered one may lead to "meltdown." If you are not sure what size resistor you need, perhaps the following example will help.

EXAMPLE

Suppose that you want to send a 10-mA (0.01-A) current through a 3300-Ω resistor. What kind of power rating should the resistor have?

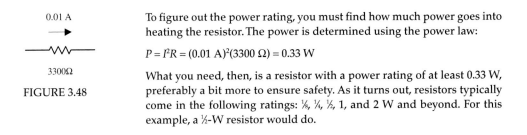

0.01 A

3300Ω

FIGURE 3.48

To figure out the power rating, you must find how much power goes into heating the resistor. The power is determined using the power law:

$$P = I^2R = (0.01 \text{ A})^2(3300 \ \Omega) = 0.33 \text{ W}$$

What you need, then, is a resistor with a power rating of at least 0.33 W, preferably a bit more to ensure safety. As it turns out, resistors typically come in the following ratings: ⅛, ¼, ½, 1, and 2 W and beyond. For this example, a ½-W resistor would do.

Notably, when dealing with ac currents and voltages, make sure you use the rms values for currents and voltages in the power equation (e.g., $P = I_{rms}^2R = V_{rms}^2/R$).

3.5.6 Variable Resistors

Variable resistors provide varying degrees of resistance that can be set with the turn of a knob.

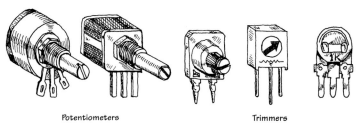

Potentiometers   Trimmers

FIGURE 3.49

Special kinds of variable resistors include potentiometers, rheostats, and trimmers. Potentiometers and rheostats are essentially the same thing, but rheostats are used specifically for high-power ac electricity, whereas potentiometers typically are used with lower-level dc electricity. Both potentiometers and rheostats are designed for frequent adjustment. Trimmers, on the other hand, are miniature potentiometers that are adjusted infrequently and usually come with pins that can be inserted into printed-circuit boards. They are used for fine-tuning circuits (e.g., fine-tuning a circuit that goes astray as it ages), and they are usually hidden within a circuit's enclosure box.

Variable resistors come with two or three terminals.

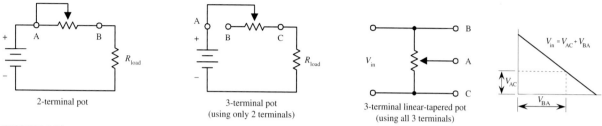

FIGURE 3.50

When purchasing variable resistors, make sure that you understand the distinction between linear-tapered and nonlinear-tapered variable resistors. The "taper" describes the way in which the resistance changes as the control knob is twisted. Figure 3.51 shows how the resistance changes as the control knob is turned for both a linear and nonlinear-tapered variable resistor.

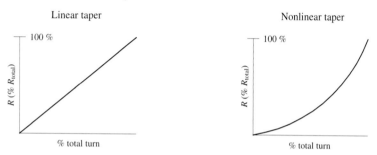

FIGURE 3.51

Why do variable resistors come with nonlinear tapers? Well, as it turns out, human physiology has a weird way of perceiving changes in signal intensity, such as sound and light intensities. For example, you may think that if you doubled the intensity of sound or light, you would perceive a doubling in sound and light. Unfortunately—at least in terms of intuition (not in terms of safety control for our brains)—humans do not work this way. In fact, our perceptions of sight and sound work as follows: Perceived loudness/brightness is proportional to $\log_{10}$ (actual intensity measured with a nonhuman instrument). Thus, if you are building an amplifier for a set of speakers or building a home light-dimming circuit, it would be wise to use a variable resistor with a nonlinear taper.

3.6 Capacitors

As you discovered in Chap. 2, capacitors act as temporary charge-storage units, whose behavior can be described by $I = C dV/dt$. A simple explanation of what this equation tells us is this: If you supply a 1-μF capacitor with a 1-mA, 1-s-long pulse of current, the voltage across its leads will increase 1000 V ($dV = Idt/C$). More generally, this equation states that a capacitor "likes" to pass current when the voltage across its leads is changing with time (e.g., high-frequency ac signals) but "hates" to pass current when the applied voltage is constant (e.g., dc signals). A capacitor's "dislike" for passing a current is given by its capacitive reactance $X_c = 1/\omega C$ (or $Z_c = -j/\omega C$ in complex form). As the applied voltage's frequency approaches infinity, the capacitor's reactance goes to zero, and it acts like a perfect conductor. However, as the frequency approaches zero, the capacitor's reactance goes to infinity, and it acts like an infinitely large resistor. Changing the value of C also affects the reactance. As C gets large, the reactance decreases, and the displacement current increases.

In terms of applications, the ability of a capacitor to vary its reactance as the voltage across its leads fluctuates makes it a particularly useful device in frequency-sensitive applications. For example, capacitors are used in frequency-sensitive voltage dividers, bypassing and blocking networks, filtering networks, transient noise suppressors, differentiator circuits, and integrator circuits. Capacitors are also used in voltage-multiplier circuits, oscillator circuits, and photoflash circuits.

3.6.1   How a Capacitor Works

A simple capacitor consists of two parallel plates. When the two plates are connected to a dc voltage source (e.g., a battery), electrons are "pushed" onto one plate by the negative terminal of the battery, while electrons are "pulled" from the other plate by the positive terminal of the battery. If the charge difference between the two plates become excessively large, a spark may jump across the gap between them and discharge the capacitor. To increase the amount of charge that can be stored on the plates, a nonconducting dielectric material is placed between them. The dielectric acts as a "spark blocker" and consequently increases the charge capacity of the capacitor. Other factors that affect capacitance levels include the capacitor's plate surface area and the distance between the parallel plates. Commercially, a capacitor's dielectric may be either paper, plastic film, mica, glass, ceramic, or air, while the plates may be either aluminum disks, aluminum foil, or a thin film of metal applied to opposite sides of a solid dielectric. The conductor-dielectric-conductor sandwich may be left flat or rolled into a cylinder. The figure below shows some examples.

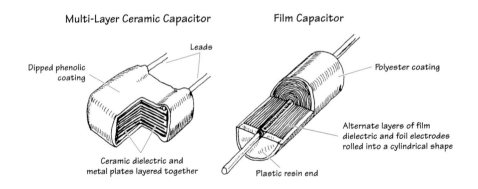

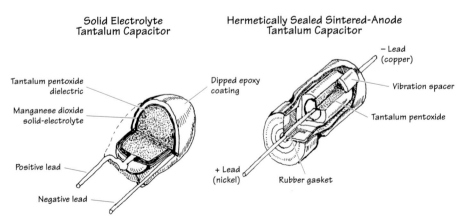

FIGURE 3.52

3.6.2 A Note about I = CdV/dt

Note that according to the laws of physics, no individual charges (electrons) ever make it across the gap separating the capacitor plates. However, according to $I = CdV/dt$, it appears as if there is a current flow across the gap. How can both these statements be true? Well, as it turns out, both are true. The misleading thing to do is to treat I as if it were a conventional current, such as a current through a resistor or wire. Instead, I represents a *displacement current*. A displacement current represents an apparent current through the capacitor that results from the act of the plates charging up with time and produces magnetic fields that induce electrons to move in the opposite plates. Current never goes across the gap—instead, charges on one plate are "shoved" by the changing magnetic fields produced by the opposing plate. The overall effect makes things appear as if current is flowing across the gap.

3.6.3 Water Analogy of a Capacitor

Here's the best thing I could come up with in terms of a water analogy for a capacitor. Pretend that electrons are water molecules and that voltage is water pressure.

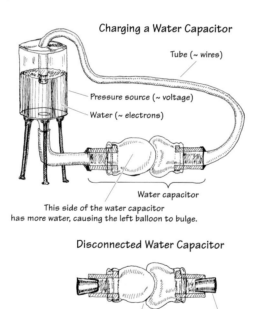

FIGURE 3.53

The water capacitor is built from two balloons. Normally, the two balloons are filled with the same amount of water; the pressure within each one is the same (analogous to an uncharged capacitor). In the figure, a real capacitor is charged by a battery, whereas the water capacitor is "charged" using a pump or pressurized water source. The real capacitor has a voltage across its plates, whereas in the water capacitor there is a difference in pressure between the two balloons. When the real capacitor is removed from the battery, it retains its charge; there is no conductive path through which charges can escape. If the water capacitor is removed from the pressurized water source—you have to pretend that corks are placed in its lead pipes—it too retains its stored-up pressure. When an alternating voltage is applied across a real capacitor, it appears as if an alternating current (displacement current) flows through the capacitor due to the changing magnetic fields. In the water capacitor, if an alternating pressure is applied across its lead tubes, one balloon will fill with water and push against the other semifilled balloon, causing water to flow out it. As the frequency of the applied pressure increases, the water capacitor resembles a rubber membrane that fluctuates very rapidly back and forth, making it appear as if it is a short circuit (at least in ac terms). In reality, this analogy is overly simplistic and does not address the subtleties involved in a real capacitor's operation. Take this analogy lightheartedly.

3.6.4   Basic Capacitor Functions

FIGURE 3.54

Getting a Sense of What $I = CdV/dt$ Means

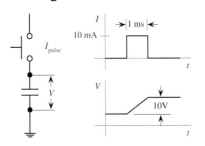

If you supply a 1-μF capacitor with a 10-mA, 1-ms pulse of current, the voltage across its leads will increase by 10 V. Here's how to figure it out:

$$I = C\frac{dV}{dt}$$

$$dV = \frac{Idt}{C} = \frac{(10\ \text{mA})(1\ \text{ms})}{1\ \mu F} = 10\ V$$

Charging/Discharging a Capacitor through a Resistor

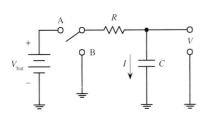

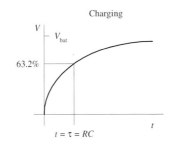

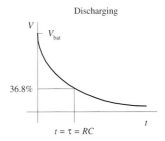

When the switch is thrown to position A, the capacitor charges through the resistor, and the voltage across its leads increases according to the far-left graph. If the switch is thrown to position B after the capacitor has been charged, the capacitor will discharge, and the voltage will decay according to the near-right graph. See Section 2.18 for details.

Signal Filtering

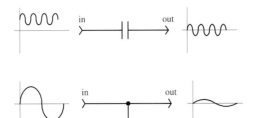

When a fluctuating signal with a dc component is sent through a capacitor, the capacitor removes the dc component yet allows the ac component through.

A capacitor can be used to divert unwanted fluctuating signals to ground.

3.6.5 Kinds of Capacitors

There are a number of different capacitor families available, each of which has defining characteristic features. Some families are good for storing large amounts of charge yet may have high leakage currents and bad tolerances. Other families may have great tolerances and low leakage currents but may not have the ability to store large amounts of charge. Some families are designed to handle high voltages yet may be bulky and expensive. Other families may not be able to handle high voltages but may have good tolerances and good temperature performance. Some families may contain members that are polarized or nonpolarized in nature. Polarized capacitors, unlike nonpolarized capacitors, are specifically designed for use with dc fluctuating voltages (a nonpolarized capacitor can handle both dc and ac voltages). A polarized capacitor has a positive lead that must be placed at a higher potential in a circuit and has a negative lead that must be placed at a lower potential. Placing a polarized capacitor in the wrong direction may destroy it. (Polarized capacitors' limitation to use in dc fluctuating circuits is counterbalanced by extremely large capacitances.) Capacitors also come in fixed or variable forms. Variable capacitors have a knob that can be rotated to adjust the capacitance level. The symbols for fixed, polarized, and variable capacitors are shown below.

Fixed capacitor   Polarized capacitor   Variable capacitor

FIGURE 3.55

FIGURE 3.56   Now let's take a closer look at the capacitor families.

Electrolytic

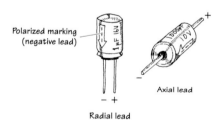

Polarized marking (negative lead)

Axial lead

Radial lead

These capacitors include both aluminum and tantalum electrolytics. They are manufactured by an electrochemical formation of an oxide film onto a metal (aluminum or tantalum) surface. The metal on which the oxide film is formed serves as the anode or positive terminal, the oxide film acts as the dielectric, and a conducting liquid or gel acts as the cathode or negative terminal. Tantalum electrolytic capacitors have larger capacitance per volume ratios when compared with aluminum electrolytics. A majority of electrolytic capacitors are polarized. Electrolytic capacitors, when compared with nonelectrolytic capacitors, typically have greater capacitances but have poor tolerances (as large as ±100 percent for aluminum and about ±5 to ±20 percent for tantalum), bad temperature stability, high leakage, and short lives. Capacitances range from about 1 μF to 1 F for aluminum and 0.001 to 1000 μF for tantalum, with maximum voltage ratings from 6 to 450 V.

Ceramic

This is very popular nonpolarized capacitor that is small and inexpensive but has poor temperature stability and poor accuracy. It contains a ceramic dielectric and a phenolic coating. It is often used for bypass and coupling applications. Tolerances range from ±5 to ±100 percent, while capacitances range from 1 pF to 2.2 μF, with maximum voltages rating from 3 V to 6 kV.

Mylar

This is a very popular nonpolarized capacitor that is reliable, inexpensive, and has low leakage current but poor temperature stability. Capacitances range from 0.001 to 10 µF, with voltages ratings from 50 to 600 V.

Mica

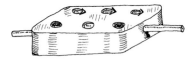

FIGURE 3.56   (*Continued*)

This is an extremely accurate device with very low leakage currents. It is constructed with alternate layers of metal foil and mica insulation, stacked and encapsulated. These capacitors have small capacitances and are often used in high-frequency circuits (e.g., RF circuits). They are very stable under variable voltage and temperature conditions. Tolerances range from ±0.25 to ±5 percent. Capacitances range from 1 pF to 0.01 µF, with maximum voltage ratings from 100 V to 2.5 KV

Other kinds of capacitors include paper, polystyrene, polycarbonate, polyester, glass, and oil capacitors; their characteristics are covered in Table 3.7. Note that in Table 3.7 a large *insulation resistance* means that a capacitor has good leakage protection.

TABLE 3.7   Characteristics of Various Capacitors

TYPE	CAPACITANCE RANGE	MAXIMUM VOLTAGE	MAXIMUM OPERATING TEMPERATURE (°C)	TOLERANCE (%)	INSULATION RESISTANCE (MΩ)	COMMENTS
Electrolytics Aluminum Tantalum	1 µF–1 F 0.001–1000 µF	3–600 V 6–100 V	85 125	+100 to −20 ±5 to 20	<1 >1	Popular, large capacitance, awful leakage, horrible tolerances
Ceramic	10 pF–1 µF	50–1000 V	125	±5 to 100	1000	Popular, small, inexpensive, poor tolerances.
Mica	1 pF–0.1 µF	100–600 V	150	±0.25 to ±5	100,000	Excellent performance; used in high-frequency applications
Mylar	0.001–10 µF	50–600 V		Good	Good	Popular, good performance, inexpensive
Paper	500 pF–50 µF	100,000 V	125	±10 to ±20	100	—
Polystyrene	10 pF–10 µF	100–600 V	85	±0.5	10,000	High quality, very accurate; used in signal filters
Polycarbonate	100 pF–10 µF	50–400 V	140	±1	10,000	High quality, very accurate
Polyester	500 pF–10 µF	600 V	125	±10	10,000	—
Glass	10–1000 pF	100–600 V	125	±1 to ±20	100,000	Long-term stability
Oil	0.1–20 µF	200 V–10 kV			Good	Large, high-voltage filters, long life

3.6.6 Variable Capacitors

Variable capacitors are devices that can be made to change capacitance values with the twist of a knob. These devices come in either air-variable or trimmer forms. Air-variable capacitors consist of two sets of aluminum plates (stator and rotor) that mesh together but do not touch. Rotating the rotor plates with respect to the stator varies the capacitor's effective plate surface area, thus changing the capacitance. Air-variable capacitors typically are mounted on panels and are used in frequently adjusted tuning applications (e.g., tuning communication receivers over a wide band of frequencies). A trimmer capacitor is a smaller unit that is designed for infrequent fine-tuning adjustments (e.g., fine-tuning fixed-frequency communications receivers, crystal frequency adjustments, adjusting filter characteristics). Trimmers may use a mica, air, ceramic, or glass dielectric and may use either a pair of rotating plates or a compression-like mechanism that forces the plates closer together.

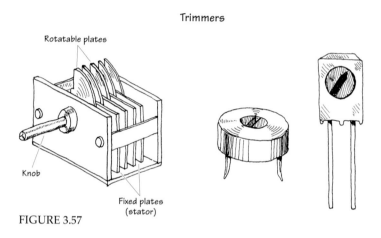

FIGURE 3.57

3.6.7 Reading Capacitor Labels

Reading capacitor labels is tricky business. Each family of capacitors uses its own unique labeling system. Some systems are easy to understand, whereas others make use of misleading letters and symbols. The best way to figure out what a capacitor label means is to first figure out what family the capacitor belongs to. After that, try seeing if the capacitor label follows one of the conventions in Fig. 3.58.

3.6.8 Important Things to Know about Capacitors

Even though two capacitors may have the same capacitance values, they may have different voltage ratings. If a smaller-voltage capacitor is substituted in place of a higher-voltage capacitor, the voltage level across the replacement may "zap" its dielectric, turning it into a low-level resistor. Also remember that with polarized capacitors, the positive lead must go to the more positive connection; otherwise, it may get zapped as well.

As a practical note, capacitor tolerances can be atrocious. For example, an aluminum electrolytic capacitor's capacitance may be as much as 20 to 100 percent off the actual value. If an application specifies a low-tolerance capacitor, it is usually safe to substitute a near-value capacitor in place of the specified one.

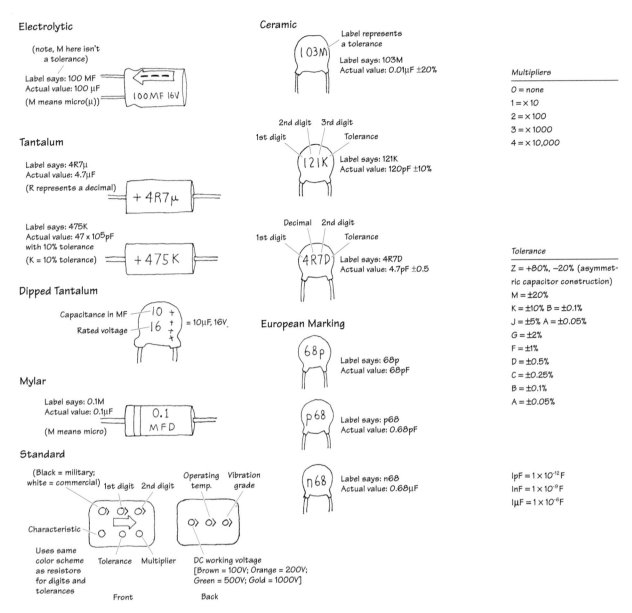

Multipliers

0 = none
1 = × 10
2 = × 100
3 = × 1000
4 = × 10,000

Tolerance

Z = +80%, −20% (asymmetric capacitor construction)
M = ±20%
K = ±10% B = ±0.1%
J = ±5% A = ±0.05%
G = ±2%
F = ±1%
D = ±0.5%
C = ±0.25%
B = ±0.1%
A = ±0.05%

$1pF = 1 \times 10^{-12}F$
$1nF = 1 \times 10^{-9}F$
$1\mu F = 1 \times 10^{-6}F$

FIGURE 3.58

As a final note, capacitors with small capactances (less than 0.01 μF) do not pose much danger to humans. However, when the capacitances start exceeding 0.1 μF, touching capacitor leads can be a shocking experience. For example, large electrolytic capacitors found in television sets and photoflashes can store a lethal charge. As a rule, never touch the leads of large capacitors. If in question, discharge the capacitor by shorting the leads together with a screwdriver tip before handling it.

FIGURE 3.59

3.6.9 Applications

Power Supply Filtering

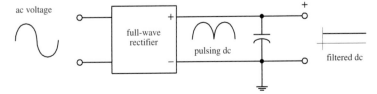

Capacitors are often used to smooth out pulsing dc signals generated by rectifier sections within power supplies by diverting the ac portion of the wavelength to ground, while passing the dc portion.

Filter Circuits

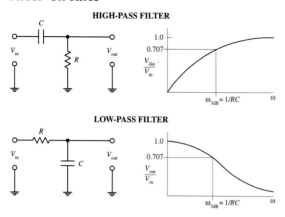

High-pass and low-pass filters can be made from simple *RC* networks. In a high-pass filter, low frequencies are blocked by the capacitor, but high-frequency signals get through to the output. In a low-pass filter, high-frequency signals are routed to ground, while low-frequency signals are sent to the output. See Chaps. 2 (section 2.30) and 8 for more on filters.

Passive Integrator

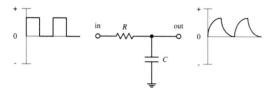

The input signal is integrated at the output. *RC* is the time constant, and it must be at least 10 times the period of the input signal. If not, the amplitude of the output will then be a low-pass filter that blocks high frequencies.

Passive Differentiator

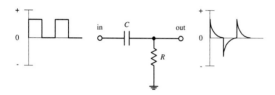

The input signal is differentiated at the output. *RC* network transforms incoming square waves into a pulsed or spiked waveform. The *RC* time constant should be ¹⁄₁₀ (or less) of the duration of the incoming pulses. Differentiators are often used to create trigger pulses.

Spike and Noise Suppression

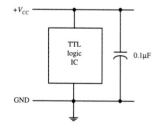

With certain types of ICs (e.g., TTL logic ICs), sudden changes in input and output states can cause brief but drastic changes in power supply current which can result in sharp, high-frequency current spikes within the power supply line. If a number of other devices are linked to the same supply line, the unwanted spike can cause false triggering of these devices. The spikes can also generate unwanted electromagnetic radiation. To avoid unwanted spikes, decoupling capacitors can be used. A decoupling capacitor [typically 0.1 to 1 μF (>5V) for TTL logic ICs] is placed directly across the positive (V_{CC}) and ground pins of each IC within a system. The capacitors help absorb the spikes and keep the V_{CC} level constant, thus reducing the likelihood of false triggering and electromagnetic radiation.

Simple Oscillator

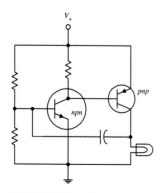

A capacitor can be used like a "spring" in an oscillator circuit. Here, a capacitor alters the biasing voltage at the *npn* transistor's base. At one moment, the *npn* transistor receives a biasing voltage large enough to turn it on. This causes the *pnp* transistor to turn on too (its base becomes properly biased). Current will flow through the bulb, causing it to glow. However, sometime later, the capacitor stores enough charge to bring the voltage level at the *npn* transistor's base down, turning it off. The *pnp* transistor then turns off, and the light turns off. But shortly afterward, the charge from the capacitor flows through lower resistor to ground, and the voltage at the *npn* transistor's base goes back up. It turns on again, and the process repeats itself over and over again. The resistance values of the voltage-divider resistors and the capacitance value of the capacitor control the oscillatory frequency.

FIGURE 3.59　*(Continued)*

3.7   Inductors

As you learned in Chap. 2, inductors act to resist changes in current flow while freely passing steady (dc) current. The behavior of an inductor can be described by $V = LdI/dt$. This equation tells you that if the current flow through a 1-H inductor is changing at a rate of 1 A/s, the induced voltage across the inductor will be 1 V. Now, the direction (polarity) of the voltage will be in the direction that opposes the change in current. For example, if the current increases, the voltage inducted across the inductor will be more negative at the end through which current enters. If the current decreases, the voltage at the end of the inductor through which current enters will become more positive in an attempt to keep current flowing.

More generally, $V = LdI/dt$ states that an inductor "likes" to pass current when the voltage across its leads does not change (e.g., dc signals) but "hates" to pass current when the voltage across its leads is changing with time (e.g., high-frequency ac signals). An inductor's "dislike" for passing a current is given by its inductive reactance $X_L = \omega L$ (or $Z_L = j\omega L$ in complex form). As the applied voltage's frequency approaches zero, the inductor's reactance goes to zero, and it acts like a perfect conductor. However, as the frequency approaches infinity, the inductor's reactance goes to infinity, and it acts like an infinitely large resistor. (Using $X_L = \omega L$, you can see that if you take a 20-mH coil and apply a 100-kHz signal across it, the inductive reactance will be 2000 Ω.) Notice that L also influences the inductor's reactance. As L increases, the inductor's reactance increases, while the current flow through it decreases.

In terms of applications, the ability of an inductor to vary its reactance as the voltage across its leads fluctuates makes it a particularly useful device in frequency-sensitive applications. For example, inductors are used in frequency-sensitive voltage dividers, blocking networks, and filtering networks (e.g., RF transmitter and receiver circuits). Inductors are also used in oscillator circuits, transformers, and magnetic relays—where they act as magnets.

3.7.1   How an Inductor Works

The physical explanation of how an inductor resists changes in current flow can be explained by the following figure. Assume that you have a steady (dc) current flowing through an inductor's coil, as shown at left below.

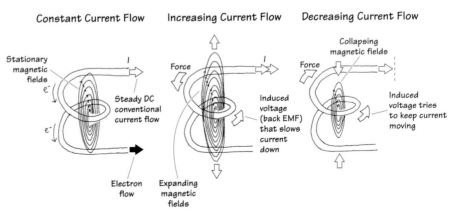

FIGURE 3.60

Within the dc current, electrons are collectively moving at a constant velocity, and as a result of their movement, a steady flux of magnetic fields encircles them. The direction of the magnetic fields is found by using the left-hand rule. The left-hand rule tells us that if we pretend to be hitch-hiking with our left hand, the direction in which our thumb points represents the direction in which electrons are moving, whereas the direction in which our four curled fingers point represents the direction of the magnetic field lines. When the magnetic field lines produced by the flowing electrons within every segment of the coil are added vectorally, the resulting magnetic field appears to emanate from the center of the coil, as shown in Fig. 3.60. Now, as long as the electron flow is constant (steady flow), the centralized magnetic field will not change. The magnetic field lines may pass through various sections of the wire's coil, but they will not affect the current flow through these sections. However, if the current flow suddenly increases, the magnetic field lines expand, as shown in Fig. 3.60. According to Faraday's law of induction (Faraday's law states that a changing magnetic field induces a force on a charged object), the expanding magnetic field will induce a force on the electrons within the coil. The force applied by the expanding magnetic fields on the electrons is such that the electrons are forced to slow down. Now, if you start with a steady (dc) current and suddenly decrease the current flow, the magnetic fields collapse, and they induce a force on the electrons in such a way as to keep them moving, as shown in Fig. 3.60.

In terms of $V = L dI/dt$, the dI/dt term represents the change in current with time, and V represents the inducted voltage (the work done on electrons to reduce changes in their motions). Other names for V include self-inducted voltage, counter emf, and back emf. Now, the only thing left in the equation is the inductance L. L is the proportionality constant between the induced voltage and the changing current; it tells us how good an inductor is at resisting changes in current. The value of L depends on a number of factors, such as the number of coil turns, coil size, coil spacing, winding arrangement, core material, and overall shape of the inductor (see Fig. 3.61).

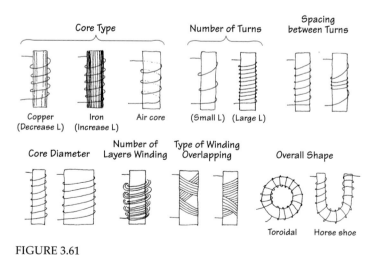

FIGURE 3.61

For a simple, single-layer, closely wound coil, the inductance is given by

$$L = \frac{\mu N^2 A}{l}$$

where μ is the permeability of the material on which the coil is wound (in free space, $\mu = 4\pi \times 10^{-7}\ \text{N}/\text{A}^2$), N is the number of coil windings, A is the cross-sectional area of the coil, and l is the length of the coil.

The following are approximation methods for finding the inductances of closely spaced single and multilayer air-core coils.

Single-Layer
Air-Core Coil

$$L = \frac{(N \times r)^2}{9r + 10l}$$

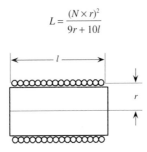

Multi-Layer
Air-Core Coil

$$L = \frac{0.8(N \times 10l)}{6r + 9l + 10b}$$

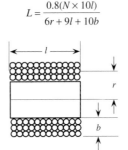

FIGURE 3.62

3.7.2   Basic Inductor Operation

Current/Voltage Behavior

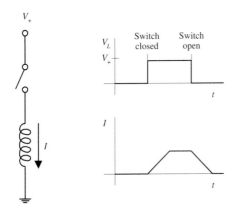

If the switch in a circuit to the left is opened for a specific amount of time and then closed, the inductor initially will "fight" the applied voltage, and the current will rise with a slope related to the coil's inductance. Once steady state is reached, the coil passes a current that is equal to the applied voltage divided by the resistance of the coil; the inductor acts like a low-resistive wire. When the switch is opened, the inductor "fights" the sudden change, and current will fall with a negative slope related to the coil's inductance. The current is related to the applied voltage by $V = L\, dI/dt$.

Resistor/Inductor Behavior

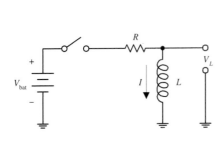

When the switch is closed, the inductor fights to keep the current flow down by inducing a voltage across its leads that opposes the applied voltage. The current reaches 63 percent of max at $t = L/R$. As time passes, the current flow levels off and assumes a dc state. At this point, the inductor acts like a wire, and the current is simply equal to V/R. When the switch is then opened, the opposite effect occurs; the inductor fights to keep current flow up by inducting a voltage in the opposite direction.

FIGURE 3.63

Signal Filtering

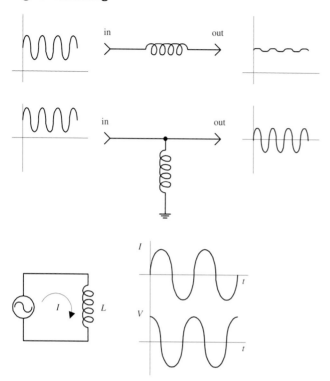

An inductor blocks fluctuating currents yet allows steady (dc) currents to pass.

Here, an inductor acts to block dc current (dc current is routed through inductor to ground) yet allows ac current to pass.

If an alternating voltage is applied across an inductor, the voltage and current will be 90° out of phase (the current will lag the voltage by 90°). As the voltage across the inductor goes to its maximum value, the current goes to zero (energy is being stored in the magnetic fields). As the voltage across the inductor goes to zero, the current through the inductor goes to its maximum value (energy is released from the magnetic fields).

FIGURE 3.63   (*Continued*)

3.7.3   Kinds of Coils

Coils typically have either an air, iron, or powered-iron core. An inductor with an iron core has a greater inductance than an inductor with an air core; the permeability of iron is much greater than the permeability of air. Some inductors come with variable cores that slide in and out of the coil's center. The distance that the core travels into the coil determines the inductance.

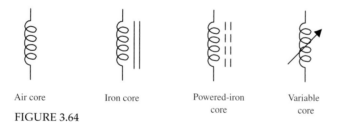

Air core   Iron core   Powered-iron   Variable
                                          core   core

FIGURE 3.64

Here's a rundown of some common coils you will find at the store.

Chokes

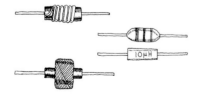

These are general-purpose inductors that act to limit or suppress fluctuating current while passing steady (dc) current. They come in an assortment of shapes, winding arrangements, package types, and inductance and tolerance ratings. Typical inductances range from about 1 to about 100,000 μH, with tolerances from 5 to 20 percent. Some use a resistor-like color code to specify inductance values.

FIGURE 3.65

Tuning Coil

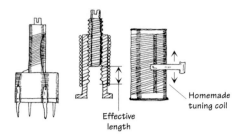

Homemade
tuning coil

Effective
length

This type of coil often contains a screwlike "magnetic field blocker" that can be adjusted to select the desired inductance value. Tuning coils often come with a series of taps. They are frequently used in radio receivers to select a desired frequency (see wireless electronics). A homemade tuning coil can be constructed by wrapping wire around a plastic cylinder and then attaching a movable wiper, as shown at left. The wire can be lacquer-covered magnetic wire. To provide a conductive link between the wiper and the coil, a lacquer strip can be removed with a file.

Toroidal Coil

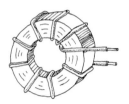

This type of coil resembles a donut with a wire wrapping. They have high inductance per volume ratios, have high quality factors, and are self-shielding. Also, they can be operated at extremely high frequencies.

Antenna Coil

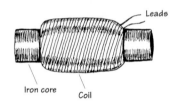

Leads

Iron core   Coil

This type of coil contains an iron core that magnifies magnetic field effects, making it very sensitive to very small changes in current. Such coils are used to tune in ultra-high-frequency signals (e.g., RF signals).

FIGURE 3.65   (*Continued*)

3.8   Transformers

A basic transformer is a four-terminal device that is capable of transforming an ac input voltage into a higher or lower ac output voltage (transformers are not designed to raise or lower dc voltages). A typical transformer consists of two or more coils that share a common laminated iron core, as shown in Fig. 3.66. One of the coils is called the *primary* (containing N_p turns), while the other coil is called the *secondary* (containing N_S turns).

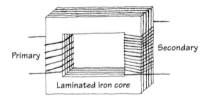

Primary   Secondary

Laminated iron core

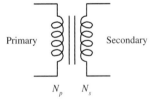

Primary   Secondary

N_p N_s

FIGURE 3.66

When an ac voltage is applied across the primary coil, an alternating magnetic flux of $\Phi = \int (V_{in}/N_p)\, dt$ emanates from the primary, propagates through the iron-laminated core, and passes through the secondary coil. (The iron core increases the inductance, and the laminations decrease power-consuming eddy currents.) According to Faraday's law of induction, the changing magnetic flux induces a voltage of $V_S = N_S d\Phi/dt$ within the secondary coil. Combining the flux equation and the secondary-induced-voltage equation results in the following useful expression:

$$V_S = \frac{N_S}{N_p} V_p$$

This equation says that if the number of turns in the primary coil is larger than the number of turns in the secondary coil, the secondary voltage will be smaller than the primary voltage. Conversely, if the number of turns in the primary coil is less than the number of turns in the secondary, the secondary voltage will be larger than the primary voltage.

When a source voltage is applied across a transformer's primary terminals while the secondary terminals are open circuited (see Fig. 3.67), the source treats the transformer as if it were a simple inductor with an impedance of $Z_p = j\omega L_p$, where L_p represents the inductance of the primary coil. This means that the primary current will lag the primary voltage (source voltage) by 90°, and the primary current will be equal to V_p/Z_p, according to Ohm's law. At the same time, a voltage of $(N_S/N_p)V_p$ will be present across the secondary. Because of the polarity of the induced voltage, the secondary voltage will be 180° out of phase with the primary voltage.

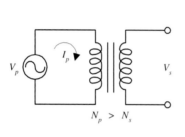 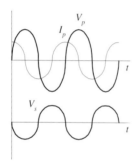

FIGURE 3.67

Now, let's take a look at what happens when you attach a load to the secondary, as shown in Fig. 3.68. When a load R_L is placed across the secondary, the induced voltage acts to move current through the load. Since the magnetic flux from the primary is now being used to induce a current in the secondary, the primary current and voltage move toward being in phase with each other. At the same time, the secondary voltage and the induced secondary current I_s move toward being in phase with each other, but both of them will be 180° out of phase with the primary voltage and current (the physics of induced voltages again).

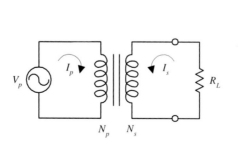 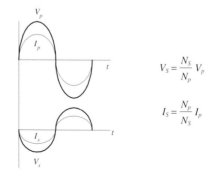

$$V_S = \frac{N_S}{N_p} V_p$$

$$I_S = \frac{N_p}{N_S} I_p$$

FIGURE 3.68

To figure out the relationship between the primary and secondary currents, consider that an ideal transformer is 100 percent efficient (real transformers are around 95 to 99 percent efficient), and then infer that all the power disipated by the load will be equal to the power supplied by the primary source. By combining the power law ($P = V^2/R$) and the transformer voltage relation, it can be seen that both the power dissipated by the load and the power supplied in the primary are equal to

$$P = \frac{V_S^2}{R_L} = \frac{N_S^2 V_p^2}{N_p^2 R_L}$$

By using the $P = IV$ form of the power law, the primary current is

$$I_p = \frac{P_R}{V_p} = \frac{N_S^2 V_p}{N_p^2 R_L}$$

Now, Ohm's law can be used to come up with the equivalent resistance the source feels from the combined transformer and load sections:

$$R_{eq} = \frac{V_p}{I_p} = \left(\frac{N_p}{N_S}\right)^2 R_L$$

In terms of circuit reduction, you can then replace the transformer/load section with R_{eq} (see Fig. 3.69).

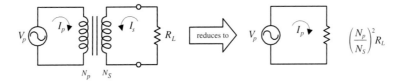

FIGURE 3.69

This reduction trick also can be applied to loads that have complex impedances (e.g., *RLC* networks). The only alteration is that R_L must be replaced with the complex load impedance Z_L.

To finish up with the theory, suppose that you are simply interested in figuring out the relationship between the primary current and the secondary current. Since the power in the primary and the power in the secondary must be equal, you can make use of $P = IV$ and the transformer voltage relation to come up with

$$I_S = \frac{N_p}{N_S} I_p$$

This equation tells you that if you increase the number of turns in the secondary (which results in a larger secondary voltage relative to the primary voltage), the secondary current will be smaller than the primary current. Conversely, if you decrease the number of turns in the secondary (which results in a smaller secondary voltage relative to the primary voltage), the secondary current will be larger than the primary current.

3.8.1 Basic Operation

Voltage/Current Relationships

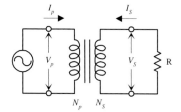

$$V_S = \frac{N_S}{N_p} V_p$$

$$I_S = \frac{N_p}{N_S} I_p$$

To obtain a secondary voltage that's larger than the primary voltage, the number of turns in the secondary must be greater than the number of turns in the primary. Increasing the secondary voltage results in a smaller secondary current relative to the primary current. To obtain a secondary voltage that's smaller than the primary voltage, the number of turns in the secondary must be smaller than the number of turns in the primary. Decreasing the secondary voltage relative to the primary voltage will result in a larger secondary current.

Step-Down Voltage/Step-Up Current

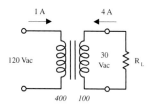

$$V_S = \frac{100}{400} (120 \text{ V}) = 30 \text{ V}$$

$$I_S = \frac{400}{100} (1 \text{ A}) = 4 \text{ A}$$

A transformer with 400 turns in the primary and 100 turns in the secondary will convert a 120-V ac, 1-A primary voltage and current into a 30-V ac, 4-A secondary voltage and current. A transformer that decreases the secondary voltage relative to the primary voltage is called a *step-down transformer.*

Step-Up Voltage/Step-Down Current

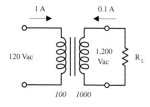

$$V_S = \frac{1000}{100} (120 \text{ V}) = 1200 \text{ V}$$

$$I_S = \frac{100}{1000} (1 \text{ A}) = 0.1 \text{ A}$$

A transformer with 100 turns in the primary and 1000 turns in the secondary will convert a 120-V ac, 1-A primary voltage and current into a 1200-V ac, 0.1-A secondary voltage and current. A transformer that increases the secondary voltage relative to the primary voltage is called a *step-up transformer.*

Circuit Analysis Problem

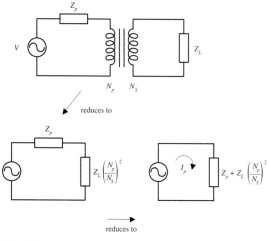

Here's an example of how to treat transformers in circuit analysis. Say that you have a complex network of components that contains resistive, capacitive, and inductive elements (e.g., *RLC* network) in both the primary and secondary sides. To figure out the equivalent impedance the source feels, you can reduce the circuit by finding the equivalent impedance of the load (Z_L) and transformer combined. From what you learned earlier, this is given by $Z_L(N_p/N_S)^2$. After that, you simply have to deal with the impedance in the primary (Z_p) and $Z_L(N_p/N_S)^2$. You apply a final reduction by adding Z_p and $Z_L(N_p/N_S)^2$ in series. The effective primary current is found by using ac Ohm's law ($I = V/Z$).

FIGURE 3.70

3.8.2 Special Kinds of Transformers

Tapped Transformer

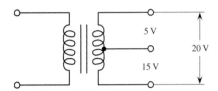

Tapped transformers have an additional connection, or tap, on their primary and/or secondary windings. An additional tap lead on the secondary gives a transformer three possible output voltages.

Multiple-Winding Transformer

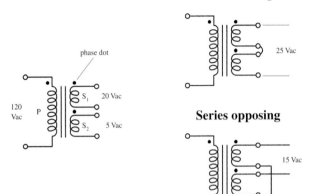

It is often useful to have a number of different secondary windings, each of which is electrically isolated from others (unlike a taped transformer). Each of the secondary coils will have a voltage that is proportional to its number of turns. The secondary windings can be connected in series-aiding (voltage is summed) or series-opposing (the voltage is the difference) configurations. Dots are often used to indicate the terminals that have the same phase.

Autotransformer

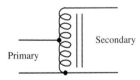

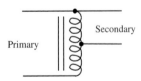

This device only uses a single coil and a tap to make a primary and secondary. Autotransformers can be used to step up or step down voltages; however, they are not used for isolation application because the primary and secondary are on the same coil (there is no electrical isolation between the two). These devices are frequently used in impedance-matching applications.

Continuous Tapped Transformer (Variac)

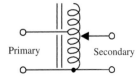

This device contains a variable tap that slides up and down the secondary coil to control the secondary coil length and hence the secondary voltage.

FIGURE 3.71

3.8.3 Applications

Three fundamental applications for transformers include isolation protection, voltage conversion, and impedance matching. Here are a few examples showing application for all three.

FIGURE 3.72

Isolation

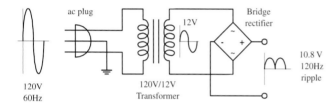

1:1
120 Vac
60Hz

Primary-line plug

Isolated receptacle

Transformers with a 1:1 ratio between primary and secondary windings are often used to protect secondary circuits (or individuals touching secondary elements, e.g., knobs, panels, etc.) from electrical shocks. The reason why the protection works is that the secondary is only magnetically coupled—not electrically coupled—with the high-current utility line. All test equipment, especially ones that are "floated" (ground is removed), must use isolation protection to eliminate shock hazards. Another advantage in using an isolation transformer is that there are no dc connections between circuit elements in the primary and secondary sides; ac devices can be coupled with other devices through the transformer without any dc signals getting through. The circuit in the figure shows a simple example of how a power outlet can be isolated from a utility outlet. The isolated receptacle is then used to power test equipment.

Power Conversion

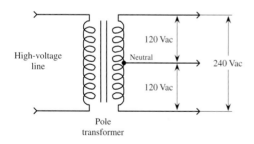

ac plug

12V

Bridge rectifier

10.8 V
120Hz
ripple

120V
60Hz

120V/12V
Transformer

Transformers are essential ingredients in power supply design. Here a 120-to-12-V transformer steps down a 120-V, 60-Hz line voltage to a 12-V, 60-Hz secondary voltage. A bridge rectifier (network of four diodes) then rectifies the secondary voltage into a pulsed dc voltage with a peak of 10.8 V and a frequency of 120 Hz. (1.2 V is lost during rectification due to diode biasing voltages, and it appears that the output is double in frequency due to the negative swings being converted into positive swings by the rectifier.) The average dc voltage at the output is equal to 0.636 of the peak rectified voltage.

Tapped Transformer Application

High-voltage line

120 Vac
Neutral
240 Vac
120 Vac

Pole transformer

In the United States, main power lines carry ac voltages of upwards of 1000 V. A center-tapped pole transformer is used to step down the line voltage to 240 V. The tap then acts to break this voltage up into 120-V portions. Small appliances, such as TVs, lights, and hairdryers, can use either the top line and the neutral line or the bottom line and the neutral line. Larger appliances, such as stoves, refrigerators, and clothes dryers, make use of the 240-V terminals and often use the neutral terminal as well. See Appendix A for more on power distribution and home wiring.

Impedance Matching

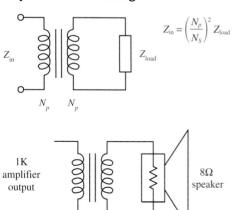

Z_{in}

Z_{load}

$$Z_{in} = \left(\frac{N_p}{N_S}\right)^2 Z_{load}$$

N_p N_p

1K amplifier output

8Ω speaker

$N_p/N_S = 11.2$

Earlier you discovered that any device connected to the primary side of a transformer with a secondary load feels an equivalent impedance equal to $(N_p/N_S)^2 Z_{load}$. By manipulating the N_p/N_S ratio, you can trick an input device into thinking it is connected to an output device with a similar impedance, even though the output device may have an entirely different impedance. For example, if you want to use a transformer to match impedances between an 8-Ω speaker and a 1-k amplifier, you plug 8 Ω into Z_{load} and 1000 Ω into Z_{in} and solve for N_p/N_S. This provides the ratio between the primary and secondary windings that is needed to match the two devices. N_p/N_S turns out to be 11.2. This means the primary must have 11.2 more winding than the secondary.

3.8.4 Real Kinds of Transformers

Isolation

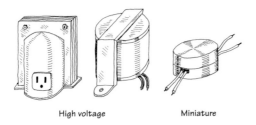

High voltage   Miniature

This type of transformer acts exclusively as an isolation device; it does not increase or decrease the secondary voltage. The primary coil of an ac power-line isolation transformer (in the United States) accepts a 120-V ac, 60-Hz input voltage and outputs a similar but electrically isolated voltage at the secondary. Such transformers usually come with an electrostatic shield between the primary and secondary. They often come with a three-wire plug and receptacle that can be plugged directly into a power outlet.

Power Conversion

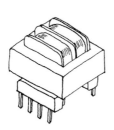

PCB Mount

These transformers are used primarily to reduce line voltages. They come in a variety of different shapes, sizes, and primary and secondary winding ratios. They often come with taps and multiple secondary windings. Color-coded wires are frequently used to indicate primary and secondary leads (e.g., black wires for primary, green for secondary, and yellow for tap lead is one possibility). Other transformers use pins for primary, secondary, and taped leads, allowing them to be mounted on a PC board.

Audio

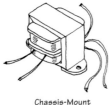

Chassis-Mount   Miniature, Epoxy-Coated

Audio transformers are used primarily to match impedances between audio devices (e.g., between microphone and amplifier or amplifier and speaker). They work best at audio frequencies from 150 Hz to 15 kHz. They come in a variety of shapes and sizes and typically contain a center tap in both the primary and secondary windings. Some use color-coded wires to specify leads (e.g., blue and brown as primary leads, green and yellow as secondary leads, and red and black for primary and secondary taps is one possibility). Other audio transformers have pinlike terminals that can be mounted on PC boards. Spec tables provide dc resistance values for primary and secondary windings to help you select the appropriate transformer for the particular matching application.

Miniature

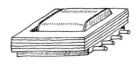

Miniature transformers are used primarily for impedance matching. They usually come with pinlike leads that can be mounted on PC boards. Spec tables provide dc resistance values for both primary and secondary windings and come with primary/secondary turn ratios.

High-Frequency Transformers

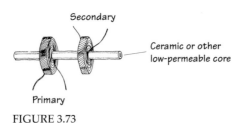

Secondary

Ceramic or other low-permeable core

Primary

These often come with air or powered-iron cores instead of laminated-iron cores. They are used for high-frequency applications, such as matching RF transmission lines to other devices (e.g., transmission line to antenna). One particularly useful high-frequency rated transformer is the broad-band transmission-line transformer (see the section on matching impedances).

FIGURE 3.73

3.9 Fuses and Circuit Breakers

Fuses and circuit breakers are devices designed to protect circuits from excessive current flows, which are often a result of large currents that result from shorts or sudden power surges. A fuse contains a narrow strip of metal that is designed to melt when current flow exceeds its current rating, thereby interrupting power to the circuit. Once a fuse blows (wire melts), it must be replaced with a new one. A circuit breaker is a mechanical device that can be reset after it "blows." It contains a spring-loaded contact that is latched into position against another contact. When the current flow exceeds a breaker's current rating, a bimetallic strip heats up and bends. As the strip bends, it "trips" the latch, and the spring pulls the contacts apart. To reset the breaker, a button or rockerlike switch is pressed to compress the spring and reset the latch.

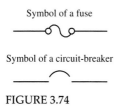

Symbol of a fuse

Symbol of a circuit-breaker

FIGURE 3.74

In homes, fuses/circuit breakers are used to prevent the wires within the walls from melting under excessive current flow (typically upwards of 15 A); they are not designed to protect devices, such as dc power supplies, oscilloscopes, and other line-powered devices, from damage. For example, if an important current-limiting component within a test instrument (powered from the main line) shorts out, or if it is connected to an extremely large test current, the circuit within the device may be flooded with, say, 10 A instead of 0.1 A. According to $P = I^2R$, the increase in wattage will be 10,000 times larger, and as a result, the components within the circuit will fry. As the circuit is melting away, no help will come from the main 15-A breaker—the surge in current through the device may be large, but not large enough to trip the breaker. For this reason, it is essential that each individual device contain its own properly rated fuse.

Fuses come in *fast-action* (quick-blow) and *time-lag* (slow-blow) types. A fast-acting fuse will turn off with just a brief surge in current, whereas a time-lag fuse will take a while longer (a second or so). Time-lag fuses are used in circuits that have large turn-on currents, such as motor circuits, and other inductive-type circuits.

In practice, a fuse's current rating should be around 50 percent larger than the expected nominal current rating. The additional leeway allows for unexpected, slight variations in current and also provides compensation for fuses whose current ratings decrease as they age.

Fuses and breakers that are used with 120-V ac line power must be placed in the hot line (black wire) and must be placed before the device they are designed to protect. If a fuse or circuit breaker is placed in the neutral line (white wire), the full line-voltage will be present at the input, even if the fuse/circuit breaker blows. Circuit breakers that are used to protect 240-V ac appliances (e.g., stoves, clothes dryers) have fuses on all three input wires (both the hot wires and the neutral wire) (see Fig. 3.75). Power distribution and home wiring are covered in detail in Appendix A.

Home wiring

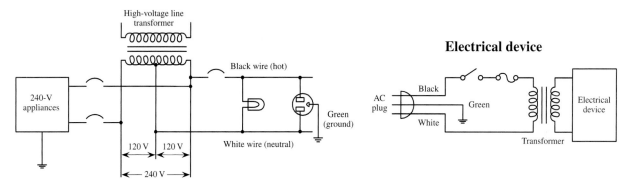

Electrical device

FIGURE 3.75

3.9.1   Types of Fuses and Circuit Breakers

Glass and Ceramic

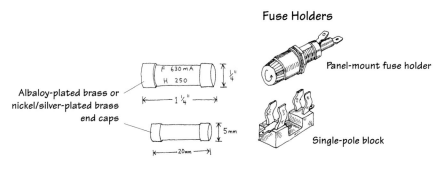

Fuse Holders

Panel-mount fuse holder

Single-pole block

Albaloy-plated brass or nickel/silver-plated brass end caps

These are made by encapsulating a current-sensitive wire or ceramic element within a glass cylinder. Each end of the cylinder contains a metal cap that acts as a contact lead. Fuses may be fast-acting or time-lagging. They are used in instruments, electric circuits, and small appliances. Typical cylinders come in $\frac{1}{4} \times 1\frac{1}{4}$ in or 5×20 mm sizes. Current ratings range from around $\frac{1}{4}$ to 20 A, with voltage ratings of 32, 125, and 250 V.

Blade

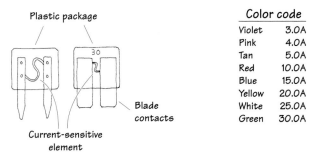

Plastic package

Blade contacts

Current-sensitive element

Color code

Violet	3.0A
Pink	4.0A
Tan	5.0A
Red	10.0A
Blue	15.0A
Yellow	20.0A
White	25.0A
Green	30.0A

These are fast-acting fuses with bladelike contacts. They are easy to install and remove from their sockets. Current ratings range from 3 to 30 A, with voltage ratings of 32 and 36 V. They are color-coded according to current rating and are used primarily as automobile fuses.

Miscellaneous Fuses

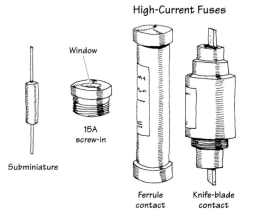

High-Current Fuses

Window

Subminiature

15A screw-in

Ferrule contact

Knife-blade contact

FIGURE 3.76

Other types of fuses include subminiature fuses and high-current screw-in and cartridge fuses. Subminiature fuses are small devices with two wire leads that can be mounted on PC boards. Current ratings range from 0.05 to 10 A. They are used primarily in miniature circuits. Cartridge fuses are designed to handle very large currents. They are typically used as main power shutoffs and in subpanels for 240-V applications such as electric dryers and air conditioners. Cartridge fuses are wrapped in paper, like shotgun shells, and come with either ferrule or knife-blade contacts. Ferrule-contact fuses protect from up to 60 A, while knife-blade contact fuses are rated at 60 A or higher.

Circuit Breakers

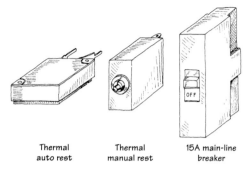

Thermal
auto rest

Thermal
manual rest

15A main-line
breaker

FIGURE 3.76   (*Continued*)

These come in both rocker and push-button forms. Some have manual resets, while other have thermally actuated resets (they reset themselves as they cool down). Main-line circuit breakers are rated at 15 to 20 A. Smaller circuit breakers may be rated as low as 1 A.

CHAPTER 4

Semiconductors

4.1 Semiconductor Technology

The most important and perhaps most exciting electrical devices used today are built from semiconductive materials. Electronic devices, such as diodes, transistors, thyristors, thermistors, photovoltaic cells, phototransistors, photoresistors, lasers, and integrated circuits, are all made from semiconductive materials, or semiconductors.

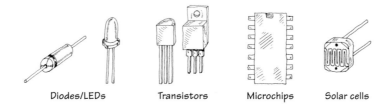

FIGURE 4.1   Diodes/LEDs   Transistors   Microchips   Solar cells

4.1.1 What Is a Semiconductor?

Materials are classified by their ability to conduct electricity. Substances that easily pass an electric current, such as silver and copper, are referred to as *conductors*. Materials that have a difficult time passing an electric current, such as rubber, glass, and Teflon, are called *insulators*. There is a third category of material whose conductivity lies between those of conductors and insulators. This third category of material is referred to as a *semiconductor*. A semiconductor has a kind of neutral conductivity when taken as a group. Technically speaking, semiconductors are defined as those materials that have a conductivity σ in the range between 10^{-7} and 10^3 mho/cm (see Fig. 4.2). Some semiconductors are pure-elemental structures (e.g., silicon, germanium), others are alloys (e.g., nichrome, brass), and still others are liquids.

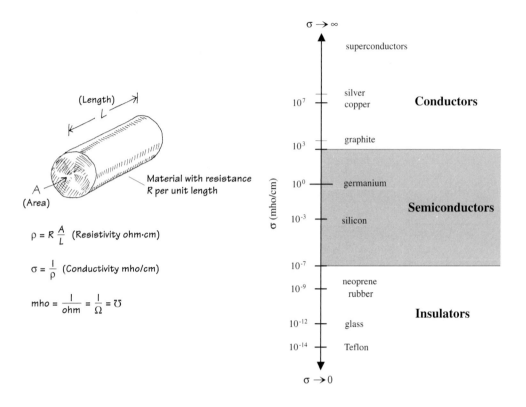

FIGURE 4.2

Silicon

Silicon is the most important semiconductor used in building electrical devices. Other materials such as germanium and selenium are sometimes used, too, but they are less popular. In pure form, silicon has a unique atomic structure with very important properties useful in making electrical devices.

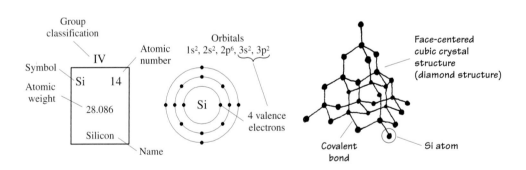

FIGURE 4.3

Silicon is ranked second in the order of elements appearing in the earth's crust, an average of 27 percent occurring in igneous rocks. It is estimated that a cubic mile of seawater contains about 15,000 tons of silicon. It is extremely rare to find silicon in its pure crystalline form in nature, and before it can be used in making electronic devices, it must be separated from its binding elements. After individuals—chemists, material scientists, etc.—perform the purification process, the silicon is melted and spun into a large "seed" crystal. This long crystal can then be cut up into slices or wafers that semiconductor-device designers use in making electrical contraptions.

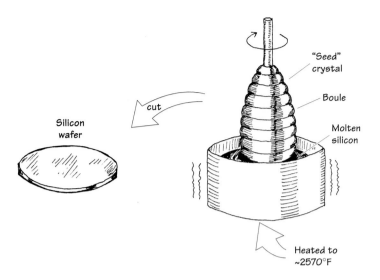

"Seed" crystal

Boule

Molten silicon

cut

Silicon wafer

Heated to ~2570°F

FIGURE 4.4

For the semiconductor-device designer, a silicon wafer alone does not prove very useful. A designer would not use the silicon wafer in its pure form to build a device; it just does not have quite the right properties needed to be useful. A semiconductor-device designer is looking for a material that can be made to alter its conductive state, acting as a conductor at one moment and as an insulator at another moment. For the material to change states, it must be able to respond to some external force applied at will, such as an externally applied voltage. A silicon wafer alone is not going to do the trick. In fact, a pure silicon wafer acts more as an insulator than a conductor, and it does not have the ability to change conductive states when an external force is applied. Every designer today knows today that a silicon wafer can be transformed and combined with other transformed silicon wafers to make devices that have the ability to alter their conductive states when an external force is applied. The transforming process is referred to as *doping*.

Doping

Doping refers to the process of "spicing up" or adding ingredients to a silicon wafer in such a way that it becomes useful to the semiconductor-device designer. Many ingredients can be added in the doping process, such as antimony, arsenic, aluminum, and gallium. These ingredients provide specialized characteristics such as frequency response to applied voltages, strength, and thermal integrity, to name a few. By far, however, the two most important ingredients that are of fundamental importance to the semiconductor-device designer are boron and phosphorus.

When a silicon wafer is doped with either boron or phosphorus, its electrical conductivity is altered dramatically. Normally, a pure silicon wafer contains no free electrons; all four of its valence electrons are locked up in covalent bonds with neighboring silicon atoms (see Fig. 4.5). Without any free electrons, an applied voltage will have little effect on producing an electron flow through the wafer.

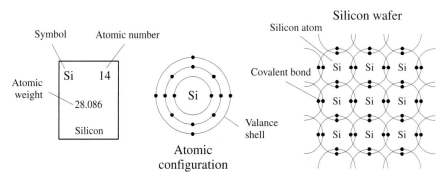

FIGURE 4.5

A silicon wafer in pure form doesn't contain any free charge carriers; all the electrons are locked up into covalent bonds between neighboring atoms.

However, if phosphorus is added to the silicon wafer, something very interesting occurs. Unlike silicon, phosphorus has five valance electrons instead of four. Four of its valance electrons will form covalent bonds with the valance electrons of four neighboring silicon atoms (see Fig. 4.6). However, the fifth valance electron will not have a "home" (binding site) and will be loosely floating about the atoms. If a voltage is applied across the silicon-phosphorus mixture, the unbound electron will migrate through the doped silicon toward the positive voltage end. By supplying more phosphorus to the mixture, a larger flow of electrons will result. Silicon that is doped with phosphorus is referred to as *n-type silicon*, or negative-charge-carrier-type silicon.

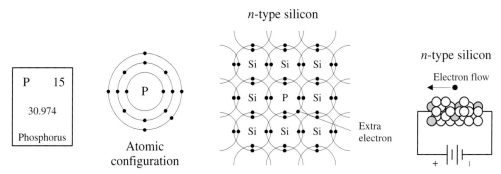

FIGURE 4.6

A phosphorus atom added to a silicon wafer provides an extra unbound electron that aids in conduction. Silicon doped with phosphorus is called *n*-type silicon.

Now, if you take pure silicon and add some boron, you will see a different kind conduction effect. Boron, unlike silicon or phosphorus, contains only three valance electrons. When it is mixed with silicon, all three of its valance electrons will bind with neighboring silicon atoms (see Fig. 4.7). However, there will be a vacant spot—called a *hole*—within the covalent bond between one boron and one silicon atom. If a voltage is applied across the doped wafer, the hole will move toward the negative voltage end, while a neighboring electron will fill in its place. Holes are considered positive charge carriers even though they do not contain a physical charge per se. Instead, it only appears as if a hole has a positive charge because of the charge imbalance between the protons within the nucleus of the silicon atom that receives the hole and the electrons in the outer orbital. The net charge on a particular silicon atom with a hole will appear to be positive by an amount of charge equivalent to one proton (or a "negative electron"). Silicon that is doped with boron is referred to as *p-type silicon*, or positive-charge-carrier-type silicon.

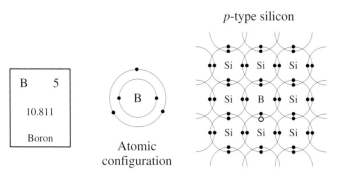

p-type silicon

Atomic
configuration

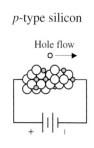

p-type silicon

Hole flow

When boron is added
to silicon, a hole is
formed. This hole acts
like a positive charge
(see text) that aids in
conduction. Silicon
doped with boron is
called *p*-type silicon.

FIGURE 4.7

As you can see, both *n*-type and *p*-type silicon have the ability to conduct electricity; one does it with extra unbound electrons (*n*-type silicon), and the other does it with holes (*p*-type silicon).

A Note to Avoid Confusion

Boron atoms have three valance electrons, not four like silicon. This means that the combined lattice structure has fewer free valance electrons as a whole. However, this does not mean that a *p*-type silicon semiconductor has an overall positive charge; the missing electrons in the structure are counterbalanced by the missing protons in the nuclei of the boron atoms. The same idea goes for *n*-type silicon, but now the extra electrons within the semiconductor are counterbalanced by the extra protons within the phosphorus nuclei.

Another Note to Avoid Confusion (Charge Carriers)

What does it mean for a hole to flow? I mean, a hole is nothing, right? How can nothing flow? Well, it is perhaps misleading, but when you hear the phrase "hole flow" or "flow of positive charge carriers in *p*-type silicon," electrons are in fact flowing. You may say, doesn't that make this just like the electron flow in *n*-type silicon? No. Think about tipping a sealed bottle of water upside down and then right side up (see Fig. 4.8). The bubble trapped in the bottle will move in the opposite direction of the water. For the bubble to proceed, water has to move out of its way. In this analogy, the water represents the electrons in the *p*-type silicon, and the holes represent the bubble. When a voltage is applied across a *p*-type silicon semiconductor, the electrons around the boron atom will be forced toward the direction of the positive terminal. Now here is where it gets tricky. A hole about a boron atom will be pointing toward the negative terminal. This hole is just waiting for an electron from a neighboring atom to fall into it, due in part to the lower energy configuration there. Once an electron, say, from a neighboring silicon atom, falls into the hole in the boron atom's valance shell, a hole is briefly created in that silicon atom's valance shell. The electrons in the silicon atom lean toward the positive terminal, and the newly created hole leans toward the negative terminal. The next silicon atom over will let go of one of its electrons, and the electron will fall into the hole, and the hole will move over again—the process continues, and it appears as if the hole flows in a continuous motion through the *p*-type semiconductor.

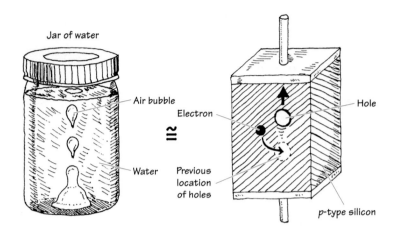

FIGURE 4.8

A Final Note to Avoid Confusion

And finally, why is a hole called a positive charge carrier? How can "nothing" carry a positive charge? Well, what's going on here is this: A hole, as it moves through the mostly silicon-based crystal, causes a brief alteration in the electrical field strength around the silicon atom in the crystal where it happens to be situated. When an electron moves out of the way, thus creating a new hole, this silicon atom as a whole will be missing an electron, and hence the positive charge from the nucleus of the silicon atom will be felt (one of the protons is not counterbalanced). The "positive charge carrier" attributed to holes comes from this effective positive nuclear charge of the protons fixed within the nucleus.

4.1.2 Applications of Silicon

You may be asking yourself, Why are these two new types of silicon (*n*-type and *p*-type) so useful and interesting? What good are they for semiconductor-device designers? Why is there such a fuss over them? These doped silicon crystals are now conductors, big deal, right? Yes, we now have two new conductors, but the two new conductors have two unique ways of passing an electric current—one does it with holes, the other with electrons. This is very important.

The manners in which *n*-type and *p*-type silicon conduct electricity (electron flow and hole flow) are very important in designing electronic devices such as diodes, transistors, and solar cells. Some clever people figured out ways to arrange slabs, chucks, strings, etc. made of *n*-type and *p*-type silicon in such a way that when an external voltage or current is applied to these structures, unique and very useful features result. These unique features are made possible by the interplay between hole flow and electron flow between the *n*-type and *p*-type semiconductors. With these new *n*-type/*p*-type contraptions, designers began building one-way gates for current flow, opening and closing channels for current flow controlled by an external electrical voltage and/or current. Folks figured out that when an *n*-type and a *p*-type semiconductor were placed together and a particular voltage was applied across the slabs, light, or photons, could be produced as the electrons jumped across the junction between the interface. It was noticed that this process could work backward as well. That is, when light was exposed at the *np* junction, electrons were made to flow, thus resulting in an electric current. A number of clever contraptions have been built using *n*-type and *p*-type semiconductor combinations. The following chapters describe some of the major devices people came up with.

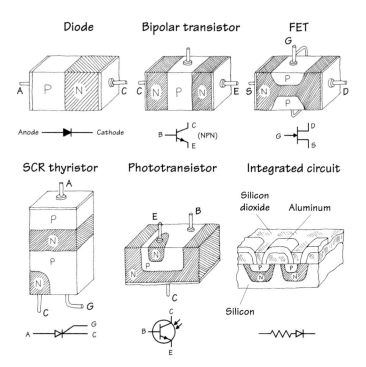

FIGURE 4.9

4.2 Diodes

A *diode* is a two-lead semiconductor device that acts as a one-way gate to electric current flow. When a diode's *anode* lead is made more positive in voltage than its *cathode* lead—a condition referred to as *forward biasing*—current is permitted to flow through the device. However, if the polarities are reversed (the anode is made more negative in voltage than the cathode)—a condition referred to as *reversed biasing*—the diode acts to block current flow.

FIGURE 4.10

anode   cathode

Diodes are used most commonly in circuits that convert ac voltages and current into dc voltages and currents (e.g., ac/dc power supply). Diodes are also used in voltage-multiplier circuits, voltage-shifting circuits, voltage-limiting circuits, and voltage-regulator circuits.

4.2.1 How pn-Junction Diodes Work

A *pn-junction diode* (*rectifier diode*) is formed by sandwiching together *n*-type and *p*-type silicon. In practice, manufacturers grow an *n*-type silicon crystal and then abruptly change it to a *p*-type crystal. Then either a glass or plastic coating is placed around the joined crystals. The *n* side is the cathode end, and the *p* side is the anode end.

The trick to making a one-way gate from these combined pieces of silicon is getting the charge carriers in both the *n*-type and *p*-type silicon to interact in such a way that when a voltage is applied across the device, current will flow in only one direction. Both *n*-type and *p*-type silicon conducts electric current; one does it with electrons (*n*-type), and the other does it with holes (*p*-type). Now the important feature to note here, which makes a diode work (act as a one-way gate), is the manner in which the two types of charge carriers interact with each other and how they interact with an applied electrical field supplied by an external voltage across its leads. Below is an explanation describing how the charge carriers interact with each other and with the electrical field to create an electrically controlled one-way gate.

Forward-Biased ("Open Door")

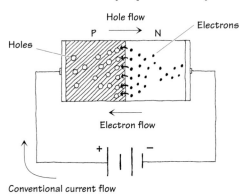

When a diode is connected to a battery, as shown here, electrons from the *n* side and holes from the *p* side are forced toward the center (*pn* interface) by the electrical field supplied by the battery. The electrons and holes combine, and current passes through the diode. When a diode is arranged in this way, it is said to be *forward-biased*.

Reverse-Biased ("Closed Door")

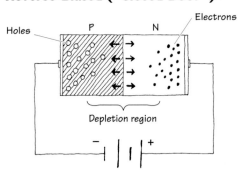

When a diode is connected to a battery, as shown here, holes in the *n* side are forced to the left, while electrons in the *p* side are forced to the right. This results in an empty zone around the *pn* junction that is free of charge carriers, better known as the *depletion region*. This depletion region has an insulative quality that prevents current from flowing through the diode. When a diode is arranged in this way, it is said to be *reverse-biased*.

FIGURE 4.11

A diode's one-way gate feature does not work all the time. That is, it takes a minimal voltage to turn it on when it is placed in forward-biased direction. Typically for silicon diodes, an applied voltage of 0.6 V or greater is needed; otherwise, the diode will not conduct. This feature of requiring a specific voltage to turn the diode on may seem like a drawback, but in fact, this feature becomes very useful in terms of acting as a voltage-sensitive switch. Germanium diodes, unlike silicon diodes, often require a forward-biasing voltage of only 0.2 V or greater for conduction to occur. Figure 4.12 shows how the current and voltage are related for silicon and germanium diodes.

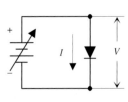
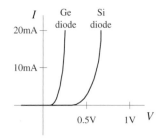

FIGURE 4.12

Another fundamental difference between silicon diodes and germanium diodes, besides the forward-biasing voltages, is their ability to dissipate heat. Silicon diodes do a better job of dissipating heat than germanium diodes. When germanium diodes get hot—temperatures exceeding 85°C—the thermal vibrations affect the physics inside the crystalline structure to a point where normal diode operation becomes unreliable. Above 85°C, germanium diodes become worthless.

4.2.2 Diode Water Analogy

In the following water analogy, a diode is treated like a one-way gate that has a "forward-biasing spring."

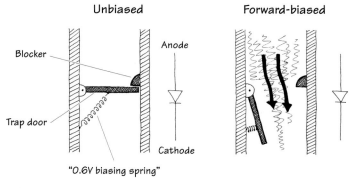

FIGURE 4.13

In this water analogy, a spring holds the one-way gate shut. Water pressure on top of the gate must apply a force large enough to overcome the restoring spring force. This spring force is analogous to the 0.6 V needed to forward bias a silicon diode. A germanium diode with a 0.2-V biasing voltage can be represented with a similar arrangement, but the water analogy will contain a weaker spring, one that can be compressed more easily. Note that if pressure is applied to the bottom of the trap door, a blocker will prevent the trap door from swinging upward, preventing current flow in the upward direction, which is analogous to reverse-biasing a diode.

4.2.3 Basic Applications

Diodes are used most frequently in rectifier circuits—circuits that convert ac to dc. Diodes are used in a number of other applications as well. Here's a rundown of some of the most common applications.

Half-Wave Rectifier

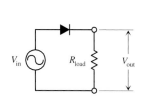
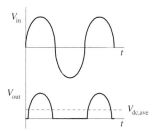

In this circuit, the diode acts to convert an ac input voltage into a pulsed dc output voltage. Whenever the voltage attempts to swing negative at the anode, the diode acts to block current flow, thus causing the output voltage (voltage across the resistor) to go to zero. This circuit is called a *half-wave rectifier*, since only half the input waveform is supplied to the output. Note that there will be a 0.6 V drop across the diode, so the output peak voltage will be 0.6 V less than the peak voltage of V_{in}. The output frequency is the same as the input frequency, and the average dc voltage at the output is 0.318 times zero-to-peak output voltage. A transformer is typically used to step down or step up the voltage before it reaches the rectifier section.

Full-Wave Bridge Rectifier

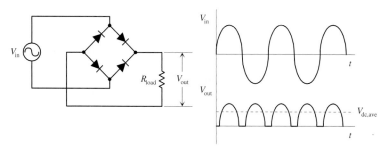

FIGURE 4.14

This circuit is called a *full-wave rectifier*, or *bridge rectifier*. Unlike the half-wave rectifier, a full-wave rectifier does not merely block negative swings in voltage but also converts them into positive swings at the output. To understand how the device works, just follow the current flow through the diode one-way gates. Note that there will be a 1.2-V drop from zero-to-peak input voltage to zero-to-peak output voltage (there are two 0.6-V drops across a pair of diodes during a half cycle). The output frequency is twice the input frequency, and the average dc voltage at the output is 0.636 times the zero-to-peak output voltage.

Basic AC-to-DC Power Supply

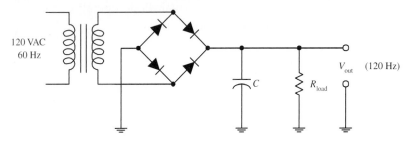

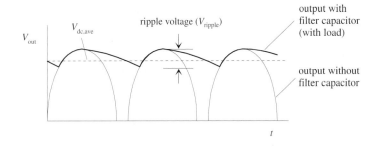

By using a transformer and a full-wave bridge rectifier, a simple ac-to-dc power supply can be constructed. The transformer acts to step down the voltage, and the bridge rectifier acts to convert the ac input into a pulsed dc output. A filter capacitor is then used to delay the discharge time and hence "smooth" out the pulses. The capacitor must be large enough to store a sufficient amount of energy to provide a steady supply of current to the load. If the capacitor is not large enough or is not being charged fast enough, the voltage will drop as the load demands more current. A general rule for choosing C is to use the following relation:

$$R_{\text{load}}C \gg 1/f$$

where f is the rectified signal's frequency (120 Hz). The ripple voltage (deviation from dc) is approximated by

$$V_{\text{ripple}} = \frac{I_{\text{load}}}{fC}$$

Voltage Dropper

DC application   AC application

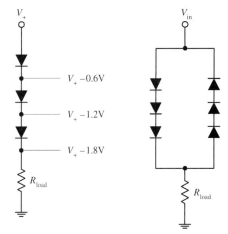

When current passes through a diode, the voltage is lowered by 0.6 V (silicon diodes). By placing a number of diodes in series, the total voltage drop is the sum of the individual voltage drops of each diode. In ac applications, two sets of diodes are placed in opposite directions.

Voltage Regulator

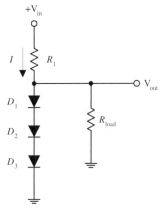

This circuit will supply a steady output voltage equal to the sum of the forward-biasing voltages of the diodes. For example, if D_1, D_2, and D_3 are silicon diodes, the voltage drop across each one will be 0.6 V; the voltage drop across all three is then 1.8 V. This means that the voltage applied to the load (V_{out}) will remain at 1.8 V. R_1 is designed to prevent the diodes from "frying" if the resistance of the load gets very large or the load is removed. The value of R_1 should be approximately equal to

$$R_1 = \frac{(V_{\text{in}} - V_{\text{out}})}{I}$$

FIGURE 4.14   (*Continued*)

Voltage-Multiplier Circuits

Conventional doubler

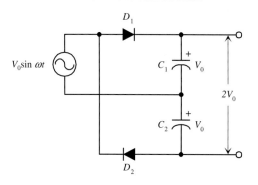

Charge pump doubler

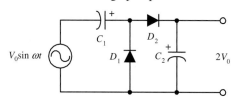

Voltage tripler

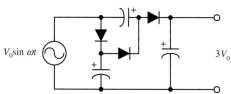

Voltage quadrupler

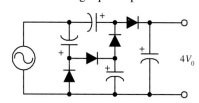

It is often useful to have a rectifier circuit that is capable of producing a dc voltage that is higher than the zero-to-peak voltage of the ac source. Although the usual procedure is to use a transformer to step up the voltage before the rectifier stage, an alternative approach is to use a voltage-multiplier circuit. The top circuit shown here represents a conventional voltage doubler. During a positive half cycle, D_1 is forward biased, allowing C_1 to charge to a dc voltage of V_0 (the peak source voltage). During the negative half cycle, D_2 is forward-biased, and C_2 charges to V_0. Because C_1 and C_2 are connected in series with their polarities aiding, the output voltage is the sum of the capacitors' voltages, or is equal to $2V_0$. The second circuit shown is a variation of the first and is called a *charge pump.* During the negative half cycle, the source pumps charge into C_1 through D_1 with D_2 open-circuited, and then during the positive half cycle, D_1 becomes an open circuit and D_2 becomes a short circuit, and some of the charge in C_1 flows into C_2. The process continues until enough charge is pumped into C_2 to raise its voltage to $2V_0$. One advantage of the charge pump over the conventional voltage doubler is that one side of the source and one side of the output are common and hence can be grounded. To obtain a larger voltage, additional stages can be added, as shown within the bottom two circuits. A serious limitation of voltage-multiplier circuits is their relatively poor voltage regulation and low-current capability.

FIGURE 4.14 *(Continued)*

Diode Clamp

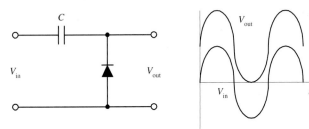

The diode clamp circuit shown here takes a periodic ac signal that oscillates between positive and negative values and displaces it so that it is either always positive or always negative. The capacitor charges up to a dc voltage equal to the zero-to-peak value of V_{in}. The capacitor is made large enough so that it looks like a short circuit for the ac components of V_{in}. If, for example, V_{in} is a sine wave, V_{out} will equal the sum of V_{in} and the dc voltage on the capacitor. By placing the diode in the opposite position (pointing down), V_{out} will be displaced downward so that it is always negative.

Waveform Clipper

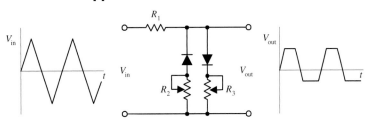

This circuit is often used to protect circuit components from damage due to overvoltage and is used for generating special waveforms. R_2 controls the lower-level clipping at the output, while R_3 controls the upper-level clipping. R_1 is used as a safety resistor to prevent a large current from flowing through a diode if a variable resistor happens to be set to zero resistance.

Reverse-Polarity Protector

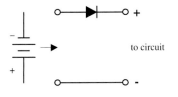

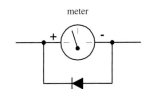

A single diode can be used to protect a circuit that could be damaged if the polarity of the power source were reversed. In the leftmost network, the diode acts to block the current flow to a circuit if the battery is accidentally placed the wrong way around. In the rightmost circuit, the diode prevents a large current from entering the negative terminal of a meter.

Transient Protector

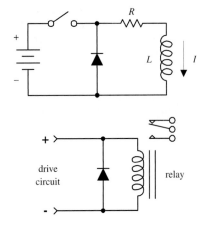

Placing a diode in the reverse-biased direction across an inductive load eliminates voltage spikes by going into conduction before a large voltage can form across the load. The diode must be rated to handle the current, equivalent to the maximum current that would have been flowing through the load before the current supply was interrupted. The lower network shows how a diode can be used to protect a circuit from voltage spikes that are produced when a dc-actuated relay suddenly switches states.

FIGURE 4.14   (*Continued*)

Battery-Powered Backup

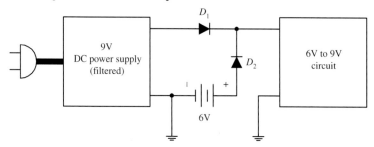

FIGURE 4.14   (*Continued*)

This circuit shows how a battery along with two diodes can be used to provide backup power to a circuit that is normally powered by an ac-driven dc power supply (with filtered output). Here you have a circuit that is designed to run off a voltage from 6 to 9 V. The voltage supplied by the power supply is at 9 V, while the battery voltage is 6 V. As long as power is supplied by the main power supply, the voltage that the circuit receives will be 8.4 V (there is a 0.6-V drop across D_1). At the same time, D_2's anode will be more negative in voltage than its cathode by 2.6 V (8.4 V minus 6.0 V), which means D_2 will block current flow from the battery, hence preventing battery drain. Now, if a lightning bolt strikes, causing the power supply to fail, D_2's anode will become more positive in voltage than its cathode, and it will allow current to flow from the battery to the load. At the same time, D_1 acts to block current from flowing from the battery into the power supply.

4.2.4   *Important Things to Know about Diodes*

Diodes come in many shapes and sizes. High-current diodes are often mounted on a heat-sink device to reduce their operating temperature. It is possible to place diodes in parallel to increase the current-carrying capacity, but the *VI* characteristics of both diodes must be closely matched to ensure that current divides evenly (although a small resistor can be placed in series with each diode to help equalize the currents).

All diodes have some leakage current (current that gets through when a diode is reverse-biased). This leakage current—better known as the *reverse current* (I_R)—is very small, typically within the nanoampere range. Diodes also have a maximum allowable *reverse voltage, peak reverse voltage* (PRV), or *peak inverse voltage* (PIV), above which a large current will flow in the wrong direction. If the PIV is exceeded, the diode may get zapped and may become permanently damaged. The PIV for diodes varies from a few volts to as much as several thousand volts. One method for achieving an effectively higher PIV is to place diodes in series. Again, it is important that diodes are matched to ensure that the reverse voltage divides equally (although a small resistor placed in parallel with each diode can be used to equalize the reverse voltages).

Other things to consider about diodes include maximum forward current (I_F), capacitance (formed across the *pn* junction), and reverse recovery time.

Most diodes have a 1-prefix designation (e.g., 1N4003). The two ends of a diode are usually distinguished from each other by a mark. For glass-encapsulated diodes, the cathode is designated with a black band, whereas black-plastic-encapsulated diodes use a white band (see Fig. 4.15). If no symbols are present (as seen with many power diodes), the cathode may be a boltlike piece. This piece is inserted through a heat-sink device (piece of metal with a hole) and is fastened down by a nut. A fiber or mica washer is used to isolate the cathode electrically from the metal heat sink, and a special silicone grease is placed between the washer and heat sink to enhance thermal conductivity.

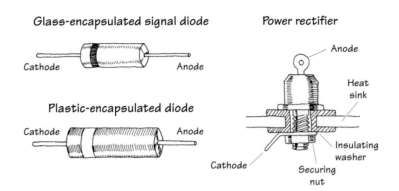

FIGURE 4.15

4.2.5   Zener Diodes

A zener diode is a device that acts as a typical *pn*-junction diode when it comes to forward biasing, but it also has the ability to conduct in the reverse-biased direction when a specific breakdown voltage (V_B) is reached. Zener diodes typically have breakdown voltages in the range of a few volts to a few hundred volts (although larger effective breakdown voltages can be reached by placing zener diodes in series). Figure 4.16 shows the symbol for a zener diode and the forward and reverse current characteristics as a function of applied voltage.

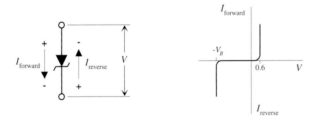

FIGURE 4.16

4.2.6   Zener Diode Water Analogy

In the following water analogy, a zener diode is treated like a two-way gate that has a "forward-biasing spring" and a stronger "reverse-biasing spring."

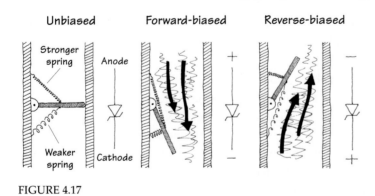

FIGURE 4.17

This water analogy is much like the water analogy used for the *pn*-junction diode. The only difference is that there is no block to prevent the gate from swinging in the "reverse-biasing direction." Instead, a "reverse-biasing spring" is used to hold the gate shut. If the water pressure is applied in the reverse-biasing direction, the gate will only open if the pressure is greater than the force applied by the reverse-biasing spring. Placing a stronger reverse-biasing spring into the system is analogous to choosing a zener diode with a larger reverse-biasing voltage.

4.2.7   Basic Applications for Zener Diodes

Zener diodes are used most frequently in voltage-regulator applications. Here are a few applications.

Voltage Regulator

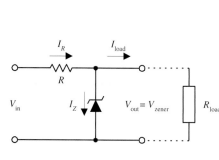

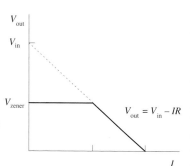

$$I_{\text{load}} = \frac{V_{\text{in}}}{R_{\text{load}}} \approx \frac{V_{\text{zener}}}{R_{\text{load}}}$$

$$I_R = \frac{V_{\text{in}} - V_{\text{out}}}{R} \approx \frac{V_{\text{in}} - V_{\text{zener}}}{R}$$

$$I_{\text{zener}} = I_R - I_{\text{load}}$$

Here, a zener diode is used to regulate the voltage supplied to a load. When V_{in} attempts to push V_{out} above the zener diode's breakdown voltage (V_{zener}), the zener diode draws as much current through itself in the reverse-biased direction as is necessary to keep V_{out} at V_{zener}, even if the input voltage V_{in} varies considerably. The resistor in the circuit is used to limit current through the zener diode in case the load is removed, protecting it from excessive current flow. The value of the resistor can be found by using

$$R = \frac{V_{\text{in}} - V_{\text{zener}}}{I_{\text{max,zener}}}$$

where $I_{\text{max,zener}}$ is the maximum current rating of the zener diode. The power rating of the resistor should be

$$P_R = IV_R = I_{\text{max,zener}}(V_{\text{in}} - V_{\text{zener}})$$

I_{load}, I_R, and I_{zener} are given by the formulas accompanying the figure.

Example 1

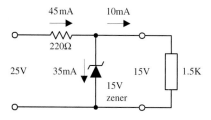

$$I_{\text{load}} = \frac{15\text{V}}{15\text{K}} = 10\text{mA}$$

$$I_R = \frac{25\text{V} - 15\text{V}}{220\Omega} = 45\text{mA}$$

$$I_{\text{zener}} = 45\text{mA} - 10\text{mA} = 35\text{mA}$$

Example 2

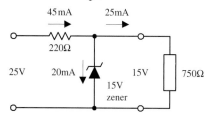

$$I_{\text{load}} = \frac{15\text{V}}{750\Omega} = 20\text{mA}$$

$$I_R = \frac{25\text{V} - 15\text{V}}{220\Omega} = 45\text{mA}$$

$$I_{\text{zener}} = 45\text{mA} - 20\text{mA} = 25\text{mA}$$

Voltage Shifter

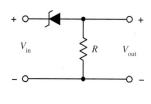

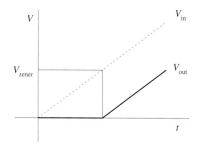

This circuit shifts the input voltage (V_{in}) down by an amount equal to the breakdown voltage of the zener diode.

Waveform Clipper

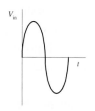

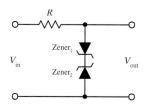

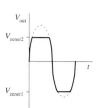

Two opposing zener diodes placed in series act to clip both halves of an input signal. Here, a sine wave is converted to a near square wave. Besides acting as a signal reshaper, this arrangement also can be placed across a power supply's output terminals to prevent voltage spikes from reaching a circuit that is attached to the power supply. The breakdown voltages of both diodes must be chosen according to the particular application.

FIGURE 4.18

4.3   Transistors

Transistors are semiconductor devices that act as either electrically controlled switches or amplifier controls. The beauty of transistors is the way they can control electric current flow in a manner similar to the way a faucet controls the flow of water. With a faucet, the flow of water is controlled by a control knob. With a transistor, a small voltage and/or current applied to a control lead acts to control a larger electric flow through its other two leads.

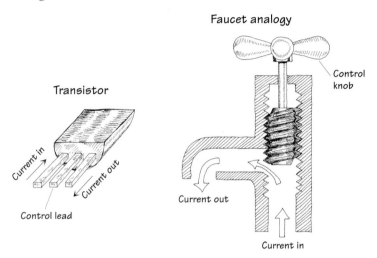

FIGURE 4.19

Transistors are used in almost every electric circuit you can imagine. For example, you find transistors in switching circuits, amplifier circuits, oscillator circuits, current-source circuits, voltage-regulator circuits, power-supply circuits, digital logic ICs, and almost any circuit that uses small control signals to control larger currents.

4.3.1   Introduction to Transistors

Transistors come in a variety of designs and come with unique control and current-flow features. Most transistors have a variable current-control feature, but a few do not. Some transistors are normally off until a voltage is applied to the base or gate, whereas others are normally on until a voltage is applied. (Here, *normally* refers to the condition when the control lead is open circuited. Also, *on* can represent a variable amount of current flow.) Some transistors require both a small current and a small voltage applied to their control lead to function, whereas others only require a voltage. Some transistors require a negative voltage and/or output current at their base lead (relative to one of their other two leads) to function, whereas others require a positive voltage and/or input current at their base.

The two major families of transistors include *bipolar transistors* and *field-effect transistors* (FETs). The major difference between these two families is that bipolar transistors require a biasing input (or output) current at their control leads, whereas FETs require only a voltage—practically no current. [Physically speaking, bipolar transistors require both positive (holes) and negative (electrons) carriers to operate, whereas FETs only require one charge carrier.] Because FETs draw little or no current, they have high input impedances (~10^{14} Ω). This high input impedance means that the FET's control lead will not have much influence on the current dynamics within the circuit that controls the FET. With a bipolar transistor, the control lead may draw a small amount of current from the control circuit, which then combines with the main current flowing through its other two leads, thus altering the dynamics of the control circuit.

In reality, FETs are definitely more popular in circuit design today than bipolar transistors. Besides drawing essentially zero input-output current at their control leads, they are easier to manufacture, cheaper to make (require less silicon), and can be made extremely small—making them useful elements in integrated circuits. One drawback with FETs is in amplifier circuits, where their transconductance is much lower than that of bipolar transistors at the same current. This means that the voltage gain will not be as large. For simple amplifier circuits, FETs are seldom used unless extremely high input impedances and low input currents are required.

TABLE 4.1 Overview of Transistors

TRANSISTOR TYPE	SYMBOL	MODE OF OPERATION
Bipolar	*npn*	Normally off, but a small input current and a small positive voltage at its base (B)—relative to its emitter (E)—turns it on (permits a large collector-emitter current). Operates with $V_C > V_E$. Used in switching and amplifying applications.
	pnp	Normally off, but a small output current and negative voltage at its base (B)—relative to its emitter (E)—turns it on (permits a large emitter-collector current). Operates with $V_E > V_C$. Used in switching and amplifying applications.
Junction FET	*n-channel*	Normally on, but a small negative voltage at its gate (G)—relative to its source (S)—turns it off (stops a large drain-source current). Operates with $V_D > V_S$. Does not require a gate current. Used in switching and amplifying applications.
	p-channel	Normally on, but a small positive voltage at its gate (G)—relative to its source (S)—turns it off (stops a large source-drain current). Operates with $V_S > V_D$. Does not require a gate current. Used in switching and amplifying applications.
Metal oxide semiconductor FET (MOSFET) (depletion)	*n-channel*	Normally on, but a small negative voltage at its gate (G)—relative to its source (S)—turns it off (stops a large drain-source current). Operates with $V_D > V_S$. Does not require a gate current. Used in switching and amplifying applications.
	p-channel	Normally on, but a small positive voltage at its gate (G)—relative to its source (S)—turns it off (stops a large source-drain current). Operates with $V_S > V_D$. Does not require a gate current. Used in switching and amplifying applications.
Metal oxide semiconductor FET (MOSFET) (enhancement)	*n-channel*	Normally off, but a small positive voltage at its gate (G)—relative to its source (S)—turns it on (permits a large drain-source current). Operates with $V_D > V_S$. Does not require a gate current. Used in switching and amplifying applications.
	p-channel	Normally off, but a small negative voltage at its gate (G)—relative to its source (S)—turns it on (permits a large source-drain current). Operates with $V_S > V_D$. Does not require a gate current. Used in switching and amplifying applications.
Unijunction FET (UJT)	*UJT*	Normally a very small current flows from base 2 (B_2) to base 1 (B_1), but a positive voltage at its emitter (E)—relative to B_1 or B_2—increases current flow. Operates with $V_{B2} > V_{B1}$. Does not require a gate current. Only acts as a switch.

Table 4.1 provides an overview of some of the most popular transistors. Note that the term *normally* used in this chart refers to conditions where the control lead (e.g., base, gate) is shorted (is at the same potential) with one of its channel leads (e.g., emitter, source). Also, the terms *on* and *off* used in this chart are not to be taken too literally; the amount of current flow through a device is usually a variable quantity, set by the magnitude of the control voltage. The transistors described in this chart will be discussed in greater detail later on in this chapter.

4.3.2 Bipolar Transistors

Bipolar transistors are three-terminal devices that act as electrically controlled switches or as amplifier controls. These devices come in either *npn* or *pnp* configurations, as shown in Fig. 4.20. An *npn* bipolar transistor uses a small input current and positive voltage at its base (relative to its emitter) to control a much larger collector-to-emitter current. Conversely, a *pnp* transistor uses a small output base current and negative base voltage (relative its emitter) to control a larger emitter-to-collector current.

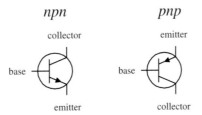

FIGURE 4.20

Bipolar transistors are incredibly useful devices. Their ability to control current flow by means of applied control signals makes them essential elements in electrically controlled switching circuits, current-regulator circuits, voltage-regulator circuits, amplifier circuits, oscillator circuits, and memory circuits.

How Bipolar Transistors Work

Here is a simple model of how an *npn* bipolar transistor works. (For a *pnp* bipolar transistor, all ingredients, polarities, and currents are reversed.)

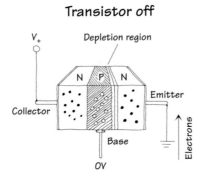

FIGURE 4.21

An *npn* bipolar transistor is made by sandwiching a thin slice of *p* semiconductor between two *n*-type semiconductors. When no voltage is applied at the transistor's base, electrons in the emitter are prevented from passing to the collector side because of the *pn* junction. (Remember that for electrons to flow across a *pn* junction, a biasing voltage is needed to give the electrons enough energy to "escape" the atomic forces holding them to the *n* side.) Notice that if a negative voltage is applied to the base, things get even worse—the *pn* junction between the base and emitter becomes reverse-biased. As a result, a depletion region forms and prevents current flow.

Transistor on

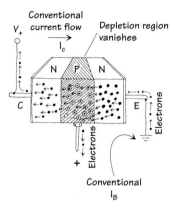

FIGURE 4.21   *(Continued)*

If a positive voltage (of at least 0.6 V) is applied to the base of an *npn* transistor, the *pn* junction between the base and emitter is forward-biased. During forward bias, escaping electrons are drawn to the positive base. Some electrons exit through the base, but—this is the trick—because the *p*-type base is so thin, the onslaught of electrons that leave the emitter get close enough to the collector side that they begin jumping into the collector. Increasing the base voltage increases this jumping effect and hence increases the emitter-to-collector electron flow. Remember that conventional currents are moving in the opposite direction to the electron flow. Thus, in terms of conventional currents, a positive voltage and input current applied at the base cause a "positive" current *I* to flow from the collector to the emitter.

Theory

Figure 4.22 shows a typical characteristic curve for a bipolar transistor. This characteristic curve describes the effects the base current I_B and the emitter-to-collector voltage V_{EC} have on the emitter/collector currents I_E and I_C. (As you will see in a second, I_C is practically equal to I_E.)

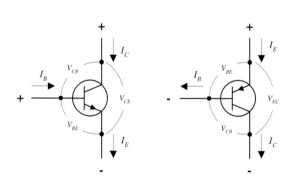
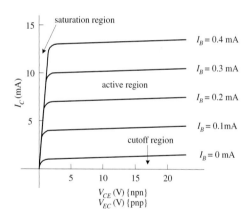

FIGURE 4.22

Some important terms used to describe a transistor's operation include saturation region, cutoff region, active mode/region, bias, and quiescent point (Q-point). *Saturation region* refers to a region of operation where maximum collector current flows and the transistor acts much like a closed switch from collector to emitter. *Cutoff region* refers to the region of operation near the voltage axis of the collector characteristics graph, where the transistor acts like an open switch—only a very small leakage current flows in this mode of operation. *Active mode/region* describes transistor operation in the region to the right of saturation and above cutoff, where a near-linear relationship exists between terminal currents (I_B, I_C, I_E). *Bias* refers to the specific dc terminal voltages and current of the transistor to set a desired point of active-mode operation, called the *quiescent point,* or *Q-point.*

SOME IMPORTANT RULES

Rule 1 For an *npn* transistor, the voltage at the collector V_C must be greater than the voltage at the emitter V_E by at least a few tenths of a volt; otherwise, current will not flow through the collector-emitter junction, no matter what the applied voltage is at the base. For *pnp* transistors, the emitter voltage must be greater than the collector voltage by a similar amount.

Rule 2 For an *npn* transistor, there is a voltage drop from the base to the emitter of 0.6 V. For a *pnp* transistor, there is a 0.6-V rise from base to emitter. In terms of operation, this means that the base voltage V_B of an *npn* transistor must be at least 0.6 V greater than the emitter voltage V_E; otherwise, the transistor will pass an emitter-to-collector current. For a *pnp* transistor, V_B must be at least 0.6 V less than V_E; otherwise, it will not pass a collector-to-emitter current.

The Formulas

The fundamental formula used to describe the behavior of a bipolar transistor (at least within the active region) is

$$I_C = h_{FE}I_B = \beta I_B$$

where I_B is the base current, I_C is the collector current, and h_{FE} (also referred to as β) is the *current gain.* Every transistor has its own unique h_{FE}. The h_{FE} of a transistor is often taken to be a constant, typically around 10 to 500, but it may change slightly with temperature and with changes in collector-to-emitter voltage. (A transistor's h_{FE} is given in transistor spec tables.) A simple explanation of what the current-gain formula tells you is this: If you take a bipolar transistor with, say, an h_{FE} of 100 and then feed (*npn*) or sink (*pnp*) a 1-mA current into (*npn*) or out of (*pnp*) its base, a collector current of 100 mA will result. Now, it is important to note that the current-gain formula applies only if rules 1 and 2 are met, i.e., assuming the transistor is within the active region. Also, there is a limit to how much current can flow through a transistor's terminals and a limit to the size of voltage that can be applied across them. I will discuss these limits later in the chapter (Fig. 4.23).

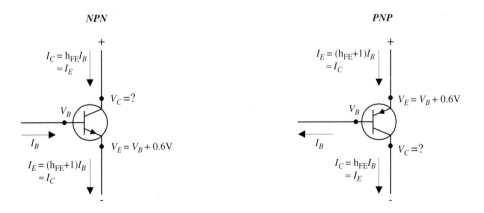

FIGURE 4.23

Now, if you apply the law of conservation of current (follow the arrows in Fig. 4.23), you get the following useful expression relating the emitter, collector, and base currents:

$$I_E = I_C + I_B$$

If you combine this equation with the current-gain equation, you can come up with an equation relating the emitter and base currents:

$$I_E = (h_{FE} + 1)I_B$$

As you can see, this equation is almost identical to the current-gain equation ($I_C = h_{FE}I_B$), with the exception of the +1 term. In practice, the +1 is insignificant as long as h_{FE} is large (which is almost always the case). This means that you can make the following approximation:

$$I_E \approx I_C$$

Finally, the second equation below is simply rule 2 expressed in mathematical form:

$$V_{BE} = V_B - V_E = +0.6 \text{ V } (npn)$$
$$V_{BE} = V_B - V_E = -0.6 \text{ V } (pnp)$$

Figure 4.23 shows how all the terminal currents and voltages are related. In the figure, notice that the collector voltage has a question mark next to it. As it turns out, the value of V_C cannot be determined directly by applying the formulas. Instead, V_C's value depends on the network that is connected to it. For example, if you consider the setup shown in Fig. 4.24, you must find the voltage drop across the resistor in order to find the collector voltage. By applying Ohm's law and using the current-gain relation, you can calculate V_C. The results are shown in the figure.

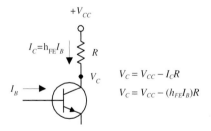

FIGURE 4.24

It is important to note that the equations here are idealistic in form. In reality, these equations may result in "unreal" answers. For instance, they tend to "screw up" when the currents and voltages are not within the bounds provided by the characteristic curves. If you apply the equations blindly, without considering the operating characteristics, you could end up with some wild results that are physically impossible.

One final note with regard to bipolar transistor theory involves what is called *transresiststance* r_{tr}. Transresistance represents a small resistance that is inherently present within the emitter junction region of a transistor. Two things that determine the transresistance of a transistor are temperature and emitter current flow. The following equation provides a rough approximation of the r_{tr}:

$$r_{tr} \approx \frac{0.026 \text{ V}}{I_E}$$

In many cases, r_{tr} is insignificantly small (usually well below 1000 Ω) and does not pose a major threat to the overall operation of a circuit. However, in certain types of circuits, treating r_{tr} as being insignificant will not do. In fact, its presence may be the major factor determining the overall behavior of a circuit. We will take a closer look at transresistance later on in this chapter.

Here are a couple of problems that should help explain how the equations work. The first example deals with an *npn* transistor; the second deals with a *pnp* transistor.

EXAMPLE 1   Given $V_{CC} = +20$ V, $V_B = 5.6$ V, $R_1 = 4.7$ kΩ, $R_2 = 3.3$ kΩ, and $h_{FE} = 100$, find V_E, I_E, I_B, I_C, and V_C.

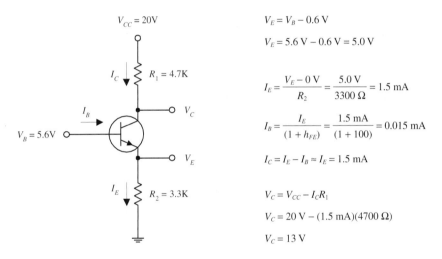

$$V_E = V_B - 0.6 \text{ V}$$

$$V_E = 5.6 \text{ V} - 0.6 \text{ V} = 5.0 \text{ V}$$

$$I_E = \frac{V_E - 0 \text{ V}}{R_2} = \frac{5.0 \text{ V}}{3300 \text{ }\Omega} = 1.5 \text{ mA}$$

$$I_B = \frac{I_E}{(1 + h_{FE})} = \frac{1.5 \text{ mA}}{(1 + 100)} = 0.015 \text{ mA}$$

$$I_C = I_E - I_B \approx I_E = 1.5 \text{ mA}$$

$$V_C = V_{CC} - I_C R_1$$

$$V_C = 20 \text{ V} - (1.5 \text{ mA})(4700 \text{ }\Omega)$$

$$V_C = 13 \text{ V}$$

FIGURE 4.25

EXAMPLE 2   Given $V_{CC} = +10$ V, $V_B = 8.2$ V, $R_1 = 560$ Ω, $R_2 = 2.8$ kΩ, and $h_{FE} = 100$, find V_E, I_E, I_B, I_C, and V_C.

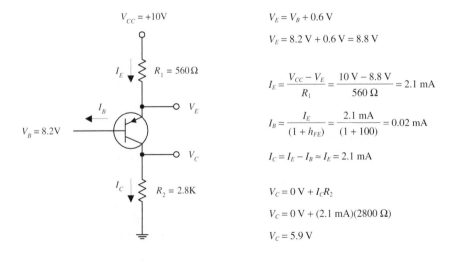

$$V_E = V_B + 0.6 \text{ V}$$

$$V_E = 8.2 \text{ V} + 0.6 \text{ V} = 8.8 \text{ V}$$

$$I_E = \frac{V_{CC} - V_E}{R_1} = \frac{10 \text{ V} - 8.8 \text{ V}}{560 \text{ }\Omega} = 2.1 \text{ mA}$$

$$I_B = \frac{I_E}{(1 + h_{FE})} = \frac{2.1 \text{ mA}}{(1 + 100)} = 0.02 \text{ mA}$$

$$I_C = I_E - I_B \approx I_E = 2.1 \text{ mA}$$

$$V_C = 0 \text{ V} + I_C R_2$$

$$V_C = 0 \text{ V} + (2.1 \text{ mA})(2800 \text{ }\Omega)$$

$$V_C = 5.9 \text{ V}$$

FIGURE 4.26

Bipolar Transistor Water Analogy

NPN WATER ANALOGY

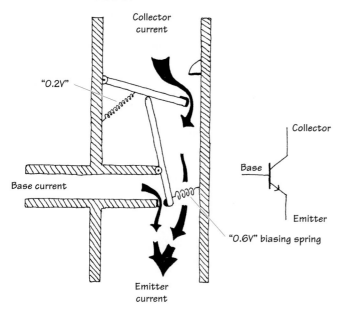

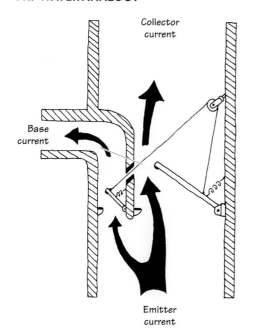

The base of the *npn* water transistor represents the smaller tube entering the main device from the left side. The collector is represented by the upper portion of the vertical tube, while the emitter is represented by the lower portion of the vertical tube. When no pressure or current is applied through the "base" tube (analogous to an *npn* transistor's base being open circuited), the lower lever arm remains vertical while the top of this arm holds the upper main door shut. This state is analogous to a real bipolar *npn* transistor off state. In the water analogy, when a small current and pressure are applied to the base tube, the vertical lever is pushed by the entering current and swings counterclockwise. When this lever arm swings, the upper main door is permitted to swing open a certain amount that is dependent on the amount of swing of the lever arm. In this state, water can make its way from the collector tube to the emitter tube, provided there is enough pressure to overcome the force of the spring holding the door shut. This spring force is analogous to the 0.6 V biasing voltage needed to allow current through the collector-emitter channel. Notice that in this analogy, the small base water current combines with the collector current.

PNP WATER ANALOGY

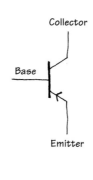

The main feature to note here is the need for a lower pressure at the base for the *pnp* water transistor to turn on. By allowing current to flow out the base tube, the lever moves, allowing the emitter-collector door to open. The degree of openness varies with the amount of swing in the lever arm, which corresponds to the amount of current escaping through the base tube. Again, note the 0.6 V biasing spring.

FIGURE 4.27

Basic Operation

TRANSISTOR SWITCH

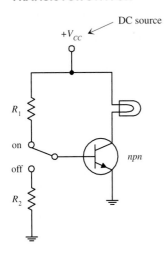

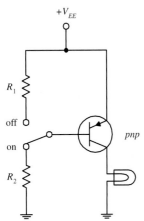

FIGURE 4.28

Here, an *npn* transistor is used to control current flow through a light bulb. When the switch is thrown to the on position, the transistor is properly biased, and the collector-to-emitter channel opens, allowing current to flow from V_{CC} through the light bulb and into ground. The amount of current entering the base is determined by

$$I_B = \frac{V_E + 0.6\,\text{V}}{R_1} = \frac{0\,\text{V} + 0.6\,\text{V}}{R_1}$$

To find the collector current, you can use the current-gain relation ($I_C = h_{FE}I_B$), provided that there is not too large a voltage drop across the light bulb (it shouldn't cause V_C to drop below $0.6\,\text{V} + V_E$). When the switch is thrown to the off position, the base is set to ground, and the transistor turns off, cutting current flow to the light bulb. R_2 should be large (e.g., 10 kΩ) so that very little current flows to ground.

In the *pnp* circuit, everything is reversed; current must leave the base in order for a collector current to flow.

CURRENT SOURCE

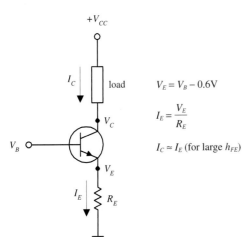

$$V_E = V_B - 0.6\,\text{V}$$

$$I_E = \frac{V_E}{R_E}$$

$$I_C \approx I_E \text{ (for large } h_{FE})$$

FIGURE 4.29

The circuit here shows how an *npn* transistor can be used to make a simple current source. By applying a small input voltage and current at the transistor's base, a larger collector/load current can be controlled. The collector/load current is related to the base voltage by

$$I_C = I_{\text{load}} = \frac{V_B - 0.6\,\text{V}}{R_E}$$

The derivation of this equation is shown with the figure.

CURRENT BIASING METHODS

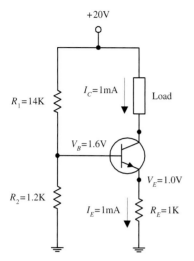

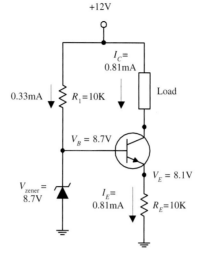

Two common methods for biasing a current source are to use either a voltage-divider circuit (shown in the leftmost circuit) or a zener diode regulator (shown in the rightmost circuit). In the voltage-divider circuit, the base voltage is set by R_1 and R_2 and is equal to

$$V_B = \frac{R_2}{R_1 + R_2} V_{CC}$$

In the zener diode circuit, the base voltage is set by the zener diode's breakdown voltage such that

$$V_B = V_{zener}$$

FIGURE 4.30

EMITTER FOLLOWER

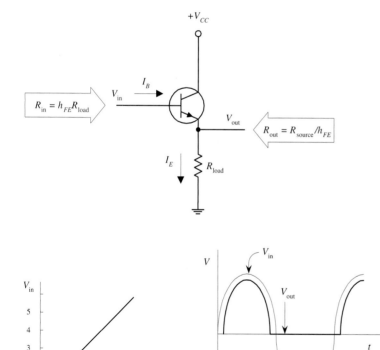

FIGURE 4.31

The network shown here is called an *emitter follower*. In this circuit, the output voltage (tapped at the emitter) is almost a mirror image of the input (output "follows" input), with the exception of a 0.6 V drop in the output relative to the input (caused by base-emitter *pn* junction). Also, whenever $V_B \leq V_E + 0.6$ V (during negative swings in input), the transistor will turn off (the *pn* junction is reversed-biased). This effect results in clipping of the output (see graph). At first glance, it may appear that the emitter follower is useless—it has no voltage gain. However, if you look at the circuit more closely, you will see that it has a much larger input impedance than an output impedance, or more precisely, it has a much larger output current (I_E) relative to an input current (I_B). In other words, the emitter follower has current gain, a feature that is just as important in applications as voltage gain. This means that this circuit requires less power from the signal source (applied to V_{in}) to drive a load than would otherwise be required if the load were to be powered directly by the source. By manipulating the transistor gain equation and using Ohm's law, the input resistance and output resistance are:

$$R_{in} = h_{FE}R_{load}$$

$$R_{out} = \frac{R_{source}}{h_{FE}}$$

EMITTER-FOLLOWER (COMMON-COLLECTOR) AMPLIFIER

The circuit shown here is called a *common-collector amplifier*, which has current gain but no voltage gain. It makes use of the emitter-follower arrangement but is modified to avoid clipping during negative input swings. The voltage divider (R_1 and R_2) is used to give the input signal (after passing through the capacitor) a positive dc level or operating point (known as the *quiescent point*). Both the input and output capacitors are included so that an ac input-output signal can be added without disturbing the dc operating point. The capacitors, as you will see, also act as filtering elements.

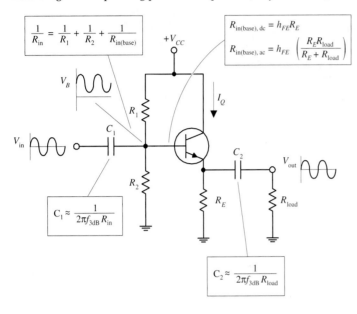

$$\frac{1}{R_{in}} = \frac{1}{R_1} + \frac{1}{R_2} + \frac{1}{R_{in(base)}}$$

$$R_{in(base), dc} = h_{FE}R_E$$

$$R_{in(base), ac} = h_{FE}\left(\frac{R_E R_{load}}{R_E + R_{load}}\right)$$

$$C_1 \approx \frac{1}{2\pi f_{3dB} R_{in}}$$

$$C_2 \approx \frac{1}{2\pi f_{3dB} R_{load}}$$

To design a common-collector amplifier used to power a 3-kΩ load, which has a supply voltage V_{CC} = +10 V, a transistor h_{FE} of 100, and a desired f_{3dB} point of 100 Hz, you

1. Choose a quiescent current $I_Q = I_C$. For this problem, pick $I_Q = 1$ mA.

2. Next, select $V_E = \frac{1}{2}V_{CC}$ to allow for the largest possible symmetric output swing without clipping, which in this case, is 5 V. To set $V_E = 5$ V and still get $I_Q = 1$ mA, make use of R_E, whose value you find by applying Ohm's law:

$$R_E = \frac{\frac{1}{2}V_{CC}}{I_Q} = \frac{5\,\text{V}}{1\,\text{mA}} = 5\,\text{k}\Omega$$

3. Next, set the $V_B = V_{EE} + 0.6$ V for quiescent conditions (to match up V_{EE} so as to avoid clipping). To set the base voltage, use the voltage divider (R_1 and R_2). The ratio between R_1 and R_2 is determined by rearranging the voltage-divider relation and substituting into it $V_B = V_{EE} + 0.6$ V:

$$\frac{R_2}{R_1} = \frac{V_B}{V_{CC} - V_B} = \frac{V_E + 0.6\,\text{V}}{V_{CC} - (V_E + 0.6\,\text{V})}$$

Fortunately, you can make an approximation and simply let $R_1 = R_2$. This approximation "forgets" the 0.6-V drop but usually isn't too dramatic. The actual sizes of R_2 and R_1 should be such that their parallel resistance is less than or equal to one-tenth the dc (quiescent) input resistance at the base (this prevents the voltage divider's output voltage from lowering under loading conditions):

$$\frac{R_1 R_2}{R_1 + R_2} \leq \frac{1}{10} R_{in(base),dc}$$

$$\frac{R}{2} \leq \frac{1}{10} R_{in(base),dc} \quad \text{(using the approximation } R = R_1 = R_2\text{)}$$

Here, $R_{in(base),dc} = h_{FE}R_E$, or specially, $R_{in(base),dc} = 100(5\,\text{k}) = 500\,\text{k}$. Using the approximation above, R_1 and R_2 are calculated to be 100 k each. (Here you did not have to worry about the ac coupler load; it did not influence the voltage divider because you assumed quiescent setup conditions; C_2 acts as an open circuit, hence "eliminating" the presence of the load.)

EXAMPLE

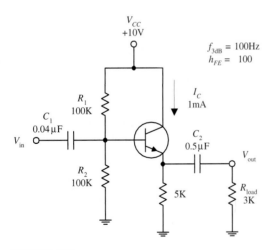

f_{3dB} = 100Hz
h_{FE} = 100

FIGURE 4.32

4. Next, choose the ac coupling capacitors so as to block out dc levels and other undesired frequencies. C_1 forms a high-pass filter with R_{in} (see diagram). To find R_{in}, treat the voltage divider and $R_{in(base),ac}$ as being in parallel:

$$\frac{1}{R_{in}} = \frac{1}{R_1} + \frac{1}{R_2} + \frac{1}{R_{in(base),ac}}$$

Notice that $R_{in(base),ac}$ is used, not $R_{in(base),dc}$. This is so because you can no longer treat the load as being absent when fluctuating signals are applied to the input; the capacitor begins to pass a displacement current. This means that you must take R_E and R_{load} in parallel and multiply by h_{FE} to find $R_{in(base,ac)}$:

$$R_{in(base),ac} = h_{FE}\left(\frac{R_E R_{load}}{F_E + R_{load}}\right) = 100\left[\frac{5\,\text{k}\Omega(3\,\text{k}\Omega)}{5\,\text{k}\Omega + 3\,\text{k}\Omega}\right] = 190\,\text{k}\Omega$$

Now you can find R_{in}:

$$\frac{1}{R_{\text{in}}} = \frac{1}{100 \text{ k}\Omega} + \frac{1}{100 \text{ k}\Omega} + \frac{1}{190 \text{ k}\Omega}$$

$$R_{\text{in}} = 40 \text{ k}\Omega$$

Once you have found R_{in}, choose C_1 to set the f_{3dB} point (C_1 and R_{in} form a high-pass filter.) The f_{3dB} point is found by using the following formula:

$$C_1 = \frac{1}{2\pi f_{\text{3dB}} R_{\text{in}}} = \frac{1}{2\pi (100 \text{ Hz})(40 \text{ k}\Omega)} = 0.04 \text{ }\mu\text{F}$$

C_2 forms a high-pass filter with the load. It is chosen by using

$$C_2 = \frac{1}{2\pi f_{\text{3dB}} R_{\text{load}}} = \frac{1}{2\pi (100 \text{ Hz})(3 \text{ k}\Omega)} = 0.5 \text{ }\mu\text{F}$$

COMMON-EMITTER CONFIGURATION

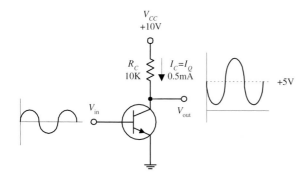

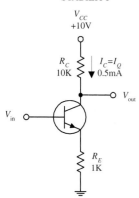

INCLUDING R_E FOR TEMPERATURE
STABILITY

ACHIEVING HIGH-GAIN WITH
TEMPERATURE STABILITY

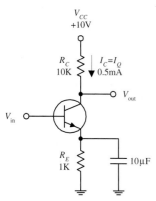

FIGURE 4.33

The transistor configuration here is referred to as the *common-emitter configuration.* Unlike the emitter follower, the common emitter has voltage gain. To figure out how this circuit works, first set $V_C = \frac{1}{2}V_{CC}$ to allow for maximum swing without clipping. Like the emitter follower, again pick a quiescent current I_Q to start with. To set $V_C = \frac{1}{2}V_{CC}$ with a desired I_Q, use R_C, which is found by Ohm's law:

$$R_C = \frac{V_{CC} - V_C}{I_C} = \frac{V_{CC} - \frac{1}{2}V_{CC}}{I_Q} = \frac{\frac{1}{2}V_{CC}}{I_Q}$$

For example, if V_{CC} is 10V and I_Q is 0.5 mA, R_C is then 10 k. The gain of this circuit is found by realizing that $\Delta V_E = \Delta V_B$ (where Δ represents a small fluctuation). The emitter current is found using Ohm's law:

$$\Delta I_E = \frac{\Delta V_E}{R_E} = \frac{\Delta V_B}{R_E} \approx \Delta I_C$$

Using $V_C = V_{CC} - I_C R_C$ and the last expression, you get

$$\Delta V_C = -\Delta I_C R_C = -\frac{\Delta V_B}{R_E} R_C$$

Since V_C is V_{out} and V_B is V_{in}, the gain is

$$\text{Gain} = \frac{V_{\text{out}}}{V_{\text{in}}} = \frac{\Delta V_C}{\Delta V_B} = -\frac{R_C}{R_E}$$

But what about R_E? According to the circuit, there's no emitter resistor. If you use the gain formula, it would appear that $R_E = 0 \text{ }\Omega$, making the gain infinite. However, as mentioned earlier, bipolar transistors have a trans-resistance (small internal resistance) in the emitter region, which is approximated by using

$$r_{\text{tr}} \approx \frac{0.026 \text{ V}}{I_E}$$

Applying this formula to the example, taking $I_Q = 0.5$ mA $= I_C \approx I_E$, the R_E term in the gain equation, or r_{tr} equals 52 Ω. This means the gain is actually equal to

$$\text{Gain} = -\frac{R_C}{R_E} = -\frac{R_C}{r_{\text{tr}}} = -\frac{10 \text{ k}\Omega}{52 \text{ }\Omega} = -192$$

Notice that the gain is negative (output is inverted). This results in the fact that as V_{in} increases, I_C increases, while V_C (V_{out}) decreases (Ohm's law). Now, there is one problem with this circuit. The r_{tr} term happens to be very unstable, which in effect makes the gain unstable. The instability stems from r_{tr} dependence on temperature. As the temperature rises, V_E

and I_C increase, V_{BE} decreases, but V_B remains fixed. This means that the biasing-voltage range narrows, which in effect turns the transistor's "valve" off. To eliminate this pinch, an emitter resistor is placed from emitter to ground (see second circuit). Treating R_E and r_{tr} as series resistors, the gain becomes

$$\text{Gain} = \frac{R_C}{R_E + r_{tr}}$$

By adding R_E, variations in the denominator are reduced, and therefore, the variations in gain are reduced as well. In practice, R_E should be chosen to place V_E around 1 V (for temperature stability and maximum swing in output). Using Ohm's law (and applying it to the example), choose $R_E = V_E/I_E = V_E/I_Q = 1\,\text{V}/1\,\text{mA} = 1\,\text{k}$. One drawback that arises when R_E is added to the circuit is a reduction in gain. However, there is a trick you can use to eliminate this reduction in voltage gain and at the same time maintain the temperature stability. If you bypass R_E with a capacitor (see third circuit), you can make R_E "disappear" when high-frequency input signals are applied. (Recall that a capacitor behaves like an infinitely large resistor to dc signals but becomes less "resistive," or reactive to ac signals.) In terms of the gain equation, the R_E term goes to zero because the capacitor diverts current away from it toward ground. The only resistance left in the gain equation is r_{tr}.

COMMON-EMITTER AMPLIFIER

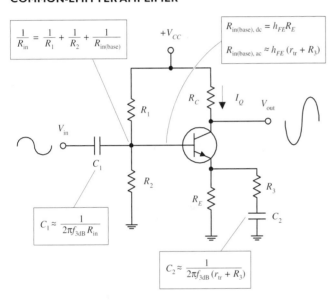

$$\frac{1}{R_{in}} = \frac{1}{R_1} + \frac{1}{R_2} + \frac{1}{R_{in(base)}}$$

$$R_{in(base),\,dc} = h_{FE} R_E$$

$$R_{in(base),\,ac} \approx h_{FE}\,(r_{tr} + R_3)$$

$$C_1 \approx \frac{1}{2\pi f_{3dB} R_{in}}$$

$$C_2 \approx \frac{1}{2\pi f_{3dB}\,(r_{tr} + R_3)}$$

EXAMPLE

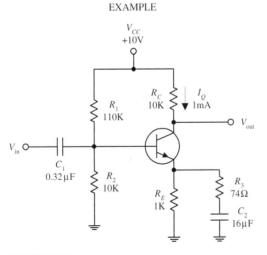

FIGURE 4.34

The circuit shown here is known as a *common-emitter amplifier*. Unlike the common-collector amplifier, the common-emitter amplifier provides voltage gain. This amplifier makes use of the common-emitter arrangement and is modified to allow for ac coupling. To understand how the amplifier works, let's go through the following example.

To design a common-emitter amplifier with a voltage gain of −100, an f_{3dB} point of 100 Hz, and a quiescent current $I_Q = 1$ mA, where $h_{FE} = 100$ and $V_{CC} = 20$ V:

1. Choose R_C to center V_{out} (or V_C) to $\frac{1}{2}V_{CC}$ to allow for maximum symmetrical swings in the output. In this example, this means V_C should be set to 10 V. Using Ohm's law, you find R_C:

$$R_C = -\frac{V_C - V_{CC}}{I_C} = -\frac{0.5V_{CC} - V_{CC}}{I_Q} = \frac{10\,\text{V}}{1\,\text{mA}} = 10\,\text{k}\Omega$$

2. Next we select R_E to set $V_E = 1$ V for temperature stability. Using Ohm's law, and taking $I_Q = I_E = 1$ mA, we get $R_E = V_E/I_E = 1\text{V}/1\,\text{mA} = 1\,\text{k}\Omega$.

3. Now, choose R_1 and R_2 to set the voltage divider to establish the quiescent base voltage of $V_B = V_E + 0.6$ V, or 1.6 V. To find the proper ratio between R_1 and R_2, use the voltage divider (rearranged a bit):

$$\frac{R_2}{R_1} = \frac{V_B}{V_{CC} - V_B} = \frac{1.6\,\text{V}}{20\,\text{V} - 1.6\,\text{V}} = \frac{1}{11.5}$$

This means $R_1 = 11.5R_2$. The size of these resistors is found using the similar procedure you used for the common-collector amplifier; their parallel resistance should be less than or equal to $\frac{1}{10}R_{in(base),dc}$.

$$\frac{R_1 R_2}{R_1 + R_2} \leq \frac{1}{10} R_{in(base),dc}$$

After plugging $R_1 = 11.5R_2$ into this expression and using $R_{in(base),dc} = h_{FE}R_E$, you find that $R_2 = 10$ kΩ, which in turn means $R_1 = 115$ kΩ (let's say, 110 kΩ is close enough for R_1).

4. Next, choose R_3 for the desired gain, where

$$\text{Gain} = -\frac{R_C}{R_E \| (r_{tr} + R_3)} = -100$$

(The double line means to take R_E and $(r_{tr} + R_3)$ in parallel.) To find r_{tr}, use $r_{tr} = 0.026\,\text{V}/I_E = 0.026\,\text{V}/IC = 0.026\,\text{V}/1\,\text{mA} = 26\,\Omega$. Now you can simplify the gain expression by assuming R_E "disappears" when ac signals are applied. This means the gain can be simplified to

$$\text{Gain} = -\frac{R_C}{r_{tr} + R_3} = -\frac{10\,\text{k}}{26\,\Omega + R_3} = -100$$

Solving this equation for R_3, you get $R_3 = 74\,\Omega$.

5. Next, choose C_1 for filtering purposes such that $C_1 = \frac{1}{2}\pi f_{3dB} R_{in}$. Here, R_{in} is the combined parallel resistance of the voltage-divider resistors, and $R_{in(base),ac}$ looking in from the left into the voltage divider:

$$\frac{1}{R_{in}} = \frac{1}{R_1} + \frac{1}{R_2} + \frac{1}{h_{FE}(r_{tr} + R_3)} = \frac{1}{110\,\text{k}\Omega} + \frac{1}{10\,\text{k}\Omega} + \frac{1}{100(26\,\Omega + 74\,\Omega)}$$

Solving this equation, you get $R_{in} = 5\,\text{k}$. This means

$$C_1 = \frac{1}{2\pi(100\,\text{Hz})(5\,\text{k}\Omega)} = 0.32\,\mu\text{F}$$

6. To choose C_2, treat C_2 and $r_{tr} + R_3$ as a high-pass filter (again, treat R_F as being negligible during ac conditions). C_2 is given by

$$C_2 = \frac{1}{2\pi f_{3dB}(r_{tr} + R_3)} = \frac{1}{2\pi(100\,\text{Hz})(26\,\Omega + 74\,\Omega)} = 16\,\mu\text{F}$$

VOLTAGE REGULATOR

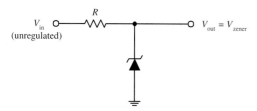

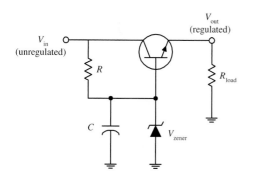

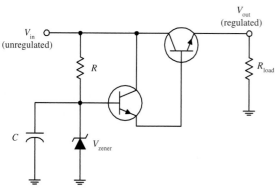

The zener diode circuit here can be used to make a simple voltage regulator. However, in many applications, the simple regulator has problems; V_{out} isn't adjustable to a precise value, and the zener diode provides only moderate protection against ripple voltages. Also, the simple zener diode regulator does not work particularly well when the load impedance varies. Accommodating large load variations requires a zener diode with a large power rating—which can be costly.

The second circuit in the figure, unlike the first circuit, does a better job of regulating; it provides regulation with load variations and provides high-current output and somewhat better stability. This circuit closely resembles the preceding circuit, except that the zener diode is connected to the base of an *npn* transistor and is used to regulate the collect-to-emitter current. The transistor is configured in the emitter-follower configuration. This means that the emitter will follow the base (except there is the 0.6-V drop). Using a zener diode to regulate the base voltage results in a regulated emitter voltage. According to the transistor rules, the current required by the base is only $1/h_{FE}$ times the emitter-to-collector current. Therefore, a low-power zener diode can regulate the base voltage of of a transistor that can pass a considerable amount of current. The capacitor is added to reduce the noise from the zener diode and also forms an RC filter with the resistor that is used to reduce ripple voltages.

In some instances, the preceding zener diode circuit may not be able to supply enough base current. One way to fix this problem is to add a second transistor, as shown in the third circuit. The extra transistor (the one whose base is connected to the zener diode) acts to amplify current sent to the base of the upper transistor.

FIGURE 4.35

DARLINGTON PAIR

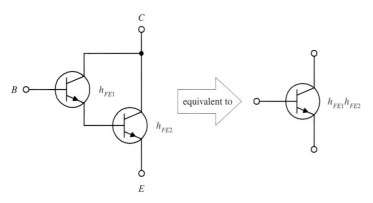

FIGURE 4.36

By attaching two transistors together as shown here, a larger current-handling, larger h_{FE} equivalent transistor circuit is formed. The combination is referred to as a *Darlington pair*. The equivalent h_{FE} for the pair is equal to the product of the individual transistor's h_{FE} values ($h_{FE} = h_{FE1}h_{FE2}$). Darlington pairs are used for large current applications and as input stages for amplifiers, where big input impedances are required. Unlike single transistors, however, Darlington pairs have slower response times (it takes longer for the top transistor to turn the lower transistor on and off) and have twice the base-to-emitter voltage drop (1.2V instead of 0.6V) as compared with single transistors. Darlington pairs can be purchased in single packages.

Types of Bipolar Transistors

SMALL SIGNAL

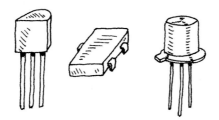

This type of transistor is used to amplify low-level signals but also can be used as a switch. Typical h_{FE} values range from 10 to 500, with maximum I_C ratings from about 80 to 600 mA. They come in both *npn* and *pnp* forms. Maximum operating frequencies range from about 1 to 300 MHz.

SMALL SWITCHING

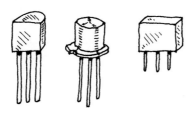

These transistors are used primarily as switches but also can be used as amplifiers. Typical h_{FE} values range from around 10 to 200, with maximum I_C ratings from around 10 to 1000 mA. They come in both *npn* and *pnp* forms. Maximum switching rates range between 10 and 2000 MHz.

HIGH FREQUENCY (RF)

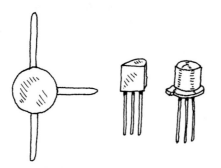

These transistors are used for small signals that run at high frequencies for high-speed switching applications. The base region is very thin, and the actual chip is very small. They are used in HF, VHF, UHF, CATV, and MATV amplifier and oscillator applications. They have a maximum frequency rating of around 2000 MHz and maximum I_C currents from 10 to 600 mA. They come in both *npn* and *pnp* forms.

POWER

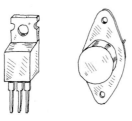

FIGURE 4.37

These transistors are used in high-power amplifiers and power supplies. The collector is connected to a metal base that acts as a heat sink. Typical power ratings range from around 10 to 300 W, with frequency ratings from about 1 to 100 MHz. Maximum I_C values range between 1 to 100 A. They come in *npn, pnp,* and Darlington (*npn* or *pnp*) forms.

DARLINGTON PAIR

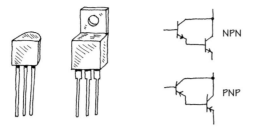

NPN

PNP

These are two transistors in one. They provide more stability at high current levels. The effective h_{FE} for the device is much larger than that of a single transistor, hence allowing for a larger current gain. They come in *npn* (*D-npn*) and *pnp* (*D-pnp*) Darlington packages.

PHOTOTRANSISTOR

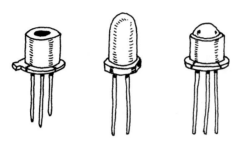

This transistor acts as a light-sensitive bipolar transistor (base is exposed to light). When light comes in contact with the base region, a base current results. Depending on the type of phototransistor, the light may act exclusively as a biasing agent (two-lead phototransistor) or may simply alter an already present base current (three-lead phototransistor). See Chap. 5 for more details.

TRANSISTOR ARRAY

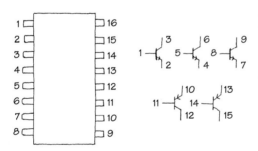

This consists of a number of transistors combined into a single integrated package. For example, the transistor array shown here is made of three *npn* transistors and two *pnp* transistors.

FIGURE 4.37   (*Continued*)

Important Things to Know about Bipolar Transistors

The current gain of a transistor (h_{FE}) is not a very good parameter to go by. It can vary from, say, 50 to 500 within same transistor group family and varies with changes in collector current, collector-to-emitter voltage, and temperature. Because h_{FE} is somewhat unpredictable, one should avoid building circuits that depend specifically on h_{FE} values.

All transistors have maximum collector-current ratings ($I_{C,max}$), maximum collector-to-base (BV_{CBO}), collector-to-emitter (BV_{CEO}), and emitter-to-base (V_{EBO}) breakdown voltages, and maximum collector power dissipation (P_D) ratings. If these rating are exceeded, the transistor may get zapped. One method to safeguard against BV_{EB} is to place a diode from the emitter to the base, as shown in Fig. 4.38*a*. The diode prevents emitter-to-base conduction whenever the emitter becomes more positive than the base (e.g., input at base swings negative while emitter is grounded). To avoid exceeding BV_{CBO}, a diode placed in series with the collector (Fig. 4.38*b*) can be used to prevent collector-base conduction from occurring when the base voltage becomes excessively larger than the collector voltage. To prevent exceeding BV_{CEO}, which may be an issue if the collector holds an inductive load, a diode placed in parallel with the load (see Fig. 4.38c) will go into conduction before a collector-voltage spike, created by the inductive load, reaches the breakdown voltage.

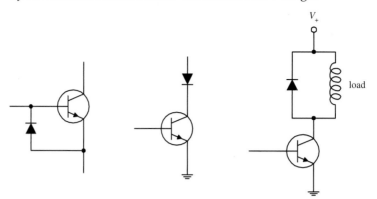

FIGURE 4.38

Pinouts for Bipolar Transistors

Bipolar transistors come in a variety of different package types. Some transistors come with plastic housings; others come with metal can-like housings. When attempting to isolate the leads that correspond to the base, emitter, and collector terminals, first check to see if the package that housed the transistor has a pinout diagram. If no pinout diagram is provided, a good cross-reference catalog (e.g., NTE Cross-Reference Catalog for Semiconductors) can be used. However, as is often the case, simple switching transistors that come in bulk cannot be "looked up"—they may not have a label. Also, these bulk suppliers often will throw together a bunch of transistors that all look alike but may have entirely different pinout designations and may include both *pnp* and *npn* polarities. If you anticipate using transistors often, it may be in your best interest to purchase a digital multimeter that comes with a transistor tester. These multimeters are relatively inexpensive and are easy to use. Such a meter comes with a number of breadboard-like slots. To test a transistor, the pins of the transistor are placed into the slots. By simply pressing a button, the multimeter then tests the transistor and displays whether the device is an *npn* or *pnp* transistor, provides you with the pinout designations (e.g., "ebc," "cbe," etc.), and will give you the transistor's h_{FE}.

Applications

RELAY DRIVER

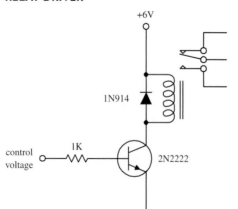

Here, an *npn* transistor is used to control a relay. When the transistor's base receives a control voltage/current, the transistor will turn on, allowing current to flow through the relay coil and causing the relay to switch states. The diode is used to eliminate voltage spikes created by the relay's coil. The relay must be chosen according to the proper voltage rating, etc.

FIGURE 4.39

DIFFERENTIAL AMPLIFIER

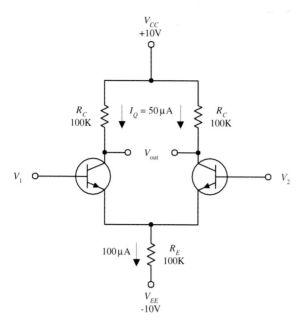

FIGURE 4.40

The differential amplifier shown here is a device that compares two separate input signals, takes the difference between them, and then amplifies this difference. To understand how the circuit works, treat both transistors as identical, and then notice that both transistors are set in the common-emitter configuration. Now, if you apply identical input signals to both V_1 and V_2, identical currents flow through each transistor. This means that (by using $V_C = V_{CC} - I_C R_C$) both transistors' collector voltages are the same. Since the output terminals are simply the left and right transistors' collector voltages, the output voltage (potential difference) is zero. Now, assume the signals applied to the inputs are different, say V_1 is larger than V_2. In this case, the current flow through the right transistor will be larger than the current flow through the left transistor. This means that right transistor's V_C will decrease relative to the left transistor's V_C. Because the transistors are set in the common-emitter configuration, the effect is amplified. The relationship between the input and output voltages is given by

$$V_{\text{out}} \approx \frac{R_C}{r_{\text{tr}}}(V_1 - V_2)$$

Rearranging this expression, you find that the gain is equal to R_C/r_{tr}.

Understanding what resistor values to choose can be explained by examining the circuit shown here. First, choose R_C to center V_C to $\frac{1}{2}V_{CC}$, or 5 V, to maximize the dynamic range. At the same time, you must choose a quiescent current (when no signals are applied), say, $I_Q = I_C = 50$ μA. By Ohm's law $R_C = 10$ V $- 5$ V/50 μA $= 100$ k. R_E is chosen to set the transistor's emitters (point A) as close to 0 V as possible. R_E is found by adding both the right and left branch's 50 μA and taking the sum to be the current flow through it, which is 100 μA. Now, apply Ohm's law: $R_E = 0$ V $- 10$ V/100 μA $= 100$ kΩ. Next, find the transresistance: $r_{\text{tr}} \approx 0.026$ V/$I_E = 0.026$ V/50 μA $= 520$ Ω. The gain then is equal to 100 kΩ/520 Ω $= 192$.

In terms of applications, differential amplifiers can be used to extract a signal that has become weak and which has picked up considerable noise during transmission through a cable (differential amplifier is placed at the receiving end). Unlike a filter circuit, which can only extract a signal from noise if the noise frequency and signal frequency are different, a differential amplifier does not require this condition. The only requirement is that the noise be common in both wires.

When dealing with differential amplifiers, the term *common-mode rejection ratio* (CMRR) is frequently used to describe the quality of the amplifier. A good differential amplifier has a high CMRR (theoretically infinite). CMRR is the ratio of the voltage that must be applied at the two inputs in parallel (V_1 and V_2) to the difference voltage ($V_1 - V_2$) for the output to have the same magnitude.

COMPLEMENTARY-SYMMETRY AMPLIFIER

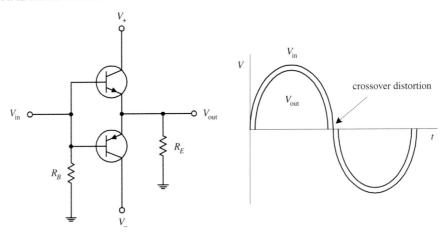

FIGURE 4.41

Recall that an *npn* emitter follower acts to clip the output during negative swings in the input (the transistor turns off when $V_B \leq V_E + 0.6\,V$). Likewise, a *pnp* follower will clip the output during positive input swings. But now, if you combine an *npn* and *pnp* transistor, as shown in the circuit shown here, you get what is called a *push-pull follower,* or *complementary-symmetry amplifier,* an amplifier that provides current gain and that is capable of conducting during both positive and negative input swings. For $V_{in} = 0\,V$, both transistors are biased to cutoff ($I_B = 0$). For $V_{in} > 0\,V$, the upper transistor conducts and behaves like an emitter follower, while the lower transistor is cut off. For $V_{in} < 0\,V$ the lower transistor conducts, while the upper transistor is cut off. In addition to being useful as a dc amplifier, this circuit also conserves power because the operating point for both transistors is near $I_C = 0$. However, at $I_C = 0$, the characteristics of h_{FE} and r_{tr} are not very constant, so the circuit is not very linear for small signals or for near-zero crossing points of large signals (crossover distortion occurs).

CURRENT MIRROR

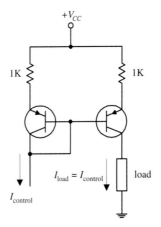

FIGURE 4.42

Here, two matched *pnp* transistors can be used to make what is called a *current mirror.* In this circuit, the load current is a "mirror image" of the control current that is sunk out of the leftmost transistor's collector. Since the same amount of biasing current leaves both transistors' bases, it follows that both transistors' collector-to-emitter currents should be the same. The control current can be set by, say, a resistor connected from the collector to a lower potential. Current mirrors can be made with *npn* transistors, too. However, you must flip this circuit upside down, replace the *pnp* transistors with *npn* transistors, reverse current directions, and swap the supply voltage with ground.

MULTIPLE CURRENT SOURCES

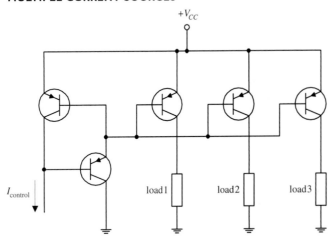

FIGURE 4.43

The circuit here is an expanded version of the previous circuit, which is used to supply a "mirror image" of control current to a number of different loads. (Again, you can design such a circuit with *npn* transistor, too, taking into consideration what was mentioned in the last example.) Note the addition of an extra transistor in the control side of the circuit. This transistor is included to help prevent one transistor that saturates (e.g., its load is removed) from stealing current from the common base line and hence reducing the other output currents.

MULTIVIBRATORS (FLIP-FLOPS)

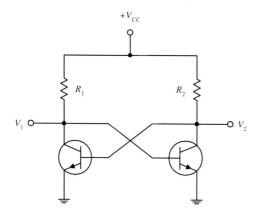

BISTABLE MULTIVIBRATOR (1ST CIRCUIT) A *bistable multivibrator* is a circuit that is designed to remain in either of two states indefinitely until a control signal is applied that causes it to change states. After the circuit switches states, another signal is required to switch it back to its previous state. To understand how this circuit works, initially assume that $V_1 = 0$ V. This means that the transistor on the right has no base current and hence no collector current. Therefore, all the current that flows through R_2 flows into the base of the left-hand transistor, driving it into saturation. In the saturation state, $V_1 = 0$, as assumed initially. Now, because the circuit is symmetric, you can say it is equally stable with $V_2 = 0$ and the right-hand transistor saturated. The bistable multivibrator can be made to switch from one state to another by simply grounding either V_1 or V_2 as needed. Bistable multivibrators can be used as memory devices or as frequency dividers, since alternate pulses restore the circuit to its initial state.

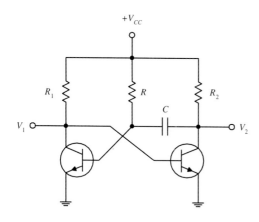

MONOSTABLE MULTIVIBRATOR (2D CIRCUIT) A monostable multivibrator is a circuit that is stable in only one state. It can be thrown into its unstable state by applying an external signal, but it will automatically return to its unstable state afterward. When $V_1 = 0$ V, the circuit is in its stable state. However, if you momentarily ground V_2, the capacitor suddenly behaves like a short circuit (a capacitor likes to pass current when the voltage across it changes suddenly) and causes the base current and hence the collector current of the left-hand transistor to go to zero. Then, all the current through R_1 flows into the base of the right-hand transistor, holding it in its saturation state until the capacitor can recharge through R. This in turn causes the circuit to switch back to its initial state. This kind of circuit produces a square pulse of voltage at V_1 with a duration determined by the RC time constant and which is independent of the duration and amplitude of the pulse that caused it to change states.

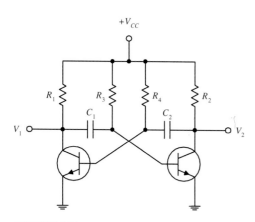

FIGURE 4.44

ASTABLE MULTIVIBRATOR (3D CIRCUIT) This circuit is not stable in either state and will spontaneously switch back and forth at a prescribed rate, even when no input signals are present. To understand how this circuit works, initially assume that V_1 is grounded. This means that the base of the right-hand transistor also will be at ground, at least until C_1 can charge up through R_2 to a high enough voltage to cause the right-hand transistor to saturate. At this time, V_2 then goes to zero, causing the base of the left-hand transistor to go to zero. V_1 then rises to a positive value, at least until C_2 can charge up through R_4 to a high enough voltage to cause the left-hand transistor to saturate. The cycle repeats itself over and over again. The time spent in each state can be controlled by the RC networks in the base section (R_3C_1 and R_4C_2 time constants set the time duration). As you can see, an astable multivibrator is basically a simple oscillator with an adjustable wave pattern (time spent in each state).

TRANSISTOR LOGIC GATES

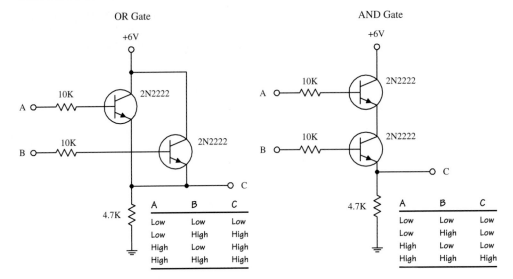

FIGURE 4.45

The two circuits here form logic gates. The OR circuit allows the output (*C*) to swing to a high voltage when either *A* or *B* or both *A* and *B* are high. In other words, as long as at least one of the transistors is biased (turned on), a high voltage will appear at the output. In the AND gate circuit, both *A* and *B* must be high in order for *C* to go high. In other words, both transistors must be biased (turned on) for a high voltage to appear at the output.

4.3.3   Junction Field-Effect Transistors

Junction field-effect transistors (JFETs) are three-lead semiconductive devices that are used as electrically controlled switches, amplifier controls, and voltage-controlled resistors. Unlike bipolar transistors, JFETs are exclusively voltage-controlled—they do not require a biasing current. Another unique trait of a JFET is that it is normally on when there is no voltage difference between its gate and source leads. However, if a voltage difference forms between these leads, the JFET becomes more resistive to current flow (less current will flow through the drain-source leads). For this reason, JFETs are referred to as *depletion devices,* unlike bipolar transistors, which are enhancement devices (bipolar transistors become less resistive when a current/voltage is applied to their base leads).

JFETs come in either *n-channel* or *p-channel* configurations. With an *n*-channel JFET, a negative voltage applied to its gate (relative to its source lead) reduces current flow from its drain to source lead. (It operates with $V_C > V_S$.) With a *p*-channel JFET, a positive voltage applied to its gate reduces current flow from its source to drain lead. (It operates with $V_S > V_C$.) The symbols for both types of JFETs are shown below.

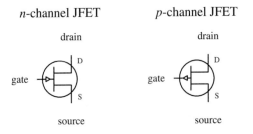

FIGURE 4.46

An important characteristic of a JFET that is useful in terms of applications is its extremely large input impedance (typically around 10^{10} Ω). This high input impedance means that the JFET draws little or no input current (lower pA range) and therefore has little or no effect on external components or circuits connected to its gate—no current is drawn away from the control circuit, and no unwanted current enters the control circuit. The ability for a JFET to control current flow while maintaining an extremely high input impedance makes it a useful device used in the construction of bidirectional analog switching circuits, input stages for amplifiers, simple two-terminal current sources, amplifier circuits, oscillators circuits, electronic gain-control logic switches, audio mixing circuits, etc.

How a JFET Works

An n-channel JFET is made with an n-type silicon channel that contains two p-type silicon "bumps" placed on either side. The gate lead is connected to the p-type bumps, while the drain and source leads are connected to either end of the n-type channel (see Fig. 4.47).

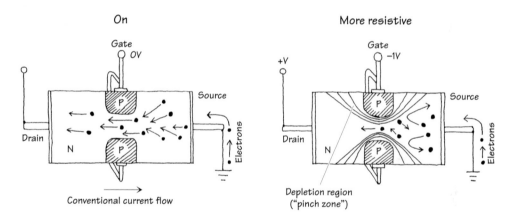

FIGURE 4.47

When no voltage is applied to the gate of an n-channel JFET, current flows freely through the central n-channel—electrons have no problem going through an n-channel; there are a lot of negative charger carriers already in there just waiting to help out with conduction. However, if the gate is set to a negative voltage—relative to the source—the area in between the p-type semiconductor bumps and the center of the n-channel will form two reverse-biased junctions (one about the upper bump, another about the lower bump). This reverse-biased condition forms a depletion region that extends into the channel. The more negative the gate voltage, the larger is the depletion region, and hence the harder it is for electrons to make it through the channel. For a p-channel JFET, everything is reversed, meaning you replace the negative gate voltage with a positive voltage, replace the n-channel with a p-channel semiconductor, replace the p-type semiconductor bumps with n-type semiconductors, and replace negative charge carriers (electrons) with positive charge carriers (holes).

JFET Water Analogies

Here are water analogies for an n-channel and p-channel JFET. Pretend water flow is conventional current flow and water pressure is voltage.

N-CHANNEL JFET WATER ANALOGY

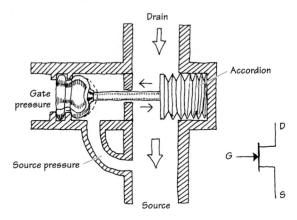

When no pressure exists between the gate and source of the *n*-channel water JFET, the device is fully on; water can flow from the drain pipe to the source pipe. To account for a real JFET's high input impedance, the JFET water analogy uses a plunger mechanism attached to a moving flood gate. (The plunger prevents current from entering the drain source channel, while at the same time it allows a pressure to control the flood gate.) When the gate of the *n*-channel JFET is made more negative in pressure relative to the source tube, the plunger is forced to the left. This in turn pulls the accordion-like flood gate across the drain-source channel, thus decreasing the current flow.

P-CHANNEL JFET WATER ANALOGY

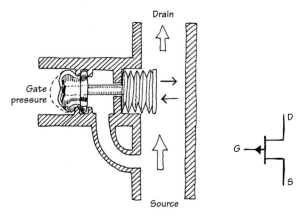

The *p*-channel water JFET is similar to the *n*-channel water JFET, except that all currents and pressures are reversed. The *p*-channel JFET is fully on until a positive pressure, relative to the source, is applied to the gate tube. The positive pressure forces the accordion across the drain-source channel, hence reducing the current flow.

FIGURE 4.48

Technical Stuff

The following graph describes how a typical *n*-channel JFET works. In particular, the graph describes how the drain current (I_D) is influenced by the gate-to-source voltage (V_{GS}) and the drain-to-source voltage (V_{DS}). The graph for a *p*-channel JFET is similar to that of the *n*-channel graph, except that I_D decreases with an increasing positive V_{GS}. In other words, V_{GS} is positive in voltage, and V_{DS} is negative in voltage.

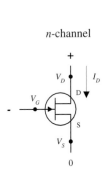

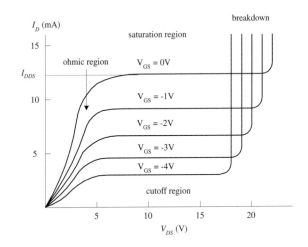

FIGURE 4.49

When the gate voltage V_G is set to the same voltage as the source ($V_{GS} = V_G - V_S = 0$ V), maximum current flows through the JFET. Technically speaking, people call this current (when $V_{GS} = 0$ V) the *drain current for zero bias*, or I_{DSS}. I_{DSS} is a constant and varies from JFET to JFET. Now notice how the I_D current depends on the drain-source voltage ($V_{DS} = V_D - V_S$). When V_{DS} is small, the drain current I_D varies nearly linearly with V_{DS} (looking at a particular curve for fixed V_{GS}). The region of the graph in which this occurs is called the *ohmic region*, or *linear region*. In this region, the JFET behaves like a voltage-controlled resistor.

Now notice the section of the graph were the curves flatten out. This region is called the *active region*, and here the drain current I_D is strongly influenced by the gate-source voltage V_{GS} but hardly at all influenced by the drain-to-source voltage V_{DS} (you have to move up and down between curves to see it).

Another thing to note is the value of V_{GS} that causes the JFET to turn off (point where practically no current flows through device). The particular V_{GS} voltage that causes the JFET to turn off is called the *cutoff voltage* (sometimes called the *pinch-off voltage V_P*), and it is expressed as $V_{GS,\text{off}}$.

Moving on with the graph analysis, you can see that when V_{DS} increases, there is a point where I_D skyrockets. At this point, the JFET loses its ability to resist current because too much voltage is applied across its drain-source terminals. In JFET lingo, this effect is referred to as *drain-source breakdown*, and the breakdown voltage is expressed as BV_{DS}.

For a typical JFET, I_{DSS} values range from about 1 mA to 1 A, $V_{GS,\text{off}}$ values range from around −0.5 to −10 V for an *n*-channel JFET (or from +0.5 to +10 V for a *p*-channel JFET), and BV_{DS} values range from about 6 to 50 V.

Like bipolar transistors, JFETs have internal resistance within their channels that varies with drain current and temperature. The reciprocal of this resistance is referred to as the *transconductance g_m*. A typical JFET transconductance is around a few thousand Ω^{-1}, where $\Omega^{-1} = 1/\Omega$ or ℧.

Another one of the JFET's built-in parameters is its *on resistance*, or $R_{DS,\text{on}}$. This resistance represents the internal resistance of a JFET when in its fully conducting state (when $V_{GS} = 0$). The $R_{DS,\text{on}}$ of a JFET is provided in the specification tables and typically ranges from 10 to 1000 Ω.

Useful Formulas

N-CHANNEL JFET

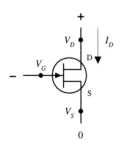

FIGURE 4.50

N-CHAN CURVES

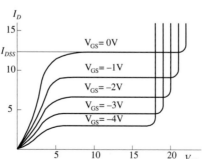

OHMIC REGION JFET is just beginning to resist. It acts like a variable resistor.

SATURATION REGION JFET is most strongly influenced by gate-source voltage, hardly at all influenced by the drain-source voltage.

CUTOFF VOLTAGE ($V_{GS,\text{off}}$) Particular gate-source voltage where JFET acts like an open circuit (channel resistance is at its maximum).

BREAKDOWN VOLTAGE (BV_{DS}) The voltage across the drain and source that caused current to "break through" the JFET's resistive channel.

P-CHANNEL JFET

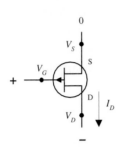

V_S

V_G

+

V_D

I_D

0

S

D

FIGURE 4.50   *(Continued)*

P-CHAN CURVES

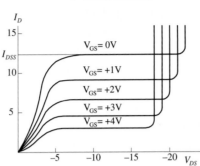

$V_{GS}= 0V$

$V_{GS}= +1V$

$V_{GS}= +2V$

$V_{GS}= +3V$

$V_{GS}= +4V$

DRAIN-CURRENT FOR ZERO BIAS (I_{DSS}) Represents the drain current when gate-source voltage is zero volts (or gate is connected to source, $V_{GS} = 0V$).

TRANSCONDUCTANCE (g_m) Represents the rate of change in the drain current with the gate-source voltage when the drain-to-source voltage is fixed for a particular V_{DS}. It is analogous to the transconductance ($1/R_{tr}$) for bipolar transistors.

**DRAIN CURRENT
(OHMIC REGION)**

$$I_D = I_{DSS}\left[2\left(1 - \frac{V_{GS}}{V_{GS,off}}\right)\frac{V_{DS}}{-V_{GS,off}} - \left(\frac{V_{DS}}{V_{GS,off}}\right)^2\right]$$

**DRAIN CURRENT
(ACTIVE REGION)**

$$I_D = I_{DSS}\left(1 - \frac{V_{GS}}{V_{GS,off}}\right)^2$$

**DRAIN-SOURCE
RESISTANCE**

$$R_{DS} = \frac{V_{DS}}{i_D} \approx \frac{V_{GS,off}}{2I_{DSS}(V_{GS} - V_{GS,off})} = \frac{1}{g_m}$$

ON RESISTANCE

$$R_{DS,on} = \text{constant}$$

**DRAIN-SOURCE
VOLTAGE**

$$V_{DS} = V_D - V_S$$

TRANSCONDUCTANCE

$$g_m = \frac{\partial I_D}{\partial V_{GS}}\bigg|_{V_{DS}} = \frac{1}{R_{DS}}$$

$$= g_{m_0}\left(1 - \frac{V_{GS}}{V_{GS,off}}\right) = g_{m_0}\sqrt{\frac{I_D}{I_{DSS}}}$$

**TRANSCONDUCTANCE
FOR SHORTED GATE**

$$g_{m_0} = -\frac{2I_{DSS}}{V_{GS,off}}$$

An *n*-channel JFET's $V_{GS,off}$ is negative. A *p*-channel JFET's $V_{GS,off}$ is positive.

$V_{GS,off}$, I_{DSS} are typically the knowns (you get their values by looking them up in a data table or on the package).

Typical JFET values:

I_{DSS}: 1 mA to 1 A

$V_{GS,off}$:
−0.5 to −10V (*n*-channel)
0.5 to 10V (*p*-channel)

$R_{DS,on}$: 10 to 1000 Ω

BV_{DS}: 6 to 50V

g_m at 1 mA:
500 to 3000 μmho

Sample Problems

PROBLEM 1

V_{DD}
+18V

I_D

G

D

S

R

FIGURE 4.51

If an *n*-channel JFET has a $I_{DSS} = 8$ mA and $V_{GS,off} = −4$V, what will be the drain current I_D if $R = 1$ kΩ and $V_{DD} = +18$V? Assume that the JFET is in the active region.

In the active region, the drain current is given by

$$I_D = I_{DSS}\left(1 - \frac{V_{GS}}{V_{GS,off}}\right)^2$$

$$= 8\text{ mA}\left(1 - \frac{V_{GS}}{−4\text{V}}\right)^2 = 8\text{ mA}\left(1 + \frac{V_{GS}}{2} + \frac{V_{GS}^2}{16}\right)$$

Unfortunately, there is one equation and two unknowns. This means that you have to come up with another equation. Here's how you get the other equation. First, you can assume the gate voltage is 0V because it's ground. This means that

$$V_{GS} = V_G - V_S = 0\text{V} - V_S = -V_S$$

From this, you can come up with another equation for the drain current by using Ohm's law and treating $I_D = I_S$:

$$I_D = \frac{V_S}{R} = -\frac{V_{GS}}{R} = -\frac{V_{GS}}{1 \text{ k}\Omega}$$

This equation is then combined with the first equation to yield

$$-\frac{V_{GS}}{1 \text{ k}\Omega} = 8 \text{ mA}\left(1 + \frac{V_{GS}}{2} + \frac{V_{GS}^2}{16}\right)$$

which simplifies to

$$V_{GS}^2 + 10V_{GS} + 16 = 0$$

The solutions to this equation are $V_{GS} = -2\,\text{V}$ and $V_{GS} = -8\,\text{V}$. But to be in the active region, V_{GS} must be between -4 and $0\,\text{V}$. This means that $V_{GS} = -2\,\text{V}$ is the correct solution, so you disregard the -8-V solution. Now you substitute V_{GS} back into one of the $I_{D(\text{active})}$ equations to get

$$I_D = -\frac{V_{GS}}{R} = -\frac{(-2\,\text{V})}{1 \text{ k}\Omega} = 2 \text{ mA}$$

PROBLEM 2

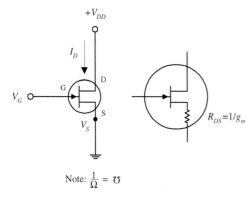

Note: $\frac{1}{\Omega} = \mho$

FIGURE 4.52

If $V_{GS,\text{off}} = -4\,\text{V}$ and $I_{DSS} = 12\,\text{mA}$, find the values of I_D and g_m and R_{DS} when $V_{GS} = -2\,\text{V}$ and when $V_{GS} = +1\,\text{V}$ Assume that the JFET is in the active region.
 When $V_{GS} = -2\,\text{V}$,

$$I_D = I_{DDS}\left(1 - \frac{V_{GS}}{V_{GS,\text{off}}}\right)^2$$

$$= 12 \text{ mA}\left(1 - \frac{(-2\,\text{V})}{(-4\,\text{V})}\right)^2 = 3.0 \text{ mA}$$

To find g_m, you first must find g_{m_0} (transconductance for shorted gate):

$$g_{m_0} = -\frac{2I_{DSS}}{V_{GS,\text{off}}} = -\frac{2(12 \text{ mA})}{(-4\,\text{V})} = 0.006 \, \mho = 6000 \text{ μmhos}$$

Now you can find g_m:

$$g_m = g_{m_0}\sqrt{\frac{I_D}{I_{DSS}}} = (0.006 \, \mho)\sqrt{\frac{3.0 \text{ mA}}{12.0 \text{ mA}}} = 0.003 \, \mho = 3000 \text{ μmhos}$$

To find the drain-source resistance (R_{DS}), use

$$R_{DS} = \frac{1}{g_m} = \frac{1}{0.003 \, \mho} = 333 \, \Omega$$

By applying the same formulas as above, you can find that when $V_{GS} = +1$ V, $I_D = 15.6$ mA, $g_m = 0.0075 \, \mho = 7500$ μmhos, and $R_{DS} = 133 \, \Omega$.

Basic Operations

LIGHT DIMMER

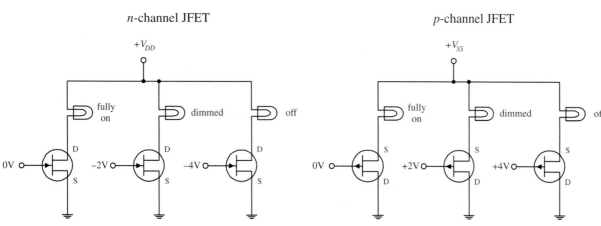

FIGURE 4.53

The two circuits here demonstrate how a JFET acts like a voltage-controlled light dimmer. In the n-channel circuit, a more negative gate voltage causes a larger source-to-drain resistance, hence causing the light bulb to receive less current. In the p-channel circuit, a more positive gate voltage causes a greater drain-to-source resistance.

BASIC CURRENT SOURCE AND BASIC AMPLIFIER

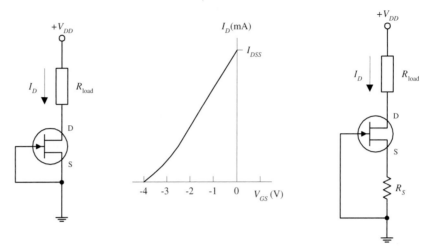

FIGURE 4.54

A simple current source can be constructed by shorting the source and gate terminals together (this is referred to as *self-biasing*), as shown in the left-most circuit. This means that $V_{GS} = V_G - V_S = 0\,\text{V}$, which means the drain current is simply equal to I_{DDS}. One obvious drawback of this circuit is that the I_{DDS} for a particular JFET is unpredictable (each JFET has its own unique I_{DDS} that is acquired during manufacturing). Also, this source is not adjustable. However, if you place a resistor between the source and ground, as shown in the right-most circuit, you can make the current source adjustable. By increasing R_S, you can decrease I_D, and vice versa (see Problem 2). Besides being adjustable, this circuit's I_D current will not vary as much as the left circuit for changes in V_{DS}. Though these simple JFET current sources are simple to construct, they are not as stable as a good bipolar or op amp current source.

SOURCE FOLLOWER

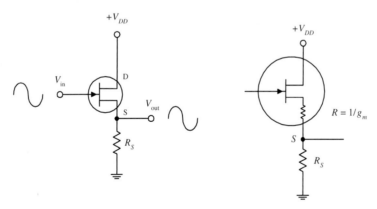

FIGURE 4.55

The JFET circuit here is called a *source follower*, which is analogous to the bipolar emitter follower; it provides current gain but not voltage gain. The amplitude of the output signal is found by applying Ohm's law: $V_S = R_S I_D$, where $I_D = g_m V_{GS} = g_m(V_G - V_S)$. Using these equations, you get

$$V_S = \frac{R_S g_m}{1 + R_S g_m} V_G$$

Since $V_S = V_{out}$ and $V_G = V_{in}$, the gain is simply $R_S g_m/(1 + R_G g_m)$. The output impedance, as you saw in Problem 2, is $1/g_m$. Unlike the emitter follower, the source follower has an extremely larger input impedance and therefore draws practically no input current. However, at the same time, the JFET's transconductance happens to be smaller than that of a bipolar transistor, meaning the output will be more attenuated. This makes sense if you treat the $1/g_m$ term as being a small internal resistance within the drain-source channel (see rightmost circuit). Also, as the drain current changes due to an applied waveform, g_m and therefore the output impedance will vary, resulting in output distortion. Another problem with this follower circuit is that V_{GS} is a poorly controlled parameter (a result of manufacturing), which gives it an unpredictable dc offset.

IMPROVED SOURCE FOLLOWER

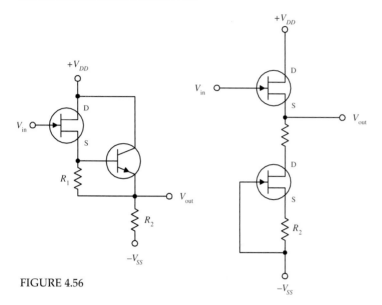

FIGURE 4.56

The source follower circuit from the preceding example had poor linearity and an unpredictable dc offset. However, you can eliminate these problems by using one of the two arrangements shown here. In the far-left circuit, you replace the source resistor with a bipolar current source. The bipolar source acts to fix V_{GS} to a constant value, which in turn eliminates the nonlinearities. To set the dc offset, you adjust R_1. (R_2 acts like R_S in the preceding circuit; it sets the gain.) The near-left circuit uses a JFET current source instead of a bipolar source. Unlike the bipolar circuit, this circuit requires no adjusting and has better temperature stability. The two JFETs used here are matched (matched JFETs can be found in pairs, assembled together within a single package). The lower transistor sinks as much current as needed to make $V_{GS} = 0$ (shorted gate). This means that both JFETs' V_{GS} values are zero,

making the upper transistor a follower with zero dc offset. Also, since the lower JFET responds directly to the upper JFET, any temperature variations will be compensated. When R_1 and R_2 are set equal, $V_{out} = V_{in}$. The resistors help give the circuit better I_D linearity, allow you to set the drain current to some value other than I_{DSS}, and help to improve the linearity. In terms of applications, JFET followers are often used as input stages to amplifiers, test instruments, or other equipment that is connected to sources with high source impedance.

JFET AMPLIFIERS

SOURCE-FOLLOWER AMP

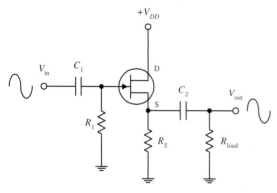

FIGURE 4.57

Recall the emitter-follower and common-emitter bipolar transistor amplifiers from the last chapter. These two amplifiers have JFET counterparts, namely, the source-follower and the common-source amplifier shown here. (The source-follower amplifier provides current gain; the common-source amplifier provides voltage gain.) If you were to set up the equations and do the math, you would find that the gain for the amplifiers would be

$$\text{Gain} = \frac{V_{out}}{V_{in}} = \frac{R_S}{R_S + 1/g_m} \qquad \text{(source-follower amp.)}$$

$$\text{Gain} = \frac{V_{out}}{V_{in}} = g_m \frac{R_D R_1}{R_D + R_1} \qquad \text{(common-source amp.)}$$

where the transconductance is given by

$$g_m = g_{m_0} \sqrt{\frac{I_D}{I_{DDS}}}, g_{m_0} = -\frac{2I_{DDS}}{V_{GS,\text{off}}}$$

As with bipolar amplifiers, the resistors are used to set the gate voltages and set the quiescent currents, while the capacitors act as ac couplers/high-pass filters. Notice, however, that both JFET amplifiers only require one self-biasing resistor.

COMMON-SOURCE AMP

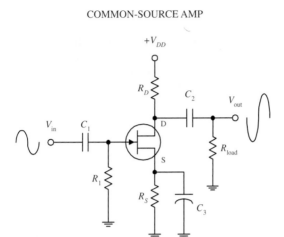

FIGURE 4.57   (*Continued*)

Now, an important question to ask at this point is, Why would you choose a JFET amplifier over a bipolar amplifier? The answer is that a JFET provides increased input impedance and low input current. However, if extremely high input impedances are not required, it is better to used a simple bipolar amplifier or op amp. In fact, bipolar amplifiers have fewer nonlinearity problems, and they tend to have higher gains when compared with JFET amplifiers. This stems from the fact that a JFET has a lower transconductance than a bipolar transistor for the same current. The difference between a bipolar's transconductance and JFET's transconductance may be as large as a factor of 100. In turn, this means that a JFET amplifier will have a significantly smaller gain.

VOLTAGE-CONTROLLED RESISTOR

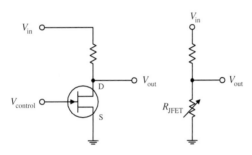

Electronic gain control

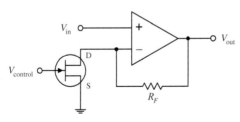

FIGURE 4.58

According to the graphs you saw earlier, if V_{DS} drops low enough, the JFET will operate within the linear (ohmic) region. In this region, the I_D versus V_{DS} curves follow approximate straight lines for V_{DS} smaller than $V_{GS} - V_{GS,off}$. This means that the JFET behaves like a voltage-controlled resistor for small signals of either polarity. For example, if you take a voltage-divider network and replace one of the resistors with a JFET, you get a voltage-controlled voltage divider (see upper left-hand figure). The range over which a JFET behaves like a traditional resistor depends of the particular JFET and is roughly proportional to the amount by which the gate voltage exceeds $V_{GS,off}$. For a JFET to be effective as a linearly responding resistor, it is important to limit V_{DS} to a value that is small compared with $V_{GS,off}$, and it is important to keep V_{GS} below $V_{GS,off}$. JFETs that are used in this manner are frequently used in electronic gain-control circuits, electronic attenuators, electronically variable filters, and oscillator amplitude-control circuits. A simple electronic gain-control circuit is shown here. The voltage gain for this circuit is given by gain $= 1 + R_F/R_{DS(on)}$, where R_{DS} is the drain-source channel resistance. If $R_F = 29$ kΩ and $R_{DS(on)} = 1$ kΩ, the maximum gain will be 30. As V_{GS} approaches $V_{GS,off}$, R_{DS} will increase and become very large such that $R_{DS} \gg R_F$, causing the gain to decrease to a minimum value close to unity. As you can see, the gain for this circuit can be varied over a 30:1 ratio margin.

Practical Considerations

JFETs typically are grouped into the following categories: small-signal and switching JFETs, high-frequency JFETs, and dual JFETs. Small-signal and switching JFETs are frequently used to couple a high-impedance source with an amplifier or other device such as an oscilloscope. These devices are also used as voltage-controlled switches. High-frequency JFETs are used primarily to amplify high-frequency signals (with the RF range) or are used as high-frequency switches. Dual JFETs contain two matched JFETs in one package. As you saw earlier, dual JFETs can be used to improve source-follower circuit performance.

TYPES OF JFET PACKAGES

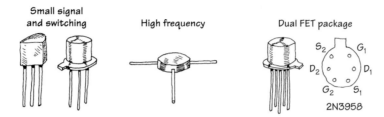

FIGURE 4.59

Like bipolar transistors, JFETs also can be destroyed with excess current and voltage. Make sure that you do not exceed maximum current and breakdown voltages. Table 4.2 is a sample of a JFET specification table designed to give you a feel for what to except when you start searching for parts.

TABLE 4.2   Portion of a JFET Specification Table

| TYPE | POLARITY | BV_{GS} (V) | I_{DSS} (mA) | | $V_{GS,OFF}$ (V) | | G_M TYPICAL (μmhos) | C_{ISS} (pF) | C_{RSS} (pF) |
			MIN (mA)	MAX (mA)	MIN (V)	MAX (V)			
2N5457	n-ch	25	1	5	−0.5	−6	3000	7	3
2N5460	p-ch	40	1	5	1	6	3000	7	2
2N5045	Matched-pair n-ch	50	0.5	8	−0.5	−4.5	3500	6	2

Applications

RELAY DRIVER

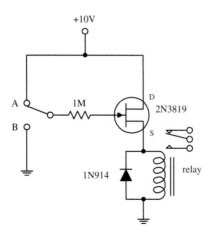

FIGURE 4.60

Here, an *n*-channel JFET is used to switch a relay. When the switch is set to position *A,* the JFET is on (gate isn't properly biased for a depletion effect to occur). Current then passes through the JFET's drain-source region and through the relay's coil, causing the relay to switch states. When the switch is thrown to position *B,* a negative voltage—relative to the source—is set at the gate. This in turn causes the JFET to block current flow from reaching the relay's coil, thus forcing the relay to switch states.

AUDIO MIXER/AMPLIFIER

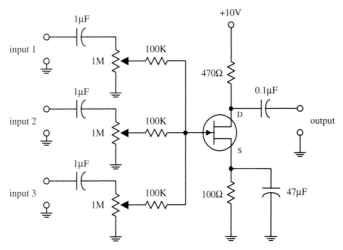

FIGURE 4.61

This circuit uses a JFET—set in the common-source arrangement—to combine (mix) signals from a number of different sources, such as microphones, preamplifiers, etc. All inputs are applied through ac coupling capacitors/filters. The source and drain resistors are used to set the overall amplification, while the 1-MΩ potentiometers are used to control the individual gains of the input signals.

ELECTRICAL FIELD METER

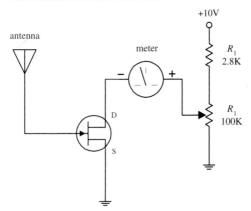

FIGURE 4.62

Here, a JFET is used to construct a simple static electricity detector. When the antenna (simple wire) is placed near a charged object, the electrons in the antenna will be draw toward or away from the JFET's gate, depending of weather the object is positively or negatively charged. The repositioning of the electrons sets up a gate voltage that is proportional to the charge placed on the object. In turn, the JFET will either begin to resist or allow current to flow through its drain-source channel, hence resulting in ammeter needle deflection. R_1 is used to protect the ammeter, and R_2 is used to calibrate it.

4.3.4   Metal Oxide Semiconductor Field-Effect Transistors

Metal oxide semiconductor field-effect transistors (MOSFETs) are incredibly popular transistors that in some ways resemble JFETs. For instance, when a small voltage is applied at its gate lead, the current flow through its drain-source channel is altered. However, unlike JFETS, MOSFETs have larger gate lead input impedances ($\geq 10^{14}$ Ω, as compared with ~10^9 Ω for JFETs), which means that they draw almost no gate current whatsoever. This increased input impedance is made possible by placing a metal oxide insulator between the gate-drain/source channel. There is a price to pay for this increased amount of input impedance, which amounts to a very low gate-to-channel capacitance (a few pF). If too much static electricity builds up on the gate of

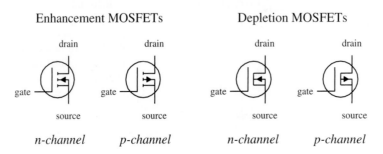

FIGURE 4.63

certain types of MOSFETs during handling, the accumulated charge may break through the gate and destroy the MOSFET. (Some MOSFETs are designed with safeguards against this breakdown—but not all.)

The two major kinds of MOSFETs include *enhancement-type MOSFETs* and *depletion-type MOSFETs* (see Fig. 4.63). A depletion-type MOSFET is normally on (maximum current flows from drain to source) when no difference in voltage exists between the gate and source terminals ($V_{GS} = V_G - V_S = 0$ V). However, if a voltage is applied to its gate lead, the drain-source channel becomes more resistive—a behavior similar to a JFET. An enhancement-type MOSFET is normally off (minimum current flows from drain to source) when $V_{GS} = 0$ V. However, if a voltage is applied to its gate lead, the drain-source channel becomes less resistive.

Both enhancement-type and depletion-type MOSFETs come in either *n*-channel or *p*-channel forms. For an *n*-channel depletion-type MOSFET, a negative gate-source voltage ($V_G < V_S$) increases the drain-source channel resistance, whereas for a *p*-channel depletion-type MOSFET, a positive gate-source voltage ($V_G > V_S$) increases the channel resistance. For an *n*-channel enhancement-type MOSFET, a positive gate-source voltage ($V_G > V_S$) decreases the drain-source channel resistance, whereas for a *p*-channel enhancement-type MOSFET, a negative gate-source voltage ($V_G < V_S$) decreases the channel resistance.

MOSFETs are perhaps the most popular transistors used today; they draw very little input current, are easy to make (require few ingredients), can be made extremely small, and consume very little power. In terms of applications, MOSFETs are used in ultrahigh input impedance amplifier circuits, voltage-controlled "resistor" circuits, switching circuits, and found with large-scale integrated digital ICs.

Like JFETs, MOSFETs have small transconductance values when compared with bipolar transistors. In terms of amplifier applications, this can lead to decreased gain values. For this reason, you will rarely see MOSFETs in simple amplifier circuits, unless there is a need for ultrahigh input impedance and low input current features.

How MOSFETs Work

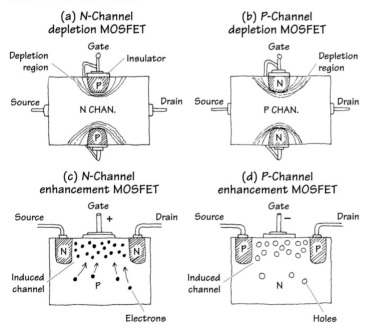

Both depletion and enhancement MOSFETs use an electrical field—produced by a gate voltage—to alter the flow of charge carriers through the semiconductive drain-source channel. With depletion-type MOSFETs, the drain-source channel is inherently conductive; charge carriers such as electrons (*n*-channel) or holes (*p*-channel) are already present within the *n*-type or *p*-type channel. If a negative gate-source voltage is applied to an *n*-channel depletion-type MOSFET, the resulting electrical field acts to "pinch off" the flow of electrons through the channel (see Fig. 4.64*a*). A *p*-channel depletion-type MOSFET uses a positive gate-source voltage to "pinch off" the flow of holes through its channel (see Fig. 4.64*b*). (The pinching-off effect results from depletion regions forming about the upper and lower gate contacts.) Enhancement MOSFETs, unlike depletion MOSFETs, have a normally resistive channel; there are few charge carriers within it. If a positive gate-source voltage is applied to an

FIGURE 4.64

n-channel enhancement-type MOSFET, electrons within the *p*-type semiconductor region migrate into the channel and thereby increase the conductance of the channel (see Fig. 4.64*c*). For a *p*-channel enhancement MOSFET, a negative gate-source voltage draws holes into the channel to increase the conductivity (see Fig. 4.64*d*).

Basic Operation

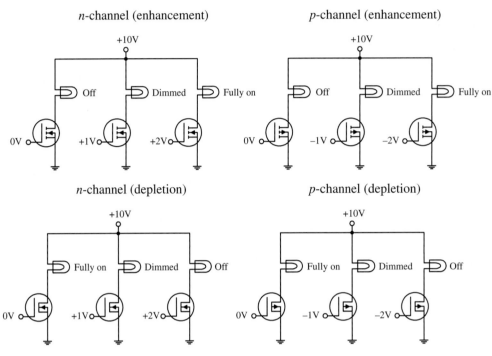

The circuits shown here demonstrate how MOSFETs can be used to control current flow through a light bulb. The desired dimming effects produced by the gate voltages may vary depending on the specific MOSFET you are working with.

FIGURE 4.65

Theory

FIGURE 4.66

In terms of theory, you can treat depletion-type MOSFETs like JFETs, except you must give them larger input impedances. The following graphs, definitions, and formulas summarize the theory.

N-CHANNEL DEPLETION-TYPE MOSFET

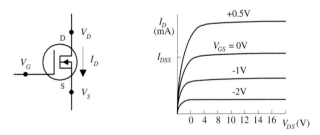

OHMIC REGION   MOSFET is just beginning to resist. In this region, the MOSFET behaves like a resistor.

ACTIVE REGION   MOSFET is most strongly influenced by gate-source voltage (V_{GS}) but hardly at all influenced by drain-source voltage (V_{DS}).

CUTOFF VOLTAGE ($V_{GS,off}$)   Often referred to as the *pinch-off voltage* (V_p). Represents the particular gate-source voltage that causes the MOSFET to block most all drain-source current flow.

BREAKDOWN VOLTAGE (BV_{DS})   The drain-source voltage (V_{DS}) that causes current to "break through" MOSFET's resistive channel.

DRAIN CURRENT FOR ZERO BIAS (I_{DSS})   Represents the drain current when gate-source voltage is zero volts (or when gate is shorted to source).

P-CHANNEL DEPLETION-TYPE MOSFET

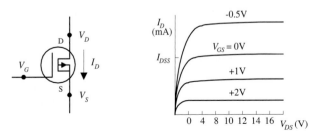

TRANSCONDUCTANCE   (g_m)   Represents the rate of change in the drain current with change in gate-source voltage when drain-source voltage is fixed for a particular V_{DS}. It is analogous to the transconductance (I/R_{tr}) for bipolar transistors.

Useful Formulas for Depletion-Type MOSFETs

DRAIN CURRENT
(OHMIC REGION)

$$I_D = I_{DSS}\left[2\left(1 - \frac{V_{GS}}{V_{GS,off}}\right)\frac{V_{DS}}{-V_{GS,off}} - \left(\frac{V_{DS}}{V_{GS,off}}\right)^2\right]$$

> An n-channel JFET's $V_{GS,off}$ is negative. A p-channel JFET's $V_{GS,off}$ is positive.

DRAIN CURRENT
(ACTIVE REGION)

$$I_D = I_{DSS}\left(1 - \frac{V_{GS}}{V_{GS,off}}\right)^2$$

DRAIN-SOURCE
RESISTANCE

$$R_{DS} = \frac{V_{DS}}{I_D} \approx \frac{V_{GS,off}}{2I_{DSS}(V_{GS} - V_{GS,off})} = \frac{1}{g_m}$$

> $V_{GS,off}$, I_{DSS} are typically the knowns (you get their values by looking them up in a data table or on the package).

ON RESISTANCE

$$R_{DS,on} = \text{constant}$$

DRAIN-SOURCE
VOLTAGE

$$V_{DS} = V_D - V_S$$

TRANSCONDUCTANCE

$$g_m = \frac{\partial I_D}{\partial V_{GS}}\bigg|_{V_{DS}} = \frac{1}{R_{DS}}$$

$$= g_{m_0}\left(1 - \frac{V_{GS}}{V_{GS,off}}\right) = g_{m_0}\sqrt{\frac{I_D}{I_{DSS}}}$$

> Typical JFET values:
> I_{DSS}: 1 mA to 1 A
> $V_{GS,off}$:
> –0.5 to –10 V (n-channel)
> 0.5 to 10 V (p-channel)
> $R_{DS,on}$: 10 to 1000 Ω
> BV_{DS}: 6 to 50 V
> g_m at 1 mA:
> 500 to 3000 µmho

TRANSCONDUCTANCE
FOR SHORTED GATE

$$g_{m_0} = \frac{2I_{DSS}}{V_{GS,off}}$$

Technical Info and Formulas for Enhancement-Type MOSFETs

FIGURE 4.67

Predicting how enhancement-type MOSFETs will behave requires learning some new concepts and formulas. Here's an overview of the theory.

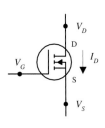

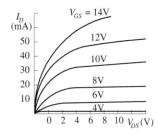

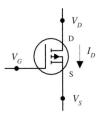

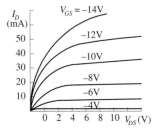

OHMIC REGION   MOSFET is just beginning to conduct. Acts like a variable resistor.

ACTIVE REGION   MOSFET is most strongly influenced by gate-source voltage V_{GS} but hardly at all influenced by drain-source voltage V_{DS}.

THRESHOLD VOLTAGE ($V_{GS,th}$) Particular gate-source voltage where MOSFET is just beginning to conduct.

BREAKDOWN VOLTAGE (BV_{DS}) The voltage across drain source (V_{DS}) that causes current to "break through" MOSFET's resistance channel.

DRAIN-CURRENT FOR GIVEN BIAS ($I_{D,on}$) Represents the amount of current I_D at a particular V_{GS}, which is given on data sheets, etc.

TRANSCONDUCTANCE (g_m) Represents the rate of change in the drain current with the change is gate-source voltage when drain-source voltage is fixed. It is analogous to the transconductance (I/R_{tr}) for bipolar transistors.

DRAIN CURRENT
(OHMIC REGION)

$$I_D = k[2(V_{GS} - V_{GS,th})V_{DS} - \tfrac{1}{2}V_{DS}{}^2]$$

> The value of the construction parameter k is proportional to the width/length ratio of the transistor's channel and is dependent on temperature. It can be determined by using the construction parameter equations to the left.

DRAIN CURRENT
(ACTIVE REGION)

$$I_D = k(V_{GS} - V_{GS,th})^2$$

> $V_{GS,th}$ is positive for n-channel enhancement MOSFETs.
> $V_{GS,th}$ is negative for p-channel enhancement MOSFETs.

CONSTRUCTION
PARAMETER

$$k = \frac{I_D}{(V_{GS} - V_{GS,th})^2}$$

$$= \frac{I_{D,on}}{(V_{GS,on} - V_{GS,th})^2}$$

> Typical values
> $I_{D,on}$: 1 mA to 1 A
> $R_{DS(on)}$: 1 Ω to 10 kΩ
> $V_{GS,off}$: 0.5 to 10 V
> $BV_{DS(off)}$: 6 to 50 V
> $BV_{GS(off)}$: 6 to 50 V

TRANSCONDUCTANCE

$$g_m = \frac{\partial I_D}{\partial V_{GS}}\bigg|_{V_{DS}} = \frac{1}{R_{DS}}$$

$$= 2k(V_{GS} - V_{GS,th}) = 2\sqrt{kI_D}$$

$$= g_{m_0}\sqrt{\frac{I_D}{I_{D_o}}}$$

> $V_{GS,th}$, $I_{D,on}$, g_m at a particular I_D are typically "knowns" you can find in the data tables or on package labels.

RESISTANCE OF
DRAIN-SOURCE CHANNEL

$$R_{DS} = 1/g_m$$

$$R_{DS_2} = \frac{V_{G_1} - V_{GS,th}}{V_{G_2} - V_{GS,th}} R_{DS_1}$$

> R_{DS_1} is the known resistance at a given voltage V_{G_1}. R_{DS_2} is the resistance you calculate at another gate voltage V_{G_2}.

Sample Problems

PROBLEM 1

An n-channel depletion-type MOSFET has an $I_{DDS} = 10$ mA and a $V_{GS,off} = -4$ V. Find the values of I_D, g_m, and R_{DS} when $V_{GS} = -2$ V and when $V_{GS} = +1$ V. Assume that the MOSFET is in the active region.

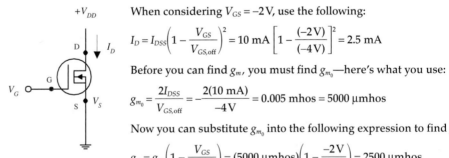

FIGURE 4.68

When considering $V_{GS} = -2$ V, use the following:

$$I_D = I_{DSS}\left(1 - \frac{V_{GS}}{V_{GS,off}}\right)^2 = 10 \text{ mA}\left[1 - \frac{(-2\text{ V})}{(-4\text{ V})}\right]^2 = 2.5 \text{ mA}$$

Before you can find g_m, you must find g_{m_0}—here's what you use:

$$g_{m_0} = \frac{2I_{DSS}}{V_{GS,off}} = -\frac{2(10 \text{ mA})}{-4\text{ V}} = 0.005 \text{ mhos} = 5000 \text{ μmhos}$$

Now you can substitute g_{m_0} into the following expression to find g_m:

$$g_m = g_{m_0}\left(1 - \frac{V_{GS}}{V_{GS,off}}\right) = (5000 \text{ μmhos})\left(1 - \frac{-2\text{ V}}{-4\text{ V}}\right) = 2500 \text{ μmhos}$$

The drain-source resistance is then found by using $R_{DS} = 1/g_m = 400 \text{ Ω}$. If you do the same calculations for $V_{GS} = +1$ V, you get $I_D = 15.6$ mA, $g_m = 6250$ μmhos, and $R_{DS} = 160$ Ω.

PROBLEM 2

An n-channel enhancement-type MOSFET has a $V_{GS,th} = +2$ V and an $I_D = 12$ mA. When $V_{GS} = +4$ V, find parameters k, g_m, and R_{DS}. Assume that the MOSFET is in the active region.

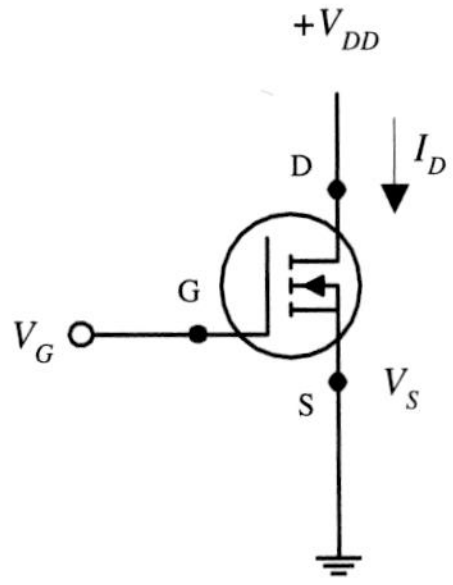

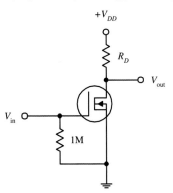

FIGURE 4.69

To find k, use the expression for the drain current in the active region:

$$I_D = k(V_{GS} - V_{GS,th})^2$$

Rearranging this equation and plugging in the knowns, you get

$$k = \frac{I_D}{(V_{GS} - V_{GS,th})^2} = \frac{12 \text{ mA}}{(4\text{V} - 2\text{V})^2} = 0.003 \text{ mhos/V} = 3000 \text{ }\mu\text{mhos/V}$$

To find g_m, use the following:

$$g_m = 2k(V_{GS} - V_{GS,th}) = 2\sqrt{kI_D}$$
$$= 2\sqrt{(3000 \text{ }\mu\text{mhos/V})(12 \text{ mA})} = 0.012 \text{ mhos} = 12{,}000 \text{ }\mu\text{mhos}$$

The drain-source resistance is then found by using $R_{DS} = 1/g_m = 83 \text{ }\Omega$.

PROBLEM 3

In the following n-channel depletion-type MOSFET circuit, $I_{DSS} = 10$ mA, $V_{GS,off} = -4$ V, $R_D = 1$ kΩ, and $V_{DD} = +20$ V, find V_D and the gain V_{out}/V_{in}.

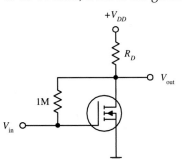

FIGURE 4.70

Applying Ohm's and Kirchhoff's laws, you can come up with the following expressions:

$$V_{DD} = V_{DS} + I_D R_D$$
$$V_{DD} = V_D + I_D R_D$$

where the last expression takes into account the grounded source terminal. (Note the 1-MΩ resistor. It is a self-biasing resistor and is used to compensate leakage currents and other parameters that can lead to MOSFET instability. The voltage drop across this resistor can be neglected because the gate leakage current is very small, typically in the nA or pA range.) Now, if you assume that there is no input voltage, you can say that $I_D = I_{DDS}$. This means that

$$V_D = V_{DD} - I_{DDS}R_D$$
$$= 20\text{V} - (10 \text{ mA})(1 \text{ k}\Omega) = 10\text{V}$$

To find the gain, use the following formula:

$$\text{Gain} = \frac{V_{out}}{V_{in}} = g_{m_0}R_D$$

where

$$g_{m_0} = -\frac{2I_{DDS}}{V_{GS,off}} = -\frac{2(10 \text{ mA})}{-4\text{V}} = 5000 \text{ }\mu\text{mhos}$$

Substituting g_{m_0} back into the gain formula, you get a resulting gain of 5.

PROBLEM 4

In the following n-channel enhancement-type MOSFET circuit, if $k = 1000$ μmhos/V, $V_{DD} = 20$ V, $V_{GS,th} = 2$ V, and $V_{GS} = 5$ V, what should the resistance of R_D be to center V_D to 10 V? Also, what is the gain for this circuit?

FIGURE 4.71

First, determine the drain current:

$$I_D = k(V_{GS} - V_{GS,th})^2$$
$$= (1000 \text{ }\mu\text{mhos/V})(5\text{V} - 2\text{V})^2 = 9 \text{ mA}$$

Next, to determine the size of R_D that is needed to set V_D to 10V, use Ohm's law:

$$R_D = \frac{V_{DD} - V_D}{I_D} = \frac{20\text{V} - 10\text{V}}{9 \text{ mA}} = 1100 \text{ }\Omega$$

(The 1-MΩ resistor has the same role as the 1-MΩ resistor in the last example.)

To find the gain, first find the transconductance:

$$g_m = 2k(V_{GS} - V_{GS,th}) = 2(1000 \ \mu\text{mhos/V})(5\,\text{V} - 2\,\text{V}) = 6000 \ \mu\text{mhos}$$

Next, substitute g_m into the gain expression:

$$\text{Gain} = \frac{V_{\text{out}}}{V_{\text{in}}} = g_m R_D = 6.6$$

Important Things to Know about MOSFETs

MOSFETs may come with a fourth lead, called the *body terminal.* This terminal forms a diode junction with the drain-source channel. It must be held at a nonconducting voltage [say, to the source or to a point in a circuit that is more negative than the source (*n*-channel devices) or more positive than the source (*p*-channel devices)]. If the base is taken away from the source (for enhancement-type MOSFETs) and set to a different voltage than that of the source, the effect shifts the threshold voltage $V_{GS,th}$ by an amount equal to $\frac{1}{2}V_{BS}^{1/2}$ in the direction that tends to decrease drain current for a given V_{GS}. Some instances when shifting the threshold voltage becomes important are when leakage effects, capacitance effects, and signal polarities must be counterbalanced. The body terminal of a MOSFET is often used to determine the operating point of a MOSFET by applying an incremental ac signal to its gate.

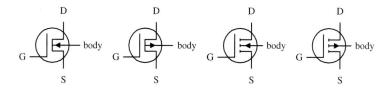

FIGURE 4.72

Damaging a MOSFET Is Easy

MOSFETs are extremely fragile. Their delicate gate-channel oxide insulators are subject to electron bombardment from statically charged objects. For example, it is possible for you to blow a hole through one of these insulators simply by walking across a carpet and then touching the gate of the MOSFET. The charge you pick up during your walk may be large enough to set yourself at a potential of a few thousand volts. Although the amount of current discharged during an interaction is not incredibly large, it does not matter; the oxide insulator is so thin (the gate-channel capacitance is so small, typically a few pF) that a small current can be fatal to a MOSFET. When installing MOSFETs, it is essential to eliminate all static electricity from your work area. In Chap. 14 you'll find guidelines for working with components subject to electrostatic discharge.

Kinds of MOSFETs

Like the other transistors, MOSFETs come in either metal can-like containers or plastic packages. High-power MOSFETs come with metal tabs that can be fastened to heat sinks. High/low MOSFET driver ICs also are available. These drivers (typically DIP in form) come with a number of independent MOSFETs built in and operate with logic signals.

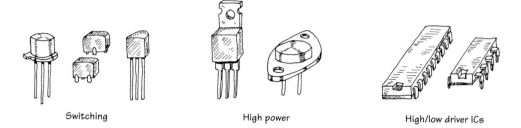

FIGURE 4.73   Switching   High power   High/low driver ICs

Things to consider when purchasing a MOSFET include breakdown voltages, $I_{D,\text{max}}$, $R_{DS(\text{on}),\text{max}}$, power dissipation, switching speed, and electrostatic discharge protection.

Applications

LIGHT DIMMER

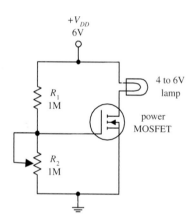

FIGURE 4.74

Here, an *n*-channel enhancement-type power MOSFET is used to control the current flow through a lamp. The voltage-divider resistor R_2 sets the gate voltage, which in turn sets the drain current through the lamp.

CURRENT SOURCE

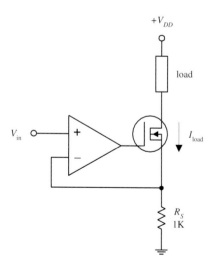

FIGURE 4.75

In the circuit shown here, an op amp is combined with an *n*-channel depletion-type MOSFET to make a highly reliable current source (less than 1 percent error). The MOSFET passes the load current, while the inverting input of the op amp samples the voltage across R_S and then compares it with the voltage applied to the noninverting input. If the drain current attempts to increase or decrease, the op amp will respond by decreasing or increasing its output, hence altering the MOSFETs gate voltage in the process. This in turn controls the load current. This op amp/MOSFET current source is more reliable than a simple bipolar transistor-driven source. The amount of leakage current is extremely small. The load current for this circuit is determined by applying Ohm's law (and applying the rules for op amps discussed in Chap. 7):

$$I_{\text{load}} = V_{\text{in}}/R_S$$

AMPLIFIERS

common-source amplifier
(depletion MOSFET)

source-follower amplifier
(depletion MOSFET)

Common-source and source-follower amplifiers can be constructed using both depletion-
and enhancement-type MOSFETs. The depletion-type amplifiers are similar to the
JFET amplifiers discussed earlier, except that
they have higher input impedances. The
enhancement-type MOSFET amplifiers essentially perform the same operations as the
depletion-type MOSFET amplifiers, but they
require a voltage divider (as compared with a
single resistor) to set the quiescent gate voltage. Also, the output for the enhancement-type
common-source MOSFET amplifier is inverted.
The role of the resistors and capacitors within
these circuits can be better understood by
referring to the amplifier circuits discussed
earlier.

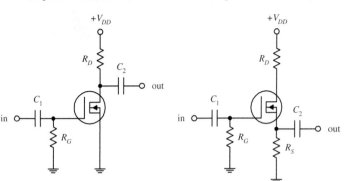

common-source amplifier
(enhancement MOSFET)

source-follower amplifier
(enhancement MOSFET)

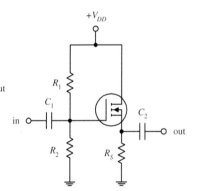

FIGURE 4.76

AUDIO AMPLIFIER

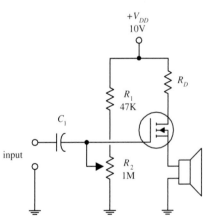

In this circuit, an n-channel enhancement-type MOSFET is used to
amply an audio signal generated by a high-impedance microphone and
then uses the amplified signal to drive a speaker. C_1 acts as an ac coupling capacitor, and the R_2 voltage divider resistor acts to control the
gain (the volume).

FIGURE 4.77

RELAY DRIVER (DIGITAL-TO-ANALOG CONVERSION)

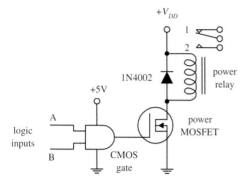

FIGURE 4.78

The circuit shown here uses an *n*-channel depletion-type MOSFET as an interface between a logic circuit and an analog circuit. In this example, an AND gate is used to drive a MOSFET into conduction, which in turn activates the relay. If inputs *A* and *B* are both high, the relay is switched to position 2. Any other combination (high/low, low/high, low/low) will put the relay into position 1. The MOSFET is a good choice to use as a digital-to-analog interface; its extremely high input resistance and low input current make it a good choice for powering high-voltage or high-current analog circuits without worrying about drawing current from the driving logic.

DIRECTION CONTROL OF A DC MOTOR

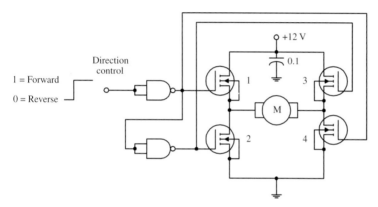

FIGURE 4.79

Logic input signals applied to this circuit act to control the direction of rotation of a dc motor. When the input is set high, the upper NAND gate outputs a low in response, turning transistors 1 and 4 on. At the same time, the high output from this gate is sent to the input of the lower NAND gate. The lower gate responses by outputting a low, thereby turning off transistors 2 and 3. Now, the only direction in which current can flow through the circuit is from the power supply through transistor 1, through the motor, and through transistor 4 to ground. This in turn causes the motor to turn in one direction. However, if you now apply a low to the input, transistors 2 and 3 turn on, while transistors 4 and 1 remain off. This causes current to flow through the motor in the opposite direction, thereby reversing the motor's direction of rotation.

4.3.5 Unijunction Transistors

Unijunction transistors (UJTs) are three-lead devices that act exclusively as electrically controlled switches (they are not used as amplifier controls). The basic operation of a UJT is relatively simple. When no potential difference exists between its emitter and either of its base leads (B_1 or B_2), only a very small current flows from B_2 to B_1. However, if a sufficiently large positive *trigger voltage*—relative to its base leads—is applied to the emitter, a larger current flows from the emitter and combines with the small B_2-to-B_1 current, thus giving rise to a larger B_1 output current. Unlike the other transistors covered earlier—where the control leads (e.g., emitter, gate) provided little or no additional current—the UJT is just the opposite; its emitter current is the primary source of additional current.

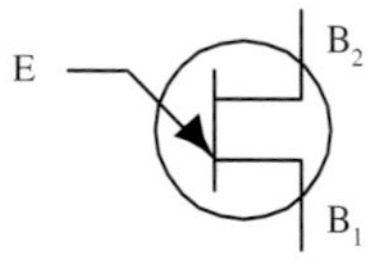

FIGURE 4.80

How UJTs Work

A simple model of a UJT is shown in Fig. 4.80. It consists of a single bar of *n*-type semiconductor material with a *p*-type semiconductor "bump" in the middle. One end of the bar makes up the base 1 terminals, the other end the base 2 terminal, and the "bump" represents the emitter terminal. Below is a simple "how it works" explanation.

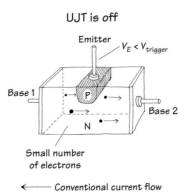

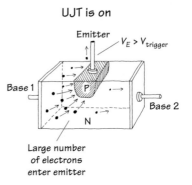

FIGURE 4.81

With no voltage applied to the emitter, only a relatively small number of electrons makes it through the *n*-region between base 1 and base 2. Normally, both connectors to bases 1 and 2 are resistive (each around a few thousand ohms).

When a sufficiently large voltage is applied to the emitter, the emitter-channel *pn* junction is forward-biased (similar to forward-biasing a diode). This in turn allows a large number of base 1 electrons to exit through the emitter. Now, since conventional currents are defined to be flowing in the opposite direction of electron flow, you would say that a positive current flows from the emitter and combines with channel current to produce a larger base 1 output current.

Technical Info

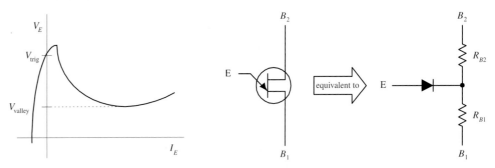

FIGURE 4.82

Figure 4.82 shows a typical V_E versus I_E graph of an UJT, as well as a UJT equivalent circuit. In terms of the UJT theory, if B_1 is grounded, a voltage applied to the emitter will have no effect (does not increase conductance from one base to another) until it exceeds a critical voltage, known as the *triggering voltage*. The triggering voltage is given by the following expression:

$$V_{\text{trig}} = \frac{R_{B1}}{R_{B1} + R_{B2}} V_{B2} = \eta \, V_{B2}$$

In this equation, R_{B1} and R_{B2} represent the inherent resistance within the region between each base terminal and the *n*-channel. When the emitter is open-circuited, the combined channel resistance is typically around a few thousand ohms, where R_{B1} is somewhat larger than R_{B2}. Once the trigger voltage is reached, the *pn* junction is forward-biased (the diode in the equivalent circuit begins to conduct), and current then flows from the emitter into the channel. But how do we determine R_{B1} and R_{B2}? Will the manufacturers give you these resistances? They most likely will not. Instead, they typically give you a parameter called the *intrinsic standoff ratio* η. This intrinsic standoff ratio is equal to the $R_{B1}/(R_{B1} + R_{B2})$ term in the preceding expression, provided the emitter is not conducting. The value of η is between 0 to 1, but typically it hangs out at a value around 0.5.

TYPICAL APPLICATION (RELAXATION OSCILLATOR)

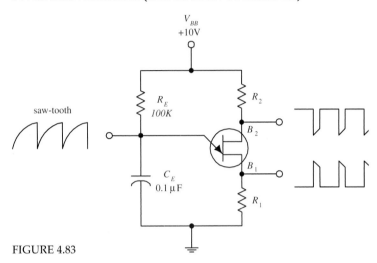

FIGURE 4.83

Most frequently, UJTs are used in oscillator circuits. Here, a UJT, along with some resistors and a capacitor, makes up a relaxation oscillator that is capable of generating three different output waveforms. During operation, at one instant in time, C_E charges through R_E until the voltage present on the emitter reaches the triggering voltage. Once the triggering voltage is exceeded, the E-to-B_1 conductivity increases sharply, which allows current to pass from the capacitor-emitter region, through the emitter-base 1 region, and then to ground. C_E suddenly loses its charges, and the emitter voltage suddenly falls below the triggering voltage. The cycle then repeats itself. The resulting waveforms generated during this process are shown in the figure. The frequency of oscillation is determined by the RC charge-discharge period and is given by

$$f = \frac{1}{R_E C_E \ln [1/(1 - \eta)]}$$

For example, if $R_E = 100$ kΩ, $C_E = 0.1$ μF, and η = 0.61, then $f = 106$ Hz.

Kinds of UJTs

BASIC SWITCHING

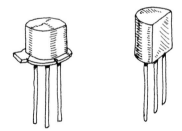

FIGURE 4.84

These UJTs are used in oscillatory, timing, and level-detecting circuits. Typical maximum ratings include 50 mA for I_E, 35 to 55 V for the interbase voltage (V_{BB}), and 300 to 500 mW for power dissipation.

PROGRAMMABLE (PUTs)

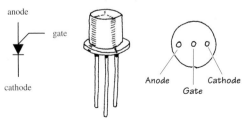

FIGURE 4.85

A PUT is similar to a UJT, except that R_{BB}, I_V (valley current level), I_P (peak current level), and η (intrinsic standoff ratio) can be programmed by means of an external voltage divider. Being able to alter these parameters is often essential in order to eliminate circuit instability. The electronic symbol for a PUT looks radically different when compared with a UJT (see figure). The lead names are also different; there is a gate lead, a cathode lead, and an anode lead. PUTs are used to construct timer circuits, high-gain phase-control circuits, and oscillator circuits. I have included a simple PUT application in the applications section that follows.

Applications

TIMER/RELAY DRIVER

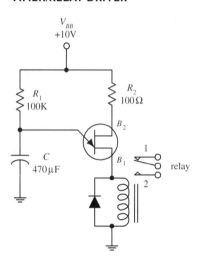

The circuit here causes a relay to throw its switch from one position to another in a repetitive manner. The positive supply voltage charges up the capacitor. When the voltage across the capacitor reaches the UJT's triggering voltage, the UJT goes into conduction. This causes the relay to throw its switch to position 2. When the capacitor's charge runs out, the voltage falls below the triggering voltage, and the UJT turns off. The relay then switches to position 1. R_1 controls the charging rate of the capacitor, and the size of the capacitor determines the amount of voltage used to trigger the UJT. C also determines the charge rate.

FIGURE 4.86

RAMP GENERATOR WITH AMPLIFIER

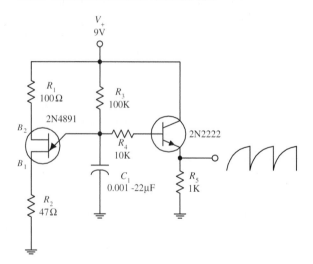

FIGURE 4.87

Here, a UJT is combined with a few resistors, a bipolar transistor, and a capacitor to make an oscillatory sawtooth generator that has controlled amplification (set by the bipolar transistor). Like the preceding oscillators, C_1 and R_3 set the frequency. The bipolar transistor samples the voltage on the capacitor and outputs a ramp or sawtooth waveform.

PUT RELAXATION OSCILLATOR

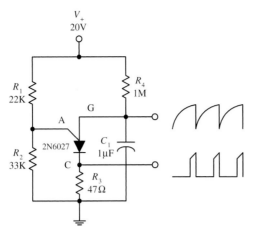

FIGURE 4.88

Here, a PUT is programmed by R_1 and R_2 to set the desired triggering voltage and anode current. These two resistors form a voltage divider that sets the gate voltage V_G (terminal used to turn PUT on or off). For a PUT to conduct, the anode voltage must exceed the gate voltage by at least 0.7 V. At a moment when the capacitor is discharged, the gate is reverse-biased, and the PUT is turned off. Over time, the capacitor begins charging through R_4, and when enough charge is collected, a large enough voltage will be present to forward bias the gate. This then turns the PUT on (i.e., if the anode current I_A exceeds the peak current I_P). Next, the capacitor discharges through the PUT and through R_3. (*Note:* When a PUT is conducting, the voltage from the anode to the cathode is about 1 V.) As the capacitor reaches full discharge, the anode current decreases and finally stops when the gate no longer has a sufficient voltage applied to it. After that, the charging begins again, and the cycle repeats itself, over and over again. By taping the circuit at the gate and source terminals, you can output both a spiked and sawtooth wave pattern.

CALCULATIONS

PUT begins to conduct when

$$V_A = V_S + 0.7 \text{ V}$$

where V_S is set by the voltage divider:

$$V_S = \frac{R_2}{R_2 + R_1} V_+$$

When V_A is reached, the anode current becomes:

$$I_A = \frac{V_+ - V_A}{R_1 + R_2}$$

4.4 Thyristors

4.4.1 Introduction

Thyristors are two- to four-lead semiconductor devices that act exclusively as switches—they are not used to amplify signals, like transistors. A three-lead thyristor uses a small current/voltage applied to one of its leads to control a much larger current flow through its other two leads. A two-lead thyristor, on the other hand, does not use a control lead but instead is designed to switch on when the voltage across its leads reaches a specific level, known as the *breakdown voltage*. Below this breakdown voltage, the two-lead thyristor remains off.

You may be wondering at this point, Why not simply use a transistor instead of a thyristor for switching applications? Well, you could—often transistors are indeed used as switches—but compared with thyristors, they are trickier to use because they require exacting control currents/voltages to operate properly. If the control current/voltage is not exact, the transistor may lay in between on and off states. And according to common sense, a switch that lies in between states is not a good switch. Thyristors, on the other hand, are not designed to operate in between states. For these devices, it is all or nothing—they are either on or off.

In terms of applications, thyristors are used in speed-control circuits, power-switching circuits, relay-replacement circuits, low-cost timer circuits, oscillator circuits, level-detector circuits, phase-control circuits, inverter circuits, chopper circuits, logic circuits, light-dimming circuits, motor speed-control circuits, etc.

TABLE 4.3   Major Kinds of Thyristors

TYPE	SYMBOL	MODE OF OPERATION
Silicon-controlled rectifier (SCR)		Normally off, but when a small current enters its gate (G), it turns on. Even when the gate current is removed, the SCR remains on. To turn it off, the anode-to-cathode current flow must be removed, or the anode must be set to a more negative voltage than the cathode. Current flows in only one direction, from anode (A) to cathode (C).
Silicon-controlled switch (SCS)		Similar to an SCR, but it can be made to turn off by applying a positive voltage pulse to a four-lead, called the *anode gate*. This device also can be made to trigger on when a negative voltage is applied to the anode-gate lead. Current flows in one direction, from anode (A) to cathode (C).
Triac		Similar to a SCR, but it can switch in both directions, meaning it can switch ac as well as dc currents. A triac remains on only when the gate is receiving current, and it turns off when the gate current is removed. Current flows in both directions, through *MT1* and *MT2*.
Four-layer diode		It has only two leads. When placed between two points in a circuit, it acts as a voltage-sensitive switch. As long as the voltage difference across its leads is below a specific breakdown voltage, it remains off. However, when the voltage difference exceeds the breakdown point, it turns on. Conducts in one direction, from anode (A) to cathode (C).
Diac		Similar to the four-layer diode but can conduct in both directions. Designed to switch either ac or dc.

Table 4.3 provides an overview of the major kinds of thyristors. When you see the phrase *turns it on,* this means a conductive path is made between the two conducting leads [e.g., anode (A) to cathode (C), MT1 to MT2]. *Normally off* refers to the condition when no voltage is applied to the gate (the gate is open-circuited). I will present a closer look at these thyristors in the subsections that follow.

4.4.2   Silicon-Controlled Rectifiers

SCRs are three-lead semiconductor devices that act as electrically controlled switches. When a specific positive trigger voltage/current is applied to the SCR's gate lead (G), a conductive channel forms between the anode (A) and the cathode (C) leads. Current flows in only one direction through the SCR, from anode to cathode (like a diode).

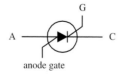

FIGURE 4.89

Another unique feature of an SCR, besides its current-controlled switching, has to do with its conduction state after the gate current is removed. After an SCR is triggered into conduction, removing the gate current has no effect. That is, the SCR will remain on even when the gate current/voltage is removed. The only way to turn the device off is to remove the anode-to-cathode current or to reverse the anode and cathodes polarities.

In terms of applications, SCRs are used in switching circuits, phase-control circuits, inverting circuits, clipper circuits, and relay-control circuits, to name a few.

How SCRs Work

An SCR is essentially just an *npn* and a *pnp* bipolar transistor sandwiched together, as shown in Fig. 4.90. The bipolar transistor equivalent circuit works well in describing how the SCR works.

FIGURE 4.90

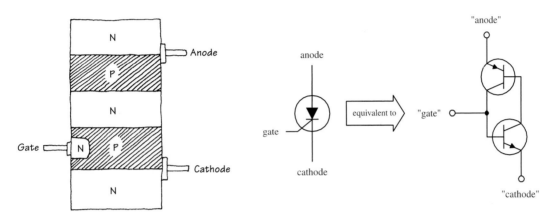

THE SCR IS OFF

Using the bipolar equivalent circuit, if the gate is not set to a specific positive voltage needed to turn the *npn* transistor on, the *pnp* transistor will not be able to "sink" current from its own base. This means that neither transistor will conduct, and hence current will not flow from anode to cathode.

THE SCR IS ON

If a positive voltage is applied to the gate, the *npn* transistor's base is properly biased, and it turns on. Once on, the *pnp* transistor's base can now "sink" current though the *npn* transistor's collector—which is what a *pnp* transistor needs in order to turn on. Since both transistors are on, current flows freely between anode and cathode. Notice that the SCR will remain on even after the gate current is removed. This—according to the bipolar equivalent circuit—results from the fact that both transistors are in a state of conduction when the gate current is removed. Because current is already in motion through the *pnp* transistors base, there is no reason for the transistors to turn off.

Basic SCR Applications

BASIC LATCHING SWITCH

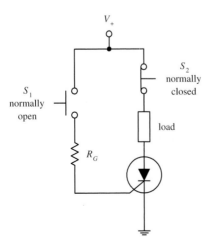

Here, an SCR is used to construct a simple latching circuit. S_1 is a momentary contact, normally open pushbutton switch, while S_2 is a momentary contact, normally closed pushbutton switch. When S_1 is pushed in and released, a small pulse of current enters the gate of the SCR, thus turning it on. Current will then flow through the load. The load will continue to receive current until the moment S_2 is pushed, at which time the SCR turns off. The gate resistor acts to set the SCR's triggering voltage/current. We'll take a closer look at the triggering specifications in a second.

FIGURE 4.91

ADJUSTABLE RECTIFIER

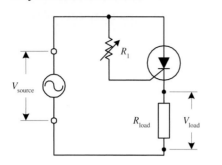
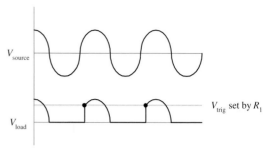

FIGURE 4.92

Here, an SCR is used to rectify a sinusoidal signal that is to be used to power a load. When a sinusoidal waveform is applied to the gate, the SCR turns on when the anode and gate receive the positive going portion of the waveform (provided the triggering voltage is exceeded). Once the SCR is on, the waveform passes through the anode and cathode, powering the load in the process. During the negative going portion of the waveform, the SCR acts like a reverse-biased diode; the SCR turns off. Increasing R_1 has the effect of lowering the current/voltage supplied to the SCR's gate. This in turn causes a lag in anode-to-cathode conduction time. As a result, the fraction of the cycle over which the device conducts can be controlled (see graph), which means that the average power dissipated by R_{load} can be adjusted. The advantage of using an SCR over a simple series variable resistor to control current flow is that essentially no power is lost to resistive heating.

DC MOTOR SPEED CONTROLLER

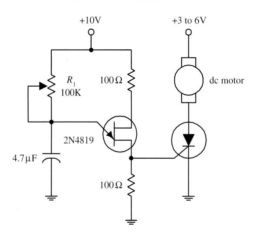

FIGURE 4.93

An SCR along with a few resistors, a capacitor, and a UJT can be connected together to make a variable-speed control circuit used to run a dc motor. The UJT, the capacitor, and the resistors make up an oscillator that supplies an ac voltage to the SCR's gate. When the voltage at the gate exceeds the SCR's triggering voltage, the SCR turns on, thus allowing current to flow through the motor. Changing the resistance of R_1 changes the frequency of the oscillator and hence determines the number of times the SCR's gate is triggered over time, which in turn controls the speed of the motor. (The motor appears to turn continuously, even though it is receiving a series of on/off pulses. The number of on cycles averaged over time determines the speed of the motor.) Using such a circuit over a simple series variable resistor to control the speed of the motor wastes less energy.

Kinds of SCRs

Some SCRs are designed specifically for phase-control applications, while others are designed for high-speed switching applications. Perhaps the most distinguishing feature of SCRs is the amount of current they can handle. Low-current SCRs typically come with maximum current/voltage ratings approximately no bigger than 1 A/100 V. Medium-current SCRs, on the other hand, come with maximum current/voltage ratings typically no bigger than 10 A/100 V. The maximum ratings for high-current SCRs may be several thousand amps at several thousand volts. Low-current SCRs come in plastic or metal can-like packages, while medium and high-current SCRs come with heat sinks built in.

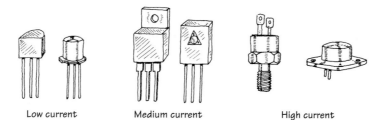

FIGURE 4.94   Low current   Medium current   High current

Technical Stuff

Here are some common terms used by the manufacturers to describe their SCRs:

V_T *On state-voltage.* The anode-to-cathode voltage present when the SCR is on.

I_{GT} *Gate trigger current.* The minimum gate current needed to switch the SCR on.

V_{GT} *Gate trigger voltage.* The minimum gate voltage required to trigger the gate trigger current.

I_H *Holding current.* The minimum current through the anode-to-cathode terminal required to maintain the SCR's on state.

P_{GM} *Peak gate power dissipation.* The maximum power that may be dissipated between the gate and the cathode region.

V_{DRM} *Repetitive peak off-state voltage.* The maximum instantaneous value of the off-state voltage that occurs across an SCR, including all repetitive transient voltages but excluding all nonrepetitive transient voltages.

I_{DRM} *Repetitive peak off-state current.* The maximum instantaneous value of the off-state current that results from the application of repetitive peak off-state voltage.

V_{RMM} *Repetitive peak reverse voltage.* The maximum instantaneous value of the reverse voltage that occurs across an SCR, including all repetitive transient voltages but excluding all nonrepetitive transient voltages.

I_{RMM} *Repetitive peak reverse current.* Maximum instantaneous value of the reverse current that results from the application of repetitive peak reverse voltage.

Here's a sample section of an SCR specifications table to give you an idea of what to expect (Table 4.4).

TABLE 4.4   Sample Section of an SCR Specifications Table

MNFR #	V_{DRM} (MIN) (V)	I_{DRM} (MAX) (mA)	I_{RRM} (MAX) (mA)	V_T (V)	I_{GT} (TYP/MAX) (mA)	V_{GT} (TYP/MAX) (V)	I_H (TYP/MAX) (mA)	P_{GM} (W)
2N6401	100	2.0	2.0	1.7	5.0/30	0.7/1.5	6.0/40	5

4.4.3   Silicon-Controlled Switches

A silicon-controlled switch (SCS) is a device similar to an SCR, but it is designed to turn off when a positive voltage/input current pulse is applied to an additional *anode gate* lead. The device also can be triggered into conduction by applying a negative voltage/output current pulse to the same lead. Other than this, the SCS behaves just like an SCR (see last section for the details). Figure 4.95 shows the symbol for an SCS. Note that the lead names may not appear as *cathode, gate,* and *anode gate.* Instead, they may be referred to as *emitter* (cathode), *base* (gate), and *collector* (anode gate).

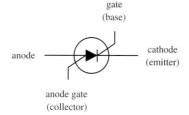

gate (base)

anode   cathode (emitter)

anode gate (collector)

FIGURE 4.95

SCSs are used in counters, lamp drivers, power-switching circuits, and logic circuits, as well as in essentially any circuit that requires a switch that can be turned on and off by two separate control pulses.

How an SCS Works

Figure 4.96 shows a basic *n*-type/*p*-type silicon model of an SCS, along with its bipolar equivalent circuit. As you can see, the equivalent circuit looks a lot that the SCR equivalent circuit, with the exception of the anode gate connection. When a positive pulse of current is applied to the gate, the *npn* transistor turns on. This allows current to exit the *pnp* transistor's base, hence turning the *pnp* transistor on. Now that both transistors are on, current can flow from anode to cathode—the SCS is turned on. The SCS will remain on until you remove the anode-to-cathode current, reverse the anode and cathode polarities, or apply a negative voltage to the anode gate. The negative anode gate voltage removes the transistor's self-sustaining biasing current.

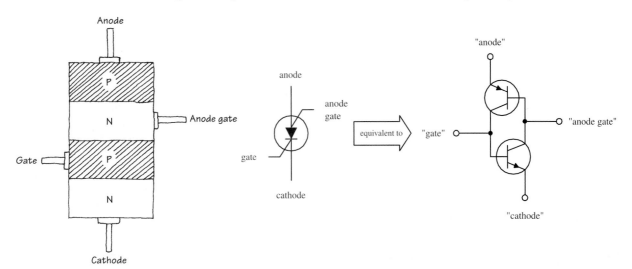

FIGURE 4.96

Specifications

When buying an SCS, make sure to select a device that has the proper breakdown voltage, current, and power-dissipation ratings. A typical specification table will provide the following ratings: BV_{CB}, BV_{EB}, BV_{CE}, I_E, I_C, I_H (holding current), and P_D (power dissipation). Here I have assumed the alternate lead name designations.

4.4.4   Triacs

Triacs are devices similar to SCRs—they act as electrically controlled switches—but unlike SCRs, they are designed to pass current in both directions, therefore making them suitable for ac applications. Triacs come with three leads, a gate lead and two conducting leads called *MT*1 and *MT*2. When no current/voltage is applied to the gate, the triac remains off. However, if a specific trigger voltage is applied to the gate, the device turns on. To turn the triac off, the gate current/voltage is removed.

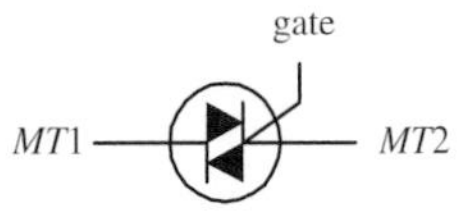

FIGURE 4.97

Triacs are used in ac motor control circuits, light-dimming circuits, phase-control circuits, and other ac power-switching circuits. They are often used as substitutes for mechanical relays.

How a Triac Works

Figure 4.98 shows a simple *n*-type/*p*-type silicon model of a triac. This device resembles two SCRs placed in reverse parallel with each other. The equivalent circuit describes how the triac works.

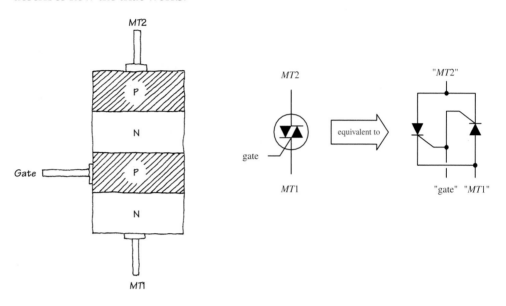

FIGURE 4.98

TRIAC IS OFF

Using the SCR equivalent circuit, when no current/voltage is applied to the gate lead, neither of the SCRs' gates receives a triggering voltage; hence current cannot flow in either direction through *MT*1 and *MT*2.

TRIAC IS ON

When a specific positive triggering current/voltage is applied to the gate, both SCRs receive sufficient voltage to trigger on. Once both SCRs are on, current can flow in either direction through *MT*1 to *MT*2 or from *MT*2 to *MT*1. If the gate voltage is removed, both SCRs will turn off when the ac waveform applied across *MT*1 and *MT*2 crosses zero volts.

Basic Applications

SIMPLE SWITCH

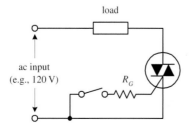

FIGURE 4.99

Here is a simple circuit showing how a triac acts to permit or prevent current from reaching a load. When the mechanical switch is open, no current enters the triac's gate; the triac remains off, and no current passes through the load. When the switch is closed, a small current slips through R_G, triggering the triac into conduction (provided the gate current and voltage exceed the triggering requirements of the triac). The alternating current can now flow through the triac and power the load. If the switch is open again, the triac turns off, and current is prevented from flowing through the load.

DUAL RECTIFIER

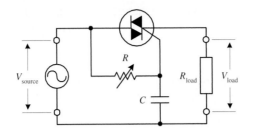

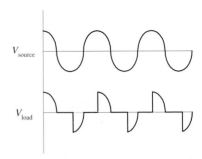

FIGURE 4.100

A triac along with a variable resistor and a capacitor can be used to construct an adjustable full-wave rectifier. The resistance R of the variable resistor sets the time at which the triac will trigger on. Increasing R causes the triac to trigger at a later time and therefore results in a larger amount of clipping (see graph). The size of C also determines the amount of clipping that will take place. (The capacitor acts to store charge until the voltage across its terminals reaches the triac's triggering voltage. At that time, the capacitor will dump its charge.) The reason why the capacitor can introduce additional clipping results from the fact that the capacitor may cause the voltage at the gate to lag the $MT2$-to-$MT1$ voltage (e.g., even if the gate receives sufficient triggering voltage, the $MT2$-to-$MT1$ voltage may be crossing zero volts). Overall, more clipping results in less power supplied to the load. Using this circuit over a simple series variable resistor connected to a load saves power. A simple series variable resistor gobbles up energy. This circuit, however, supplies energy-efficient pulses of current.

AC LIGHT DIMMER

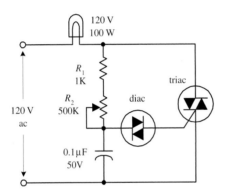

FIGURE 4.101

This circuit is used in many household dimmer switches. The diac—described in the next section—acts to ensure accurate triac triggering. (The diac acts as a switch that passes current when the voltage across its leads reaches a set breakdown value. Once the breakdown voltage is reached, the diac release a pulse of current.) In this circuit, at one moment the diac is off. However, when enough current passes through the resistors and charges up the capacitor to a voltage that exceeds the diac's triggering voltage, the diac suddenly passes all the capacitor's charge into the triac's gate. This in turn causes the triac to turn on and thus turns the lamp on. After the capacitor is discharged to a voltage below the breakdown voltage of the diac, the diac turns off, the triac turns off, and the lamp turns off. Then the cycle repeats itself, over and over again. Now, it appears that the lamp is on (or dimmed to some degree) because the on/off cycles are occurring very quickly. The lamp's brightness is controlled by R_2.

AC MOTOR CONTROLLER

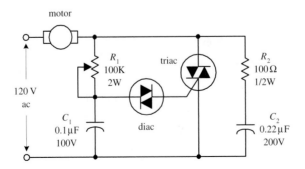

FIGURE 4.102

This circuit has the same basic structure as the light dimmer circuit, with the exception of the transient suppressor section (R_2C_2). The speed of the motor is adjusted by varying R_1.

Kinds of Triacs

Triacs come in low-current and medium-current forms. Low-current triacs typically come with maximum current/voltage ratings no bigger than 1 A/(several hundred volts). Medium-current triacs typically come with maximum current/voltage rating of up to 40 A/(few thousand volts). Triacs cannot switch as much current as high-current SCRs.

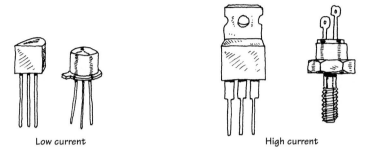

FIGURE 4.103   Low current   High current

Technical Stuff

Here are some common terms used by the manufacturers to describe their triacs:

$I_{TRMS,max}$ *RMS on-state current.* The maximum allowable *MT*1-to-*MT*2 current

$I_{GT,max}$ *DC gate trigger current.* The minimum dc gate current needed to switch the triac on

$V_{GT,max}$ *DC gate trigger voltage.* The minimum dc gate voltage required to trigger the gate trigger current

I_H *DC holding current.* The minimum *MT*1-to-*MT*2 dc current needed to keep the triac in its on state

P_{GM} *Peak gate power dissipation.* The maximum gate-to-*MT*1 power dissipation

I_{surge} *Surge current.* Maximum allowable surge current

Here's a sample section of a triac specifications table to give you an idea of what to expect (Table 4.5).

TABLE 4.5   Sample Section of a Triac Specifications Table

MNFR #	$I_{T,RMS}$ MAX. (A)	I_{GT} MAX. (mA)	V_{GT} MAX. (V)	V_{FON} (V)	I_H (mA)	I_{SURGE} (A)
NTE5600	4.0	30	2.5	2.0	30	30

4.4.5   Four-Layer Diodes and Diacs

Four-layer diodes and diacs are two-lead thyristors that switch current without the need of a gate signal. Instead, these devices turn on when the voltage across their leads reaches a particular *breakdown voltage* (or *breakover voltage*). A four-layer diode resembles an SCR without a gate lead, and it is designed to switch only dc. A diac resembles a *pnp* transistor without a base lead, and it is designed to switch only ac.

four-layer diode   diac

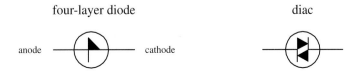

anode   cathode

FIGURE 4.104

Four-layer diodes and diacs are used most frequently to help SCRs and triacs trigger properly. For example, by using a diac to trigger a triac's gate, as shown in Fig. 4.105a, you can avoid unreliable triac triggering caused by device instability resulting from temperature variations, etc. When the voltage across the diac reaches the breakdown voltage, the diac will suddenly release a "convincing" pulse of current into the triac's gate.

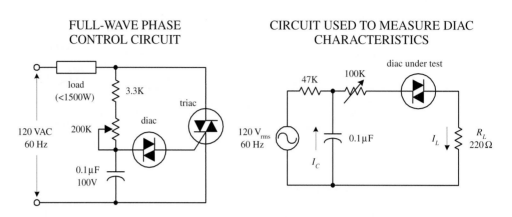

FIGURE 4.105

The circuit in Fig. 4.105 *right* is used to measure diac characteristics. The 100-kΩ variable resistor is adjusted until the diac fires once for every half-cycle.

Specifications

Here's a typical portion of a specifications table for a diac (Table 4.6).

TABLE 4.6   Sample Section of a Diac Specifications Table

MNFR #	V_{BO} (V)	I_{BO} MAX (μA)	I_{PULSE} (A)	V_{SWITCH} (V)	P_D (mW)
NTE6411	40	100	2	6	250

Here, V_{BO} is the breakover voltage, I_{BO} is the breakover current, I_{pulse} is the maximum peak pulse current, V_{switch} is the maximum switching voltage, and P_D is the maximum power dissipation.

CHAPTER 5

Optoelectronics

Optoelectrics is a branch of electronics that deals with light-emitting and light-detecting devices. Light-emitting devices, such as lamps and light-emitting diodes (LEDs), create electromagnetic energy (e.g., light) by using an electric current to excite electrons into higher energy levels (when an electron changes energy levels, a photon is emitted). Light-detecting devices such as phototransistors and photoresistors, on the other hand, are designed to take incoming electromagnetic energy and convert it into electric currents and voltages. This is usually accomplished using photons to liberate bound electrons within semiconductor materials. Light-emitting devices typically are used for illumination purposes or as indicator lights. Light-detecting devices are used primarily in light-sensing and communication devices, such as dark-activated switches and remote controls. This chapter examines the following optoelectronic devices: lamps, LEDs, photoresistors, photodiodes, solar cells, phototransistors, photothyristors, and optoisolators.

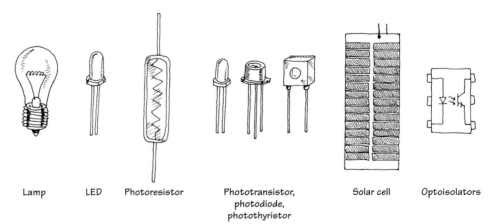

Lamp   LED   Photoresistor   Phototransistor, photodiode, photothyristor   Solar cell   Optoisolators

FIGURE 5.1

5.1   A Little Lecture on Photons

Photons are the elemental units of electromagnetic radiation. White light, for example, is composed of a number of different kinds of photons; some are blue photons, some are red photons, etc. It is important to note that there is no such thing as a white photon. Instead, when the combination of the various colored photons interacts with our eye, our brain perceives what we call *white light*.

Photons are not limited to visible light alone. There are also radiofrequency photons, infrared photons, microwave photons, and other kinds of photons that out eyes cannot detect.

In terms of the physics, photons are very interesting creatures. They have no rest mass, but they do carry momentum (energy). A photon also has a distant wavelike character within its electromagnetic bundle. The wavelength of a photon (horizontal distance between consecutive electrical or magnetic field peaks) depends on the medium in which it travels and on the source that produced it. It is this wavelength that determines the color of a photon. A photon's frequency is related to its wavelength by $\lambda = v/f$, where v is the speed of the photon. In free space, v is equal to the speed of light ($c = 3.0 \times 10^8$ m/s), but in other media, such as glass, v becomes smaller than the speed of light. A photon with a large wavelength (or small frequency) is less energetic than a photon with a shorter wavelength (or a higher frequency). The energy of a photon is equal to $E = hf$, where h is Planck's constant (6.63×10^{-34} J·s).

The trick to "making" a photon is to accelerate/decelerate a charged particle. For example, an electron that is made to vibrate back and forth within an antenna will produce radiofrequency photons that have very long wavelengths (low energies) when compared with light photons. Visible light, on the other hand, is produced when outer-shell electrons within atoms are forced to make transitions between energy levels, accelerating in the process. Other frequency photons may be created by vibrating or rotating molecules very quickly, while still others, specifically those with very high energy (e.g., gamma rays), can be created by the charge accelerations within the atomic nuclei.

FIGURE 5.2

Electromagnetic Spectrum

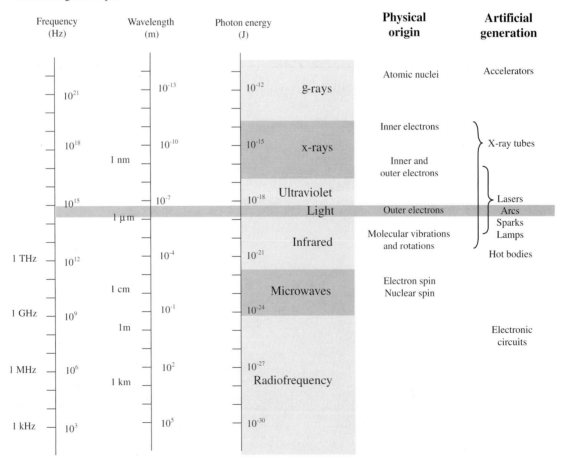

Figure 5.2 shows the breakdown of the electromagnetic spectrum. *Radiofrequency* photons extend from a few hertz to about 10^9 Hz (wavelengths from kilometers to about 0.3 m). They are often generated by alternating currents within power lines and electric circuits such as radio and television transmitters.

Microwave photons extend from about 10^9 up to 3×10^{11} Hz (wavelengths from 30 cm to 1 mm). These photons can penetrate the earth's atmosphere and hence are used in space vehicle communications, radio astronomy, and transmitting telephone conversations to satellites. They are also used for cooking food. Microwaves are often produced by atomic transitions and by electron and nuclear spins.

Infrared photons extend from about 3×10^{11} to 4×10^{14} Hz. Infrared radiation is created by molecular oscillations and is commonly emitted from incandescent sources such as electric heaters, glowing coals, the sun, human bodies (which radiate photons in the range of 3000 to 10,000 nm), and special types of semiconductor devices.

Light photons comprise a narrow frequency band from about 3.84×10^{14} to about 7.69×10^{14} Hz and are generally produced by a rearrangement of outer electrons in atoms and molecules. For example, in the filament of an incandescent light bulb, electrons are randomly accelerated by applied voltages and undergo frequent collision. These collisions result in a wide range of electron acceleration, and as a result, a broad frequency spectrum (within the light band) results, giving rise to white light.

Ultraviolet photons extend from approximately 8×10^{14} to 3.4×10^{16} Hz and are produced when an electron in an atom makes a long jump down from a highly excited state. The frequency of ultraviolet photons—unfortunately for us—tend to react badly with human cell DNA, which in turn can lead to skin cancer. The sun produces a large output of ultraviolet radiation. Fortunately for us, protective ozone molecules in the upper atmosphere can absorb most of this ultraviolet radiation by converting the photon's energy into a vibrating motion within the ozone molecule.

X-rays are highly energetic photons that extend from about 2.4×10^{16} to 5×10^{19} Hz, making their wavelengths often smaller than the diameter of an atom. One way of producing x-rays is to rapidly decelerate a high-speed charged particle. X-rays tend to act like bullets and can be used in x-ray imagery.

Gamma rays are the most energetic of the photons, whose frequency begins around 5×10^{19} Hz. These photons are produced by particles undergoing transitions within the atomic nuclei. The wavelike properties of gamma rays are extremely difficult to observe.

5.2 Lamps

Lamps are devices that convert electric current into light energy. One approach used in the conversion process is to pass a current through a special kind of wire filament. As current collides with the filament's atoms, the filament heats up, and photons are emitted. (As it turns out, this process produces a variety of different wavelength photons, so it appears that the emitted light is white in color.) Another approach used to produce light involves placing a pair of electrodes a small distance apart within a glass gas-enclosed bulb. When a voltage is set across the electrodes, the gas ionizes (electrons are striped from the gas atoms) emitting photons in the process. Here's an overview of some of the major kinds of lamps.

Incandescent

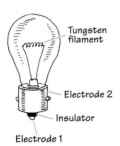

Tungsten filament

Electrode 2

Insulator

Electrode 1

These lamps use a tungsten wire filament to produce a glowing white light when current passes through it. The filament is enclosed in an evacuated glass bulb filled with a gas such as argon, krypton, or nitrogen that helps increase the brilliance of the lamp and also helps prevent the filament from burning out (as would be the case in an oxygen-rich environment). Incandescent lamps are used in flashlights, home lighting, and as indicator lights. They come in a variety of different sizes and shapes, as well as various current, voltage, and candlelight power ratings.

Halogen

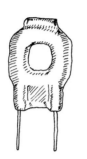

Similar to a typical incandescent lamp, these lamps provide ultrabright output light. Unlike a typical incandescent lamp, the filament is coated on the inside of a quartz bulb. Within this bulb, a halogen gas, such as bromine or iodine, is placed. These lamps are used in projector lamps, automotive headlights, strobe lights, etc.

Gas-Discharge

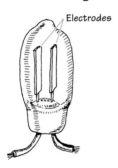

Electrodes

This lamp produces a dim, pale light that results from the ionization of neon gas molecules between two electrodes within the bulb. Types of gas-discharge lamps include neon, xenon flash, and mercury vapor lamps. Gas-discharge lamps have a tendency to suddenly switch on when a particular minimum operating voltage is met. For this reason, they are sometimes used in triggering and voltage-regulation applications. They are also used as indicator lights and for testing home ac power outlets.

Fluorescent

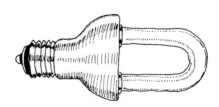

FIGURE 5.3

This is a lamp consisting of a mercury vapor–filled glass tube whose inner wall is coated with a material that fluoresces. At either end of the tube are cathode and anode incandescent filaments. When electrons emitted from an incandescent cathode electrode collide with the mercury atoms, ultraviolet (UV) radiation is emitted. The UV radiation then collides with the lamp's florescent coating, emitting visible light in the process. Such lamps require an auxiliary glow lamp with bimetallic contacts and a choke placed in parallel with the cathode and anode to initiate discharge within the lamp. These are highly efficient lamps that are often used in home lighting applications.

Xenon Flash Lamp

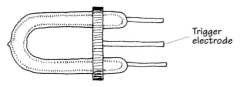

Trigger
electrode

FIGURE 5.3   (*Continued*)

This is a gas-discharge lamp that is filled with a xenon gas that ionizes when a particular voltage is applied across its electrodes. These lamps come with three leads: an anode, a cathode, and a trigger-voltage lead. Normally, a particular voltage is applied across the anode and cathode leads, and the lamp is off. However, when a particular voltage is applied to the trigger lead, the gas suddenly ionizes and releases an extremely bright flash. These lamps are used in photographic applications and in special-effects lighting projects.

Technical Stuff about Light Bulbs

A lamp's brightness is measured in what is called the *mean spherical candle power* (MSCP). Bulb manufacturers place a lamp at the center of an integrating sphere that averages the lamp's light output over its surface. The actual value of the MSCP for a lamp is a function of color temperature of the emitting surface of the lamp's filament. For a given temperature, doubling the filament's surface area doubles the MSCP. Other technical things to consider about lamps include voltage and current ratings, life expectancy, physical geometry of the bulb, and filament type. Table 5.1 shows typical specifications for an incandescent lamp. Figure 5.4 shows a number of different bulb types.

TABLE 5.1   Typical Incandescent Lamp Specifications

MNFR #	DESIGN VOLTAGE	LIFE AMPS	FILAMENT MSCP	LIFE EXPECTANCY (HOURS)	FILAMENT TYPE
PR2	2.38	0.500	0.800	15	C-2R

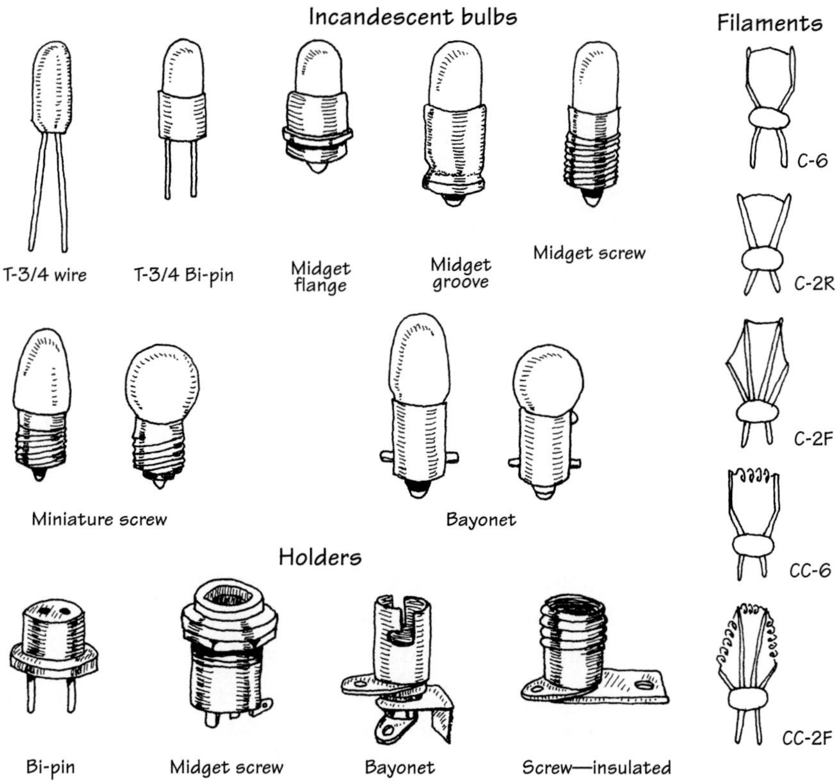

Incandescent bulbs

T-3/4 wire   T-3/4 Bi-pin   Midget flange   Midget groove   Midget screw

Filaments

C-6

C-2R

C-2F

CC-6

CC-2F

Miniature screw   Bayonet

Holders

FIGURE 5.4   Bi-pin   Midget screw   Bayonet   Screw—insulated

5.3   Light-Emitting Diodes

Light-emitting diodes (LEDs) are two-lead devices that are similar to *pn*-junction diodes, except that they are designed to emit visible or infrared light. When a LED's anode lead is made more positive in voltage than its cathode lead (by at least 0.6 to 2.2 V), current flows through the device and light is emitted. However, if the polarities are reversed (anode is made more negative than the cathode), the LED will not conduct, and hence it will not emit light. The symbol for an LED is shown below.

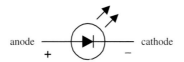

anode ——————————— cathode

\+   −

FIGURE 5.5

Unlike lamps, which give off a wide range of different colored photons—yielding white light—LEDs only give off a particular color photon. Typical emitted photon colors are red, yellow, green, and infrared. In terms of applications, LEDs are frequently used as indicator lights for displays or for low-level lighting applications (e.g., bicycle signal lights). Often, LEDs (especially infrared LEDs) are used as transmitting elements in remote-control circuits (e.g., TV remote control). The receiving element in this case may be a phototransistor that responds to changes in LED light output by altering the current flow within the receiving circuit.

5.3.1   How an LED Works

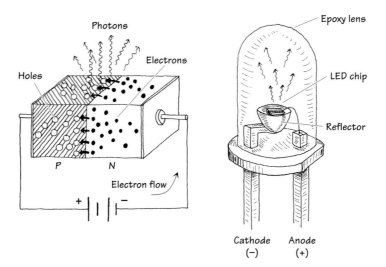

FIGURE 5.6

The light-emitting section of an LED is made by joining *n*-type and *p*-type semiconductors together to form a *pn* junction. When this *pn* junction is forward-biased, electrons in the *n* side are excited across the *pn* junction and into the *p* side, where they combine with holes. As the electrons combine with the holes, photons are emitted. Typically, the *pn*-junction section of an LED is encased in an epoxy shell that is doped with light-scattering particles to diffuse the light and make the LED appear brighter. Often a reflector placed beneath the semiconductor is used to direct light upward.

The cathode and anode leads are made from a heavy-gauge conductor to help wick heat away from the semiconductor.

FIGURE 5.7   ## 5.3.2   Kinds of LEDs

Visible-Light LEDs

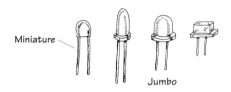

Miniature

Jumbo

These LEDs are inexpensive and durable devices that typically are used as indicator lights. Common colors include green (~565 nm), yellow (~585 nm), orange (~615 nm), and red (~650 nm). Maximum forward voltages are about 1.8 V, with typical operating currents from 1 to 3 mA. Typical brightness levels range from 1.0 to 3.0 mcd/1 mA to 3.0 mcd/2 mA. High-brightness LEDs also exist that are used in high-brightness flashers (e.g., bicycle flashers).

Infrared LEDs

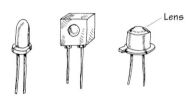

Lens

These LEDs are designed to emit infrared photons that have wavelength between approximately 880 and 940 nm. They are used in conjunction with a photosensor (e.g., photodiode, photoresistor, phototransistor) in remote-control circuits (e.g., TV remotes, intrusion alarms). They tend to have a narrower viewing angle when compared with a visible-light LED so that transmitted information can be directed efficiently. Photon output is characterized in terms of output power per specific forward current. Typical outputs range from around 0.50 mW/20 mA to 8.0 mW/50 mA. Maximum forward voltages at specific forward currents range from about 1.60 V at 20 mA to 2.0 V at 100 mA.

Blinking LEDs

These LEDs contain a miniature integrated circuit within their package that causes the LED to flash from 1 to 6 times each second. They are used primary as indicator flashers, but may be used as simple oscillators.

Tricolor LEDs

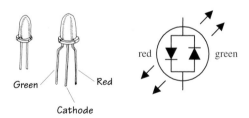

Green   Red

Cathode

red   green

These devices behave like two LEDs placed in parallel but facing opposite directions. One LED is typically red or orange in color, while the other is green. When current flows through the device in one direction, one of the LEDs turns on, while the other is reversed-biased (off). When the direction of current is reverse, the first LED turns off, and the second one turns on. They are used as a polarity indicator and usually come with a maximum voltage rating of about 3 V and operating range from 10 to 20 mA.

LED Displays

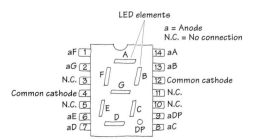

These are used for displaying numbers and other characters. In the LED display shown here, seven individual LEDs are used to make up the display. When a voltage is applied across one of the LEDs, a portion of the 8 lights up. Unlike liquid-crystal displays, LED displays tend to be more rugged, but they also consume more power. Displays are discussed in greater detail in Appendix I.

5.3.3   Technical Stuff about LEDs

Like conventional *pn*-junction diodes, LEDs are current-dependent devices. To control an LED's output intensity, the current that enters the diode (called the *forward current*, or I_F) is varied. The maximum amount of current an LED can handle under continuous drain is relatively small, perhaps as much as 100 mA. However, LEDs can handle a considerable amount of pulsed current, perhaps as large as 10 A.

To protect a diode from excessive current, a resistor is placed in series with the LED. The value of the series resistor R_S depends on the *forward voltage V_F* of the LED, the supply voltage V_+, and the desired forward current I_F. To find the value of R_S, apply Ohm's law, as shown in Fig. 5.8*a*.

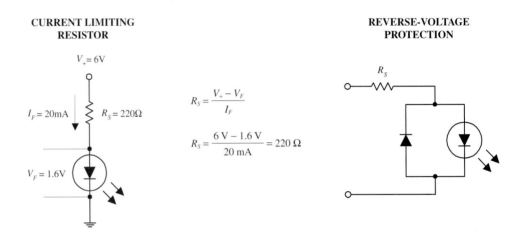

CURRENT LIMITING RESISTOR

$V_+ = 6V$

$I_F = 20mA$ $R_S = 220\Omega$

$V_F = 1.6V$

$$R_S = \frac{V_+ - V_F}{I_F}$$

$$R_S = \frac{6\,V - 1.6\,V}{20\,mA} = 220\,\Omega$$

REVERSE-VOLTAGE PROTECTION

R_S

FIGURE 5.8

Like conventional *pn*-junction diodes, LEDs have a reversed breakdown voltage V_R that, if exceeded, may result in a zapped semiconductor. The reversed breakdown voltage for LEDs is relatively small, typically around 5 V. To provide reverse polarity protection to a LED, a diode can be placed in series and in the reverse direction with respect to the LED; the diode will conduct before the reversed voltage across the LED becomes dangerously large (see Fig. 5.8*b*). It is important to note that LEDs provide a small voltage drop when placed in circuits—a result of the *pn* junction. The magnitude of this voltage drop is equal to the forward voltage V_F and may vary from 0.6 V to around 2.2 V depending of the type of semiconductors used. Table 5.2 shows a typical specification listing for a few types of LEDs.

TABLE 5.2 Typical LED Specifications

MNFR #	TYPE OF LED	PACKAGE	TYPICAL VIEWING ANGLE (DEGREES)	COLOR	LUMINOUS INTENSITY (MCD) I_V	TYPICAL FORWARD VOLTAGE DROP (V) V_F	MAX. REVERSE BREAKDOWN VOLTAGE (V) V_R	MAX. DC FORWARD CURRENT (mA) I_F	MAX. POWER DISSIPATION (mW) P_D
NTE3000	Indicator	T-3/4	80	Clear red	1.4	1.65	5	40	80
NTE3010	Indicator	T-1	90	Green	1	2.2	5	35	105
NTE3026	Tristate	T-1 3/4	50	Red/green	1.5, red	1.65, red	—	70, red	200
					0.5, green	2.2, green	—	35, green	
NTE3130	Blinker (3 Hz)	T-1 3/4	30	Yellow blinking	3	5.25	0.4	20	—
NTE3017	Infrared	—	—	900 nm	—	1.28	6	100	175

199

5.3.4   Basic LED Operations

LED Brightness Control

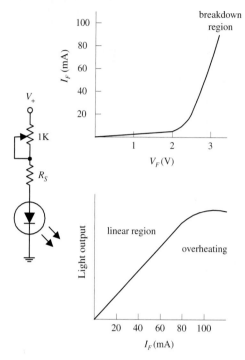

Here, a 1-k variable resistor is used to control the amount of current that passes through the LED and in turn acts to control the brightness. R_S is used to protect the LED from excessive current and is found by using

$$R_S = V_+ - V_F / I_F$$

When an LED begins to conduct, the voltage increases gradually, while the current increases rapidly. Too much current will overheat the LED.

Voltage-Level Indicator

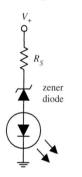

The circuit shows how an LED along with a zener diode can be used to make a voltage-level indicator circuit. Whenever the voltage V_+ exceeds the breakdown voltage of the zener diode, the zener diode conducts and allows current to pass through the LED. Zener diodes come with various breakdown voltages, so it is possible to connect a number of these types of circuits in parallel to form a voltage indicator display.

Logic Probes

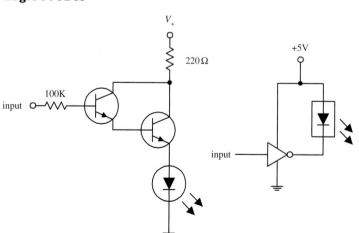

The circuit shown here indicates how an LED can be used to display the status of a logic gate. When the output of a logic gate is attached to the input of this circuit, a high will turn on the transistors and will light the LED. However, a low applied to the input will turn the LED off. The rightmost circuit shows how a flasher LED can be used to do the same thing. This circuit works with TTL gates and with high-output CMOS gates.

FIGURE 5.9

Tristate Polarity Indicator

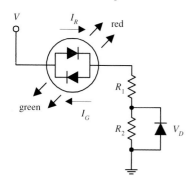

V	Color
+	red
−	green
AC	yellow

$$R_1 = \frac{V - (V_G + V_D)}{I_G}$$

$$R_2 = \frac{V - V_R}{I_R} - R_1$$

Here, a tristate LED is used to indicate the direction and type of current flow. If V is a positive dc voltage, the device emits red light. However, if V is a negative dc voltage, the device emits green light. If a high-frequency ac voltage is applied, it appears as if the device emits a yellow light. R_1 and R_2 are chosen to protect the LEDs, and the diode is used to provide reversed-voltage protection when the applied voltage exceeds the maximum reversed voltage. The formulas show how to calculate R_1 and R_2. (In the equations, V is the applied voltage, V_G and V_R are the forward voltages for the LED, and V_D is the forward voltage of the diode (0.6 V.)

LED Flasher Circuits

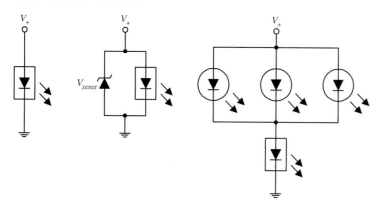

A flasher LED does not require a current-limiting resistor like the other LEDs. Typically, a voltage between 3 and 7 V is used to drive a flasher LED. To protect a flasher LED from excessive forward voltage, a zener diode placed in parallel can be added (a typical value for a zener would be around 6 V). A single flasher LED as shown in the near left circuit can be used to flash a number of ordinary LEDs.

Relay Driver

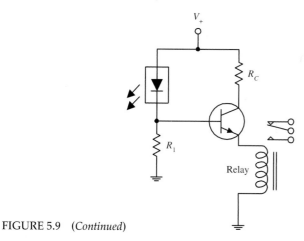

Here, a flasher LED is used to supply a series of on/off pulses of current/voltage to the base of a transistor. When the flasher LED goes into conduction, the bipolar transistor's base receives a positive voltage and input current needed to turn it on, thus providing power to drive the relay. R_1 sets the biasing voltage for the transistor, and R_C sets the collector current.

FIGURE 5.9   (*Continued*)

5.4   Photoresistors

Photoresistors are light-controlled variable resistors. In terms of operation, a photoresistor is usually very resistive (in the megaohms) when placed in the dark. However, when it is illuminated, its resistance decreases significantly; it may drop as low as a few hundred ohms, depending on the light intensity. In terms of applications, photoresistors are used in light- and dark-activated switching circuits and in light-sensitive detector circuits. Figure 5.10 shows the symbol for a photoresistor.

FIGURE 5.10

5.4.1   How a Photoresistor Works

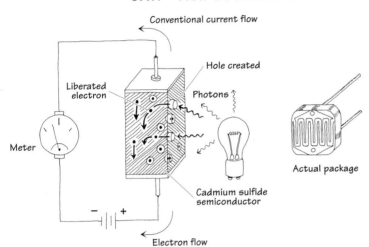

FIGURE 5.11

Photoresistors are made from a special kind of semiconductor crystal, such as cadmium sulfide (for light) or lead sulfide (for infrared). When this semiconductor is placed in the dark, electrons within its structure do not want to flow through the resistor because they are too strongly bound to the crystal's atoms. However, when illuminated, incoming photons of light collide with the bound electrons, stripping them from the binding atom, thus creating a hole in the process. These liberated electrons can now contribute to the current flowing through the device (the resistance goes down).

5.4.2   Technical Stuff

Photoresistors may require a few milliseconds or more to fully respond to changes in light intensity and may require a few seconds to return to their normal *dark resistance* once light is removed. In general, photoresistors pretty much all function in a similar manner. However, the sensitivity and resistance range of a photoresistor may vary greatly from one device to the next. Also, certain photoresistor may respond better to light that contains photons within a particular wavelength of the spectrum. For example, cadmium sulfide photoresistors respond best to light within the 400- to 800-nm range, whereas lead sulfide photoresistors respond best to infrared photons.

5.4.3   Applications

Simple Light Meter

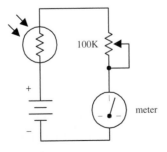

FIGURE 5.12

Here, a photoresistor acts as the light-sensing element in a simple light meter. When dark, the photoresistor is very resistive, and little current flows through the series loop; the meter is at its lowest deflection level. When an increasingly bright light source is shone on the photoresistor, the photoresistor's resistance begins to decrease, and more current begins to flow through the series loop; the meter starts to deflect. The potentiometer is used to calibrate the meter.

Light-Sensitive Voltage Divider

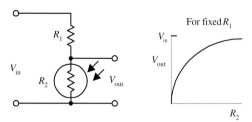

For fixed R_1

These circuits resemble the voltage-divider circuit described in Chap. 3. As before, the output voltage is given by

$$V_{out} = \frac{R_2}{R_1 + R_2} V_{in}$$

As the intensity of light increases, the resistance of the photoresistor decreases, so V_{out} in the top circuit gets smaller as more light hits it, whereas V_{out} in the lower circuit gets larger.

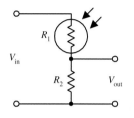
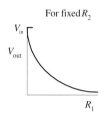

For fixed R_2

Light-Activated Relay

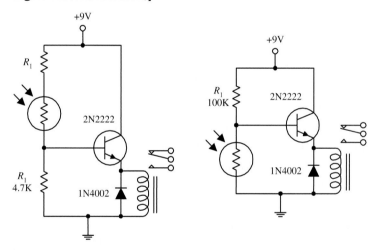

The two circuits shown here make use of light-sensitive voltage dividers to trip a relay whenever the light intensity changes. In the light-activated circuit, when light comes in contact with the photoresistor, the photoresistor's resistance decreases, causing an increase in the transistor's base current and voltage. If the base current and voltage are large enough, the transistor will allow enough current to pass from collector to emitter, triggering the relay. The dark-activated relay works in a similar but opposite manner. The value of R_1 in the light-activated circuit should be around 1 kΩ but may need some adjusting. R_1 in the dark-activated circuit (100 kΩ) also may need adjusting. A 6- to 9-V relay with a 500-Ω coil can be used in either circuit.

FIGURE 5.12 *(Continued)*

5.5 Photodiodes

Photodiodes are two-lead devices that convert light energy (photon energy) directly into electric current. If the anode and cathode leads of a photodiode are joined together by a wire and then the photodiode is placed in the dark, no current will flow through the wire. However, when the photodiode is illuminated, it suddenly becomes a small current source that pumps current from the cathode through the wire and into the anode. Figure 5.13 depicts the symbol for a photo diode.

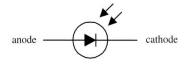

anode —— cathode

FIGURE 5.13

Photodiodes are used most commonly to detect fast pulses of near-infrared light used in wireless communications. They are often found in light-meter circuits (e.g., camera light meters, intrusion alarms, etc.) because they have very linear light/current responses.

5.5.1   How a Photodiode Works

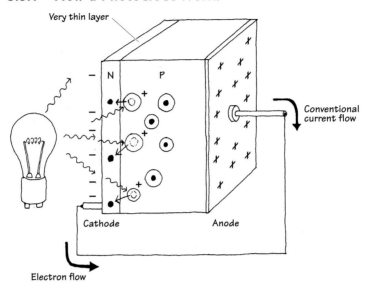

FIGURE 5.14

A photodiode is built by sandwiching a very thin *n*-type semiconductor together with a thicker *p*-type semiconductor. (The *n* side has an abundance of electrons; the *p* side has an abundance of holes.) The *n* side of the combination is considered the cathode, while the *p* side is considered the anode. Now, if you shine light on this device, a number of photons will pass through the *n*-semiconductor and into the *p*-semiconductor. Some of the photons that make it into the *p* side will then collide with bound electrons within the *p*-semiconductor, ejecting them and creating holes in the process. If these collisions are close enough to the *pn* interface, the ejected electrons will then cross the junction. What you get in the end is extra electrons on the *n* side and extra holes on the *p* side. This segregation of positive and negative charges leads to a potential difference being formed across the *pn* junction. Now if you connect a wire from the cathode (*n* side) to the anode (*p* side), electrons will flow through the wire, from the electron-abundant cathode end to the hole-abundant anode end (or if you like, a conventional positive current will flow from the anode to cathode). A commercial photodiode typically places the *pn*-semiconductor in a plastic or metal case that contains a window. The window may contain a magnifying lens and a filter.

5.5.2   Basic Operations

Photovoltaic Current Source

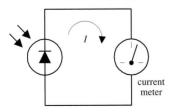

FIGURE 5.15

Here, a photodiode acts to convert light energy directly into electric current that can be measured with a meter. The input intensity of light (brightness) and the output current are nearly linear.

Photoconductive Operation

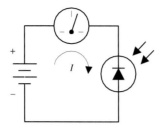

FIGURE 5.15   (*Continued*)

Individual photodiodes may not produce enough current needed to drive a particular light-sensitive circuit. Usually they are incorporated along with a voltage source. Here, a photodiode is connected in reversed-biased direction with a battery. When dark, a small current called the *dark current* (within the nA range) flows through the photodiode. When the photodiode is illuminated, a larger current flows. This circuit, unlike the preceding circuit, uses the battery for increased output current. A resistor placed in series with the diode and battery can be used to calibrate the meter. Note that if you treat the photodiode as if it were an ordinary diode, conduction will not occur; it must be pointed in the opposite direction.

5.5.3   Kinds of Photodiodes

Photodiodes come in all shapes and sizes. Some come with built-in lenses, some come with optical filters, some are designed for high-speed responses, some have large surface areas for high sensitivity, and some have small surface areas. When the surface area of a photodiode increases, the response time tends to slow down. Table 5.3 presents a sample portion of a specifications table for a photodiode.

FIGURE 5.16

TABLE 5.3   Part of a Specifications Table for a Photodiode

MNFR #	DESCRIPTION	REVERSE VOLTAGE (V) V_R	MAX. DARK CURRENT (nA) I_D	MIN. LIGHT CURRENT (μA) I_L	POWER DISSIPATION (mW) P_D	RISE TIME (ns) t_r	TYPICAL DETECTION ANGLE (°)	TYPICAL PEAK EMISSION WAVELENGTH (nm) λ_P
NTE3033	Infrared	30	50	35	100	50	65	900

5.6   Solar Cells

Solar cells are photodiodes with very large surface areas. The large surface area of a solar cell makes the device more sensitive to incoming light, as well as more powerful (larger currents and voltages) than photodiodes. For example, a single silicon solar cell may be capable of producing a 0.5-V potential that can supply up to 0.1 A when exposed to bright light.

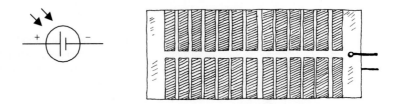

FIGURE 5.17

Solar cells can be used to power small devices such as solar-powered calculators or can be added in series to recharge nickel cadmium batteries. Often solar cells are used as light-sensitive elements in detectors of visible and near-infrared light (e.g., light meters, light-sensitive triggering mechanism for relays). Like photodiodes, solar cells have a positive and negative lead that must be connected to the more positive and more negative voltage regions within a circuit. The typical response time for a solar cell is around 20 ms.

5.6.1   Basic Operations

Power Sources

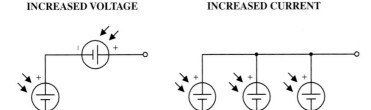

INCREASED VOLTAGE   INCREASED CURRENT

Like batteries, solar cells can be combined in series or parallel configurations. Each solar cells produces an open-circuit voltage from around 0.45 to 0.5 V and may generate as much as 0.1 A in bright light. By adding cells in series, the output voltage becomes the sum of the individual cell voltages. When cells are placed in parallel, the output current increases.

Battery Recharger

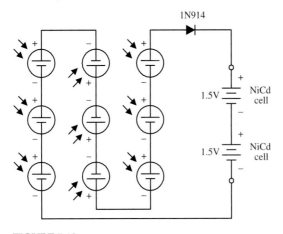

1N914

1.5V NiCd cell

1.5V NiCd cell

The circuit here shows how nine solar cells placed in series can be used to recharge two 1.5-V NiCd cells. (Each cell provides 0.5 V, so the total voltage is 4.5 V minus a 0.6-V drop due to the diode.) The diode is added to the circuit to prevent the NiCd cells from discharging through the solar cell during times of darkness. It is important not to exceed the safe charging rate of NiCd cells. To slow the charge rate, a resistor placed in series with the batteries can be added.

FIGURE 5.18

5.7   Phototransistors

Phototransistors are light-sensitive transistors. A common type of phototransistor resembles a bipolar transistor with its base lead removed and replaced with a light-sensitive surface area. When this surface area is kept dark, the device is off (practically no current flows through the collector-to-emitter region). However, when the light-sensitive region is exposed to light, a small base current is generated that controls a much larger collector-to-emitter current. Field-effect phototransistors (photoFETs) are light-sensitive field-effect transistors. Unlike photobipolar transistors, photoFETs use light to generate a gate voltage that is used to control a drain-source current. PhotoFETs are extremely sensitive to variations in light but are more fragile (electrically speaking) than bipolar phototransistors.

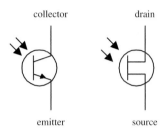

FIGURE 5.19

5.7.1 *How a Phototransistor Works*

Figure 5.20 shows a simple model of a two-lead bipolar phototransistor. The details of how this device works are given below.

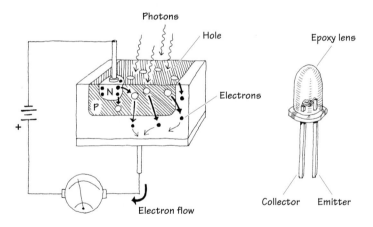

FIGURE 5.20

The bipolar phototransistor resembles a bipolar transistor (with no base lead) that has an extralarge *p*-type semiconductor region that is open for light exposure. When photons from a light source collide with electrons within the *p*-type semiconductor, they gain enough energy to jump across the *pn*-junction energy barrier—provided the photons are of the right frequency/energy. As electrons jump from the *p* region into the lower *n* region, holes are created in the *p*-type semiconductor. The extra electrons injected into the lower *n*-type slab are drawn toward the positive terminal of the battery, while electrons from the negative terminal of the battery are drawn into the upper *n*-type semiconductor and across the *np* junction, where they combine with the holes. The net result is an electron current that flows from the emitter to the collector. In terms of conventional currents, everything is backward. That is, you would say that when the base region is exposed to light, a positive current *I* flows from the collector to the emitter. Commercial phototransistors often place the *pnp* semiconductor in an epoxy case that also acts as a magnifying lens. Other phototransistors use a metal container and a plastic window to encase the chip.

5.7.2 Basic Confxigurations

EMITTER FOLLOWER   **COMMON EMITTER**

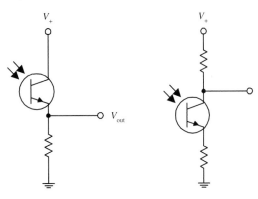

In many ways, a phototransistor is much like an ordinary bipolar transistor. Here, you can see the emitter-follower (current gain, no voltage gain) and the common-emitter amplifier (voltage gain) configurations. The emitter-follower and common-emitter circuits are discussed in Chap. 4.

FIGURE 5.21

5.7.3 Kinds of Phototransistors

Three-Lead Phototransistor

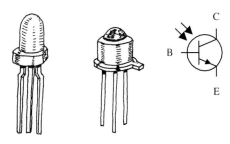

Two-lead phototransistors may not be able to inject enough electrons into the base region to promote a desired collector-to-emitter current. For this reason, a three-lead phototransistor with a base lead may be used. The extra base lead can be fed external current to help boost the number of electrons injected into the base region. In effect, the base current depends on both the light intensity and the supplied base current. With optoelectronic circuits, three-lead phototransistors are often used in place of two-lead devices, provided the base is left untouched.

Photodarlington

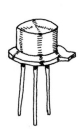

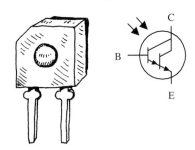

This is similar to a conventional bipolar Darlington transistor but is light-sensitive. Photodarlingtons are much more sensitive to light than ordinary phototransistors, but they tend to have slower response times. These devices may or may not come with a base lead.

FIGURE 5.22

5.7.4 Technical Stuff

Like ordinary transistors, phototransistors have maximum breakdown voltages and current and power dissipation ratings. The collector current I_C through a phototransistor depends directly on the input radiation density, the dc current gain of the device, and the external base current (for three-lead phototransistors). When a phototransistor is used to control a collector-to-emitter current, a small amount of leakage current, called the *dark current* I_D, will flow through the device even when the device is kept in the dark. This current is usually insignificant (within the nA range). Table 5.4 presents a portion of a typical data sheet for phototransistors.

TABLE 5.4   Part of a Specifications Table for Phototransistors

MNFR #	DESCRIPTION	COLLECTOR TO BASE VOLTAGE (V) BV_{CBO}	MAX. COLLECTOR CURRENT (mA) I_C	MAX. COLLECTOR DARK CURRENT (nA) I_D	MIN. LIGHT CURRENT (mA) I_L	MAX. POWER DISSIPATION (mW) P_D	TYPICAL RESPONSE TIME (μS)
NTE3031	npn, Si, visible and IR	30 (V_{CEO})	40	100 at 10V V_{CE}	1	150	6
NTE3036	npn, Si, Darlington, visible and near IR	50	250	100	12	250	151

5.7.5   Applications

Light-Activated Relay

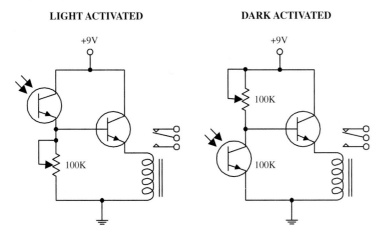

LIGHT ACTIVATED   **DARK ACTIVATED**

Here, a phototransistor is used to control the base current supplied to a power-switching transistor that is used to supply current to a relay. In the light-activated circuit, when light comes in contact with the phototransistor, the phototransistor turns on, allowing current to pass from the supply into the base of the power-switching transistor. The power-switching transistor then turns on, and current flows through the relay, triggering it to switch states. In the dark-activated circuit, an opposite effect occurs. When light is removed from the phototransistor, the phototransistor turns off, allowing more current to enter the base of the power-switching transistor. The 100-k potentiometers are used to adjust the sensitivity of both devices by controlling current flow through the phototransistor.

Receiver Circuit

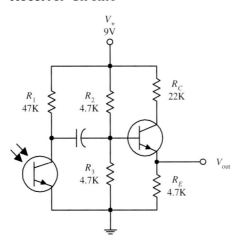

The circuit shown here illustrates how a phototransistor can be used as a modulated lightwave detector with an amplifier section (current gain amplifier). R_2 and R_3 are used to set the dc operating point of the power-switching transistor, and R_1 is used to set the sensitivity of the phototransistor. The capacitor acts to block unwanted dc signals from entering the amplifier section.

FIGURE 5.23

Tachometer

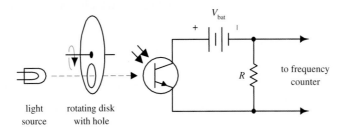

light source | rotating disk with hole

FIGURE 5.23 (*Continued*)

Here is a simple example of how a phototransistor can be used as a simple frequency counter, or tachometer. A rotating disk with a hole (connected to a rotating shaft) will pass light through its hole once every revolution. The light that passes will then trigger the phototransistor into conduction. A frequency counter is used to count the number of electrical pulses generated.

5.8 Photothyristors

Photothyristors are light-activated thyristors. Two common photothyristors include the light-activated SCR (LASCR) and the light-activated triac. A LASCR acts like a switch that changes states whenever it is exposed to a pulse of light. Even when the light is removed, the LASCR remains on until the anode and cathode polarities are reversed or the power is removed. A light-active triac is similar to a LASCR but is designed to handle ac currents. The symbol for a LASCR is shown below.

LASCR

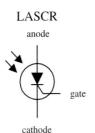

FIGURE 5.24

5.8.1 How LASCRs Work

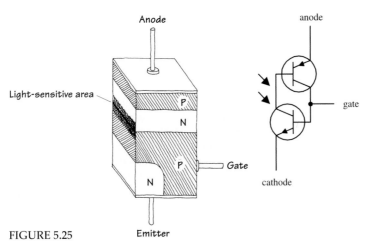

FIGURE 5.25

The equivalent circuit shown here helps explain how a LASCR works. Again, like other *pn*-junction optoelectronic device, a photon will collide with an electron in the *p*-semiconductor side, and an electron will be ejected across the *pn* junction into the *n* side. When a number of photons liberate a number of electrons across the junction, a large enough current at the base is generated to turn the transistors on. Even when the photons are eliminated, the LASCR will remain on until the polarities of the anode and cathode are reversed or the power is cut. (This results from the fact that the transistors' bases are continuously simulated by the main current flowing through the anode and cathode leads.)

5.8.2 Basic Operation

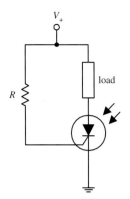

When no light is present, the LASCR is off; no current will flow through the load. However, when the LASCR is illuminated, it turns on, allowing current to flow through the load. The resistor in this circuit is used to set the triggering level of the LASCR.

FIGURE 5.26

5.9 Optoisolators

Optoisolators/optocouplers are devices that interconnect two circuits by means of an optical interface. For example, a typical optoisolator may consist of an LED and a phototransistor enclosed in a light-tight container. The LED portion of the optoisolator is connected to the source circuit, whereas the photransistor portion is connected to the detector circuit. Whenever the LED is supplied current, it emits photons that are detected by the phototransistor. There are many other kinds of source-sensor combinations, such as LED-photodiode, LED-LASCR, and lamp-photoresistor pairs.

In terms of applications, optoisolators are used frequently to provide electrical isolation between two separate circuits. This means that one circuit can be used to control another circuit without undesirable changes in voltage and current that might occur if the two circuits were connected electrically. Isolation couplers typically are enclosed in a dark container, with both source and sensor facing each other. In such an arrangement, the optoisolator is referred to as a *closed pair* (see Fig. 5.27a). Besides being used for electrical isolation applications, closed pairs are also used for level conversions and solid-state relaying. A *slotted coupler/interrupter* is a device that contains an open slot between the source and sensor through which a blocker can be placed to interrupt light signals (see Fig. 5.27b). These devices are frequently used for object detection, bounce-free switching, and vibration detection. A *reflective pair* is another kind of optoisolator configuration that uses a source to emit light and a sensor to detect that light once it has reflected off an object. Reflective pairs are used as object detectors, reflectance monitors, tachometers, and movement detectors (see Fig. 5.27c).

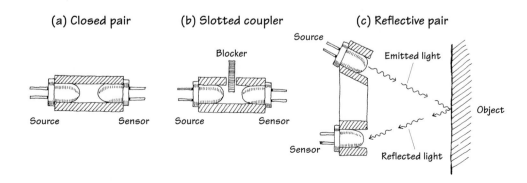

FIGURE 5.27

5.9.1   Integrated Optoisolators

Closed-pair optoisolators usually come in integrated packages. Figure 5.28 shows two sample optoisolator ICs.

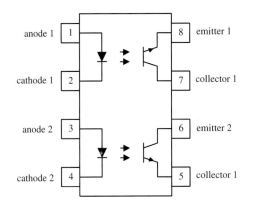
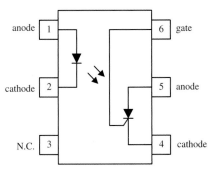

FIGURE 5.28

5.9.2   Applications

Basic Isolators/Level Shifters

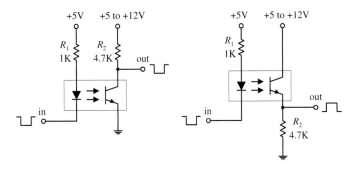

Here, a diode/phototransistor optoisolator is used to provide electrical isolation between the source circuit and the sensor circuit, as well as providing a dc level shift in the output. In the leftmost circuit, the output is noninverted, while the output in the rightmost circuit is inverted.

Optocoupler with Amplifier

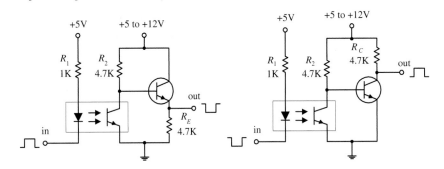

In optoelectronic applications, the phototransistor section of an optoisolator may not be able to provide enough power-handling capacity to switch large currents. The circuits to the right incorporate a power-switching transistor to solve this problem.

FIGURE 5.29

Integrated Circuits

An *integrated circuit* (IC) is a miniaturized circuit that contains a number of resistors, capacitors, diodes, and transistors stuffed together on a single chip of silicon no bigger than your fingernail. The number of resistors, capacitors, diodes, and transistors within an IC may vary from just a few to hundreds of thousands in number. The trick to cramming everything into such a small package is to make all the components out of tiny *n*-type and *p*-type silicon structures that get imbedded into the silicon chip during the production phase. To connect the little transistors, resistors, capacitors, and diodes together, aluminum plating is applied along the surface of the chip. Here is a magnified cross-sectional view of an IC showing how the various components are imbedded and linked together.

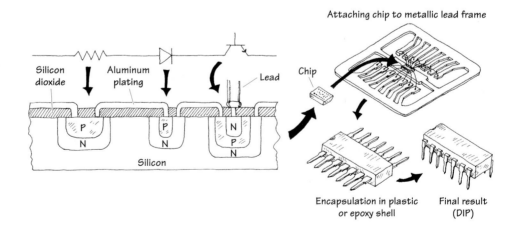

FIGURE 6.1

ICs come in either analog, digital, or analog/digital form. Analog (or linear) ICs produce, amplify, or respond to varying voltages. Digital (or logic) ICs respond to or produce signals having only high and low voltage states, whereas analog/digital ICs share properties common with both analog and digital ICs. Some common analog ICs include voltage regulators, operational amplifiers, comparators, timers, and oscillators. Common digital ICs include logic gates (e.g., AND, OR, NOR, etc.), flip-flops, memories, processors, binary counters, shift registers, multiplexers, encoders and decoders, etc. Analog/digital ICs may take on a number of different forms. For example, such an IC may be designed primarily as an analog timer but may contain a digital counter. Alternatively, the IC may be designed to read in digital information and then use this information to produce a linear output that can be used to drive, say, a stepper motor or LED display.

6.1   IC Packages

ICs are housed most frequently in dual in-line packages (DIPs). These packages consist of a plastic or ceramic box with metal pins sticking out of its sides. Each pin has a specific function and an identification number. The number 1 pin is located to the left of an index marker (see Fig. 6.2), while the succeeding pins are counted off in a counterclockwise direction. A manufacturer's logo, manufacturer's prefix, part number, and date code are almost always printed on an IC package. Other numbers and symbols found on ICs are often used to specify production process, temperature range, package type, etc. Here are a few deciphered ICs to get you going.

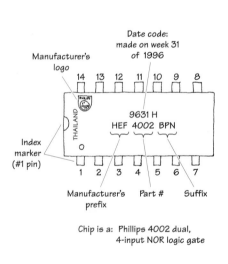

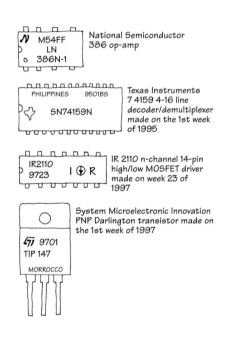

FIGURE 6.2

Notably, some IC labeling systems use an "xx44c55p-1" format. "xx" represents the manufacturer's prefix, "44" represents the chip family code, "c" represents the fabrication process (e.g., c = CMOS, f = fast, hc = high speed CMOS, hct = high speed CMOS TTL compatible, ls = low power schottky, etc.), "pp" (the suffix) represents the package code (specific to an individual manufacturer), and "1" represents the speed of the device.

Table 6.1 presents a partial list of company prefixes. It is important to note that the prefixes may not always be helpful in determining the actual manufacturer. As it turns out, a number of different manufacturers use similar prefixes. For example, the prefix "P" is used by the following companies: ASD, Harris, Hewlett-Packard, Intel, Mitsubishi, Motorola, NEC, Phillips, SGS, and Temic. If you are not sure which company the prefix belongs to, check out the manufacturer's logo.

TABLE 6.1 Company Prefixes

MANUFACTURER	PREFIX	MANUFACTURER	PREFIX
Allegro	A, μPA, UCX	Siliconix	L, LD
Analog Devices	AD	Linear Technology	LT
Advanced Micro Devices	Am, A, AO	Mitsubishi	M
Panasonic	An	Fugitsu	MB, FTU
General Instrument	AY, GIC, GP, GI, FE, GF, W, GI	MOS Technology	MCS
Sony	Bx, Cx, GI	Microsystems International	MIL
Intel	B, C, I, M, IR, A, AP AT	Mostek	MK
RCA (now Harris)	CA, CD, CDP	Plessey	MN, SL, SP
TRW	CA, TDC, MPY, CMP, DAC, MAT, OP	Signetics	N, NE, S, SE, SP
Precision Monolithics	PM, REF, SSS	Next	Nx
National Semiconductor	DM, LF, LFT, LH, LM, NH, NA, NDx	NTE	NTE
		Precision Monolithics	PM
Sanyo	DSK	Quality Semiconductor	QS
Fairchild (now National Semiconductor)	F μA, μL, Unx	Raytheon	R, RAY, RC, RM
Ferranti	FSS, ZLD, Zn	Silicon General	SG
GE	GEL	Shanghai Belling Micro	SGS
Harris	HA	Siemens	B, BB, BF
Hitachi	HA, HD, HG, HI, HZ	Texas Instruments	SN, TL, TMS, TEX
Motorola	HEP, M, MC, MCC, MCM, MFC, MM, MWM, HEF, HEED, T, CH, J	Toshiba	T, TC, HZ, M, TA, TH, JT
Intersil	ICH, ICL, ICM, IM	Sprague	ULN, ULS
IR	IR, IRB, IRxx	NEC	μP, μTD, NRA, NRB, NRx
Sharp	IR	Westinghouse	WC, WM
ITT	ITT, MIC	Exar	XR
Philips	HEF, HCF, M, AD, AJ, J, B, BB, BA, ON, OT	Yamaha	YAC, YM, YMF, YSS
Samsung	KA, IR, K Kxx	Hewlett-Packard	5082-nnnn, AT

Another problem when deciphering IC labels is figuring out what the suffixes mean. Often a manufacturer will place a suffix at the end of a part number to indicate the type of package (e.g., ceramic dual-in-line, plastic dual-in-line package, etc.) or to indicate temperature range. The frustrating thing with suffixes is that there is no standard convention used. Each manufacturer's suffix may have its own unique meaning.

A more complete and up-to-date listing of company prefixes can be found in the Chip Directory's Web site (*www.hitex.com/chipdir/index.htm*). If this link does not work, run a search using "Chip Directory" as your keyword. The Chip Directory also provides listings of company logos, company abbreviations, suffix meanings, chips by number, chips by name, chips by family, chip manufacturers, and links to manufacturers' Web sites.

6.2   Some Basic ICs to Get You Started

TL783 Three-Terminal Adjustable Voltage Regulator

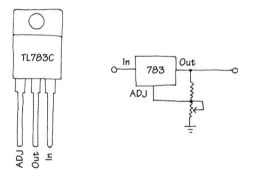

FIGURE 6.3

This device is used as an adjustable voltage regulator for high-voltage applications. Output can be adjusted from 1.25 to 125 V, with a maximum output current of 700 mA. Two external resistors are required to program the output voltage. Voltage regulators will be covered in greater detail in Chap. 10.

741CD Op Amp

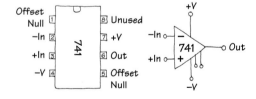

FIGURE 6.4

A 741 op amp is a very popular high-performance linear amplifier IC. When a feedback resistor is connected from its output to its inverting input, the device can be made to do some amazing things. For example, you can use op amps to build inverting and noninverting amplifiers, oscillators, integrators, differentiators, adders, and subtractors. Without feedback, the device acts as a comparator. The 741 op amp is made by a number of different manufacturers. To learn more about op amps, see Chap. 7.

555 Timer

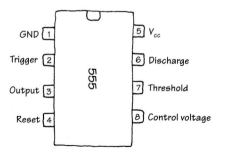

FIGURE 6.5

This is a very versatile IC that acts as a kind of smart on/off switch that can be made to turn on and off at a rate you set by wiring resistors and capacitors between its leads and to and from supply voltages. These devices are used in timer circuits, oscillator circuits, logic clock generators, and tone generators. The 555 timer is described in detail in Chap. 8.

LM7411 Three-Input AND Gate

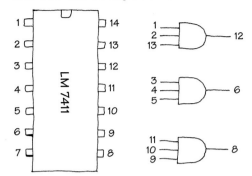

FIGURE 6.6

The 7411 contains three functionally independent three-input AND gates. All three gates are powered through a common supply pin. NAND gates, OR gates, NOR gates, memories, microprocessors, microcontrollers, and other logic ICs are discussed in Chap. 12.

MM5480 LED Display Driver

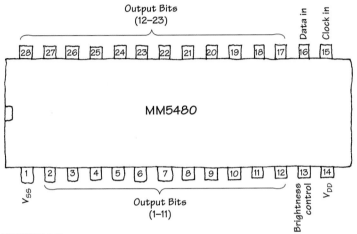

FIGURE 6.7

This IC is designed to drive common-anode, separate-cathode LED displays. A single pin connected to a supply voltage through a variable resistor controls the LED display's brightness. Some features include continuous brightness control, serial data input, no load signal requirement, TTL compatibility, 3½-digit displays, and wide dc power supply operation. The MM5480 can be used in industrial control indicators, digital clocks, thermometers, counters, and voltmeters. LED displays and display drivers are discussed in Chap. 12.

LM628 Precision Motion Controller

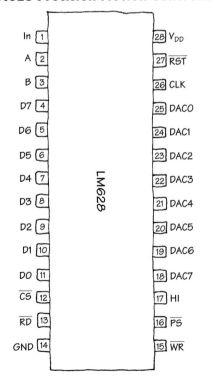

FIGURE 6.8

The LM628 is a motion-control processor that uses a quadrature incremental position feedback signal from dc motors, brushless dc servo motors, and other servo mechanisms. The device has an 8-bit output that can drive an 8-bit or 12-bit DAC, which in turn makes it possible to construct a servo system that includes using a dc motor, an actuator, an incremental encoder, a DAC, and a power amplifier. Features include 32-bit position, velocity and acceleration registers, position and velocity modes of operation, real-time programmable host interrupts, an 8-bit parallel asynchronous host interface, and a quadrature incremental encoder.

CHAPTER 7

Operational Amplifiers

Operational amplifiers (op amps) are incredibly useful high-performance differential amplifiers that can be employed in a number of amazing ways. A typical op amp is an integrated device with a noninverting input, an inverting input, two dc power supply leads (positive and negative), an output terminal, and a few other specialized leads used for fine-tuning. The positive and negative supply leads, as well as the fine-tuning leads, are often omitted from circuit schematics. If you do not see any supply leads, assume that a dual supply is being used.

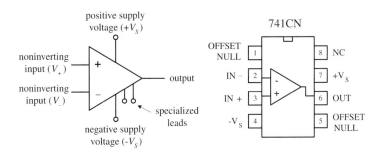

FIGURE 7.1

By itself, an op amp's operation is simple. If the voltage applied to the *inverting terminal* V_- is more positive than the voltage applied to the *noninverting terminal* V_+, the output saturates toward the *negative supply voltage* $-V_S$. Conversely, if $V_+ > V_-$, the output saturates toward the positive supply voltage $+V_S$ (see Fig. 7.2). This "maxing out" effect occurs with the slightest difference in voltage between the input terminals.

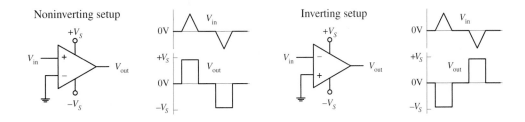

FIGURE 7.2

At first glance, it may appear that an op amp is not a very impressive device—it switches from one maximum output state to another whenever there's a voltage difference between its inputs. Big deal, right? By itself, it does indeed have limited applications. The trick to making op amps useful devices involves applying what is called *negative feedback*.

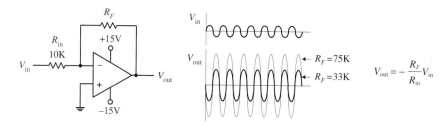

FIGURE 7.3

When voltage is "fed" back from the output terminal to the inverting terminal (this is referred to as *negative feedback*), the gain of an op amp can be controlled—the op amp's output is prevented from saturating. For example, a feedback resistor R_F placed between the output and the inverting input, as shown in Fig. 7.3, acts to convey the state of the output back to the op amp's input. This feedback information basically tells the op amp to readjust its output voltage to a value determined by the resistance of the feedback resistor. The circuit in Fig. 7.3, called an *inverting amplifier*, and has an output equal to $-V_{in}/(R_F/R_{in})$ (you will learn how to derive this formula later in this chapter). The negative sign means that the output is inverted relative to the input—a result of the inverting input. The gain is then simply the output voltage divided by the input voltage, or $-R_F/R_{in}$ (the negative sign indicates that the output is inverted relative to the input). As you can see from this equation, if you increase the resistance of the feedback resistor, there is an increase in the voltage gain. On the other hand, if you decrease the resistance of the feedback resistor, there is a decrease in the voltage gain.

By adding other components to the negative-feedback circuit, an op amp can be made to do a number of interesting things besides pure amplification. Other interesting op amp circuits include voltage-regulator circuits, current-to-voltage converters, voltage-to-current converters, oscillator circuits, mathematical circuits (adders, subtractors, multipliers, differentiators, integrators, etc.), waveform generators, active filter circuits, active rectifiers, peak detectors, sample-and-hold circuits, etc. Most of these circuits will be covered in this chapter.

Besides negative feedback, there's positive feedback, where the output is linked through a network to the noninverting input. Positive feedback has the opposite effect as negative feedback; it drives the op amp harder toward saturation. Although positive feedback is seldom used, it finds applications in special comparator circuits that are often used in oscillator circuits. Positive feedback also will be discussed in detail in this chapter.

7.1   Operational Amplifier Water Analogy

This is the closest thing I could come up with in terms of a water analogy for an op amp. To make the analogy work, you have to pretend that water pressure is analogous to voltage and water flow is analogous to current flow.

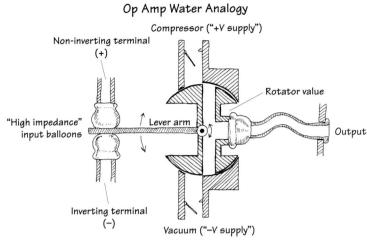

Op Amp Water Analogy

FIGURE 7.4

The inverting and noninverting terminals of the water op amp are represented by the two tubes with elastic balloon ends. When the water pressure applied to both input tubes is equal, the lever arm is centered. However, if the water pressure applied to the noninverting tube is made larger than the pressure applied to the inverting tube, the noninverting balloon expands and forces the lever arm downward. The lever arm then rotates the rotator valve counterclockwise, thus opening a canal from the compressor tube (analogous to the positive supply voltage) to the output tube. (This is analogous to an op amp saturating in the positive direction whenever the noninverting input is more positive in voltage than the inverting input.) Now, if the pressure applied at the noninverting tube becomes less than the pressure applied at the inverting tube, the lever arm is pushed upward by the inverting balloon. This causes the rotator value to rotate clockwise, thus opening the canal from the vacuum tube (analogous to the negative supply voltage) to the output. (This is analogous to an op amp saturating in the negative direction whenever the inverting input is made more positive in voltage than the noninverting input.) See what you can do with the analogy in terms of explaining negative feedback. Also note that in the analogy there is an infinite "input water impedance" at the input tubes, while there is a zero "output water impedance" at the output tube. As you will see, ideal op amps also have similar input and output impedance. In real op amps, there are always some leakage currents.

7.2 How Op Amps Work (The "Cop-out" Explanation)

An op amp is an integrated device that contains a large number of transistors, several resistors, and a few capacitors. Figure 7.5 shows a schematic diagram of a typical low-cost general-purpose bipolar operational amplifier.

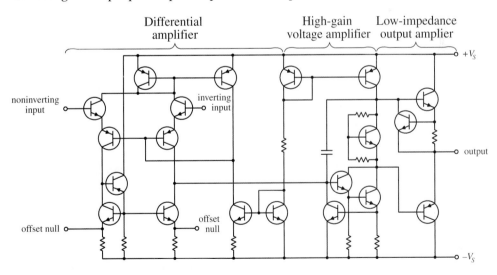

FIGURE 7.5

This op amp basically consists of three stages: a high-input-impedance differential amplifier, a high-gain voltage amplifier with a level shifter (permitting the output to swing positive and negative), and a low-impedance output amplifier. However, realizing that an op amp is composed of various stages does not help you much in terms of figuring out what will happen between the input and output leads. That is, if you attempt to figure out what the currents and voltages are doing within the complex system, you will be asking for trouble. It is just too difficult a task. What is important here is not to focus on understanding the op amp's internal circuitry but instead to focus on memorizing some rules that individuals came up with that require only working with the input and output leads. This approach seems like a "cop-out," but it works.

7.3   Theory

There is essentially only one formula you will need to know for solving op amp circuit problems. This formula is the foundation on which everything else rests. It is the expression for an op amp's output voltage as a function of its input voltages V_+ (noninverting) and V_- (inverting) and of its *open-loop voltage gain* A_o:

$$V_{out} = A_o(V_+ - V_-)$$

This expression says that an *ideal op amp* acts like an ideal voltage source that supplies an output voltage equal to $A_o(V_+ - V_-)$ (see Fig. 7.6). Things can get a little more complex when we start talking about *real op amps,* but generally, the open-loop voltage expression above pretty much remains the same, except now we have to make some slight modifications to our equivalent circuit. These modifications must take into account the nonideal features of an op amp, such as its input resistance R_{in} and output resistance R_{out}. Figure 7.6 *right* shows a more realistic equivalent circuit for an op amp.

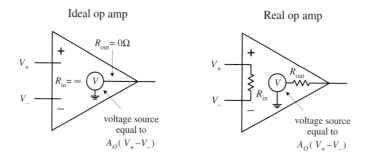

FIGURE 7.6

To give meaning to the open-loop voltage gain expression and to the ideal and real equivalent circuits, the values of A_o, R_{in}, and R_{out} are defined within the following rules:

Rule 1: For an ideal op amp, the open-loop voltage gain is infinite ($A_o = \infty$). For a real op amp, the gain is a finite value, typically between 10^4 to 10^6.

Rule 2: For an ideal op amp, the input impedance is infinite ($R_{in} = \infty$). For a real op amp, the input impedance is finite, typically between 10^6 (e.g., typical bipolar op amp) to 10^{12} Ω (e.g., typical JFET op amp). The output impedance for an ideal op amp is zero ($R_{out} = 0$). For a real op amp, R_{out} is typically between 10 to 1000 Ω.

Rule 3: The input terminals of an ideal op amp draw no current. Practically speaking, this is true for a real op amp as well—the actual amount of input current is usually (but not always) insignificantly small, typically within the picoamps (e.g., typical JFET op amp) to nanoamps (e.g., typical bipolar op amp) range.

Now that you are armed with $V_{out} = A_o(V_+ - V_-)$ and rules 1 through 3, let's apply them to a few simple example problems.

EXAMPLE I

Solve for the gain (V_{out}/V_{in}) of the circuit below.

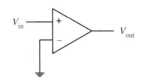

FIGURE 7.7

Since V_- is grounded (0 V) and V_+ is simply V_{in}, you can plug these values into the open-loop voltage gain expression:

$$V_{out} = A_o(V_+ - V_-)$$
$$= A_o(V_{in} - 0\,\text{V}) = A_o V_{in}$$

Rearranging this equation, you get the expression for the gain:

$$\text{Gain} = \frac{V_{out}}{V_{in}} = A_o$$

If you treat the op amp as ideal, A_o would be infinite. However, if you treat the op amp as real, A_o is finite (around 10^4 to 10^6). This circuit acts as a simple noninverting comparator that uses ground as a reference. If $V_{in} > 0$ V, the output ideally goes to $+\infty$ V; if $V_{in} < 0$ V, the output ideally goes to $-\infty$ V. With a real op amp, the output is limited by the supply voltages (which are not shown in the drawing but assumed). The exact value of the output voltage is slightly below and above the positive and negative supply voltages, respectively. These maximum output voltages are called the *positive* and *negative saturation voltages*.

EXAMPLE 2

Solve for the gain (V_{out}/V_{in}) of the circuit below.

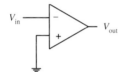

FIGURE 7.8

Since V_+ is grounded (0 V) and V_- is simply V_{in}, you can substitute these values into the open-loop voltage gain expression:

$$V_{out} = A_o(V_+ - V_-)$$
$$= A_o(0\,\text{V} - V_{in}) = -A_o V_{in}$$

Rearranging this equation, you get the expression for the gain:

$$\text{Gain} = \frac{V_{out}}{V_{in}} = -A_o$$

If you treat the op amp as ideal, $-A_o$ is negatively infinite. However, if you treat the op amp as real, $-A_o$ is finite (around -10^4 to -10^6). This circuit acts as a simple inverting comparator that uses ground as a reference. If $V_{in} > 0$ V, the output ideally goes to $-\infty$ V; if $V_{in} < 0$ V, the output ideally goes to $+\infty$ V. With a real op amp, the output swings are limited to the saturation voltages.

7.4 Negative Feedback

Negative feedback is a wiring technique where some of the output voltage is sent back to the inverting terminal. This voltage can be "sent" back through a resistor, capacitor, or complex circuit or simply can be sent back through a wire. So exactly what kind of formulas do you use now? Well, that depends on the feedback circuit, but in reality, there is nothing all that new to learn. In fact, there is really only one formula you need to know for negative-feedback circuits (you still have to use the rules, however). This formula looks a lot like our old friend $V_{out} = A_o(V_+ - V_-)$. There is, however, the V_- in the formula—this you must reconsider. V_- in the formula changes because now the output voltage from the op amp is "giving" extra voltage (positive or negative) back to the inverting terminal. What this means is that you must replace V_- with fV_{out}, where f is a fraction of the voltage "sent" back from V_{out}. That's the trick!

There are two basic kinds of negative feedback, voltage feedback and operational feedback, as shown in Fig. 7.9.

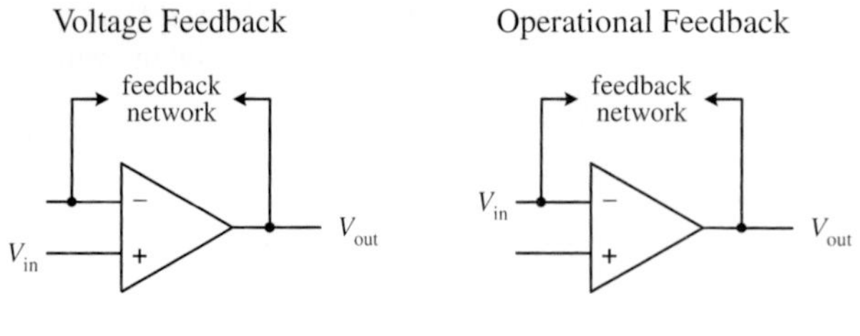

Voltage Feedback   Operational Feedback

FIGURE 7.9

$$V_{out} = A_0(V_+ - fV_{out})$$

Now, in practice, figuring out what the fraction f should be is not important. That is, you do not have to calculate it explicitly. The reason why I have introduced it in the open-loop voltage expression is to provide you with a bit of basic understanding as to how negative feedback works in theory. As it turns out, there is a simple trick for making op amp circuits with negative feedback easy to calculate. The trick is as follows: If you treat an op amp as an ideal device, you will notice that if you rearrange the open-loop voltage expression into $V_{out}/A_o = (V_+ - V_-)$, the left side of the equation goes to zero—A_o is infinite for an ideal op amp. What you get in the end is then simply $V_+ - V_- = 0$. This result is incredibly important in terms of simplifying op amp circuits with negative feedback—so important that the result receives its own rule (the fourth and final rule).

Rule 4: Whenever an op amp senses a voltage difference between its inverting and noninverting inputs, it responds by feeding back as much current/voltage through the feedback network as is necessary to keep this difference equal to zero $(V_+ - V_- = 0)$. This rule only applies for negative feedback.

The following sample problems are designed to show you how to apply rule 4 (and the other rules) to op amp circuit problems with negative feedback.

Negative Feedback Example Problems

BUFFER (UNITY GAIN AMPLIFIER)

Solve for the gain (V_{out}/V_{in}) of the circuit below.

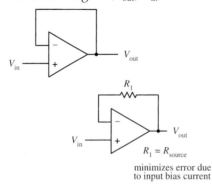

$R_1 = R_{source}$

minimizes error due
to input bias current

FIGURE 7.10

Since you are dealing with negative feedback, you can apply rule 4, which says that the output will attempt to make $V_+ - V_- = 0$. By examining the simple connections, notice that $V_{in} = V_+$ and $V_- = V_{out}$. This means that $V_{in} - V_{out} = 0$. Rearranging this expression, you get the gain:

$$\text{Gain} = \frac{V_{out}}{V_{in}} = 1$$

A gain of 1 means that there is no amplification; the op amp's output follows its input. At first glance, it may appear that this circuit is useless. However, it is important to recall that an op amp's input impedance is huge, while its output impedance is extremely small (rule 2). This feature makes this circuit useful for circuit-isolation applications. In other words, the circuit acts as a buffer. With real op amps, it may be necessary to throw in a resistor in the feedback loop (lower circuit). The resistor acts to minimize voltage offset errors caused by input bias currents (leakage). The resistance of the feedback resistor should be equal to the source resistance. I will discuss input bias currents later in this chapter.

INVERTING AMPLIFIER

Solve for the gain (V_{out}/V_{in}) of the circuit below.

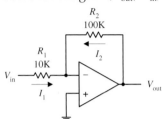

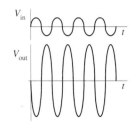

FIGURE 7.11

Because you have negative feedback, you know the output will attempt to make the difference between V_+ and V_- zero. Since V_+ is grounded (0 V), this means that V_- also will be 0 V (rule 4). To figure out the gain, you must find currents I_1 and I_2 so you can come up with an expression containing V_{out} in terms of V_{in}. Using Ohm's law, you find I_1 and I_2 to be

$$I_1 = \frac{V_{in} - V_-}{R_1} = \frac{V_{in} - 0\,V}{R_1} = \frac{V_{in}}{R_1}$$

$$I_2 = \frac{V_{out} - V_-}{R_2} = \frac{V_{out} - 0\,V}{R_2} = \frac{V_{out}}{R_2}$$

Because an ideal op amp has infinite input impedance, no current will enter its inverting terminal (rule 3). Therefore, you can apply Kirchhoff's junction rule to get $I_2 = -I_1$. Substituting the calculated values of I_1 and I_2 into this expression, you get $V_{out}/R_2 = -V_{in}/R_1$. Rearranging this expression, you find the gain:

$$\text{Gain} = \frac{V_{out}}{V_{in}} = -\frac{R_2}{R_1}$$

The negative sign tells you that the signal that enters the input will be inverted (shifted 180°). Notice that if $R_1 = R_2$, the gain is –1 (the negative sign simply means the output is inverted). In this case you get what's called a *unity-gain inverter,* or an *inverting buffer.* When using real op amps that have relatively high input bias currents (e.g., bipolar op amps), it may be necessary to place a resistor with a resistance equal to $R_1 \| R_2$ between the noninverting input and ground to minimize voltage offset errors.

NONINVERTING AMPLIFER

Solve for the gain (V_{out}/V_{in}) of the circuit below.

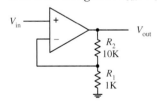

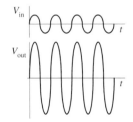

FIGURE 7.12

By inspection, you know that $V_+ = V_{in}$. By applying rule 4, you then can say that $V_- = V_+$. This means that $V_- = V_{in}$. To come up with an expression relating V_{in} and V_{out} (so that you can find the gain), the voltage divider relation is used:

$$V_- = \frac{R_1}{R_1 + R_2} V_{out} = V_{in}$$

Rearranging this equation, you find the gain:

$$\text{Gain} = \frac{V_{out}}{V_{in}} = \frac{R_1 + R_2}{R_1} = 1 + \frac{R_2}{R_1}$$

Unlike the inverting amplifier, this circuit's output is in phase with its input—the output is "noninverted." With real op amps, to minimize voltage offset errors due to input bias current, set $R_1 \| R_2 = R_{source}$.

SUMMING AMPLIFIER

Solve for V_{out} in terms of V_1 and V_2.

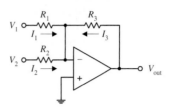

Since you know that V_+ is grounded (0 V), and since you have feedback in the circuit, you can say that $V_+ = V_- = 0\,\text{V}$ (rule 4). Now that you know V_-, solve for I_1, I_2, and I_3 in order to come up with an expression relating V_{out} with V_1 and V_2. The currents are found by applying Ohm's law:

$$I_1 = \frac{V_1 - V_-}{R_1} = \frac{V_1 - 0\,\text{V}}{R_1} = \frac{V_1}{R_1}$$

$$I_2 = \frac{V_2 - V_-}{R_2} = \frac{V_2 - 0\,\text{V}}{R_2} = \frac{V_2}{R_2}$$

$$I_3 = \frac{V_{\text{out}} - V_-}{R_3} = \frac{V_{\text{out}} - 0\,\text{V}}{R_3} = \frac{V_{\text{out}}}{R_3}$$

Like the last problem, assume that no current enters the op amp's inverting terminal (rule 3). This means that you can apply Kirchhoff's junction rule to combine I_1, I_2, and I_3 into one expression: $I_3 = -(I_1 + I_2) = -I_1 - I_2$. Plugging the results above into this expression gives the answer:

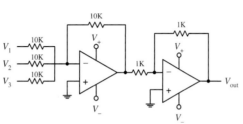

FIGURE 7.13

$$V_{\text{out}} = -\frac{R_3}{R_1}V_1 - \frac{R_3}{R_2}V_2 = -\left(\frac{R_3}{R_1}V_1 + \frac{R_3}{R_2}V_2\right)$$

If you make $R_1 = R_2 = R_3$, $V_{\text{out}} = -(V_1 + V_2)$. Notice that the sum is negative. To get a positive sum, you can add an inverting stage, as shown in the lower circuit. Here, three inputs are added together to yield the following output: $V_{\text{out}} = V_1 + V_2 + V_3$. Again, for some real op amps, an additional input-bias compensation resistor placed between the noninverting input and ground may be needed to avoid offset error caused by input bias current. Its value should be equal to the parallel resistance of all the input resistors.

DIFFERENCE AMPLIFIER

Determine V_{out}.

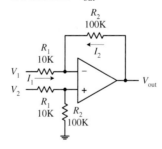

FIGURE 7.14

First, you determine the voltage at the noninverting input by using the voltage divider relation (again, assume that no current enters the inputs):

$$V_+ = \frac{R_2}{R_1 + R_2}V_2$$

Next, apply Kirchhoff's current junction law to the inverting input ($I_1 = I_2$):

$$\frac{V_1 - V_-}{R_1} = \frac{V_- - V_{\text{out}}}{R_2}$$

Using rule 4 ($V_+ = V_-$), substitute the V_+ term in for V_- in the last equation to get

$$V_{\text{out}} = \frac{R_2}{R_1}(V_2 - V_1)$$

If you set $R_1 = R_2$, then $V_{\text{out}} = V_2 - V_1$.

INTEGRATOR

Solve for V_{out} in terms of V_{in}.

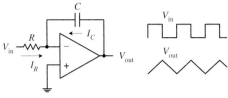

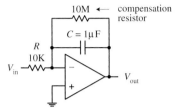

FIGURE 7.15

Because you have feedback, and because $V_+ = 0\,\text{V}$, you can say that V_- is $0\,\text{V}$ as well (rule 4). Now that you know V_-, solve for I_R and I_C so that you can come up with an expression relating V_{out} with V_{in}. Since no current enters the input of an op amp (rule 3), the displacement current I_C through the capacitor and the current I_R through the resistor must be related by $I_R + I_C = 0$. To find I_R, use Ohm's law:

$$I_R = \frac{V_{\text{in}} - V_-}{R} = \frac{V_{\text{in}} - 0\,\text{V}}{R} = \frac{V_{\text{in}}}{R}$$

I_C is found by using the displacement current relation:

$$I_C = C\frac{dV}{dt} = C\frac{d(V_{\text{out}} - V_-)}{dt} = C\frac{d(V_{\text{out}} - 0\,\text{V})}{dt} = C\frac{dV_{\text{out}}}{dt}$$

Placing these values of I_C and I_R into $I_R + I_C = 0$ and rearranging, you get the answer:

$$dV_{\text{out}} = -\frac{1}{RC}V_{\text{in}}dt$$

$$V_{\text{out}} = -\frac{1}{RC}V_{\text{in}}dt$$

Such a circuit is called an *integrator;* the input signal is integrated at the output. Now, one problem with the first circuit is that the output tends to drift, even with the input grounded, due to nonideal characteristics of real op amps such as voltage offsets and bias current. A large resistor placed across the capacitor can provide dc feedback for stable biasing. Also, a compensation resistor may be needed between the noninverting terminal and ground to correct voltage offset errors caused by input bias currents. The size of this resistor should be equal to the parallel resistance of the input resistor and the feedback compensation resistor.

DIFFERENTIATOR

Solve for V_{out} in terms of V_{in}.

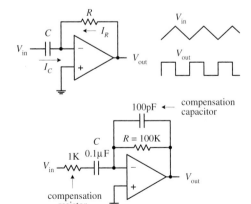

FIGURE 7.16

Since you know that V_+ is grounded ($0\,\text{V}$), and since you have feedback in the circuit, you can say that $V_- = V_+ = 0\,\text{V}$ (rule 4). Now that you know V_-, solve for I_R and I_C so that you can come up with an expression relating V_{out} with V_{in}. Since no current enters the input of an op amp (rule 3), the displacement current I_C through the capacitor and the current I_R through the resistor must be related by $I_R + I_C = 0$. To find I_C, use the displacement current equation:

$$I_C = C\frac{dV}{dt} = C\frac{d(V_{\text{in}} - V_-)}{dt} = C\frac{d(V_{\text{in}} - 0\,\text{V})}{dt} = C\frac{dV_{\text{in}}}{dt}$$

The current I_R is found using Ohm's law:

$$I_R = \frac{V_{\text{out}} - V_-}{R} = \frac{V_{\text{out}} - 0\,\text{V}}{R} = \frac{V_{\text{out}}}{R}$$

Placing these values of I_C and I_R into $I_R + I_C = 0$ and rearranging, you get the answer:

$$V_{\text{out}} = -RC\frac{dV_{\text{in}}}{dt}$$

Such a circuit is called a *differentiator;* the input signal is differentiated at the output. The first differentiator circuit shown is not in practical form. It is extremely susceptible to noise due to the op amp's high ac gain. Also, the feedback network of the differentiator acts as an *RC* low-pass filter that contributes a 90° phase lag within the loop and may cause stability problems. A more practical differentiator is shown below the first circuit. Here, both stability and noise problems are corrected with the addition of a feedback capacitor and input resistor. The additional components provide high-frequency rolloff to reduce high-frequency noise. These components also introduce a 90° lead to cancel the 90° phase lag. The effect of the additional components, however, limits the maximum frequency of operation—at very high frequencies, the differentiator becomes an integrator. Finally, an additional input-bias compensation resistor placed between the noninverting input and ground may be needed to avoid offset error caused by input bias current. Its value should be equal to the resistance of the feedback resistor.

7.5   Positive Feedback

Positive feedback involves sending output voltage back to the noninverting input. In terms of the theory, if you look at our old friend $V_{out} = A_o(V_+ - V_-)$, the V_+ term changes to fV_{out} (f is a fraction of the voltage sent back), so you get $V_{out} = A_o(fV_{out} - V_-)$. Now, an important thing to notice about this equation (and about positive feedback in general) is that the voltage fed back to the noninverting input will act to drive the op amp "harder" in the direction the output is going (toward saturation). This makes sense in terms of the equation; fV_{out} adds to the expression. Recall that negative feedback acted in the opposite way; the fV_{out} ($= V_-$) term subtracted from the expression, preventing the output from "maxing out." In electronics, positive feedback is usually a bad thing, whereas negative feedback is a good thing. For most applications, it is desirable to control the gain (negative feedback), while it is undesirable to go to the extremes (positive feedback).

There is, however, an important use for positive feedback. When using an op amp to make a comparator, positive feedback can make output swings more pronounced. Also, by adjusting the size of the feedback resistor, a comparator can be made to experience what is called *hysteresis*. In effect, hysteresis gives the comparator two thresholds. The voltage between the two thresholds is called the *hysteresis voltage*. By obtaining two thresholds (instead of merely one), the comparator circuit becomes more immune to noise that can trigger unwanted output swings. To better understand hysteresis, let's take a look at the following comparator circuit that incorporates positive feedback.

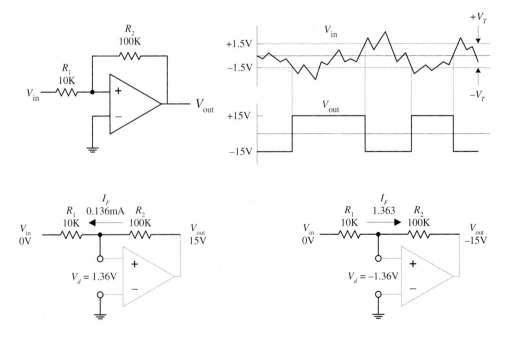

FIGURE 7.17

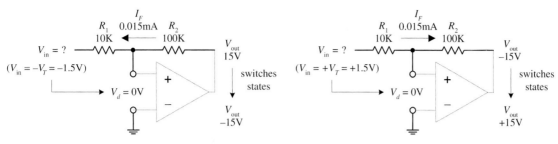

FIGURE 7.17 (*Continued*)

Assume that the op amp's output is at positive saturation, say, +15 V. If V_{in} is 0 V, the voltage difference between the inverting input and noninverting input (V_d) will be 1.36 V. You get this by using Ohm's law:

$$I_F = (V_{out} - V_{in})/(R_1 + R_2)$$
$$V_d = I_F R_1$$

This does not do anything to the output; it remains at +15 V. However, if you reduce V_{in}, there is a point when V_d goes to 0 V, at which time the output switches states. This voltage is called the *negative threshold voltage* ($-V_T$). The negative threshold voltage can be determined by using the previous two equations—the end result being $-V_T = -V_{out}/(R_2/R_1)$. In the example, $-V_T = -1.5$ V. Now, if the output is at negative saturation (-15 V) and 0 V is applied to the input, $V_d = -1.36$ V. The output remains at -15 V. However, if the input voltage is increased, there is a point where V_d goes to zero and the output switches states. This point is called the *positive threshold voltage* ($+V_T$), which is equal to $+V_{out}/(R_2/R_1)$. In the example, $+V_T = +1.5$ V. Now the difference between the two saturation voltages is the hysteresis voltage: $V_h = +V_T - (-V_T)$. In the example, $V_h = 3$ V.

7.6 Real Kinds of Op Amps

General Purpose

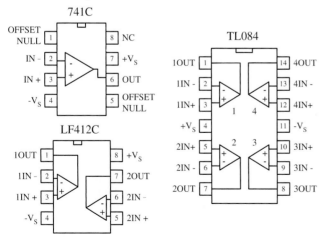

FIGURE 7.18

There is a huge selection of general-purpose and precision op amps to chose from. Precision op amps are specifically designed for high stability, low offset voltages, low bias currents, and low drift parameters. Because the selection of op amps is so incredibly large, I will leave it to you to check out the electronics catalogs to see what devices are available. When checking out these catalogs, you will find that op amps (not just general-purpose and precision) fall into one of the following categories (based on input circuitry): bipolar, JFET, MOSFET, or some hybrid thereof (e.g., BiFET). In general, bipolar op amps, like the 741 (industry standard), have higher input bias currents than either JFET or MOSFET types. This means that their input terminals have a greater tendency to "leak in" current. Input bias current results in voltage drops across resistors of feedback networks, biasing networks, or source impedances, which in turn can offset the output voltage. The amount of offset a circuit can tolerate ultimately depends on the application. Now, as I briefly mentioned earlier in this chapter, a compensation resistor placed between the noninverting terminal and ground (e.g., bipolar inverting amplifier circuit) can reduce these offset errors. (More on this in a minute.)

Precision

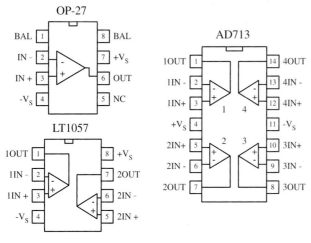

FIGURE 7.19

A simple way to avoid problems associated with input bias current is to use a FET op amp. A typical JFET op amp has a very low input bias current, typically within the lower picoamp range as compared with the nanoamp range for a typical bipolar op amp. Some MOSFET op amps come with even lower input bias currents, often as low as few tenths of a picoamp. Though FET op amps have lower input bias current than bipolar op amps, there are other features they have that are not quite as desirable. For example, JFET op amps often experience an undesired effect called *phase inversion*. If the input common-mode voltage of the JFET approaches the negative supply too closely, the inverting and noninverting input terminals may reverse directions—negative feedback becomes positive feedback, causing the op amp to latch up.

This problem can be avoided by using a bipolar op amp or by restricting the common-mode range of the signal. Here are some other general comments about bipolar and FET op amps: offset voltage (low for bipolar, medium for JFET, medium to high for MOSFET), offset drift (low for bipolar, medium for FET), bias matching (excellent for bipolar, fair for FET), bias/temperature variation (low for bipolar, fair for FET).

To avoid getting confused by the differences between the various op amp technologies, it is often easier to simply concentrate on the specifications listed in the electronics catalogs. Characteristics to look for include speed/slew rate, noise, input offset voltages and their drift, bias currents and their drift, common-mode range, gain, bandwidth, input impedance, output impedance, maximum supply voltages, supply current, power dissipation, and temperature range. Another feature to look when purchasing an op amp is whether the op amp is internally or externally frequency compensated. An externally compensated op amp requires external components to prevent the gain from dropping too quickly at high frequencies, which can lead to phase inversions and oscillations. Internally compensated op amps take care of these problems with internal circuitry. All the terms listed in this paragraph will be explained in greater detail in a minute.

Programmable Op Amp

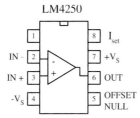

FIGURE 7.20

A programmable op amp is a versatile device that is used primarily in low-power applications (e.g., battery-powered circuits). These devices can be programmed with an external current for desired characteristics. Some of the characteristics that can be altered by applying a programming current include quiescent power dissipation, input offset and bias currents, slew rate, gain-bandwidth product, and input noise characteristics—all of which are roughly proportional to the programming current. The programming current is typically drawn from the programming pin (e.g., pin 8 of the LM4250) through a resistor and into ground. The programming current allows the op amp to be operated over a wide range of supply currents, typically from around a few microamps to a few millamps. Because a programmable op amp can be altered so as to appear as a completely different op amp for different programming currents, it is possible to use a single device for a variety of circuit functions within a system. These devices typically can operate with very low supply voltages (e.g., 1 V for the LM4250). A number of different manufacturers make programmable op amps, so check the catalogs. To learn more about how to use these devices, check out the manufacturers' literature (e.g., for National Semiconductor's LM4250, go to *www.national.com*).

Single-Supply Op Amps

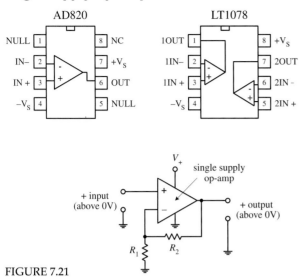

These op amps are designed to be operated from a single positive supply (e.g., +12 V) and allow input voltages all the way down to the negative rail (normally tied to ground). Figure 7.21 shows a simple dc amplifier that uses a single-supply op amp. It is important to note that the output of the amplifier shown cannot go negative; thus it cannot be used for, say, ac-coupled audio signals. These op amps are frequently used in battery-operated devices.

FIGURE 7.21

Audio Amplifiers

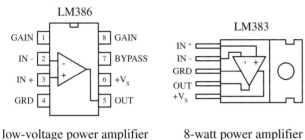

FIGURE 7.22

These are closely related to conventional op amps but designed specifically to operate best (low audio band noise, crossover distortion, etc.) within the audio-frequency spectrum (20 to 20,000 Hz). These devices are used mainly in sensitive preamplifiers, audio systems, AM-FM radio receivers, servo amplifiers, and intercom and automotive circuits. There are a number of audio amplifiers to choice from. Some of these devices contain unique features that differ when compared with those of conventional op amps. For example, the popular LM386 low-voltage audio amplifier has a gain that is internally fixed at 20 but which can be increased to up to 200 with an external capacitor and resistor place across its gain leads (pins 1 and 8). This device is also designed to drive low-impedance loads, such as an 8-Ω speaker, and runs off a single supply from +4 to +12 V—an ideal range for battery-powered applications. The LM383 is another audio amplifier designed as a power amplifier. It is a high-current device (3.5 A) designed to drive a 4-Ω load (e.g., one 4-Ω speaker or two 8-Ω speakers in parallel). This device also comes with thermal shutdown circuitry and a heat sink. We'll take a closer look at audio amplifiers in Chap. 11.

7.7   Op Amp Specifications

Common-mode rejection ratio (CMRR). The input to a difference amplifier, in general, contains two components: a common-mode and a difference-mode signal. The common-mode signal voltage is the average of the two inputs, whereas the difference-mode signal is the difference between the two inputs. Ideally, an amplifier affects the difference-mode signals only. However, the common-mode signal is also amplified to some degree. The common-mode rejection ratio (CMRR), which is defined as the ratio of the difference signal voltage gain to the common-mode signal voltage gain provides an indication of how well an op amp does at rejecting a signal applied simultaneously to both inputs. The greater the value of the CMRR, the better is the performance of the op amp.

Differential-input voltage range. Range of voltage that may be applied between input terminals without forcing the op amp to operate outside its specifications. If the inputs go beyond this range, the gain of the op amp may change drastically.

Differential input impedance. Impedance measured between the noninverting and inverting input terminals.

Input offset voltage. In theory, the output voltage of an op amp should be zero when both inputs are zero. In reality, however, a slight circuit imbalance within the internal circuitry can result in an output voltage. The input offset voltage is the amount of voltage that must be applied to one of the inputs to zero the output.

Input bias current. Theoretically, an op amp should have an infinite input impedance and therefore no input current. In reality, however, small currents, typically within the nanoamp to picoamp range, may be drawn by the inputs. The average of the two input currents is referred to as the *input bias current.* This current can result in a voltage drop across resistors in the feedback network, the bias network, or source impedance, which in turn can lead to error in the output voltage. Input bias currents depend on the input circuitry of an op amp. With FET op amps, input bias currents are usually small enough not to cause serious offset voltages. Bipolar op amps, on the other hand, may cause problems. With bipolar op amps, a compensation resistor is often required to center the output. I will discuss how this is done in a minute.

Input offset current. This represents the difference in the input currents into the two input terminals when the output is zero. What does this mean? Well, the input terminals of a real op amp tend to draw in different amounts of leakage current, even when the same voltage is applied to them. This occurs because there is always a slight difference in resistance within the input circuitry for the two terminals that originates during the manufacturing process. Therefore, if an op amp's two terminals are both connected to the same input voltage, different amounts of input current will result, causing the output to be offset. Op amps typically come with offset terminals that can be wired to a potentiometer to correct the offset current. I will discuss how this is done in a minute.

Voltage gain (A_V). A typical op amp has a voltage gain of 10^4 to 10^6 (or 80 to 120 dB; 1 dB = 20 $\log_{10} A_V$) at dc. However, the gain drops to 1 at a frequency called the *unity-gain frequency f_T,* typically from 1 to 10 MHz—a result of high-frequency limitations in the op amp's internal circuitry. I will talk more about high-frequency behavior in op amps in a minute.

Output voltage swing. This is the peak output voltage swing, referenced to zero, that can be obtained without clipping.

Slew rate. This represents the maximum rate of change of an op amp's output voltage with time. The limitation of output change with time results from internal or external frequency compensation capacitors slowing things down, which in turn results in delayed output changes with input changes (propagation delay). At high frequencies, the magnitude of an op amp's slew rate becomes more critical. A general-purpose op amp like the 741 has a 0.5 V/μs slew rate—a relatively small value when compared with the high-speed HA2539's slew rate of 600 V/μs.

Supply current. This represents the current that is required from the power supply to operate the op amp with no load present and with an output voltage of zero.

Table 7.1 is a sample op amp specifications table.

TABLE 7.1   **Sample Op Amp Specifications**

TYPE	TOTAL SUPPLY VOLTAGE MAX (V)	MIN (V)	SUPPLY CURRENT (mV)	OFFSET VOLTAGE TYPICAL (mV)	MAX (mV)	CURRENT BIAS MAX (NA)	OFFSET MAX (nA)	SLEW RATE TYPICAL (V/μS)	f_T TYPICAL (MHz)	CMRR MIN (dB)	GAIN MIN (mA)	OUTPUT CURRENT MAX (mA)
Bipolar 741C	10	36	2.8	2	6	500	200	0.5	1.2	70	86	20
MOSFET CA3420A	2	22	1	2	5	0.005	0.004	0.5	0.5	60	86	2
JFET LF411	10	36	3.4	0.8	2	0.2	0.1	15	4	70	88	30
Bipolar, precision LM10	1	45	0.4	0.3	2	20	0.7	0.12	0.1	93	102	20

7.8   Powering Op Amps

Most op amp applications require a dual-polarity power supply. A simple split ±15-V supply that uses a tapped transformer is presented in Chap. 10. If you are using batteries to power an op amp, one of the following arrangements can be used.

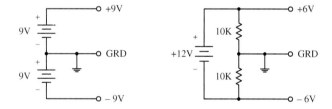

FIGURE 7.23

Now, it is often desirable to avoid split-supplies, especially with small battery-powered applications. One option in such a case is to use a single-supply op amp. However, as I pointed out a second ago, these devices will clip the output if the input attempts to go negative, making them unsuitable for ac-coupled applications. To avoid clipping while still using a single supply, it is possible to take a conventional op amp and apply a dc level to one of the inputs using a voltage-divider network. This, in turn, provides a dc offset level at the output. Both input and output offset levels are referenced to ground (the negative terminal of the battery). With the input offset voltage in place, when an input signal goes negative, the voltage applied to the input of the op amp will dip below the offset voltage but will not go below ground (provided you have set the bias voltage large enough, and provided the input signal is not too large; otherwise, clipping occurs). The output, in turn, will fluctuate about its offset level. To allow for input and output coupling, input and output capacitors are needed. The two circuits in Fig. 7.24 show noninverting and inverting ac-coupled amplifiers (designed for audio) that use conventional op amps that run off a single supply voltage.

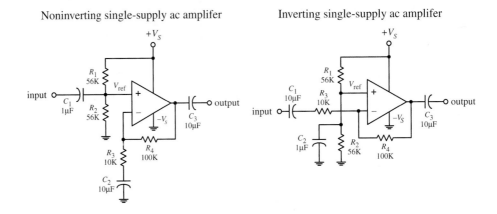

FIGURE 7.24

In the noninverting circuit, the dc offset level is set to one-half the supply voltage by R_1 and R_2 to allow for maximum symmetrical swing. C_1 (and R_2) and C_3 (and R_{load}) act as ac coupling (filtering) capacitors that block unwanted dc components and low-level frequencies. C_1 should be equal to $1/(2\pi f_{3dB}R_1)$, while C_3 should be equal to $1/(2\pi f_{3dB}R_{\text{load}})$, where f_{3dB} is the cutoff frequency (see Chaps. 8 and 11).

When using conventional op amps with single supply voltages, make sure to stay within the minimum supply voltage rating of the op amp, and also make sure to account for maximum output swing limitations and maximum common-mode input range.

7.9   Some Practical Notes

As a note of caution, never reverse an op amp's power supply leads. Doing so can result in a zapped op amp IC. One way to avoid this fate is to place a diode between the op amp's negative supply terminal and the negative supply, as shown in Fig. 7.25.

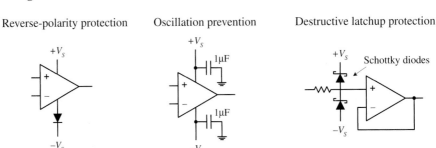

FIGURE 7.25

Also keep wires running from the power supply to the op amp's supply terminals short and direct. This helps prevent unwanted oscillations/noise from arising in the output. Disturbances also may arise from variations in supply voltage. To eliminate these effects, place bypass capacitors from the supply terminals to ground as shown in Fig. 7.25. A 0.1-μF disk capacitor or a 1.0-μF tantalum capacitor should do the trick.

Both bipolar and JFET op amps can experience a serious form of latch if the input signal becomes more positive or negative than the respective op amp power supplies. If the input terminals go more positive than $+V_s + 0.7$ V or more negative than $-V_s - 0.7$ V, current may flow in the wrong direction within the internal circuitry, short-circuiting the power supplies and destroying the device. To avoid this potentially fatal latch-up, it is important to prevent the input terminals of op amps from exceeding the power supplies. This feature has vital consequences during device turn-on; if a signal is applied to an op amp before it is powered, it may be destroyed at the moment power is applied. A "hard wire" solution to this problem involves clamping the input terminals at risk with diodes (preferably fast low-forward-voltage Schottky diodes; see Fig. 7.25). Current-limiting resistors also may be needed to prevent the diode current from becoming excessive. This protection circuitry has some problems, however. Leakage current in the diodes may increase the error. See manufacturers' literature for more information.

7.10 Voltage and Current Offset Compensation

In theory, the output voltage of an op amp should be zero when both inputs are zero. In reality, however, a slight circuit imbalance within the internal circuitry can result in an output voltage (typically within the microvolt to millivolt range). The input offset voltage is the amount of voltage that must be applied to one of the inputs to zero the output—this was discussed earlier. To zero the input offset voltage, manufacturers usually include a pair of offset null terminals. A potentiometer is placed between these two terminals, while the pot's wiper is connected to the more negative supply terminal, as shown in Fig. 7.26. To center the output, the two inputs can be shorted together and an input voltage applied. If the output saturates, the input offset needs trimming. Adjust the pot until the output approaches zero.

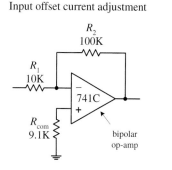

FIGURE 7.26

Notice the resistor placed between the noninverting terminal and ground within the inverting amplifier circuit shown in Fig. 7.26 *right*. What is the resistor used for? It is used to compensate for output voltage errors caused by a voltage drops across R_1 and R_2 as a result of input bias current. As discussed earlier, bipolar op amps tend to have larger input bias currents than FET op amps. With FET op amps, the input bias errors are usually so small (in the picoamp range) that the output voltage error is insignificant, and the compensation resistor is not needed. However, with bipolar op amps, this is not the case (input bias currents in the nanoamp range), and compensation is often necessary. Now, in the inverting amplifier, the bias current—assuming for now that the compensation resistor is missing—introduces a voltage drop equal to $V_{in} = I_{bias}(R_1 \| R_2)$, which is amplified by a factor of $-R_2/R_1$. In order to correct this problem, a compensation resistor with a resistance equal the $R_1 \| R_2$ is placed between the noninverting terminal and ground. This resistor makes the op amp "feel" the same input driving resistance.

7.11   Frequency Compensation

For a typical op amp, the open-loop gain is typically between 10^4 and 10^6 (80 to 120 dB). However, at a certain low frequency, called the *breakover frequency* f_B, the gain drops by 3 dB, or drops to 70.7 percent of the open-loop gain (maximum gain). As the frequency increases, the gain drops further until it reaches 1 (or 0 dB) at a frequency called the *unity-gain frequency* f_T. The unity-gain frequency for an op amp is typically around 1 MHz and is given in the manufacturer's specifications (see Fig. 7.27 *left*). The rolloff in gain as the frequency increases is caused by low-pass filter-like characteristics inherently built into the op amp's inner circuitry. If negative feedback is used, an improvement in bandwidth results; the response is flatter over a wider range of frequencies, as shown in the far-left graph in Fig. 7.27. Now op amps that exhibit an open-loop gain drop of more than 60 dB per decade at f_T are unstable due to phase shifts incurred in the filter-like regions within the interior circuitry. If these phase shifts reach 180° at some frequency at which the gain is greater than 1, negative feedback becomes positive feedback, which results in undesired oscillations (see center and right grafts in Fig. 7.27). To prevent these oscillations, frequency compensation is required. *Uncompensated* op amps can be frequency compensated by connecting an *RC* network between the op amp's frequency-compensation terminals. The network, especially the capacitor, influences the shape of the response curve. Manufacturers will supply you with the response curves, along with the component values of the compensation network for a particular desired response.

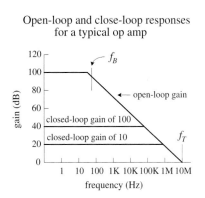
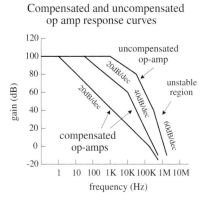
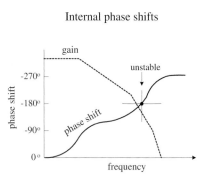

FIGURE 7.27

Perhaps the easiest way to avoid dealing with frequency compensation is to buy an op amp that is internally compensated.

7.12   Comparators

In many situations, it is desirable to know which of two signals is bigger or to know when a signal exceeds a predetermined voltage. Simple circuits that do just this can be constructed with op amps, as shown in Fig. 7.28. In the noninverting comparator circuit, the output switches from low (0 V) to high (positive saturation) when the input voltage exceeds a reference voltage applied to the inverting input. In the inverting comparator circuit, the output switches from high to low when the input exceeds the reference voltage applied to the noninverting terminal. In the far-left circuit, a voltage divider (pot) is used to set the reference voltage.

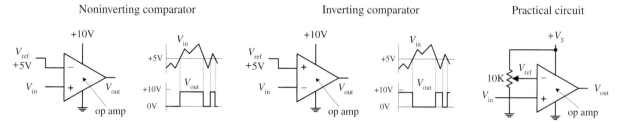

FIGURE 7.28

Another, more popular method for comparing two voltages is to use a special IC called a *comparator*. A comparator, like an op amp, has an inverting input, a noninverting input, an output, and power supply leads—its schematic looks like an op amp, too. However, unlike op amps, comparators are not frequency compensated and therefore cannot be used as linear amplifiers. In fact, comparators never use negative feedback (they often use positive feedback, as you will see). If negative feedback were used with a comparator, its output characteristics would be unstable. Comparators are specifically designed for high-speed switching—they have much larger slew rates and smaller propagation delays than op amps. Another important difference between a comparator and an op amp has to do with the output circuitry. Unlike an op amp, which typically has a push-pull output stage, a comparator uses an internal transistor whose collector is connected to the output and whose emitter is grounded. When the comparator's noninverting terminal is more positive in voltage than the inverting terminal, the output transistor turns on, grounding the output. When the noninverting terminal is made more negative than the inverting terminal, the output transistor is off. In order to give the comparator a high output state when the transistor is off ($V_- < V_+$), an external *pull-up resistor* connected from a positive voltage source to the output is used. The pull-up resistor acts like the collector resistor in a transistor amplifier. The size of the pull-up resistor should be large enough to avoid excessive power dissipation yet small enough to supply enough drive to switch whatever load circuitry is used on the comparator's output. The typical resistance for a pull-up resistor is anywhere from a few hundred to a few thousand ohms. Figure 7.29 shows a simple noninverting and inverting comparator circuit with pull-up resistors included. Both circuits have an output swing from 0 to +5 V.

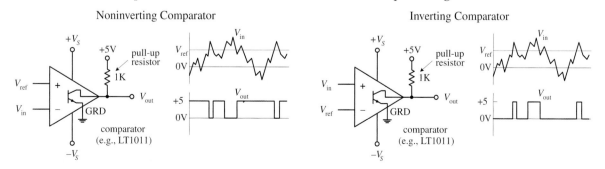

FIGURE 7.29

Comparators are commonly used in analog-to-digital conversion. A typical application might be to connect a magnetic tape sensor or photodiode to an input of a comparator (with reference voltage set at the other input) and allow the sensor to drive the comparator's output to a low or high state suitable for driving logic circuits. Analog-to-digital conversion is discussed in greater detail in Chap. 12.

7.13   Comparators with Hysteresis

Now there is a basic problem with both comparator circuits shown in Fig. 7.29. When a slowly varying signal is present that has a level near the reference voltage, the output will "jitter" or flip back and forth between high and low output states. In many situations having such a "finicky" response is undesirable. Instead, what is usually desired is a small "cushion" region that ignores small signal deviations. To provide the cushion, positive feedback can be added to the comparator to provide hysteresis, which amounts to creating two different threshold voltages or triggering points. The details of how hysteresis works within comparator circuits are given in the following two examples.

7.13.1   Inverting Comparator with Hysteresis

In the inverting comparator circuit shown in Fig. 7.30, positive feedback through R_3 provides the comparator with two threshold voltages or triggering points. The two threshold voltages result from the fact that the reference voltage applied to the noninverting terminal is different when V_{out} is high (+15 V) and when V_{out} is low (0 V)—a result of feedback current. Let's call the reference voltage when the output is high V_{ref1}, and call the reference voltage when V_{out} is low V_{ref2}. Now let's assume that the output is high (transistor is off) and $V_{in} > V_{ref1}$. In order for the output to switch high, V_{in} must be greater than V_{ref1}. However, what is V_{ref1}? It is simply the reference voltage that pops up at the noninverting terminal when the output transistor is off and the output is high (+15 V).

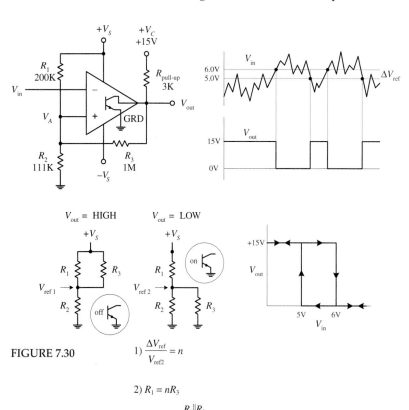

FIGURE 7.30

1) $\dfrac{\Delta V_{ref}}{V_{ref2}} = n$

2) $R_1 = nR_3$

3) $R_2 = \dfrac{R_1 \| R_3}{(+V_S / V_{ref1}) - 1}$

To calculate V_{ref1}, simply use the basic resistor network shown below the main circuit and to the left:

$$V_{ref1} = \frac{+V_S R_2}{(R_1 \| R_3) + R_2} = \frac{+V_S R_2(R_1 + R_2)}{R_1 R_2 + R_1 R_3 + R_2 R_3}$$

If $V_{in} > V_{ref1}$ when the output is already high, the output suddenly goes low—transistor turns on. With the output now low, a new reference voltage V_{ref2} is in place. To calculate V_{ref2}, use the resistor network shown to the right of the first:

$$V_{ref2} = \frac{+V_S R_2 \| R_3}{R_1 + (R_2 \| R_3)} = \frac{+V_S R_2 R_3}{R_1 R_2 + R_1 R_3 + R_2 R_3}$$

When the input voltage decreases to V_{ref2} or lower, the output suddenly goes high. The difference in the reference voltages is called the *hysteresis voltage*, or ΔV_{ref}:

$$\Delta V_{ref} = V_{ref1} - V_{ref2} = \frac{+V_S R_1 R_2}{R_1 R_2 + R_1 R_3 + R_2 R_3}$$

Now, let's try the theory out on a real-life design example.

Say we want to design a comparator circuit with a $V_{\text{ref1}} = +6\,\text{V}$, a $V_{\text{ref2}} = +5\,\text{V}$, and $+V_C = +15\,\text{V}$ that drives a 100-kΩ load. The first thing to do is pick a pull-up resistor. As a rule of thumb,

$$R_{\text{pull-up}} < R_{\text{load}}$$

$$R_3 > R_{\text{pull-up}}$$

Why? Because heavier loading on $R_{\text{pull-up}}$ (smaller values of R_3 and R_{load}) reduce the maximum output voltage, thereby reducing the amount of hysteresis by lowering the value of V_{ref1}. Pick $R_{\text{pull-up}} = 3\,\text{k}\Omega$, and choose R_3 equal to 1 MΩ. Combining equations above gives us the practical formulas below the diagrams. With equation (1), calculate n, which is (6 V − 5 V)/5 V = 0.20. Next, using equation (2), find R_1, which is simply (0.2)(1 M) = 200 kΩ. Using equation (3), find R_3, which is 166 k/(15V/6V − 1) = 111 kΩ. These are the values presented in the circuit.

7.13.2 Noninverting Comparator with Hysteresis

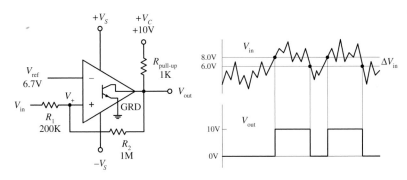

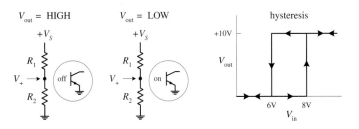

FIGURE 7.31

What's desired: $V_{\text{in1}} = 8\,\text{V}$, $V_{\text{in2}} = 6\,\text{V}$, given $+V_C = 10\,\text{V}$ and load of 100 kΩ. Question: What should V_{ref}, $R_{\text{pull-up}}$, R_2, and R_1 be?

First, chose $R_{\text{pull-up}} < R_{\text{load}}$, and $R_2 > R_{\text{pull-up}}$ to minimize the effects of loading. For $R_{\text{pull-up}}$, choose 1 kΩ, and for R_2, chose 1 M. Next, using the equations to the right, we find R_1 and V_{ref}.

$$\frac{R_1}{R_2} = \frac{\Delta V_{\text{in}}}{V_C} = \frac{10-8}{10} = 0.20$$

$$R_1 = 0.20R_2 = 0.20(1\,\text{M}) = 200\,\text{k}\Omega$$

$$V_{\text{ref}} = \frac{V_{\text{in1}}}{1 + R_1/R_2} = \frac{8\,\text{V}}{1\,\text{V} - 0.20\,\text{V}} = 6.7\,\text{V}$$

Unlike the inverting comparator, the noninverting comparator only requires two resistors for hysteresis to occur. (Extra resistors are needed if you wish to use a voltage divider to set the reference voltage. However, these resistors do not have a direct role in developing the hysteresis voltage.) Also, the terminal to which the input signal is applied is the same location where the threshold shifting occurs—a result of positive feedback. The threshold level applied to the noninverting terminal is shifted about the reference voltage as the output changes from high ($+V_C$) to low (0 V). For example, assume that V_{in} is at a low enough level to keep V_{out} low. For the output to switch high, V_{in} must rise to a triggering voltage, call it V_{in1}, which is found simply by using the resistor network shown to the far left:

$$V_{\text{in1}} = \frac{V_{\text{ref}}(R_1 + R_2)}{R_2}$$

As soon as V_{out} switches high, the voltage at the noninverting terminal will be shifted to a value that's greater than V_{ref} by

$$\Delta V_+ = V_{\text{in}} + \frac{(V_{CC} - V_{\text{in1}})R_1}{R_1 + R_2}$$

To make the comparator switch back to its low state, V_{in} must go below ΔV_+. In other words, the applied input voltage must drop below what is called the lower trip point, V_{in2}:

$$V_{\text{in2}} = \frac{V_{\text{ref}}(R_1 + R_2) - V_{CC}R_1}{R_2}$$

The hysteresis is then simply the difference between V_{in1} and V_{in2}:

$$\Delta V_{\text{in}} = V_{\text{in1}} - V_{\text{in2}} = \frac{V_{CC}R_1}{R_2}$$

A practical design example is presented to the left.

7.14   Using Single-Supply Comparators

Like op amps, comparator ICs come in both dual- and single-supply forms. With single-supply comparators, the emitter and the "negative supply" are joined internally and grounded, whereas the dual-supply comparator has separate emitter (ground) and negative supply leads. A few sample comparator ICs, along with two single-supply comparator circuits are shown below.

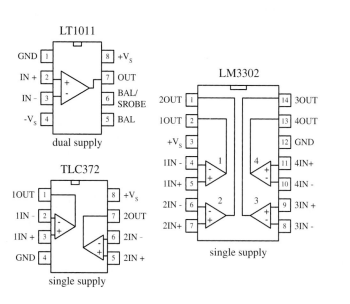

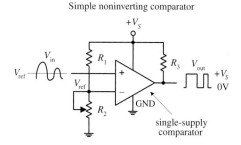

FIGURE 7.32

7.15   Window Comparator

A window comparator is a very useful circuit that changes its output state whenever the input voltage is anywhere between predetermined high and low reference voltages. The region between these two reference voltages is called the *window*. Figure 7.33 shows a simple window comparator built with two comparators (op amps also can be used). In the left-most circuit, the window is set between +3.5 V ($V_{ref,high}$) and +6.5 V($V_{ref,low}$). If V_{in} is below +3.5 V, the lower comparator's output is grounded, while the upper comparator's output floats. Only one ground is needed, however, to make $V_{out} = 0$ V. If V_{in} is above +6.5 V, the upper comparator's output is grounded, while the lower comparator's output floats—again V_{out} goes to 0 V. Only when V_{in} is between +3.5 and +6.5 V will the output go high (+5 V). The right-most circuit uses a voltage-divider network to set the reference voltages.

FIGURE 7.33

Window comparator (using comparators)

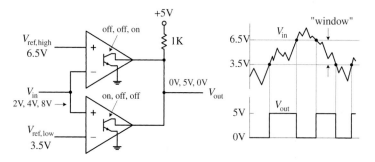

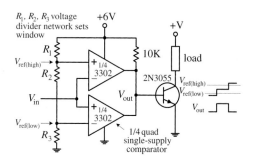

7.16   Voltage-Level Indicator

A simple way to make a voltage-level indicator is to take a number of comparators that share a common input and then supply each comparator with a different reference or triggering voltage, as shown in Fig. 7.34. In this circuit, the reference voltage applied to a comparator increases as you move up the chain of comparators (a result of the voltage-divider network). As the input voltage increases, the lower comparator's output is grounded first (diode turns on), followed in secession by the comparators (LEDs) above it. The potentiometer provides proportional control over all the reference voltages.

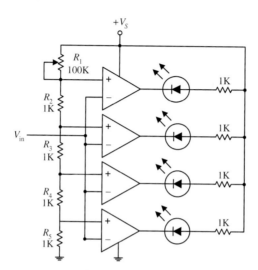

FIGURE 7.34

7.17   Applications

Op Amp Output Drivers (for Loads That Are Either On or Off)

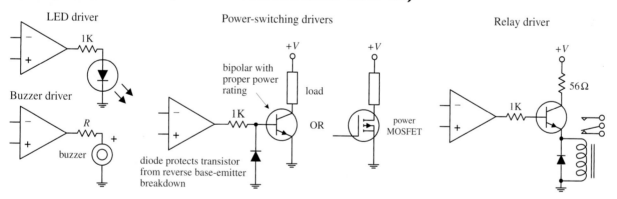

FIGURE 7.35

Comparator Output Drivers

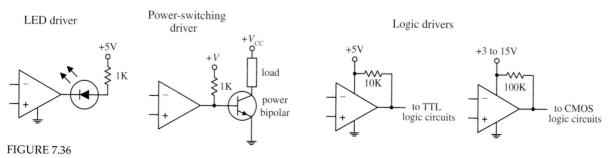

FIGURE 7.36

Op Amp Power Booster (AC Signals)

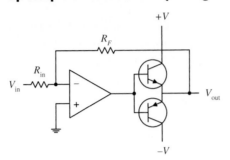

FIGURE 7.37

There are times when an op amp needs a boost in output power-handling capacity while at the same time maintaining both positive and negative output swings. A simple way to increase the output power while maintaining the swing integrity is to attach a complementary transistor push-pull circuit to the op amp's output, as shown in this circuit. At high speeds, additional biasing resistors and capacitors are needed to limit crossover distortion. At low speeds, negative feedback helps eliminate much of the crossover distortion.

Voltage-to-Current Converter

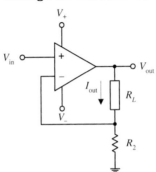

FIGURE 7.38

Here is a simple current source whose output current is determined by an input voltage applied to the noninverting terminal of the op amp. The output current and voltage are determined by the following expressions:

$$V_{out} = \frac{(R_L + R_2)}{R_2} V_{in}$$

$$I_{out} = \frac{V_{out}}{R_1 + R_2} = \frac{V_{in}}{R_2}$$

V_{in} can be set with a voltage divider.

Precision Current Source

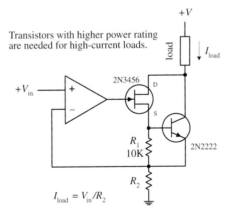

Transistors with higher power rating are needed for high-current loads.

$I_{load} = V_{in}/R_2$

FIGURE 7.39

Here, a precision current uses a JFET to drive a bipolar output transistor used to sink current through a load. Unlike the preceding current source, this circuit is less susceptible to output drift. Use of the JFET helps achieve essentially zero bias current error (a single bipolar output stage will leak base current). This circuit is accurate for output currents larger than the JFET's $I_{DS(on)}$ and provided V_{in} is greater than 0 V. For large currents, the FET-bipolar combination can be replaced with a Darlington transistor, provided its base current does not introduce significant error. The output current or load current is determined by

$$I_{load} = V_{in}/R_2$$

R_2 acts as an adjustment control. Additional compensation may be required depending on the load reactance and the transistors' parameters. Make sure to use transistors of sufficient power ratings to handle the load current in question.

Current-to-Voltage Converter

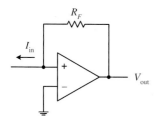

The circuit shown here transforms a current into a voltage. The feedback resistor R_F helps establish a voltage at the inverting input and controls the swing of the output. The output voltage for this circuit is given by

$$V_{out} = I_{in}R_F$$

The light-activated circuits shown below use this principle to generate an output voltage that is proportional to the amount of input current drawn through the light sensor.

photoresistor amplifier   photodiode amplifier   phototransistor amplifier and relay driver

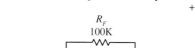
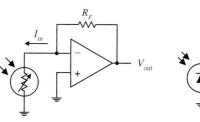
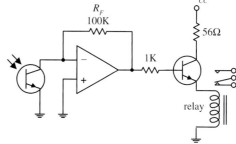

FIGURE 7.40

Overvoltage Protection (Crowbar)

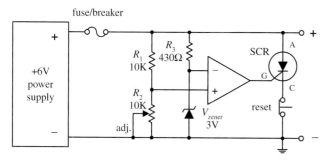

FIGURE 7.41

This circuit acts as a fast-acting overvoltage protection control used to protect sensitive loads from voltage surges generated in the power supply. Initially, let's assume that the supply is doing what it should—generating a constant +6 V. In this case, the voltage applied to the op amp's noninverting input is set to 3 V (by means of the R_1R_2 voltage divider—the pot provides fine-tuning adjustment). At the same time, the inverting input is set to 3 V by means of the 3-V zener diode. The op amp's differential input voltage in this case is therefore zero, making the op amp's output zero (op amp acts as a comparator). With the op amp's output zero, the SCR is off, and no current will pass from anode (A) to cathode to ground. Now, let's say there's a sudden surge in the supply voltage. When this occurs, the voltage at the noninverting input increases, while the voltage at the inverting input remain at 3 V (due to the 3-V zener). This causes the op amp's output to go high, triggering the SCR on and diverting all current from the load to ground in the process. As a result, the fuse (breaker) blows and the load is saved. The switch is opened to reset the SCR.

Programmable-Gain Op Amp

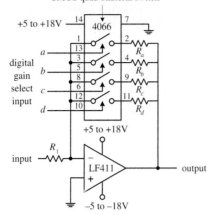

FIGURE 7.42

This circuit is simply an inverting amplifier whose feedback resistance (gain) is selected by means of a digitally controlled bilateral switch (e.g., CMOS 4066). For example, if the bilateral switch's input a is set high (+5 to +18 V) while its b through d inputs are set low (0 V), only resistor R_a will be present in the feedback loop. If you make inputs a through d high, then the effective feedback resistance is equal to the parallel resistance of resistors R_a through R_d. Bilateral switches are discussed in greater detail in Chap. 12.

Sample-and-Hold Circuits

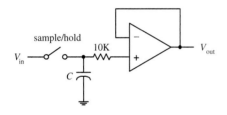

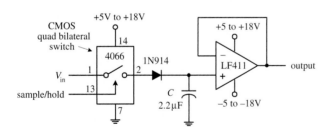

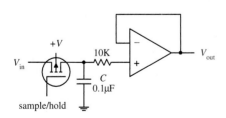

FIGURE 7.43

Sample-and-hold circuits are used to sample an analog signal and hold it so that it can be analyzed or converted into, say, a digital signal at one's leisure. In the first circuit, a switch acts as a sample/hold control. Sampling begins when the switch is closed and ends when the switch is opened. When the switch is opened, the input voltage present at that exact moment will be stored in C. The op amp acts as a unity-gain amplifier (buffer), relaying the capacitor's voltage to the output but preventing the capacitor from discharging (recall that ideally, no current enters the inputs of an op amp). The length of time a sample voltage can be held varies depending on how much current leaks out of the capacitor. To minimize leakage currents, use op amps with low input-bias currents (e.g., FET op amps). In the other two circuits, the sample/hold manual switch is replaced with an electrically controlled switch—the left-most circuit uses a bilateral switch, while the right-most circuit uses a MOSFET. Capacitors best suited for sample/hold applications include Teflon, polyethylene, and polycarbonate dielectric capacitors.

Peak Detectors

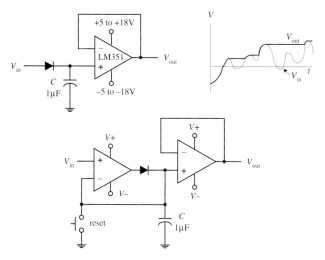

FIGURE 7.44

The circuits shown here act as peak detectors—they follow an incoming voltage signal and store its maximum voltage within C (see graph). The op amp in the upper circuit acts as a buffer—it "measures" the voltage in C, outputs that voltage, and prevents C from discharging. The diode also prevents the capacitor from discharging when the input drops below the peak voltage stored on C. The second circuit is a more practical peak detector. The additional op amp makes the detector more sensitive; it compensates for the diode voltage drop (around 0.6 V) by feeding back C's voltage to the inverting terminal. In other words, it acts as an active rectifier. Also, this circuit incorporates a switch to reset the detector. Often, peak detectors use a FET in place of the diode and use the FET's gate as a reset switch. Reducing the capacitance of C promotes faster responses times for changes in V_{in}.

Noninverting Clipper Amplifier

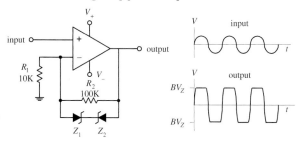

FIGURE 7.45

This simple amplifier circuit acts to clip both positive and negative going portions of the output signal. The clipping occurs in the feedback network whenever the feedback voltage exceeds a zener diode's breakdown voltage. Removing one of the zener diodes results in partial clipping (either positive or negative going, depending on which zener diode is removed). This circuit can be used to limit overloads within audio amplifiers and as a simple sinewave-to-squarewave converter.

Active Rectifiers

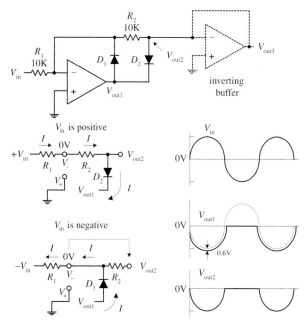

FIGURE 7.46

A single diode acts to rectify a signal—however, at the cost of a diode drop (e.g., 0.6 V). Not only does this voltage drop lower the level of the output, but it also makes it impossible to rectify low-level signals below 0.6 V. A simple solution to this problem is to construct an active rectifier like the one shown here. This circuit acts as an ideal rectifier, in so far as it rectifies signals all the way down to 0 V. To figure out how this circuit works, let's apply the rules we have learned. If V_{in} is positive, current I will flow in the direction shown in the simplified network shown below the main circuit. Since V_+ is grounded, and since we have feedback, $V_- = V_+ = 0$ V (rule 4), we can use Kirchhoff's voltage law to find V_{out1}, V_{out2}, and finally, V_{out3}:

$$0\,V - IR_2 - 0.6\,V - V_{out1} = 0$$

$$V_{out1} = 0\,V - \frac{V_{in}R_2}{R_1} - 0.6\,V = -V_{in} - 0.6\,V$$

$$V_{out2} = V_{out1} + 0.6\,V = -V_{in}$$

$$V_{out3} = V_{out2} = -V_{in}$$

Notice that there is no 0.6-V drop present at the final output; however, the output is inverted relative to the input. Now, if V_{in} is negative, the output will source current through D_1 to bring V_- to 0 V (rule 4). But since no current will pass through R_2 (due to buffer), the 0 V at V_- is present at V_{out2} and likewise present at V_{out3}. The buffer stage is used to provide low output impedance for the next stage without loading down the rectifier stage. To preserve the polarity of the input at the output, an inverting buffer (unity-gain inverter) can be attached to the output.

CHAPTER 8

Filters

A *filter* is a circuit that is capable of passing a specific range of frequencies while blocking other frequencies. As you discovered in Chap. 2, the four major types of filters include *low-pass filters, high-pass filters, bandpass filters,* and *notch filters* (or *band-reject filters*). A low-pass filter passes low-frequency components of an input signal, while a high-pass filter passes high-frequency components. A bandpass filter passes a narrow range of frequencies centered around the filter's resonant frequency, while a notch filter passes all frequencies except those within a narrow band centered around the filter's resonant frequency.

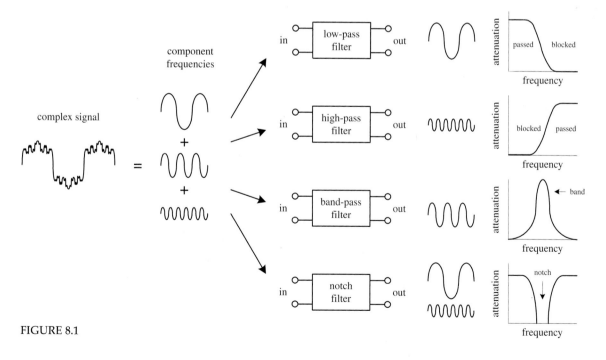

FIGURE 8.1

Filters have many practical applications in electronics. For example, within a dc power supply, filters can be used to eliminate unwanted high-frequency noise present within the ac line voltage, and they act to flatten out pulsing dc voltages generated by the supply's rectifier section. In radio communications, filters make it possible for a radio receiver to provide the listener with only the desired signal while rejecting all others. Likewise, filters allow a radio transmitter to generate only one signal while attenuating other signals that might interfere with different radio transmitters' signals. In audio electronics, filter networks called *crossover networks* are used to divert low audio signals to woofers, middle-range frequencies to midrange speakers, and high frequencies to tweeters. The list of filter applications is extensive.

There are two filter types covered in this chapter, namely, *passive filters* and *active filters*. Passive filters are designed using passive elements (e.g., resistors, capacitors, and inductors) and are most responsive to frequencies between around 100 Hz and 300 MHz. (The lower frequency limit results from the fact that at low frequencies the capacitance and inductance values become exceedingly large, meaning prohibitively large components are needed. The upper frequency limit results from the fact that at high frequencies parasitic capacitances and inductances wreak havoc.) When designing passive filters with very steep attenuation falloff responses, the number of inductor and capacitor sections increases. As more sections are added to get the desired response, the greater is the chance for signal loss to occur. Also, source and load impedances must be taken into consideration when designing passive filters.

Active filters, unlike passive filters, are constructed from op amps, resistors, and capacitors—no inductors are needed. Active filters are capable of handling very low frequency signals (approaching 0 Hz), and they can provide voltage gain if needed (unlike passive filters). Active filters can be designed to offer comparable performance to *LC* filters, and they are typically easier to make, less finicky, and can be designed without the need for large-sized components. Also, with active filters, a desired input and output impedance can be provided that is independent of frequency. One major drawback with active filters is a relatively limited high-frequency range. Above around 100 kHz or so, active filters can become unreliable (a result of the op amp's bandwidth and slew-rate requirements). At radiofrequencies, it is best to use a passive filter.

8.1 Things to Know Before You Start Designing Filters

When describing how a filter behaves, a response curve is used, which is simply an attenuation (V_{out}/V_{in}) versus frequency graph (see Fig. 8.2). As you discovered in Chap. 2, attenuation is often expressed in decibels (dB), while frequency may be expressed in either angular form ω (expressed in rad/s) or conventional form f (expressed in Hz). The two forms are related by $\omega = 2\pi f$. Filter response curves may be plotted on linear-linear, log-linear, or log-log paper. In the case of log-linear graphs, the attenuation need not be specified in decibels.

FIGURE 8.2

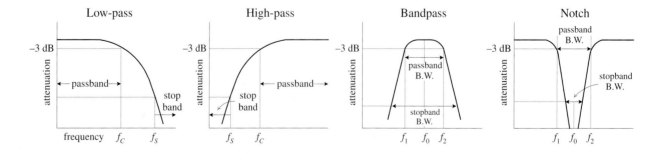

Here are some terms that are commonly used when describing filter response:

–3-dB Frequency (f_{3dB}). This represents the input frequency that causes the output signal to drop to –3 dB relative to the input signal. The –3-dB frequency is equivalent to the cutoff frequency—the point where the input-to-output power is reduced by one-half or the point where the input-to-output voltage is reduce by $1/\sqrt{2}$. For low-pass and high-pass filters, there is only one –3-dB frequency. However, for bandpass and notch filters, there are two –3-dB frequencies, typically referred to as f_1 and f_2.

Center frequency (f_0). On a linear-log graph, bandpass filters are geometrically symmetrical around the filter's resonant frequency or center frequency—provided the response is plotted on linear-log graph paper (the logarithmic axis representing the frequency). On linear-log paper, the central frequency is related to the –3-dB frequencies by the following expression:

$$f_0 = \sqrt{f_1 f_2}$$

For narrow-band bandpass filters, where the ratio of f_2 to f_1 is less than 1.1, the response shape approaches arithmetic symmetry. In this case, we can approximate f_0 by taking the average of –3-dB frequencies:

$$f_0 = \frac{f_1 + f_2}{2}$$

Passband. This represents those frequency signals that reach the output with no more than –3 dB worth of attenuation.

Stop-band frequency (f_s). This is a specific frequency where the attenuation reaches a specified value set by the designer. For low-pass and high-pass filters, the frequencies beyond the stop-band frequency are referred to as the *stop band*. For bandpass and notch filters, there are two stop-band frequencies, and the frequencies between the stop bands are also collectively called the *stop band*.

Quality factor (Q). This represents the ratio of the center frequency of a bandpass filter to the –3-dB bandwidth (distance between –3-dB points f_1 and f_2):

$$Q = \frac{f_0}{f_2 - f_1}$$

For a notch filter, use $Q = (f_2 - f_1)/f_0$, where f_0 is often referred to as the *null frequency*.

8.2 Basic Filters

In Chap. 2 you discovered that by using the reactive properties of capacitors and inductors, along with the resonant behavior of *LC* series and parallel networks, you could create simple low-pass, high-phase, bandpass, and notch filters. Here's a quick look at the basic filters covered in Chap. 2:

Low-pass filters   High-pass filters

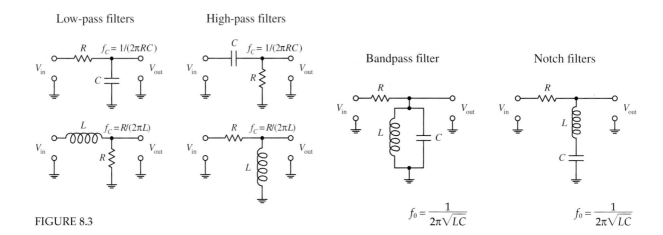

FIGURE 8.3

Now, all the filters shown in this figure have a common limiting characteristic, namely, a shallow 6-dB per octave falloff response beyond the −3-dB point(s). (You can prove this to yourself by going back to Chap. 2 and fiddling with the equations.) In certain noncritical applications, a 6-dB per ocatve falloff works fine, especially in cases where the signals you want to remove are set well beyond the −3-dB point. However, in situations where greater frequency selectivity is needed (e.g., steeper falloffs and flatter passbands), 6-dB per octave filters will not work. What is needed is a new way to design filters.

Making Filters with Sharper Falloff and Flatter Passband Responses

One approach used for getting a sharper falloff would be to combine a number of 6-dB per octave filters together. Each new section would act to filter the output of the preceding section. However, connecting one filter with another for the purpose of increasing the "dB per octave" slope is not as easy as it seems and in fact becomes impractical in certain instances (e.g., narrow-band bandpass filter design). For example, you have to contend with transient responses, phase-shift problems, signal degradation, winding capacitances, internal resistances, magnetic noise pickup, etc. Things can get nasty.

To keep things practical, what I will do is skip the hard-core filter theory (which can indeed get very nasty) and simply apply some design tricks that use basic response graphs and filter design tables. To truly understand the finer points of filter theory is by no means trivial. If you want in-depth coverage of filter theory, refer to a filter design handbook. (A comprehensive handbook written by Zverck covers almost everything you would want to know about filters.)

Let's begin by jumping straight into some practical filter design examples that require varying degrees of falloff response beyond 6 dB per octave. As you go through these examples, important new concepts will surface. First, I will discuss passive filters and then move on to active filters.

8.3   Passive Low-Pass Filter Design

Suppose that you want to design a low-pass filter that has a $f_{3dB} = 3000$ Hz (attenuation is −3 dB at 3000 Hz) and an attenuation of −25 dB at a frequency of 9000 Hz—which will be called the *stop frequency* f_s. Also, let's assume that both the signal-source impedance R_s and the load impedance R_L are equal to 50 Ω. How do you design the filter?

Step 1 (Normalization)

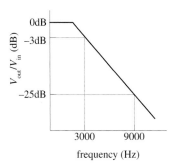

FIGURE 8.4

First, make a rough attenuation versus frequency graph to give yourself a general idea of what the response looks like (see far-left figure). Next, you must normalize the graph. This means that you set the −3-dB frequency f_{3dB} to 1 rad/s. The figure to the near left shows the normalized graph. (The reason for normalizing becomes important later on when you start applying design tricks that use normalized response curves and tables.) In order to determine the normalized stop frequency, simply use the following relation, which is also referred to as the *steepness factor:*

$$A_s = \frac{f_s}{f_{3dB}} = \frac{9000 \text{ Hz}}{3000 \text{ Hz}} = 3$$

This expression tells you that the normalized stop frequency is three times larger than the normalized −3-dB point of 1 rad/s. Therefore, the normalized stop frequency is 3 rad/s.

Step 2 (Pick Response Curve)

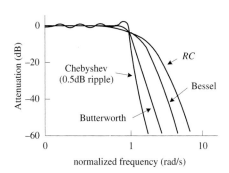

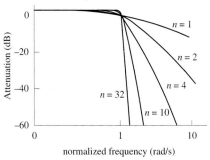

Normalized low-pass Butterworth filter reponse curves

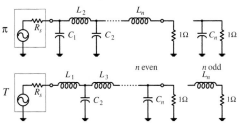

LC low-pass filter networks

FIGURE 8.5

Next, you must pick a filter response type. Three of the major kinds to choose from include the Butterworth, Chebyshev, and Bessel. Without getting too technical here, what is going on is this: Butterworth, Chebyshev, and Bessel response curves are named after individuals who where able to model *LC* filter networks after a mathematical function called the *transfer function,* given here:

$$T(S) = \frac{V_{out}}{V_{in}} = \frac{N_m S^m + N_{m-1} S^{m-1} + \cdots + N_1 S + N_0}{D_n S^n + D_{n-1} S^{n-1} + \cdots + D_1 S + D_0}$$

The N's in the equation are the numerator's coefficients, the D's are the denominator's coefficients, and $S = j\omega (j = \sqrt{-1}, \omega = 2\pi f)$. The highest power n in the denominator is referred to as the order of the filter or the number of poles. The highest power m in the numerator is referred to as the number of zeros. Now, by manipulating this function, individuals (e.g., Butterworth, Chebyshev, and Bessel) were able to generate unique graphs of the transfer function that resembled the attenuation response curves of cascaded *LC* filter networks. What is important to know, for practical purposes, is that the number of poles within the transfer function correlates with the number of *LC* sections present within the cascaded filter network and determines the overall steepness of the response curve (the decibels per octave). As the number of poles increases (number of *LC* sections increases), the falloff response becomes steeper. The coefficients of the transfer function influence the overall shape of the response curve and correlate with the specific capacitor and inductor values found within the filter network. Butterworth, Chebyshev, and Bessel came up with their own transfer functions and figured out what values to place in the coefficients and how to influence the slope of the falloff by manipulating the order of the transfer function. Butterworth figured out a way to manipulate the function to give a maximally flat passband response at the expense of steepness in the transition region between the passband and the stop band. Chebyshev figured out a way to get a very steep transition between the passband and stop band at the expense of ripples present in the passband, while Bessel figured out a way to minimize phase shifts at the expense of both flat passbands and steep falloffs. Later I will discuss the pros and cons of Butterworth, Chebyshev, and Bessell filters. For now, however, let's concentrate on Butterworth filters.

Step 3 (Determine the Number of Poles Needed)

Attenuation curves for Butterworth low-pass filter

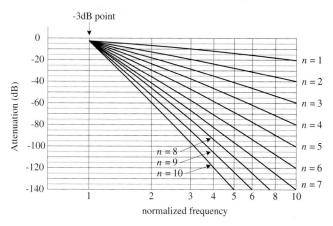

FIGURE 8.6

Continuing on with our low-pass filter problem, let's choose the Butterworth design approach, since it is one of the more popular designs used. The next step is to use a graph of attenuation versus normalized frequency curves for Butterworth low-pass filters, shown in the figure. (Response curves like this are provided in filter handbooks, along with response curves for Chebyshev and Bessel filters.) Next, pick out the single response curve from the graph that provides the desired –25 dB at 3 rad/s, as stated in the problem. If you move your finger along the curves, you will find that the $n = 3$ curve provides sufficient attenuation at 3 rad/s. Now, the filter that is needed will be a third-order low-pass filter, since there are three poles. This means that the actual filter that you will construct will have three LC sections.

Step 4 (Create a Normalized Filter)

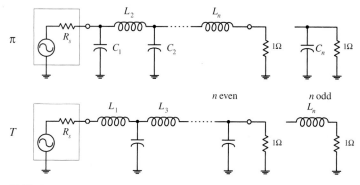

FIGURE 8.7

Now that you have determined the order of the filter, move on to the next step—creating a normalized LC filter circuit. (This circuit will not be the final filter circuit you will use—it will need to be altered.) The circuit networks that are used in this step take on either a π or the T configuration, as shown in the figure. If the source and load impedances match, either configuration can be used—though a π network is more attractive because fewer inductors are needed. However, if the load impedance is greater than the source impedance, it is better to use T configuration. If the load impedance is smaller than the source impedance, it is better to use the π configuration. Since the initial problem stated that the source and load impedances were both 50 Ω, choose the π configuration. The values of the inductors and capacitors are given in Table 8.1. (Filter handbooks will provide such tables, along with tables for Chebyshev and Bessel filters.) Since you need a third-order filter, use the values listed in the $n = 3$ row. The normalized filter circuit you get in this case is shown in Fig. 8.8.

TABLE 8.1   Butterworth Active Filter Low-Pass Values

π {T} n	R_s {$1/R_s$}	C_1 {L_1}	L_2 {C_2}	C_3 {L_3}	L_4 {C_4}	C_5 {L_5}	L_6 {C_6}	C_7 {L_7}
2	1.000	1.4142	1.4142					
3	1.000	1.0000	2.0000	1.0000				
4	1.000	0.7654	1.8478	1.8478	0.7654			
5	1.000	0.6180	1.6180	2.0000	1.6180	0.6180		
6	1.000	0.5176	1.4142	1.9319	1.9319	1.4142	0.5176	
7	1.000	0.4450	1.2470	1.8019	2.0000	1.8019	1.2470	0.4450

Note: Values of L_n and C_n are for a 1-Ω load and −3-dB frequency of 1 rad/s and have units of H and F. These values must be scaled down. See text.

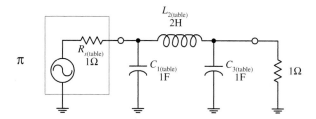

FIGURE 8.8

As mentioned a moment ago, this circuit is not the final circuit that we'll use. That is, the component values listed here will not work! This is so because the graphs and tables you used to get to this point used the normalized frequency within them. Also, you haven't considered the effects of the source and load impedances. In order to construct the final working circuit, you must frequency and impedance scale the component values listed in the circuit in Fig. 8.8. This leads us to the next step.

Step 5 (Frequency and Impedance Scaling)

$$L_{2(actual)} = \frac{R_L L_{2(table)}}{2\pi f_{3dB}} = \frac{(50\Omega)(2\text{ H})}{2\pi(3000\text{ Hz})} = 5.3 \text{ mH}$$

$$C_{1(actual)} = \frac{C_{1(table)}}{2\pi f_{3dB} R_L} = \frac{1\text{ F}}{2\pi(3000\text{ Hz})(50\ \Omega)} = 1.06\ \mu\text{F}$$

$$C_{3(actual)} = \frac{C_{3(table)}}{2\pi f_{3dB} R_L} = \frac{1\text{ F}}{2\pi(3000\text{ Hz})(50\ \Omega)} = 1.06\mu\text{F}$$

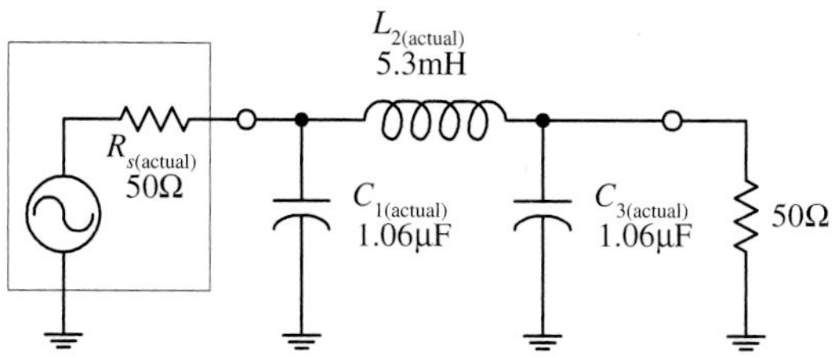

To account for impedance matching of the source and load, as well as getting rid of the normalized frequency, apply the following frequency and impedance scaling rules. To frequency scale, divide the capacitor and inductor values that you got from the table by $\omega = 2\pi f_c$. To impedance scale, multiply resistor and inductor values by the load impedance and divide the capacitor values by the load impedance. In other words, use the following two equations to get the actual component values needed:

$$L_{n(actual)} = \frac{R_L L_{n(table)}}{2\pi f_{3dB}}$$

$$C_{n(actual)} = \frac{C_{n(table)}}{2\pi f_{3dB} R_L}$$

The calculations and the final low-pass circuit are shown in the figure.

FIGURE 8.9

8.4   A Note on Filter Types

It was briefly mentioned earlier that Chebyshev and Bessel filters could be used instead of Butterworth filters. To design Chebyshev and Bessel filters, you take the same approach you used to design Butterworth filters. However, you need to use different low-pass attenuation graphs and tables to come up with the component values placed in the π and T LC networks. If you are interested in designing Chebyshev and Bessel filters, consult a filter design handbook. Now, to give you a better understanding of the differences between the various filter types, the following few paragraphs should help.

Butterworth filters are perhaps the most popular filters used. They have very flat frequency response in the middle passband region, although they have somewhat rounded bends in the region near the –3-dB point. Beyond the –3-dB point, the rate of attenuation increases and eventually reaches $n \times 6$ dB per octave (e.g., $n = 3$, attenuation = 18 dB/octave). Butterworth filters are relatively easy to construct, and the components needed tend not to require as strict tolerances as those of the other filters.

Chebyshev filters (e.g., 0.5-dB ripple, 0.1-dB ripple Chebyshev filter) provide a sharper rate of descent in attenuation beyond the –3-dB point than Butterworth and Bessel filters. However, there is a price to pay for the steep descent—the cost is a ripple voltage within the passband, referred to as the *passband ripple*. The size of the passband ripple increases with order of the filter. Also, Chebyshev filters are more sensitive to component tolerances than Butterworth filters.

Now, there is a problem with Butterworth and Chebyshev filters—they both introduce varying amounts of delay time on signals of different frequencies. In other words, if an input signal consists of a multiple-frequency waveform (e.g., a modulated signal), the output signal will become distorted because different frequencies will be displaced by different delay times. The delay-time variation over the passband is called *delay distortion*, and it increases as the order of the Butterworth and Chebyshev filters increases. To avoid this effect, a Bessel filter can be used. Bessel filters, unlike Butterworth and Chebyshev filters, provide a constant delay over the passband. However, unlike the other two filters, Bessel filters do not have as sharp an attenuation falloff. Having a sharp falloff, however, is not always as important as good signal reproduction at the output. In situations where actual signal reproduction is needed, Bessel filters are more reliable.

8.5   Passive High-Pass Filter Design

Suppose that you want to design a high-pass filter that has an $f_{3dB} = 1000$ Hz and an attenuation of at least –45 dB at 300 Hz—which we call the *stop frequency f_s*. Assume that the filter is hooked up to a source and load that both have impedances of 50 Ω and that a Butterworth response is desired. How do you design the filter? The trick, as you will see in a second, involves treating the high-pass response as an inverted low-pass response, then designing a normalized low-pass filter, applying some conversion tricks on the low-pass filter's components to get a normalized high-pass filter, and then frequency and impedance scaling the normalized high-pass filter.

Frequency Response Curve

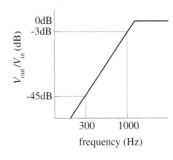

Normalized translation to low-pass filter

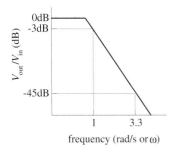

First, make a simple sketch of the response curve for the high-pass filter, as shown in the far-left graph. Next, take the high-pass curve and flip it around in the horizontal direction to get a low-pass response. Then normalize the low-pass response. (This allows you to use the low-pass design techniques. Later you will need to apply a transformation trick on the normalized component values of the low-pass filter to get the desired high-pass filter.) To find the steepness factor A_s and normalized stop-band frequency f_s, follow the same basic procedure as you used in the low-filter example, except now you must take f_{3dB} over f:

$$A_s = \frac{f_{3dB}}{f_s} = \frac{1000\ Hz}{300\ Hz} = 3.3$$

This expression tells us that the normalized stop-band frequency is 3.3 times larger than the normalized –3-dB frequency. Since the normalized graph sets f_{3dB} to 1 rad/s, f_s becomes 3.3 rad/s.

Next, take the low-pass filter response from the preceding step and determine which response curve in Fig. 8.6 provides an attenuation of at least –45 dB at 3.3 rad/s. The $n = 5$ curve does the trick, so you create a fifth-order LC network. Now, the question to ask is, Do you use the π or the T network? Initially, you might assume the π network would be best, since the load and source impedances are equal and since fewer inductors are needed. However, when you apply the transformational trick to get the low-pass filter back to a high-pass filter, you will need to interchange inductors for capacitors and capacitors for inductors. Therefore, if you choose the low-pass T network now, you will get fewer inductors in the final high-pass circuit. The fifth-order normalized low-pass filter network is shown in the figure.

To convert the low-pass into a high-pass filter, replace the inductors with capacitors that have value of $1/L$, and replace the capacitors with inductors that have values of $1/C$. In other words, do the following:

$$L_{2(transf)} = 1/C_{2(table)} = 1/1.6180 = 0.6180\ H$$
$$L_{4(transf)} = 1/C_{4(table)} = 1/1.6180 = 0.6180\ H$$

Start with a "T" low-pass filter...

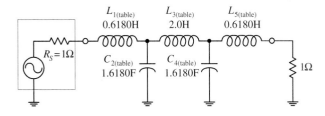

Transform low-pass filter into a high-pass filter...

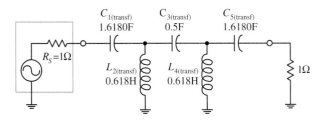

Impedance and frequency scale high-pass filter to get final circuit

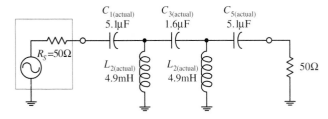

FIGURE 8.10

$$C_{1(transf)} = 1/L_{1(table)} = 1/0.6180 = 1.6180\ F$$
$$C_{3(transf)} = 1/L_{3(table)} = 1/2.0 = 0.5\ F$$
$$C_{5(transf)} = 1/L_{5(table)} = 1/0.6180 = 1.6180\ F$$

Next, frequency and impedance scale to get the actual component values:

$$C_{1(actual)} = \frac{C_{1(trans)}}{2\pi f_{3dB}R_L} = \frac{1.618\ H}{2\pi(1000\ Hz)(50\ \Omega)} = 5.1\ \mu F \qquad L_{2(actual)} = \frac{L_{2(trans)}R_L}{2\pi f_{3dB}} = \frac{(0.6180\ F)(50\ \Omega)}{2\pi(1000\ Hz)} = 4.9\ mH$$

$$C_{3(actual)} = \frac{C_{3(trans)}}{2\pi f_{3dB}R_L} = \frac{0.5\ H}{2\pi(1000\ Hz)(50\ \Omega)} = 1.6\ \mu F \qquad L_{4(actual)} = \frac{L_{4(trans)}R_L}{2\pi f_{3dB}} = \frac{(0.6180\ F)(50\ \Omega)}{2\pi(1000\ Hz)} = 4.9\ mH$$

$$C_{5(actual)} = \frac{C_{5(trans)}}{2\pi f_{3dB}R_L} = \frac{1.618\ H}{2\pi(1000\ Hz)(50\ \Omega)} = 5.1\ \mu F$$

8.6   Passive Bandpass Filter Design

Bandpass filters can be broken down into narrow-band and wide-band types. The defining difference between the two is the ratio between the upper −3-dB frequency f_1 and lower −3-dB frequency f_2. If f_2/f_1 is greater than 1.5, the bandpass filter is placed in the wide-type category. Below 1.5, the bandpass filter is placed in the narrow-band category. As you will see in a moment, the procedure used to design a wide-band bandpass filter differs from that used to design a narrow-band filter.

Wide-Band Design

The basic approach used to design wide-band bandpass filters is simply to combine a low-pass and high-pass filter together. The following example will cover the details. Suppose that you want to design a bandpass filter that has −3-dB points at $f_1 = 1000$ Hz and $f_2 = 3000$ Hz and at least −45 dB at 300 Hz and more than −25 dB at 9000 Hz. Also, again assume that the source and load impedances are both 50 Ω and a Butterworth design is desired.

Bandpass Response Curve

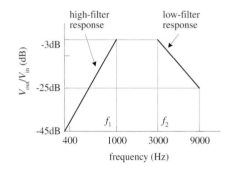

The basic sketch shown here points out the basic response desired. The ratio $f_2/f_1 = 3$, which is larger than 1.5, so you do indeed have a wide-band situation. Notice how the sketch resembles low-pass and high-pass response curves placed together on the same graph. If you break up the response into low- and high-phase curves, you get the following results:

Low-pass	−3 dB at 3000 Hz
	−25 dB at 9000 Hz
High-pass	−3 dB at 1000 Hz
	−45 dB at 300 Hz

Now, to design the wide-band bandpass filter, construct a low-pass and high-pass filter using the values above and the design technique used in the preceding two example problems. Once you have done this, simply cascade the low-pass and high-pass filters together. The nice thing about this problem is the low-pass and high-pass filters that are needed are simply the filters used in the preceding low-pass and high-pass examples. The final cascade network is shown at the bottom of the figure.

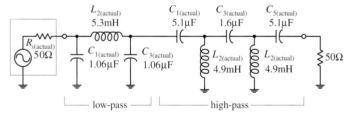

FIGURE 8.11

Narrow-Band Design

Narrow-band filters ($f_2/f_1 < 1.5$), unlike wide-band filters, cannot be made simply by cascading low-pass and high-pass filters together. Instead, you must use a new, slightly tricky procedure. This procedure involves transforming the −3-dB bandwidth ($\Delta f_{BW} = f_2 - f_1$) of a bandpass filter into the −3-dB frequency f_{3dB} of a low-pass filter. At the same time, the stop-band bandwidth of the bandpass filter is transformed into the corresponding stop-band frequency of a low-pass filter. Once this is done, a normalized low-pass filter is created. After the normalized low-pass filter is created, the filter must be frequency scaled in a special way to get the desired bandpass filter.

(The normalized circuit also must be impedance scaled, as before.) When frequency scaling the components of the normalized low-pass filter, do not divide by $\omega = 2\pi f_{3\text{dB}}$, as you would do with low-pass scaling. Instead, the normalized low-pass filter's components are divided by $2\pi(\Delta f_{BW})$. Next, the scaled circuit's branches must be resonated to the bandpass filter's center frequency f_0 by placing additional inductors in parallel with the capacitors and placing additional capacitors in series with the inductors. This creates LC resonant circuit sections. The values of the additional inductors and capacitors are determined by using the LC resonant equation (see Chap. 2 for the details):

$$f_0 = \frac{1}{2\pi\sqrt{LC}}$$

NARROW-BANDWIDTH BANDPASS FILTER EXAMPLE

Suppose that you want to design a bandpass filter with −3-dB points at $f_1 = 900$ Hz and $f_2 = 1100$ Hz and at least −20 dB worth of attenuation at 800 and 1200 Hz. Assume that both the source and load impedances are 50 Ω and that a Butterworth design is desired.

Since $f_2/f_1 = 1.2$, which is less than 1.5, a narrow-band filter is needed. The initial step in designing a narrow-band bandpass filter is to normalize the bandpass requirements. First, the geometric center frequency is determined:

$$f_0 = \sqrt{f_1 f_2} = \sqrt{(900 \text{ Hz})(1100)} = 995 \text{ Hz}$$

Next, compute the two pair of geometrically related stop-band frequencies by using

$$f_a f_b = f_0^2$$

$$f_a = 800 \text{ Hz} \qquad f_b = \frac{f_0^2}{f_a} = \frac{(995 \text{ Hz})^2}{800 \text{ Hz}} = 1237 \text{ Hz} \qquad f_b - f_a = 437 \text{ Hz}$$

$$f_b = 1200 \text{ Hz} \qquad f_a = \frac{f_0^2}{f_b} = \frac{(995 \text{ Hz})^2}{1200 \text{ Hz}} = 825 \text{ Hz} \qquad f_b - f_a = 375 \text{ Hz}$$

Notice how things are a bit confusing. For each pair of stop-band frequencies, you get two new pairs—a result of making things "geometrical" with respect to f_0. Choose the pair having the least separation, which represents the more severe requirement −375 Hz.

The steepness factor for the bandpass filter is given by

$$A_s = \frac{\text{stop-band bandwidth}}{\text{3-dB bandwidth}} = \frac{375 \text{ Hz}}{200 \text{ Hz}} = 1.88$$

Now choose a low-pass Butterworth response that provides at least −20 dB at 1.88 rad/s. According Fig. 8.6, the $n = 3$ curve does the trick. The next step is to create a third-order normalized low-pass filter using the π configuration and Table 8.1.

Next, impedance and frequency scale the normalized low-pass filter to require an impedance level of 50 Ω and a −3-dB frequency equal to the desired bandpass filter's bandwidth ($\Delta f_{BW} = f_2 - f_1$)—which in this example equals 200 Hz. Notice the frequency-scaling trick! The results follow:

$$C_{1(\text{actual})} = \frac{C_{1(\text{table})}}{2\pi(\Delta f_{BW})R_L} = \frac{1 \text{ F}}{2\pi(200 \text{ Hz})(50 \text{ }\Omega)} = 15.92 \text{ }\mu\text{F}$$

$$C_{3(\text{actual})} = \frac{C_{3(\text{table})}}{2\pi(\Delta f_{BW})R_L} = \frac{1 \text{ F}}{2\pi(200 \text{ Hz})(50 \text{ }\Omega)} = 15.92 \text{ }\mu\text{F}$$

$$L_{2(\text{actual})} = \frac{L_{2(\text{table})}R_L}{2\pi(\Delta f_{BW})} = \frac{(2 \text{ H})(50 \text{ }\Omega)}{2\pi(200 \text{ Hz})} = 79.6 \text{ mH}$$

Low-pass bandpass relationship

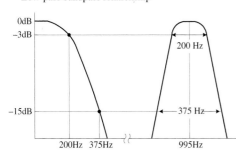

Normalized low-pass response

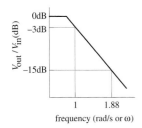

Normalized low-pass filter

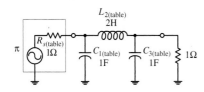

FIGURE 8.12

Impedance and frequency scaled low-pass filter

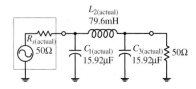

Final bandpass filter

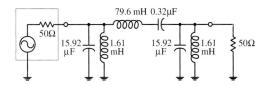

FIGURE 8.12   (*Continued*)

The important part comes now. Each circuit branch of the low-pass filter must be resonated to f_0 by adding a series capacitor to each inductor and a parallel inductor to each capacitor. The LC resonant equation is used to determine the additional component values:

$$L_{(\text{parallel with }C1)} = \frac{1}{(2\pi f_0)^2 C_{1(\text{actual})}} = \frac{1}{(2\pi \cdot 995\text{ Hz})^2 (15.92\text{ }\mu\text{F})} = 1.61\text{ mH}$$

$$L_{(\text{parallel with }C3)} = \frac{1}{(2\pi f_0)^2 C_{3(\text{actual})}} = \frac{1}{(2\pi \cdot 995\text{ Hz})^2 (15.92\text{ }\mu\text{F})} = 1.61\text{ mH}$$

$$C_{(\text{series with }L2)} = \frac{1}{(2\pi f_0)^2 L_{2(\text{actual})}} = \frac{1}{(2\pi \cdot 995\text{ Hz})^2 (79.6\text{ mH})} = 0.32\text{ }\mu\text{F}$$

The final bandpass circuit is shown at the bottom of the figure.

8.7   Passive Notch Filter Design

To design a notch filter, you can apply a technique similar to the one you used in the narrow-band bandpass example. However, now you use a high-pass filter instead of a low-pass filter as the basic building block. The idea here is to relate the notch filter's −3-dB bandwidth ($\Delta f_{BW} = f_1 - f_2$) to the −3-dB frequency of a high-pass filter and relate the notch filter's stop-band bandwidth to the stop-band frequency of a high-pass filter. After that, a normalized high-pass filter is created. This filter is then frequency scaled in a special way—all its components are divided by $2\pi\Delta f_{BW}$. (This circuit also must be impedance scaled, as before.) As with the narrow-band bandpass filter example, the scaled high-pass filter's branches must be resonated to the notch filter's center frequency f_0 by inserting additional series capacitors with existing inductors and inserting additional parallel inductors with existing capacitors.

EXAMPLE

Suppose that you want to design a notch filter with −3-dB points at $f_1 = 800$ Hz and $f_2 = 1200$ Hz and at least −20 dB at 900 and 1100 Hz. Let's assume that both the source and load impedances are 600 Ω and that a Butterworth design is desired.

High-pass bandpass relationship

FIGURE 8.13

First, you find the geometric center frequency:

$$f_0 = \sqrt{f_1 f_2} = \sqrt{(800\text{ Hz})(1200\text{ Hz})} = 980\text{ Hz}$$

Next, compute the two pairs of geometrically related stop-band frequencies:

$$f_a = 900\text{ Hz} \qquad f_b = \frac{f_0^2}{f_a} = \frac{(980\text{ Hz})^2}{900\text{ Hz}} = 1067\text{ Hz}$$

$$f_b - f_a = 1067\text{ Hz} - 900\text{ Hz} = 167\text{ Hz}$$

$$f_b = 1100\text{ Hz} \qquad f_a = \frac{f_0^2}{f_b} = \frac{(980\text{ Hz})^2}{1100\text{ Hz}} = 873\text{ Hz}$$

$$f_b - f_a = 1100\text{ Hz} - 873\text{ Hz} = 227\text{ Hz}$$

Choose the pair of frequencies that gives the more severe requirement—227 Hz.

Normalized low-pass filter

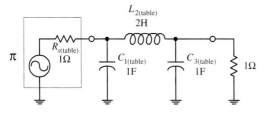

Normalized high-pass filter

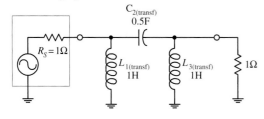

Actual high-pass filter

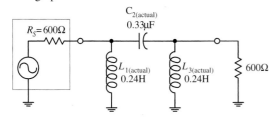

Final bandpass filter

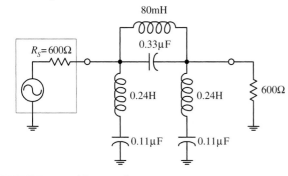

FIGURE 8.13 (*Continued*)

Next, compute the notch filter's steepness factor, which is given by

$$A_s = \frac{\text{3-dB bandwidth}}{\text{stop-band bandwidth}} = \frac{400\ \text{Hz}}{227\ \text{Hz}} = 1.7$$

To come up with the final notch filter design, start out by treating the steepness factor as the steepness factor for a high-pass filter. Next, apply the same tricks you used earlier to construct a high-pass filter. Horizontally flip the high-pass response to get a low-pass response. Then normalize the low-pass response (setting the normalized stop frequency to 1.7 rad/s) and use Fig. 8.6 ($n = 3$ provides at least –20 dB at 1.7 rad/s). Next, use Table 8.1 and the π network to come up with a normalized low-pass design. Then apply the low-pass to high-pass transformational tricks to get a normalized high-pass filter:

$$L_{1(\text{transf})} = 1/C_{1(\text{table})} = 1/1 = 1\ \text{H}$$

$$L_{3(\text{transf})} = 1/C_{3(\text{table})} = 1/1 = 1\ \text{H}$$

$$C_{2(\text{transf})} = 1/L_{2(\text{table})} = 1/2 = 0.5\ \text{F}$$

The first three circuits in the figure show the low-pass to high-pass transformational process.

Next, impedance and frequency scale the normalized high-pass filter to require an impedance level of 600 Ω and a –3-dB frequency equal to the desired notch filter's bandwidth ($\Delta f_{BW} = f_2 - f_1$)—which in the example equals 400 Hz. Notice the frequency-scaling trick! The results follow:

$$L_{1(\text{actual})} = \frac{R_L L_{1(\text{transf})}}{2\pi(\Delta f_{BW})} = \frac{(600\ \Omega)(1\ \text{H})}{2\pi(400\ \text{Hz})} = 0.24\ \text{H}$$

$$L_{3(\text{actual})} = \frac{R_L L_{3(\text{transf})}}{2\pi(\Delta f_{BW})} = \frac{(600\ \Omega)(1\ \text{H})}{2\pi(400\ \text{Hz})} = 0.24\ \text{H}$$

$$C_{2(\text{actual})} = \frac{C_{1(\text{transf})}}{2\pi(\Delta f_{BW})R_L} = \frac{(0.5\ \text{F})}{2\pi(400\ \text{Hz})(600\ \Omega)} = 0.33\ \mu\text{F}$$

And finally, the important modification—resonate each branch to the notch filter's center frequency f_0 by adding a series capacitor to each inductor and a parallel inductor to each capacitor. The values for these additional components must be

$$C_{(\text{series with }L1)} = \frac{1}{(2\pi f_0)^2 L_{1(\text{actual})}} = \frac{1}{(2\pi \cdot 400\ \text{Hz})^2(0.24\ \text{H})} = 0.11\ \mu\text{F}$$

$$C_{(\text{series with }L3)} = \frac{1}{(2\pi f_0)^2 L_{3(\text{actual})}} = \frac{1}{(2\pi \cdot 400\ \text{Hz})^2(0.24\ \text{H})} = 0.11\ \mu\text{F}$$

$$L_{(\text{parallel with }L1)} = \frac{1}{(2\pi f_0)^2 C_{2(\text{actual})}} = \frac{1}{(2\pi \cdot 400\ \text{Hz})^2(0.33\ \mu\text{F})} = 80\ \text{mH}$$

The final circuit is shown at the bottom of the figure.

8.8 Active Filter Design

This section covers some basic Butterworth active filter designs. I already discussed the pros and cons of active filter design earlier in this chapter. Here I will focus on the actual design techniques used to make unity-gain active filters. To begin, let's design a low-pass filter.

8.8.1 Active Low-Pass Filter Example

Frequency response curve

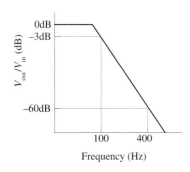

Normalized translation to low-pass filter

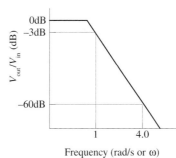

Basic two-pole section

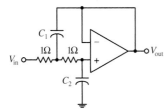

Basic three-pole section

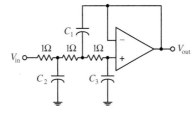

FIGURE 8.14

Suppose that you wish to design an active low-pass filter that has a 3-dB point at 100 Hz and at least 60 dB worth of attenuation at 400 Hz—which we'll call the *stop frequency* f_s.

The first step in designing the filter is to normalize low-pass requirements. The steepness factor is

$$A_s = \frac{f_s}{f_{3dB}} = \frac{400 \text{ Hz}}{100 \text{ Hz}} = 4$$

This means that the normalized position of f_s is set to 4 rad/s. See the graphs in Fig. 8.14. Next, use the Butterworth low-pass filter response curves in Fig. 8.6 to determine the order of filter you need. In this case, the $n = 5$ curve provides over −60 dB at 4 rad/s. In other words, you need a fifth-order filter.

Now, unlike passive filters design, active filter design requires the use of a different set of basic normalized filter networks and a different table to provide the components of the networks. The active filter networks are shown in Fig. 8.14—there are two of them. The one to the left is called a *two-pole section,* while the one on the right is called a *three-pole section.* To design a Butterworth low-pass normalized filter of a given order, use Table 8.2. (Filter handbooks provide Chebyshev and Bessel tables as well.) In this example, a five-pole filter is needed, so according to the table, two sections are required—a three-pole and a two-pole section. These sections are cascaded together, and the component values listed in Table 8.2 are placed next to the corresponding components within the cascaded network. The resulting normalized low-pass filter is shown in Fig. 8.15.

TABLE 8.2   Butterworth Normalized Active Low-Pass Filter Values

ORDER n	NUMBER OF SECTIONS	SECTIONS	C_1	C_2	C_3
2	1	2-pole	1.414	0.7071	
3	1	3-pole	3.546	1.392	0.2024
4	2	2-pole	1.082	0.9241	
		2-pole	2.613	0.3825	
5	2	3-pole	1.753	1.354	0.4214
		2-pole	3.235	0.3090	
6	3	2-pole	1.035	0.9660	
		2-pole	1.414	0.7071	
		2-pole	3.863	0.2588	
7	3	3-pole	1.531	1.336	0.4885
		2-pole	1.604	0.6235	
		2-pole	4.493	0.2225	
8	4	2-pole	1.020	0.9809	
		2-pole	1.202	0.8313	
		2-pole	2.000	0.5557	
		2-pole	5.758	0.1950	

Normalized low-pass filter

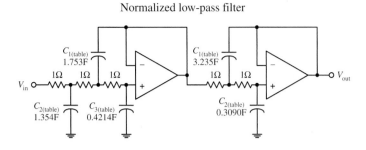

Final low-pass filter

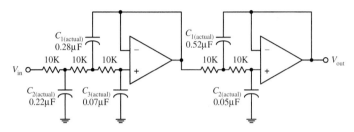

FIGURE 8.15

The normalized filter will provide the correct response, but the component values are impractical—they're too big. In order to bring these values down to size, the circuit must be frequency and impedance scaled. To frequency scale, simply divide the capacitor values by $2\pi f_{3dB}$ (you need not frequency scale the resistors—they aren't reactive). In terms of impedance scaling, you do not have to deal with source/load impedance matching. Instead, simply multiply the normalized filter circuit's resistors by a factor of Z and divide the capacitors by the same factor. The value of Z is chosen to scale the normalized filter components to more practical values. A typical value for Z is 10,000 Ω. In summary, the final scaling rules are expressed as follows:

$$C_{(actual)} = \frac{C_{(table)}}{Z \cdot 2\pi f_{3dB}}$$

$$R_{(actual)} = ZR_{(table)}$$

Taking Z to be 10,000 Ω, you get the final low-pass filter circuit shown at the bottom of the figure.

8.8.2 Active High-Pass Filter Example

The approach used to design active high-pass filters is similar to the approach used to design passive high-pass filters. Take a normalized low-pass filter, transform it into a high-pass circuit, and then frequency and impedance scale it. For example, suppose that you want to design a high-pass filter with a −3-dB frequency of 1000 Hz and 50 dB worth of attenuation at 300 Hz. What do you do?

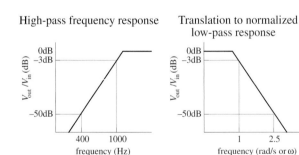

Normalized low-pass filter

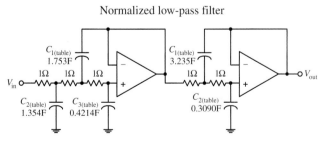

FIGURE 8.16

The first step is to convert the high-pass response into a normalized low-pass response, as shown in the figure. The steepness factor for the low-pass equivalent response is given by

$$A_s = \frac{f_{3dB}}{f_s} = \frac{1000 \text{ Hz}}{300 \text{ Hz}} = 3.3$$

This means that the stop frequency is set to 3.3 rad/s on the normalized graph. The Butterworth response curve shown in Fig. 8.6 tells you that an a fifth-order ($n = 5$) filter will provide the needed attenuation response. Like the last example, a cascaded three-pole/two-pole normalized low-pass filter is required. This filter is shown in Fig. 8.16.

Next, the normalized low-pass filter must be converted into a normalized high-pass filter. To make the conversion, exchange resistors for capacitors that have values of $1/R$ F, and exchange capacitors with resistors that have values of $1/C$ Ω. The second circuit in Fig. 8.16 shows the transformation.

Normalized high-pass filter (transformed low-pass filter)

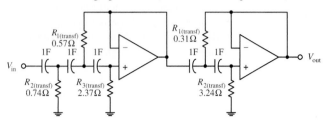

FIGURE 8.12 (*Continued*)

Like the last example problem, to construct the final circuit, the normalized high-pass filter's component values must be frequency and impedance scaled:

$$C_{(actual)} \frac{C_{(transf)}}{Z \cdot 2\pi f_{3dB}}$$

$$R_{(actual)} = ZR_{(transf)}$$

Again, let $Z = 10,000$ Ω. The final circuit is shown in Fig. 8.17.

Final high-pass filter

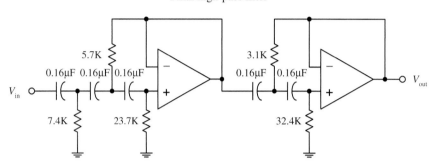

FIGURE 8.17

8.8.3 Active Bandpass Filters

To design an active bandpass filter, it is necessary to determine if a wide-band or narrow-band type is needed. If the upper 3-dB frequency divided by the lower 3-dB frequency is greater than 1.5, the bandpass filter is a wide-band type; below 1.5, it is a narrow-band type. To design a wide-band bandpass filter, simply cascade a high-pass and low-pass active filter together. To design a narrow-band bandpass filter, you have to use some special tricks.

Wide-Band Example

Suppose that you want to design a bandpass filter that has −3-dB points at $f_1 = 1000$ Hz and $f_2 = 3000$ Hz and at least −30 dB at 300 and 10,000 Hz. What do you do?

Normalized low-pass/low-pass inital setup

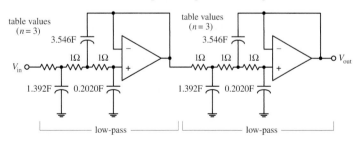

FIGURE 8.18

First, confirm that this is a wide-band situation:

$$\frac{f_2}{f_1} = \frac{3000 \text{ Hz}}{1000 \text{ Hz}} = 3$$

Yes it is—it is greater than 1.5. This means that you simply have to cascade a low-pass and high-pass filter together. Next, the response requirements for the bandpass filter are broken down into low-pass and high-pass requirements:

Low-pass: −3 dB at 3000 Hz

−30 dB at 10,000 Hz

High-pass: −3 dB at 1000 Hz

−30 dB at 300 Hz

Normalized and transformed bandbass filter

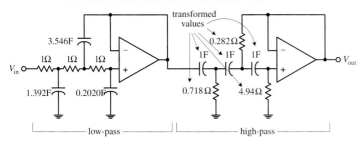

Final bandpass filter

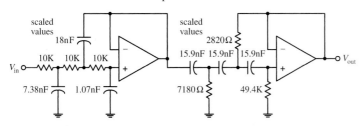

FIGURE 8.18   (*Continued*)

The steepness factor for the low-pass filter is

$$A_s = \frac{f_s}{f_{3dB}} = \frac{10{,}000\text{ Hz}}{3000\text{ Hz}} = 3.3$$

while the steepness factor for the high-pass filter is

$$A_s = \frac{f_{3dB}}{f_s} = \frac{1000\text{ Hz}}{300\text{ Hz}} = 3.3$$

This means that the normalized stop frequencies for both filters will be 3.3 rad/s. Next, use the response curves in Fig. 8.6 to determine the needed filter orders—$n = 3$ provides over -30 dB at 3.3 rad/s. To create the cascaded, normalized low-pass/high-pass filter, follow the steps in the last two examples. The upper two circuits in the figure show the steps involved in this process. To construct the final bandpass filter, the normalized bandpass filter must be frequency and impedance scaled:

Low-pass section:

$$C_{(actual)} = \frac{C_{table}}{Z \cdot 2\pi f_{3dB}} = \frac{C_{table}}{Z \cdot 2\pi(3000\text{ Hz})}$$

High-pass section:

$$C_{(actual)} = \frac{C_{table}}{Z \cdot 2\pi f_{2dB}} = \frac{C_{table}}{Z \cdot 2\pi(1000\text{ Hz})}$$

Choose $Z = 10{,}000$ Ω to provide convenient scaling of the components. In the normalized circuit, resistors are multiplied by a factor of Z. The final bandpass filter is shown at the bottom of the figure.

Narrow-Band Example

Suppose that you want to design a bandpass filter that has a center frequency $f_0 = 2000$ Hz and a -3-dB bandwidth $\Delta f_{BW} = f_2 - f_1 = 40$ Hz. How do you design the filter? Since $f_2/f_1 = 2040\text{ Hz}/1960\text{ Hz} = 1.04$, it is not possible to used the low-pass/high-pass cascading technique you used in the wide-band example. Instead, you must use a different approach. One simple approach is shown below.

Narrow-band filter circuit

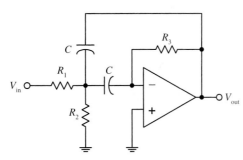

FIGURE 8.19

In this example, simply use the circuit in Fig. 8.19 and some important equations that follow. No detailed discussion will ensue.

First, find the quality factor for the desired response:

$$Q = \frac{f_0}{f_2 - f_1} = \frac{2000\text{ Hz}}{40\text{ Hz}} = 50$$

Next, use the following design equations:

$$R_1 = \frac{Q}{2\pi f_0 C} \qquad R_2 = \frac{R_1}{2Q^2 - 1} \qquad R_3 = 2R_1$$

Final filter circuit

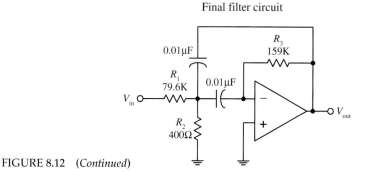

FIGURE 8.12   (*Continued*)

Picking a convenient value for *C*—which we'll set to 0.01 μF—the resistors' values become

$$R_1 = \frac{50}{2\pi(2000 \text{ Hz})(0.01 \text{ μF})} = 79.6 \text{ k}\Omega$$

$$R_2 = \frac{79.6 \text{ k}\Omega}{2(50)^2 - 1} = 400 \text{ }\Omega$$

$$R_3 = 2(79.6 \text{ k}\Omega) = 159 \text{ k}\Omega$$

The final circuit is shown at the bottom of the figure. R_2 can be replaced with a variable resistor to allow for tuning.

8.8.4   *Active Notch Filters*

Active notch filters come in narrow- and wide-band types. If the upper –3-dB frequency divided by the lower –3-dB frequency is greater than 1.5, the filter is called a *wide-band notch filter*—less than 1.5, the filter is called a *narrow-band notch filter*.

Wide-Band Notch Filter Example

To design a wide-band notch filter, simply combine a low-pass and high-pass filter together as shown in Fig. 8.20.

Basic wide-band notch filter

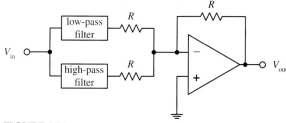

FIGURE 8.20

For example, if you need a notch filter to have – 3-dB points at 500 and 5000 Hz and at least –15 dB at 1000 and 2500 Hz, simply cascade a low-pass filter with a response of

3 dB at 500 Hz

15 dB at 1000 Hz

with a high-pass filter with a response of

3 dB at 5000 Hz

15 dB at 2500 Hz

After that, go through the same low-pass and high-pass design procedures covered earlier. Once these filters are constructed, combine them as shown in the circuit in Fig. 8.20. In this circuit, *R* = 10 k is typically used.

Narrow-Band Notch Filters Example

To design a narrow-band notch filter ($f_2/f_1 < 1.5$), an *RC* network called the *twin-T* (see Fig. 8.21) is frequently used. A deep null can be obtained at a particular frequency with this circuit, but the circuit's *Q* is only ¼. (Recall that the *Q* for a notch filter is given as the center or null frequency divided by the –3-dB bandwidth.) To increase the *Q*, use the active notch filter shown in Fig. 8.22.

Twin-T passive notch filter

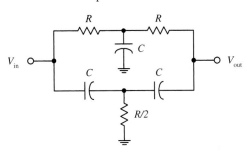

FIGURE 8.21

Improved notch filter

FIGURE 8.22

Like the narrow-bandpass example, let's simply go through the mechanics of how to pick the component values of the active notch filter. Here's an example.

Suppose that you want to make a "notch" at $f_0 = 2000$ Hz and desire a –3-dB bandwidth of $\Delta f_{BW} = 100$ Hz. To get this desired response, do the following: First determine the Q:

$$Q = \frac{\text{"notch" frequency}}{\text{–3-dB bandwidth}} = \frac{f_0}{\Delta f_{BW}} = \frac{2000 \text{ Hz}}{100 \text{ Hz}} = 20$$

The components of the active filter are found be using

$$R_1 = \frac{1}{2\pi f_0 C} \quad \text{and} \quad K = \frac{4Q - 1}{4Q}$$

Now arbitrarily choose R and C; say, let $R = 10$ k and $C = 0.01$ μF. Next, solve for R_1 and K:

$$R_1 = \frac{1}{2\pi f_0 C} = \frac{1}{2\pi (2000 \text{ Hz})(0.01 \text{ μF})} = 7961 \ \Omega$$

$$K = \frac{4Q - 1}{4Q} = \frac{4(20) - 1}{4(20)} = 0.9875$$

Substitute these values into the circuit in Fig. 8.22. Notice the variable potentiometer—it is used to fine-tune the circuit.

8.9 Integrated Filter Circuits

A number of filter ICs are available on the market today. Two of the major categories of integrated filter circuits include the state-variable and switched-capacitor filters IC. Both these filter ICs can be programmed to implement all the second-order functions described in the preceding sections. To design higher-order filters, a number of these ICs can be cascaded together. Typically, all that's needed to program these filter ICs is a few resistors. Using ICs filters allows for great versatility, somewhat simplified design, good precision, and limited design costs. Also, in most applications, frequency and selectivity factors can be adjusted independently.

An example of a state-variable filter IC is the AF100 made by National Semiconductor. This IC can provide low-pass, high-pass, bandpass, and notch filtering capabilities (see Fig. 8.23). Unlike the preceding filters covered in this chapter, the state-variable filter also can provide voltage gain.

FIGURE 8.23

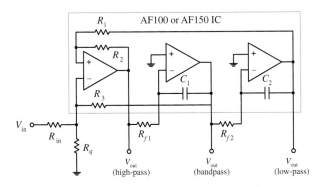

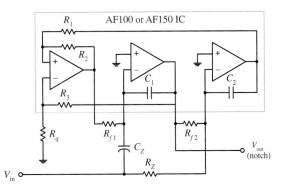

For the AF100, the low-pass gain is set using resistors R_1 and R_{in} (gain = $-R_1/R_{in}$). For the high-pass filter, the gain is set by resistors R_2 and R_{in} (gain = $-R_2/R_{in}$). (The negative sign indicates that the output is inverted relative to the input.) Setting the gain for the bandpass and notch functions is a bit more complex. Other parameters, such as Q, can be tweaked by using design formulas provided by the manufacturer. A good filter design handbook will discuss state-variable filters in detail and will provide the necessary design formulas. Also, check out the electronics catalogs to see what kinds of state-variable ICs exist besides the AF100.

Switched-capacitor filters are functionally similar to the other filters already discussed. However, instead of using external resistors to program the desired characteristics, switched-capacitor filters use a high-frequency capacitor-switching network technology. The capacitor-switching networks act like resistors whose values can be changed by changing the frequency of an externally applied clock voltage. The frequency of the clock signal determines which frequencies are passed and which frequencies get rejected. Typically, a digital clock signal is used to drive the filter—a useful feature if you are looking to design filters that can be altered by digital circuits. An example of a switched-capacitor IC is National Semiconductor's MF5 (see Fig. 8.24). By simply using a few external resistors, a power source, and an applied clock signal, you can program the filter for low-pass, high-pass, and bandpass functions. Again, like the state-variable ICs, manufacturers will provide you with necessary formulas needed for selecting the resistors and the frequency of the clock signal.

FIGURE 8.24

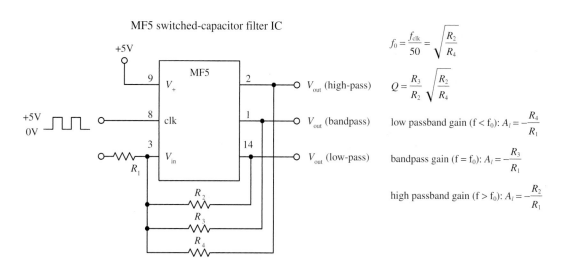

Switched-capacitor filters come in different filter orders. For example, the MF4 is a fourth-order Butterworth low-pass filter, and the MF6 is a sixth-order low-pass Butterworth filter; both are made by National Semiconductor. These two ICs have unity passband gain and require no external components, but they do require a clock input. There are a number of different kinds of switched-capacitor filters out there, made by a number of different manufacturers. Check the catalogs.

As an important note, the periodic clock signal applied to a switched-capacitor filter can generate a significant amount of noise (around 10 to 25 m V) present in the output signal. Typically, this is not of much concern because the frequency of the noise—which is the same as the frequency of the clock—is far removed from the signal band of interest. Usually a simple RC filter can be used to get rid of the problem.

CHAPTER 9

Oscillators and Timers

Within practically every electronic instrument there is an oscillator of some sort. The task of the oscillator is to generate a repetitive waveform of desired shape, frequency, and amplitude that can be used to drive other circuits. Depending on the application, the driven circuit(s) may require either a pulsed, sinusoidal, square, sawtooth, or triangular waveform.

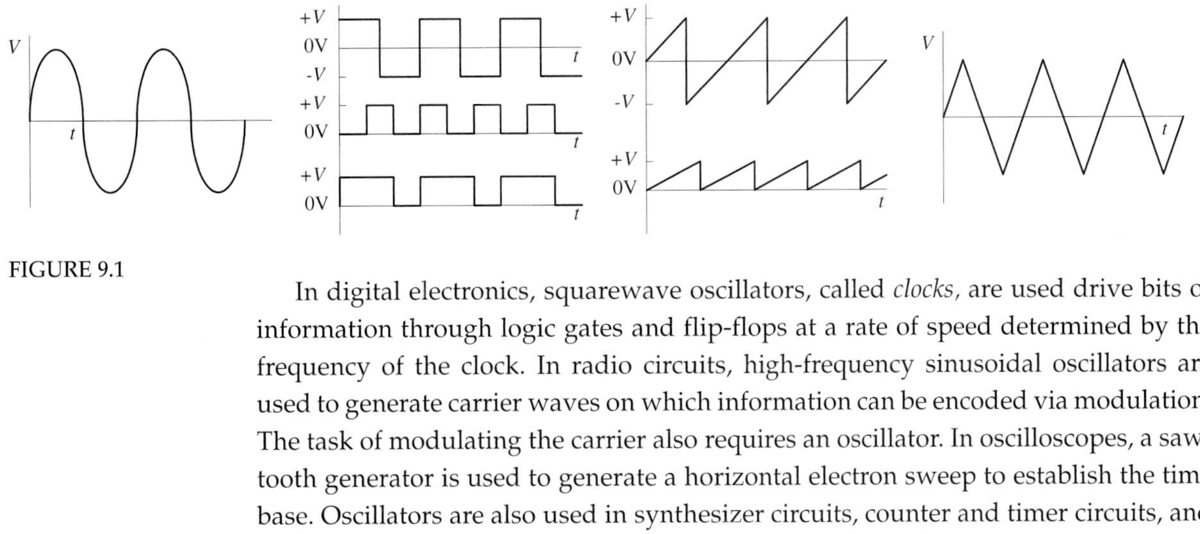

FIGURE 9.1

In digital electronics, squarewave oscillators, called *clocks,* are used drive bits of information through logic gates and flip-flops at a rate of speed determined by the frequency of the clock. In radio circuits, high-frequency sinusoidal oscillators are used to generate carrier waves on which information can be encoded via modulation. The task of modulating the carrier also requires an oscillator. In oscilloscopes, a sawtooth generator is used to generate a horizontal electron sweep to establish the time base. Oscillators are also used in synthesizer circuits, counter and timer circuits, and LED/lamp flasher circuits. The list of applications is endless.

The art of designing good oscillator circuits can be fairly complex. There are a number of designs to choose from and a number of precision design techniques required. The various designs make use of different timing principles (e.g., *RC* charge/discharge cycle, *LC* resonant tank networks, crystals), and each is best suited for use within a specific application. Some designs are simple to construct but may have limited frequency stability. Other designs may have good stability within a certain frequency range, but poor stability outside that range. The shape of the generated waveform is obviously another factor that must be considered when designing an oscillator.

This chapter discusses the major kinds of oscillators, such as the *RC* relaxation oscillator, the Wien-bridge oscillator, the *LC* oscillator, and the crystal oscillator. The chapter also takes a look at popular oscillator ICs.

9.1   *RC* Relaxation Oscillators

Perhaps the easiest type of oscillator to design is the *RC* relaxation oscillator. Its oscillatory nature is explained by the following principle: Charge a capacitor through a resistor and then rapidly discharge it when the capacitor voltage reaches a certain threshold voltage. After that, the cycle is repeated, continuously. In order to control the charge/discharge cycle of the capacitor, an amplifier wired with positive feedback is used. The amplifier acts like a charge/discharge switch—triggered by the threshold voltage—and also provides the oscillator with gain. Figure 9.2 shows a simple op amp relaxation oscillator.

Simple Square-Wave Relaxation Oscillator

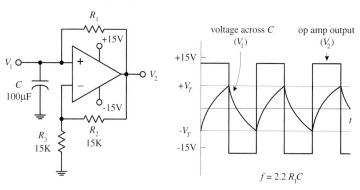

FIGURE 9.2

Assume that when power is first applied, the op amp's output goes toward positive saturation (it is equally likely that the output will go to negative saturation—see Chap. 7 for the details). The capacitor will begin to charge up toward the op amp's positive supply voltage (around +15 V) with a time constant of R_1C. When the voltage across the capacitor reaches the threshold voltage, the op amp's output suddenly switches to negative saturation (around −15 V). The threshold voltage is the voltage set at the inverting input, which is

$$V_T = \frac{R_3}{R_3 + R_2} = \frac{15\ \text{k}\Omega}{15\ \text{k}\Omega + 15\ \text{k}\Omega}(+15\ \text{V}) = +7.5\ \text{V}$$

The threshold voltage set by the voltage divider is now −7.5 V. The capacitor begins discharging toward negative saturation with the same R_1C time constant until it reaches −7.5 V, at which time the op amp's output switches back to the positive saturation voltage. The cycle repeats indefinitely, with a period equal to $2.2R_1C$.

Here's another relaxation oscillator that generates a sawtooth waveform (see Fig. 9.3). Unlike the preceding oscillator, this circuit resembles an op amp integrator network—with the exception of the PUT (programmable unijuction transistor) in the feedback loop. The PUT is the key ingredient that makes this circuit oscillate. Here's a rundown on how this circuit works.

Simple Sawtooth Generator

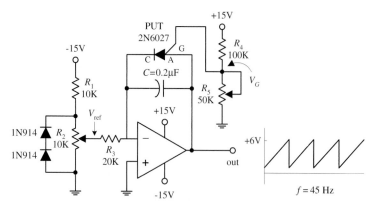

FIGURE 9.3

$$f \cong \frac{V_{\text{ref}}}{R_3C}\left(\frac{1}{V_p - 0.5\text{V}}\right)$$

Let's initially pretend the circuit shown here does not contain the PUT. In this case, the circuit would resemble a simple integrator circuit; when a negative voltage is placed at the inverting input (−), the capacitor charges up at a linear rate toward the positive saturation voltage (+15 V). The output signal would simply provide a one-shot ramp voltage—it would not generate a repetitive triangular wave. In order to generate a repetitive waveform, we must now include the PUT. The PUT introduces oscillation into the circuit by acting as an active switch that turns on (anode-to-cathode conduction) when the anode-to-cathode voltage is greater than its gate voltage. The PUT will remain on until the current through it falls below the minimum holding current rating. This switching action acts to rapidly discharge the capacitor before the output saturates. When the capacitor discharges, the PUT turns off, and the cycle repeats. The gate voltage of the PUT is set via voltage-divider resistors R_4 and R_5. The R_1 and R_2 voltage-divider resistors set the reference voltage at the inverting input, while the diodes help stabilize the voltage across R_2 when it is adjusted to vary the frequency. The output-voltage amplitude is determined by R_4, while the output frequency is approximated by the expression below the figure. (The 0.5 V represents a typical voltage drop across a PUT.)

Here's a simple dual op amp circuit that generates both triangular and square waveforms (see Fig. 9.4). This circuit combines a triangle-wave generator with a comparator.

Simple Triangle-wave/Square-wave Generator

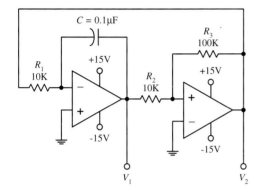

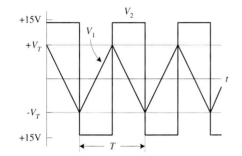

FIGURE 9.4

The rightmost op amp in the circuit acts as a comparator—it is wired with positive feedback. If there is a slight difference in voltage between the inputs of this op amp, the V_2 output voltage will saturate in either the positive or negative direction. For sake of argument, let's say the op amp saturates in the positive direction. It will remain in that saturated state until the voltage at the noninverting input (+) drops below the negative threshold voltage ($-V_T$), at which time V_2 will be driven to negative saturation. The threshold voltage is given by

$$V_T = \frac{V_{sat}}{(R_3 - R_2)}$$

where V_{sat} is a volt or so lower than the op amp's supply voltage (see Chap. 7) Now this comparator is used with a ramp generator (leftmost op amp section). The output of the ramp generator is connected to the input of the comparator, while its output is fed back to the input of the ramp generator. Each time the ramp voltage reaches the threshold voltage, the comparator changes states. This gives rise to oscillation. The period of the output waveform is determined by the R_1C time constant, the saturation voltage, and the threshold voltage:

$$T = \frac{4V_T}{V_{sat}} R_1 C$$

The frequency is $1/T$.

Now op amps are not the only active ingredient used to construct relaxation oscillators. Other components, such as transistors and digital logic gates, can take their place.

Unijunction oscillator

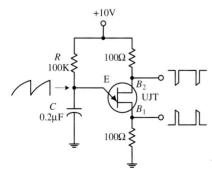

FIGURE 9.5

Here is a unijunction transistor (UJT), along with some resistors and a capacitor, that makes up a relaxation oscillator that is capable of generating three different output waveforms. During operation, at one instant in time, C charges through R until the voltage present on the emitter reaches the UJT's triggering voltage. Once the triggering voltage is exceeded, the E-to-B_1 conductivity increases sharply, which allows current to pass from the capacitor-emitter region through the emitter-base 1 region and then to ground. When this occurs, C suddenly loses its charge, and the emitter voltage suddenly falls below the triggering voltage. After that, the cycle repeats itself. The resulting waveforms generated during this process are shown in the figure. The frequency of oscillation is given by

$$f = \frac{1}{R_E C_E \ln[1/(1-\eta)]}$$

where η is the UJT's intrinsic standoff ratio, which is typically around 0.5. See Chap. 4 for more details.

Here a simple relaxation oscillator is built from a Schmitt trigger inverter IC and an *RC* network. (Schmitt triggers are used to transform slowly changing input waveforms into sharply defined, jitter-free output waveforms (see Chap. 12). When power is first applied to the circuit, the voltage across *C* is zero, and the output of the inverter is high (+5 V). The capacitor starts charging up toward the output voltage via *R*. When the capacitor voltage reaches the positive-going threshold of the inverter (e.g., 1.7 V), the output of the inverter goes low (~0 V). With the output low, *C* discharges toward 0 V. When the capacitor voltage drops below the negative-going threshold voltage of the inverter (e.g., 0.9 V), the output of the inverter goes high. The cycle repeats. The on/off times are determined by the positive- and negative-going threshold voltages and the *RC* time constant.

The third example is a pair of CMOS inverters that are used to construct a simple squarewave *RC* relaxation oscillator. The circuit can work with voltages ranging from 4 to 18 V. The frequency of oscillation is given by

$$f = \frac{1}{4\pi\sqrt{2}RC}$$

R can be adjusted to vary the frequency. I will discuss CMOS inverters in Chap. 12.

Digital oscillator
(using a Schmitt trigger inverter)

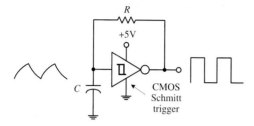

Digital oscillator (using inverters)

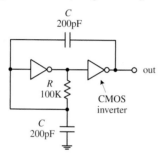

FIGURE 9.5 (*Continued*)

All the relaxation oscillators shown in this section are relatively simple to construct. Now, as it turns out, there is even an easier way to generate basic waveforms. The easy way is to use an IC especially designed for the task. An incredibly popular squarewave-generating chip that can be programmed with resistors and a capacitor is the 555 timer IC.

9.2   The 555 Timer IC

The 555 timer IC is an incredibly useful precision timer that can act as either a timer or an oscillator. In timer mode—better known as *monostable mode*—the 555 simply acts as a "one-shot" timer; when a trigger voltage is applied to its trigger lead, the chip's output goes from low to high for a duration set by an external *RC* circuit. In oscillator mode—better know as *astable mode*—the 555 acts as a rectangular-wave generator whose output waveform (low duration, high duration, frequency, etc.) can be adjusted by means of two external *RC* charge/discharge circuits.

The 555 timer IC is easy to use (requires few components and calculations) and inexpensive and can be used in an amazing number of applications. For example, with the aid of a 555, it is possible to create digital clock waveform generators, LED and lamp flasher circuits, tone-generator circuits (sirens, metronomes, etc.), one-shot timer circuits, bounce-free switches, triangular-waveform generators, frequency dividers, etc.

9.2.1   How a 555 Works (Astable Operation)

Figure 9.6 is a simplified block diagram showing what is inside a typical 555 timer IC. The overall circuit configuration shown here (with external components included) represents the astable 555 configuration.

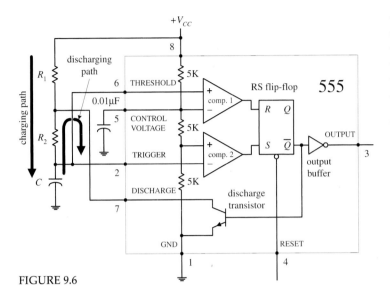

The 555 gets its name from the three 5-kΩ resistors shown in the block diagram. These resistors act as a three-step voltage divider between the supply voltage (V_{CC}) and ground. The top of the lower 5-kΩ resistor (+ input to comparator 2) is set to $\frac{1}{3}V_{CC}$ while the top of the middle 5-kΩ resistor (− input to comparator 2) is set to $\frac{2}{3}V_{CC}$. The two comparators output either a high or low voltage based on the analog voltages being compared at their inputs. If one of the comparator's positive inputs is more positive than its negative input, its output logic level goes high; if the positive input voltage is less than the negative input voltage, the output logic level goes low. The outputs of the comparators are sent to the inputs of an SR flip-flop. The flip-flop looks at the R and S inputs and produces either a high or a low based on the voltage states at the inputs (see Chap. 12).

FIGURE 9.6

Pin 1 (ground).   IC ground.

Pin 2 (trigger).   Input to comparator 2, which is used to set the flip-flop. When the voltage at pin 2 crosses from above to below $\frac{1}{3}V_{CC}$, the comparator switches to high, setting the flip-flop.

Pin 3 (output).   The output of the 555 is driven by an inverting buffer capable of sinking or sourcing around 200 mA. The output voltage levels depend on the output current but are approximately $V_{\text{out(high)}} = V_{CC} - 1.5$ V and $V_{\text{out(low)}} = 0.1$ V.

Pin 4 (reset).   Active-low reset, which forces $\overline{Q}$ high and pin 3 (output) low.

Pin 5 (control).   Used to override the $\frac{2}{3}V_{CC}$ level, if needed, but is usually grounded via a 0.01-μF bypass capacitor (the capacitor helps eliminate V_{CC} supply noise). An external voltage applied here will set a new trigger voltage level.

Pin 6 (threshold).   Input to the upper comparator, which is used to reset the flip-flop. When the voltage at pin 6 crosses from below to above $\frac{2}{3}V_{CC}$, the comparator switches to a high, resetting the flip-flop.

Pin 7 (discharge).   Connected to the open collector of the *npn* transistor. It is used to short pin 7 to ground when $\overline{Q}$ is high (pin 3 low). This causes the capacitor to discharge.

Pin 8 (Supply voltage V_{CC}).   Typically between 4.5 and 16 V for general-purpose TTL 555 timers. (For CMOS versions, the supply voltage may be as low as 1 V.)

In the astable configuration, when power is first applied to the system, the capacitor is uncharged. This means that 0 V is placed on pin 2, forcing comparator 2 high. This in turn sets the flip-flop so that $\overline{Q}$ is high and the 555's output is low (a result of the inverting buffer). With $\overline{Q}$ high, the discharge transistor is turned on, which allows the capacitor to charge toward V_{CC} through R_1 and R_2. When the capacitor voltage exceeds $\frac{1}{3}V_{CC}$, comparator 2 goes low, which has no effect on the SR flip-flop. However, when the capacitor voltage exceeds $\frac{2}{3}V_{CC}$, comparator 1 goes high, resetting the flip-flop and forcing $\overline{Q}$ high and the output low. At this point, the discharge transistor turns on and shorts pin 7 to ground, discharging the capacitor through R_2. When the capacitor's voltage drops below $\frac{1}{3}V_{CC}$, comparator 2's output jumps back to a high level, setting the flip-flop and making $\overline{Q}$ low and the output high. With $\overline{Q}$ low, the transistor turns on, allowing the capacitor to start charging again. The cycle repeats over and over again. The net result is a squarewave output pattern whose voltage level is approximately $V_{CC} - 1.5$ V and whose on/off periods are determined by the C, R_1 and R_2.

9.2.2   Basic Astable Operation

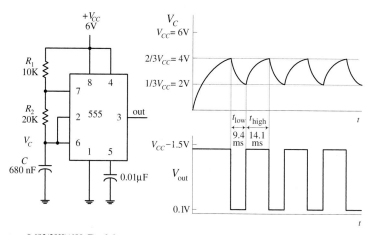

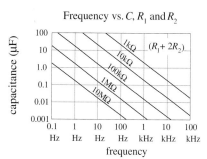

$t_{\text{low}} = 0.693(20\text{K})(680\text{nF}) = 9.6\text{ms}$

$t_{\text{high}} = 0.693(10\text{K} + 20\text{K})(680\text{nF}) = 14.1\text{ms}$

$f = \dfrac{1}{9.6\text{ms} + 14.1\text{ms}} = 42\text{Hz}$

duty cycle $= \dfrac{14.1\text{ms}}{14.1\text{ms} + 9.6\text{ms}} = 0.6$

Frequency vs. C, R_1 and R_2

FIGURE 9.7

When a 555 is set up in astable mode, it has no stable states; the output jumps back and forth. The time duration V_{out} remains low (around 0.1V) is set by the R_1C time constant and the $\frac{1}{3}V_{CC}$ and $\frac{2}{3}V_{CC}$ levels; the time duration V_{out} stays high (around $V_{CC} - 1.5$ V) is determined by the $(R_1 + R_2)C$ time constant and the two voltage levels (see graphs). After doing some basic calculations, the following two practical expression arise:

$t_{\text{low}} = 0.693R_2C$

$t_{\text{high}} = 0.693(R_1 + R_2)C$

The duty cycle (the fraction of the time the output is high) is given by

$$\text{Duty cycle} = \frac{t_{\text{high}}}{t_{\text{high}} - t_{\text{low}}}$$

The frequency of the output waveform is

$$f = \frac{1}{t_{\text{high}} + t_{\text{low}}}$$

For reliable operation, the resistors should be between approximately 10 kΩ and 14 MΩ, and the timing capacitor should be from around 100 pF to 1000 μF. The graph will give you a general idea of how the frequency responds to the component values.

Low-Duty-Cycle Operation (Astable Mode)

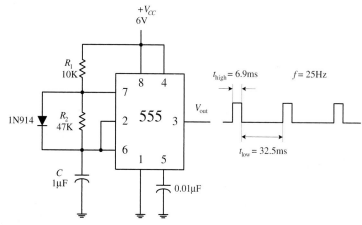

FIGURE 9.8

Now there is a slight problem with the last circuit—you cannot get a duty cycle that is below 0.5 (or 50 percent). In other words, you cannot make t_{high} shorter than t_{low}. For this to occur, the R_1C network (used to generate t_{low}) would have to be larger the $(R_1 + R_2)C$ network (used to generate t_{high}). Simple arithmetic tells us that this is impossible; $(R_1 + R_2)C$ is always greater than R_1C. How do you remedy this situation? You attach a diode across R_2, as shown in the figure. With the diode in place, as the capacitor is charging (generating t_{high}), the preceding time constant $(R_1 + R_2)C$ is reduced to R_1C because the charging current is diverted around R_2 through the diode. With the diode in place, the high and low times become

$$t_{high} = 0.693(10K)(1\mu F) = 6.9ms$$

$$t_{low} = 0.693(47K)(1\mu F) = 32.5ms$$

$$f = \frac{1}{6.9ms + 32.5ms} = 25Hz$$

$$\text{duty cycle} = \frac{6.9ms}{6.9ms + 32.5ms} = 0.18$$

$$t_{high} = 0.693R_1C$$

$$t_{low} = 0.693R_2C$$

To generate a duty cycle of less than 0.5, or 50 percent, simply make R_1 less than R_2.

9.2.3   How a 555 Works (Monostable Operation)

Figure 9.9 shows a 555 hooked up in the monostable configuration (one-shot mode). Unlike the astable mode, the monostable mode has only one stable state. This means that for the output to switch states, an externally applied signal is needed.

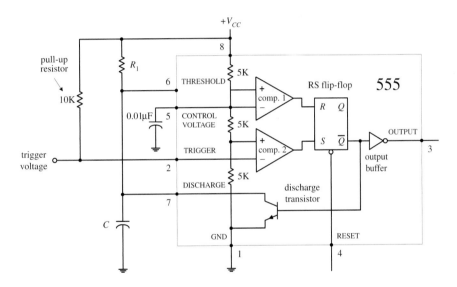

FIGURE 9.9

In the monostable configuration, initially (before a trigger pulse is applied) the 555's output is low, while the discharge transistor in on, shorting pin 7 to ground and keeping C discharged. Also, pin 2 is normally held high by the 10-k pull-up resistor. Now, when a negative-going trigger pulse (less than $\frac{1}{3}V_{CC}$) is applied to pin 2, comparator 2 is forced high, which sets the flip-flop's $\overline{Q}$ to low, making the output high (due to the inverting buffer), while turning off the discharge transistor. This allows C to charge up via R_1 from 0 V toward V_{CC}. However, when the voltage across the capacitor reaches $\frac{2}{3}V_{CC}$, comparator 1's output goes high, resetting the flip-flop and making the output low, while turning on the discharge transistor, allowing the capacitor to quickly discharge toward 0 V. The output will be held in this stable state (low) until another trigger is applied.

9.2.4 Basic Monostable Operation

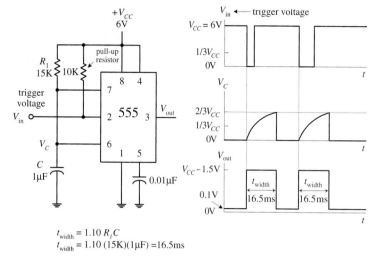

$t_{width} = 1.10\,R_1C$
$t_{width} = 1.10\,(15K)(1\mu F) = 16.5ms$

FIGURE 9.10

The monostable circuit only has one stable state. That is, the output rests at 0 V (in reality, more like 0.1 V) until a negative-going trigger pulse is applied to the trigger lead—pin 2. (The negative-going pulse can be implemented by momentarily grounding pin 2, say, by using a pushbutton switch attached from pin 2 to ground.) After the trigger pulse is applied, the output will go high (around $V_{CC} - 1.5$ V) for the duration set by the R_1C network. Without going through the derivations, the width of the high output pulse is

$$t_{width} = 1.10R_1C$$

For reliable operation, the timing resistor R_1 should be between around 10 kΩ and 14 MΩ, and the timing capacitor should be from around 100 pF to 1000 μF.

9.2.5 Some Important Notes About 555 Timers

555 ICs are available in both bipolar and CMOS types. Bipolar 555s, like the ones you used in the preceding examples, use bipolar transistors inside, while CMOS 555s use MOSFET transistors instead. These two types of 555s also differ in terms of maximum output current, minimum supply voltage/current, minimum triggering current, and maximum switching speed. With the exception of maximum output current, the CMOS 555 surpasses the bipolar 555 in all regards. A CMOS 555 IC can be distinguished from a bipolar 555 by noting whether the part number contains a C somewhere within it (e.g., ICL7555, TLC555, LMC555, etc.). (Note that there are hybrid versions of the 555 that incorporate the best features of both the bipolar and CMOS technologies.) Table 9.1 shows specifications for a few 555 devices.

TABLE 9.1 Sample Specifications for Some 555 Devices

TYPE	SUPPLY VOLTAGE MIN. (V)	SUPPLY VOLTAGE MAX. (V)	SUPPLY CURRENT (V_{CC} = 5 V) TYP. (μA)	SUPPLY CURRENT (V_{CC} = 5 V) MAX. (μA)	TRIG. CURRENT (THRES. CURRENT) TYP. (nA)	TRIG. CURRENT (THRES. CURRENT) MAX. (nA)	TYPICAL FREQUENCY (MHz)	$I_{out,max}$ (V_{CC} = 5 V) SOURCE (mA)	$I_{out,max}$ (V_{CC} = 5 V) SINK (mA)
SN555	4.5	18	3000	5000	100	500	0.5	200	200
ICL7555	2	18	60	300	—	10	1	4	25
TLC555	2	18	170	—	0.01	—	2.1	10	100
LMC555	1.5	15	100	250	0.01	—	3	—	—
NE555	4.5	15	—	6000	—	—	—	—	200

If you need more than one 555 timer per IC, check out the 556 (dual version) and 558 (quad version). The 556 contains two functionally independent 555 timers that share a common supply lead, while the 558 contains four slightly simplified 555 timers. In the 558, not all functions are brought out to the pins, and in fact, this device is intended to be used in monstable mode—although it can be tricked into astable mode with a few alterations (see manufacturer's literature for more information).

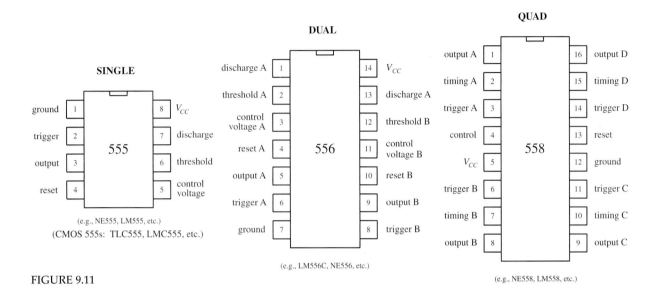

FIGURE 9.11

PRACTICAL TIP

To avoid problems associated with false triggering, connect the 555's pin 5 to ground through a 0.1-µF capacitor (we applied this trick already in this section). Also, if the power supply lead becomes long or the timer does not seem to function for some unknown reason, try attaching a 0.1-µF or larger capacitor between pins 8 and 1.

9.2.6 Simple 555 Applications

Relay Driver (Delay Timer)

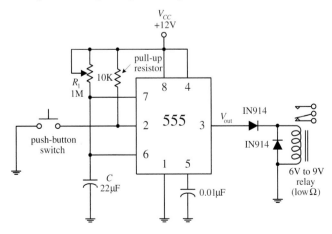

FIGURE 9.12

The monostable circuit shown here acts as a delay timer that is used to actuate a relay for a given duration. With the pushbutton switch open, the output is low (around 0.1 V), and the relay is at rest. However, when the switch is momentarily closed, the 555 begins its timing cycle; the output goes high (in this case ~10.5 V) for a duration equal to

$$t_{\text{delay}} = 1.10 R_1 C$$

The relay will be actuated for the same time duration. The diodes help prevent damaging current surges—generated when the relay switches states—from damaging the 555 IC, as well as the relay's switch contacts.

LED and Lamp Flasher and Metronome

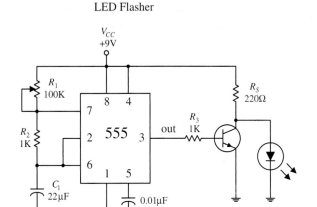

LED Flasher

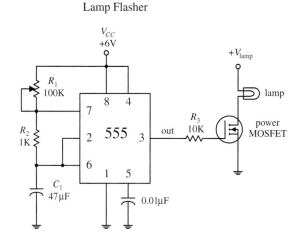

Lamp Flasher

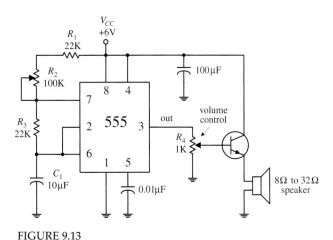

Metronome

All these circuits are oscillator circuits (astable multivibrators). In the LED flasher circuit, a transistor is used to amplify the 555's output in order to provide sufficient current to drive the LED, while R_S is used to prevent excessive current from damaging the LED. In the lamp-flasher circuit, a MOSFET amplifier is used to control current flow through the lamp. A power MOSFET may be needed if the lamp draws a considerable amount of current. The metronome circuit produces a series of "clicks" at a rate determined by R_2. To control the volume of the clicks, R_4 can be adjusted.

FIGURE 9.13

9.3   Voltage-Controlled Oscillators

Besides the 555 timer IC, there are number other voltage-controlled oscillators (VCOs) on the market—some of which provide more than just a squarewave output. For example, the NE566 function generator is a very stable, easy-to-use triangular-wave and squarewave generator. In the 556 circuit below, R_1 and C_1 set the center frequency, while a control voltage at pin 5 varies the frequency; the control voltage is applied by means of a voltage-divider network (R_2, R_3, R_4). The output frequency of the 556 can be determined by using the formula shown in Fig. 9.14.

FIGURE 9.14

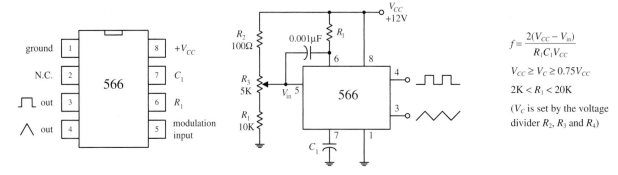

$$f = \frac{2(V_{CC} - V_{in})}{R_1 C_1 V_{CC}}$$

$$V_{CC} \geq V_C \geq 0.75 V_{CC}$$

$$2K < R_1 < 20K$$

(V_C is set by the voltage divider R_2, R_3 and R_4)

Other VCOs, such as the 8038 and the XR2206, can create a trio of output waveforms, including a sine wave (approximation of one, at any rate), a square wave, and triangular wave. Some VCOs are designed specifically for digital waveform generation and may use an external crystal in place of a capacitor for improved stability. To get a feel for what kinds of VCOs are out there, check the electronics catalogs.

9.4 Wien-Bridge and Twin-T Oscillators

A popular RC-type circuit used to generate low-distortion sinusoidal waves at low to moderate frequencies is the Wien-bridge oscillator. Unlike the oscillator circuits discussed already in this chapter, this oscillator uses a different kind of mechanism to provide oscillation, namely, a frequency-selective filter network.

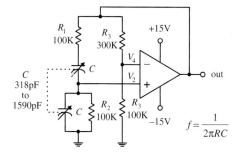

The heart of the Wien-bridge oscillator is its frequency-selective feedback network. The op amp's output is fed back to the inputs in phase. Part of the feedback is positive (makes its way through the frequency-selective RC branch to the noninverting terminal), while the other part is negative (is sent through the resistor branch to the inverting input of the op amp). At a particular frequency $f_0 = 1/(2\pi RC)$, the inverting input voltage (V_4) and the noninverting input voltage (V_2) will be equal and in phase—the positive feedback will cancel the negative feedback, and the circuit will oscillate. At any other frequency, V_2 will be too small to cancel V_4, and the circuit will not oscillate. In this circuit, the gain must be set to +3 V (R_1 and R_2 set the gain). Anything less than this value will cause oscillations to cease; anything more will cause the output to saturate. With the component values listed in the figure, this oscillator can cover a frequency range of 1 to 5 kHz. The frequency can be adjusted by means of a two-ganged variable-capacitor unit.

The second circuit shown in the figure is a slight variation of the first. Unlike the first circuit, the positive feedback must be greater than the negative feedback to sustain oscillations. The potentiometer is used to adjust the amount of negative feedback, while the RC branch controls the amount of positive feedback based on the operating frequency. Now, since the positive feedback is larger than the negative feedback, you have to contend with the "saturation problem," as encountered in the last example. To prevent saturation, two zener diodes placed face to face (or back to back) are connected across the upper 22-kΩ resistor. When the output voltage rises above the zener's breakdown voltage, one or the other zener diode conducts, depending on the polarity of the feedback. The conducting zener diode shunts the 22-kΩ resistor, causing the resistance of the negative feedback circuit to decrease. More negative feedback is applied to the op amp, and the output voltage is controlled to a certain degree.

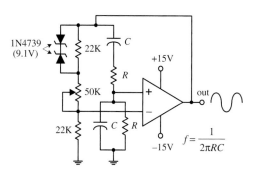

FIGURE 9.15

9.5 *LC* Oscillators (Sinusoidal Oscillators)

When it comes to generating high-frequency sinusoidal waves, commonly used in radiofrequency applications, the most common approach is to use an LC oscillator. The RC oscillators discussed so far have difficulty handling high frequencies, mainly because it is difficult to control the phase shifts of feedback signals send to the amplifier input and because, at high frequencies, the capacitor and resistor values often become impractical to work with. LC oscillators, on the other hand, can use small inductances in conjunction with capacitance to create feedback oscillators that can reach frequencies up to around 500 MHz. However, it is important to note that at low frequencies (e.g., audio range), LC oscillators become highly unwieldy.

LC oscillators basically consist of an amplifier that incorporates positive feedback through a frequency-selective *LC* circuit (or tank). The *LC* tank acts to eliminate from the amplifier's input any frequencies significantly different from its natural resonant frequency. The positive feedback, along with the tank's resonant behavior, acts to promote sustained oscillation within the overall circuit. If this is a bit confusing, envision shock exciting a parallel *LC* tank circuit. This action will set the tank circuit into sinusoidal oscillation at the *LC*'s resonant frequency—the capacitor and inductor will "toss" the charge back and forth. However, these oscillations will die out naturally due to internal resistance and loading. To sustain the oscillation, the amplifier is used. The amplifier acts to supply additional energy to the tank circuit at just the right moment to sustain oscillations. Here is a simple example to illustrate the point.

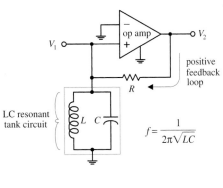

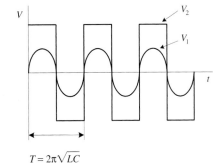

$$T = 2\pi\sqrt{LC}$$

FIGURE 9.16

Here, an op amp incorporates positive feedback that is altered by an *LC* resonant filter or tank circuit. The tank eliminates from the noninverting input of the op amp any frequencies significantly different from the tank's natural resonant frequency:

$$f = \frac{1}{2\pi\sqrt{LC}}$$

(Recall from Chap. 2 that a parallel *LC* resonant circuit's impedance becomes large at the resonant frequency but falls off on either side, allowing the feedback signal to be filtered out to ground.) If a sinusoidal voltage set at the resonant frequency is present at V_1, the amplifier is alternately driven to saturation in the positive and negative directions, resulting in a square wave at the output V_2. This square wave has a strong fundamental Fourier component at the resonant frequency, part of which is fed back to the noninverting input through the resistor to keep oscillations from dying out. If the initially applied sinusoidal voltage at V_1 is removed, oscillations will continue, and the voltage at V_1 will be sinusoidal. Now in practice (considering real-life components and not theoretical models), it is not necessary to apply a sine wave to V_1 to get things going (this is a fundamentally important point to note). Instead, due to imperfections in the amplifier, the oscillator will self-start. Why? With real amplifiers, there is always some inherent noise present at the output even when if the inputs of the amplifier are grounded (see Chap. 7). This noise has a Fourier component at the resonant frequency, and because of the positive feedback, it rapidly grows in amplitude (perhaps in just a few cycles) until the output amplitude saturates.

Now, in practice, *LC* oscillators usually do not incorporate op amps into their designs. At very high frequencies (e.g., RF range), op amps tend to become unreliable due to slew-rate and bandwidth limitations. When frequencies above around 100 kHz are needed, it is essential to use another kind of amplifier arrangement. For high-frequency applications, what is typically used is a transistor amplifier (e.g., bipolar or FET type). The switching speeds for transistors can be incredibly high—a 2000-MHz ceiling is not uncommon for special RF transistors. However, when using a transistor amplifier within an oscillator, there may be a slight problem to contend with—one that you did not have to deal with when you used the op amp. The problem stems from the fact that transistor-like amplifiers often take their outputs at a location where the output happens to be 180° out of phase with its input (see Chap. 4). However, for the feedback to sustain oscillations, the output must be in phase with the input. In certain *LC* oscillators this must be remedied by incorporating a special phase-shifting network between the output and input of the amplifier. Let's take a look at a few popular *LC* oscillator circuits.

Hartley *LC* Oscillator

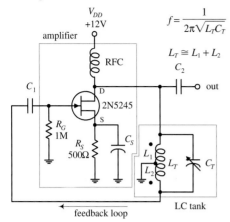

$$f = \frac{1}{2\pi\sqrt{L_T C_T}}$$

$$L_T \cong L_1 + L_2$$

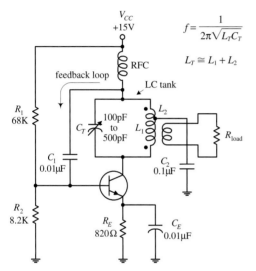

$$f = \frac{1}{2\pi\sqrt{L_T C_T}}$$

$$L_T \cong L_1 + L_2$$

FIGURE 9.17

A Hartley oscillator uses an inductive voltage divider to determine the feedback ratio. The Hartley oscillator can take on a number of forms (FET, bipolar, etc.)—a JFET version is shown here. This oscillator achieves a 180° phase shift needed for positive feedback by means of a tapped inductor in the tank circuit. The phase voltage at the two ends of the inductor differ by 180° with respect to the ground tap. Feedback via L_2 is coupled through C_1 to the base of the transistor amplifier. (The tapped inductor is basically an autotransformer, where L_1 is the primary and L_2 is the secondary.) The frequency of the Hartley is determined by the tank's resonant frequency:

$$f = \frac{1}{2\pi\sqrt{L_T C_T}}$$

This frequency can be adjusted by varying C_T. R_G acts as a gate-biasing resistor to set the gate voltage. R_S is the source resistor. C_S is used to improve amplifier stability, while C_1 and C_2 act as a dc-blocking capacitor that provides low impedance at the oscillator's operating frequency while preventing the transistor's dc operating point from being disturbed. The radiofrequency choke (RFC) aids in providing the amplifier with a steady dc supply while eliminating unwanted ac disturbances.

The second circuit is another form of the Hartley oscillator that uses a bipolar transistor instead of a JFET as the amplifier element. The frequency of operation is again determined by the resonant frequency of the *LC* tank. Notice that in this circuit the load is coupled to the oscillator via a transformer's secondary.

Colpitts *LC* Oscillator

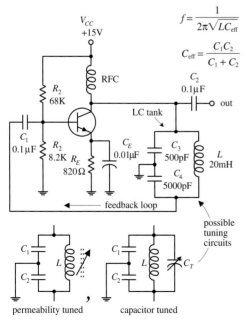

$$f = \frac{1}{2\pi\sqrt{LC_{\text{eff}}}}$$

$$C_{\text{eff}} = \frac{C_1 C_2}{C_1 + C_2}$$

FIGURE 9.18

The Colpitts oscillator is adaptable to a wide range of frequencies and can have better stability than the Hartley. Unlike the Hartley, feedback is obtained by means of a tap between two capacitors connected in series. The 180° phase shift required for sustained oscillation is achieved by using the fact that the two capacitors are in series; the ac circulating current in the *LC* circuit (see Chap. 2) produces voltage drops across each capacitor that are of opposite signs—relative to ground—at any instant in time. As the tank circuit oscillates, its two ends are at equal and opposite voltages, and this voltage is divided across the two capacitors. The signal voltage across C_4 is then connected to the transistor's base via coupling capacitor C_1, which is part of the signal from the collector. The collector signal is applied across C_3 as a feedback signal whose energy is coupled into the tank circuit to compensate for losses. The operating frequency of the oscillator is determined by the resonant frequency of the *LC* tank:

$$f = \frac{1}{2\pi\sqrt{LC_{\text{eff}}}}$$

where C_{eff} is the series capacitance of C_3 and C_4:

$$\frac{1}{C_{\text{eff}}} = \frac{1}{C_1} + \frac{1}{C_2}$$

C_1 and C_2 are dc-blocking capacitors, while R_1 and R_2 act to set the bias level of the transistor. The RFC choke is used to supply steady dc to the amplifier. This circuit's tank can be exchanged for one of the two adjustable tank networks. One tank uses permeability tuning (variable inductor), while the other uses a tuning capacitor placed across the inductor to vary the resonant frequency of the tank.

Clapp Oscillator

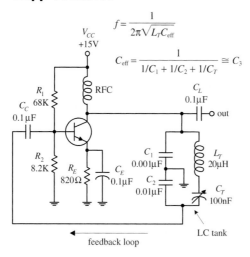

$$f = \frac{1}{2\pi\sqrt{L_T C_{\text{eff}}}}$$

$$C_{\text{eff}} = \frac{1}{1/C_1 + 1/C_2 + 1/C_T} \cong C_3$$

FIGURE 9.19

The Clapp oscillator has exceptional frequency stability. It is a simple variation of the Colpitts oscillator. The total tank capacitance is the series combination of C_1 and C_2. The effective inductance L of the tank is varied by changing the net reactance by adding and subtracting capacitive reactance via C_T from inductive reactance of L_T. Usually C_1 and C_2 are much larger than C_T, while L_T and C_T are series resonant at the desired frequency of operation. C_1 and C_2 determine the feedback ratio, and they are so large compared with C_T that adjusting C_T has almost no effect on feedback. The Clapp oscillator achieves its reputation for stability since stray capacitances are swamped out by C_1 and C_2 meaning that the frequency is almost entirely determined by L_T and C_T. The frequency of operation is determined by

$$f = \frac{1}{2\pi\sqrt{L_T C_{\text{eff}}}}$$

where C_{eff} is

$$C_{\text{eff}} = \frac{1}{1/C_1 + 1/C_2 + 1/C_T} \approx C_3$$

9.6   Crystal Oscillators

When stability and accuracy become critical in oscillator design—which is often the case in high-quality radio and microprocessor applications—one of the best approaches is to use a crystal oscillator. The stability of a crystal oscillator (from around 0.01 to 0.001 percent) is much greater than that of an *RC* oscillator (around 0.1 percent) or an *LC* oscillator (around 0.01 percent at best).

When a quartz crystal is cut in a specific manner and placed between two conductive plates that act as leads, the resulting two-lead device resembles an *RLC* tuned resonant tank. When the crystal is shock-excited by either a physical compression or an applied voltage, it will be set into mechanical vibration at a specific frequency and will continue to vibrate for some time, while at the same time generating an ac voltage between its plates. This behavior, better know as the *piezoelectric effect*, is similar to the damped electron oscillation of a shock-excited *LC* circuit. However, unlike an *LC* circuit, the oscillation of the crystal after the initial shock excitation will last longer—a result of the crystal's naturally high *Q* value. For a high-quality crystal, a *Q* of 100,000 is not uncommon. *LC* circuits typically have a *Q* of around a few hundred.

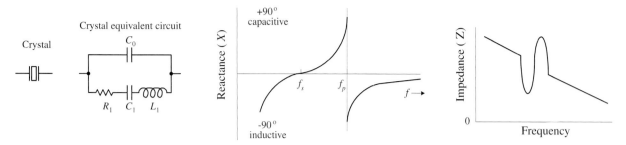

FIGURE 9.20

The *RLC* circuit shown in Fig. 9.20 is used as an equivalent circuit for a crystal. The lower branch of the equivalent circuit, consisting of R_1, C_1, and L_1 in series, is called the *motional arm*. The motional arm represents the series mechanical resonance of the crystal. The upper branch containing C_0 accounts for the stray capacitance in the crystal holder and leads. The *motional inductance* L_1 is usually many henries in size, while the motional capacitance C_1 is very small (<<1 pF). The ratio of L_1 to C_1 for a crystal is much higher than could be achieved with real inductors and capacitors. Both the internal resistance of the crystal R_1 and the value of C_0 are both fairly small. (For a 1-MHz crystal, the typical components values within the equivalent circuit would be $L_1 = 3.5$ H, $C_1 = 0.007$ pF, $R_1 = 340$ Ω, $C_0 = 3$ pF. For a 10-MHz fundamental crystal, the typical values would be $L_1 = 9.8$ mH, $C_1 = 0.026$ pF, $R_1 = 7$ Ω, $C_0 = 6.3$ pF.)

In terms of operation, a crystal can be driven at *series resonance* or *parallel resonance*. In series resonance, when the crystal is driven at a particular frequency, called the *series resonant frequency* f_s, the crystal resembles a series-tuned resonance *LC* circuit; the impedance across it goes to a minimum—only R_1 remains. In parallel resonance, when the crystal is driven at what is called the *parallel resonant frequency* f_p, the crystal resembles a parallel-tuned *LC* tank; the impedance across it peaks to a high value (see the graphs in Fig. 9.20).

Quartz crystals come in series-mode and parallel-mode forms and may either be specified as a fundamental-type or an overtone-type crystal. Fundamental-type crystals are designed for operation at the crystal's fundamental frequency, while overtone-type crystals are designed for operation at one of the crystal's overtone frequencies. (The fundamental frequency of a crystal is accompanied by harmonics or overtone modes, which are odd multiples of the fundamental frequency. For example, a crystal with a 15-MHz fundamental also will have a 45-MHz third overtone, a 75-MHz fifth overtone, a 135-MHz ninth overtone, etc. Figure 9.21 below shows an equivalent *RLC* circuit for a crystal, along with a response curve, both of which take into account the overtones.) Fundamental-type crystals are available from around 10 kHz to 30 MHz, while overtone-type crystals are available up to a few hundred megahertz. Common frequencies available are 100 kHz and 1.0, 2.0, 4, 5, 8, and 10 MHz.

FIGURE 9.21

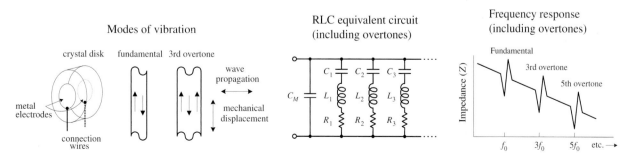

Designing crystal oscillator circuits is similar to designing LC oscillator circuits, except that now you replace the LC tank with a crystal. The crystal will supply positive feedback and gain at its series or parallel resonant frequency, hence leading to sustained oscillations. Here are a few basic crystal oscillator circuits to get you started.

Basic Crystal Oscillator

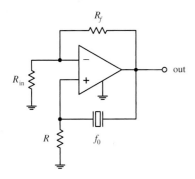

The simple op amp circuit shown here resembles the LC oscillator circuit in Fig. 9.16, except that it uses the series resonance of the crystal instead of the parallel resonance of an LC circuit to provide positive feedback at the desired frequency. Other crystal oscillators, such as the Pierce oscillator, Colpitts oscillator, and a CMOS inverter oscillator, shown below, also incorporate a crystal as a frequency-determining component. The Pierce oscillator, which uses a JFET amplifier stage, employs a crystal as a series-resonant feedback element; maximum positive feedback from drain to gate occurs only at the crystal's series-resonant frequency. The Colpitts circuit, unlike the Pierce circuit, uses a crystal in the parallel feedback arrangement; maximum base-emitter voltage signal occurs at the crystal's parallel-resonant frequency. The CMOS circuit uses a pair of CMOS inverters along with a crystal that acts as a series-resonant feedback element; maximum positive feedback occurs at the crystal's series resonant frequency.

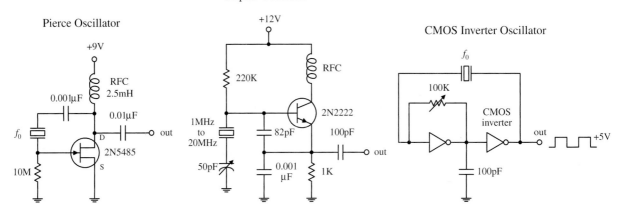

FIGURE 9.22

There are a number of ICs available that can make designing crystal oscillators a breeze. Some of these ICs, such as the 74S124 TTL VCO (squarewave generator), can be programmed by an external crystal to output a waveform whose frequency is determined by the crystal's resonant frequency. The MC12060 VCO, unlike the 74S124, outputs a pair of sine waves. Check the catalogs to see what other types of oscillator ICs are available.

Now there are also crystal oscillator modules that contain everything (crystal and all) in one single package. These modules resemble a metal-like DIP package, and they are available in many of the standard frequencies (e.g., 1, 2, 4, 5, 6, 10, 16, 24, 25, 50, and 64 MHz, etc.). Again, check out the electronics catalogs to see what is available.

CHAPTER 10

Voltage Regulators and Power Supplies

Circuits usually require a dc power supply that can maintain a fixed voltage while supplying enough current to drive a load. Batteries make good dc supplies, but their relatively small current capacities make them impractical for driving high-current, frequently used circuits. An alternative solution is to take a 120-V ac, 60-Hz line voltage and convert it into a usable dc voltage. The trick to converting the ac line voltage into a usable (typically lower-level) dc voltage is to first use a transformer to step down the ac voltage. After that, the transformed voltage is applied through a rectifier network to get rid of the negative swings (or positive swings if you are designing a negative voltage supply). Once the negative swings are eliminated, a filter network is used to flatten out the rectified signal into a nearly flat (rippled) dc voltage pattern. Figure 10.1 show the process in action.

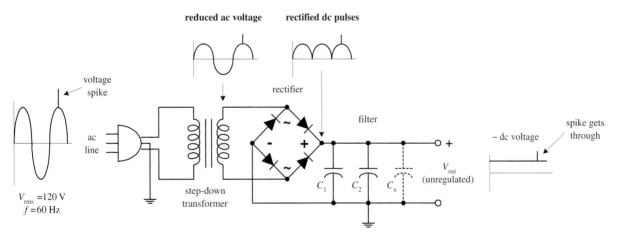

FIGURE 10.1

Now there is one problem with this supply—it is *unregulated.* This means that if there are any sudden surges within the ac input voltage (spikes, dips, etc.), these variations will be expressed at the supply's output (notice the spike that gets through in Fig. 10.1). Using an unregulated supply to run sensitive circuits (e.g., digital IC circuits) is a bad idea. The current spikes can lead to improper operating characteristics (e.g., false triggering, etc.) and may destroy the ICs in the process. An unregulated supply also has a problem maintaining a constant output voltage as the load resistance changes. If a highly resistive (low-current) load is replaced with a lower-resistance (high-current) load, the unregulated output voltage will drop (Ohm's law).

Fortunately, there is a special circuit that can be placed across the output of an unregulated supply to convert it into a regulated supply—a supply that eliminates the spikes and maintains a constant output voltage with load variations (see Fig. 10.2). This special circuit is called a *voltage regulator.*

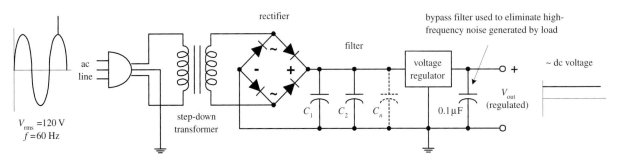

FIGURE 10.2

A voltage regulator is designed to automatically adjust the amount of current flowing through a load—so as to maintain a constant output voltage—by comparing the supply's dc output with a fixed or programmed internal reference voltage. A simple regulator consists of a sampling circuit, an error amplifier, a conduction element, and a voltage reference element (see Fig. 10.3).

The regulator's sampling circuit (voltage divider) monitors the output voltage by feeding a *sample voltage* back to the error amplifier. The reference voltage element (zener diode) acts to maintain a constant *reference voltage* that is used by the error amplifier. The error amplifier compares the output sample voltage with the reference voltage and then generates an *error voltage* if there is any difference between the two. The error amplifier's output is then fed to the current-control element (transistor), which is used to control the load current.

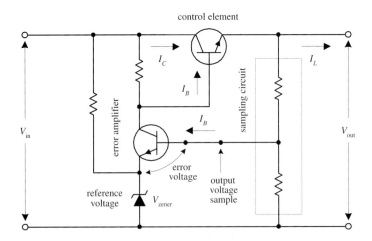

FIGURE 10.3

In practice, you do not have to worry about designing voltage-regulator circuits from scratch. Instead, what you do is spend 50 cents for a voltage-regulator IC. Let's take a closer look at these integrated devices.

10.1   Voltage-Regulator ICs

There are a number of different kinds of voltage-regulator ICs on the market today. Some of these devices are designed to output a fixed positive voltage, some are designed to output fixed negative voltage, and others are designed to be adjustable.

10.1.1   Fixed Regulator ICs

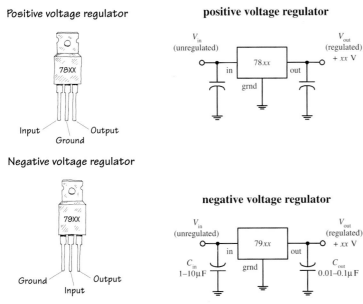

FIGURE 10.4

One popular line of regulators includes the three-terminal LM78xx series shown here. The "xx" digits represent the output voltage, e.g., 7805 (5 V), 7806 (6 V), 7808 (8 V), 7810 (10 V), 7812 (12 V), 7815 (15 V), 7818 (18 V), and 7824 (24 V). These devices can handle a maximum output current of 1.5 A if properly heat-sunk. To remove unwanted input or output spikes/noise, capacitors can be attached to the regulator's input and output terminals, as shown in the figure. A popular series of negative voltage-regulator IC is the LM79xx regulators, where "xx" represent the negative output voltage. These devices can handle a maximum output current of 1.5 A. A number of different manufacturers make their own kinds of voltage regulators. Some of the regulators can handle more current than others. Check out the catalogs to see what is available.

10.1.2   Adjustable Regulator ICs

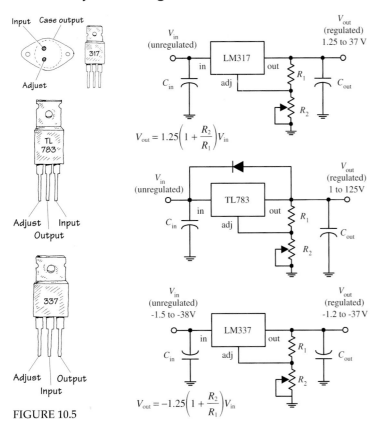

FIGURE 10.5

The LM317 regulator shown here is a popular adjustable positive voltage regulator. Unlike the 7800 fixed voltage-regulator series, the LM317's reference-voltage terminal (called the *adjust terminal*) floats. By applying a reference voltage—relative to its output voltage—to its adjust terminal, the regulator's output can be altered. The reference voltage, or *adjust voltage*, is applied by means of a voltage divider (R_1 and R_2). Increasing R_2 forces the adjust voltage upward and thus pushes the regulator's output to a higher level. The LM317 is designed to accept an unregulated input voltage of up to 37 V and can output a maximum current of 1.5 A. The TL783 is another positive adjustable regulator that can output a regulated voltage of from 1 to 125 V, with a maximum output current of 700 mA. The LM337T, unlike the previous two regulators, is an adjustable negative voltage regulator. It can output a regulated voltage of from −1.2 to −37 V, with a maximum output current of 1.5 A. Again, check the electronics catalogs to see what other kinds of adjustable regulators are available. (C_{in} should be included if the regulator is far from the power source; it should be around 0.1 μF or so. C_{out} is used to eliminate voltage spikes at the output; it should be around 0.1 μF or larger.)

10.1.3 Regulator Specifications

The specifications tables for regulators typically will provide you with the following information: output voltage, accuracy (percent), maximum output current, power dissipation, maximum and minimum input voltage, 120-Hz ripple rejection (decibels), temperature stability ($\Delta V_{out}/\Delta T$), and output impedance (at specific frequencies). A regulator's ripple rejection feature can greatly reduce voltage variations in a power supply's output, as you will discover later in this chapter.

10.2 A Quick Look at a Few Regulator Applications

Before we take a look at how voltage regulators are used in power supplies, it is worthwhile seeing how they are used in other types of applications. Here are a few examples.

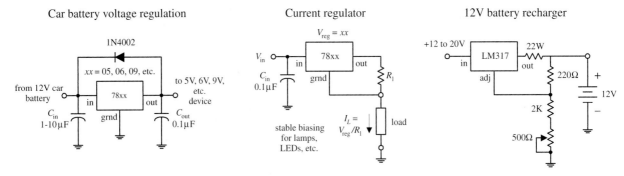

FIGURE 10.6

10.3 The Transformer

It is important that you choose the right transformer for your power supply. The transformer's secondary voltage should not be much larger than the output voltage of the regulator; otherwise, energy will be wasted because the regulator will be forced to dissipate heat. However, at the same time, the secondary voltage must not drop below the required minimum input voltage of the regulator (typically 2 to 3 V above its output voltage).

10.4 Rectifier Packages

The three basic rectifier networks used in power supply designs include the half-wave, full-wave, and bridge rectifiers, shown in Fig. 10.7. To understand how these rectifiers work, see Chap. 4.

FIGURE 10.7

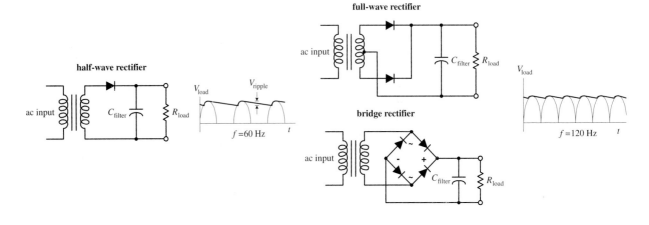

Half-wave, full-wave, and bridge rectifiers can be constructed entirely from individual diodes. However, both full-wave and bridge rectifiers also come in preassembled packages (see Fig. 10.8).

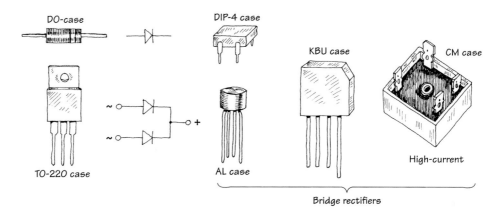

FIGURE 10.8

Make sure that the power supply's rectifier diodes have the proper current and peak-inverse-voltage (PIV) ratings. Typical rectifier diodes have current ratings from 1 to 25 A, PIV ratings from 50 to 1000 V, and surge-current ratings from around 30 to 400 A. Popular general-purpose rectifier diodes include the 1N4001 to 1N4007 series (rated at 1 A, 0.9-V forward voltage drop), the 1N5059 to 1N5062 series (rated at 2 A, 1.0-V forward voltage drop), the 1N5624 to 1N5627 series (rated at 5 A, 1.0-V forward voltage drop), and the 1N1183A-90A (rated at 40 A, 0.9-V forward voltage drop). For low-voltage applications, Schottky barrier rectifiers can be used; the voltage drop across these rectifiers is smaller than a typical rectifier (typically less than 0.4 V); however, their breakdown voltages are significantly smaller. Popular full-wave bridge rectifiers include the 3N246 to 3N252 series (rated at 1 A, 0.9-V forward voltage drop) and the 3N253 to 3N259 series (rated at 2 A, 0.85-V forward voltage drop).

10.5 A Few Simple Power Supplies

Regulated +5-V Supplies

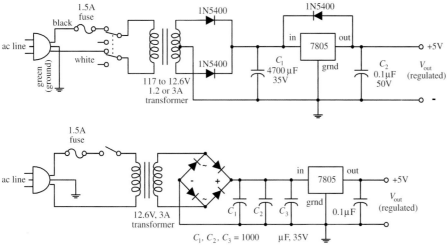

FIGURE 10.9

The first supply uses a center-tapped transformer rated at 12.6 V at 1.2 to 3 A. The voltage after rectification resides at an 8.9-V peak pulse. The filter capacitor (C_1) smoothes the pulses, and the 7805 outputs a regulated +5 V. C_2 is placed across the output of the regulator to bypass high-frequency noise that might be generated by the load. The diode placed across the 7805 helps protect the regulator from damaging reverse-current surges generated by the load. Such surges may result when the power supply is turned off. For example, the capacitance across the output may discharge more slowly than the capacitance across the input. This would reverse-bias the regulator and could damage it in the process. The diode diverts the unwanted current away from the regulator. The second power supply is similar to the first but uses a bridge rectifier.

Adjustable +1.2- to +37-V, 1.5-A Supply

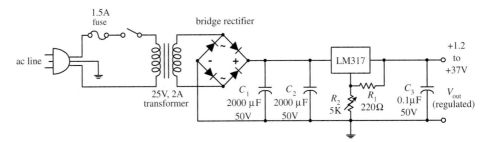

FIGURE 10.10

Here, a +1.2- to +37-V power supply uses an LM317T adjustable regulator to regulate its output. The output voltage is adjusted by varying R_2. C_1 and C_2 act as filter capacitors, and C_3 acts as a high-frequency bypass filter capacitor (see last example).

±12-V and ±15-V Power Supplies

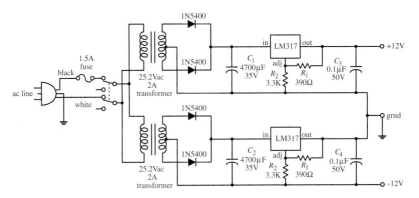

Here, two LM317 adjustable regulators and two 25.2-V center-tapped transformers are used to construct a ±12-V dual-polarity power supply that has a maximum output current of 1.5 A. Note that even though an LM317 is not specifically designed as a negative voltage regulator, it can still be "fooled" by the transformers; the transformers are connected in such a way as to set the regulators' terminal voltages to the proper polarity. (It is a relativity thing.) To change the output voltage, the programming resistors R_1 through R_4 are varied.

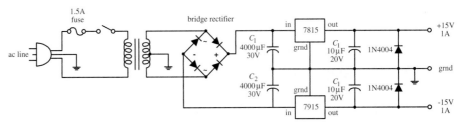

FIGURE 10.11

Here, a 7815 positive voltage regulator and a 7915 negative voltage regulator are used to construct a ±15-V supply. Notice that only one transformer is needed, as compared with the two that were needed in the preceding supply.

10.6   Technical Points about Ripple Reduction

When using a supply to power sensitive circuits, it is essential to keep the variation in output voltage as small as possible. For example, when driving digital circuits from a 5-V supply, the variation in output voltage should be no more than 5 percent, or 0.25 V, if not lower. In fact, digital logic circuits usually have a minimum 200-mV noise margin around critical logic levels. Small analog signal circuits can be especially finicky when it comes to output variations. For example, they may require a variation less than 1 percent to operate properly. How do you keep the output variation error low? You use filter capacitors and voltage regulators.

Filter capacitors act to reduce the fluctuations in output by storing charge during the positive-going rectifier cycle and then releasing charge through the load—at a slow enough rate to maintain a level output voltage—during the negative-going rectifier cycle. If the filter capacitor is too small, it will not be able to store enough charge to maintain the load current and output voltage during the negative-going cycle.

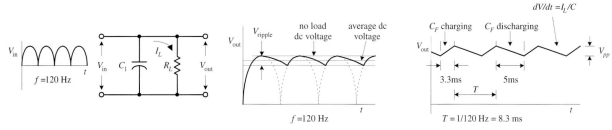

FIGURE 10.12

As it turns out, the amount of current drawn by a load influences the rate of capacitor discharge. If a low-resistance, high-current load is placed across the supply's output, the capacitor will discharge relatively quickly, causing the voltage across the capacitor and therefore the voltage across the load to drop relatively quickly. On the other hand, a highly resistive, low-current load will cause the capacitor to discharge more slowly, which means the dip in output voltage will not be as significant. To calculate the drop in voltage of the capacitor during its discharge cycle, use

$$I = C \frac{dV}{dt} \approx C \frac{\Delta V}{\Delta t}$$

where I is the load current, Δt is the discharge time, and ΔV is the fluctuation in output above and below the average dc level. ΔV is also referred to as the *peak-to-peak ripple voltage* $V_{\text{ripple(pp)}}$. (Here, I cheated by substituting straight lines for exponential decays to describe the discharge cycle; see Fig. 10.12.) Δt can be approximated by dividing 1 by the rectified output voltage frequency. For a full-wave rectifier, the period is 1/120 Hz, or 8.3×10^{-3} s. In reality, the actual amount of time the capacitor spends discharging during a peak-to-peak variation is about 5 ms. The other 3.3 ms is used up during the charge. To make life easier on you, the following equation simplifies matters:

$$V_{\text{ripple(rms)}} = (0.0024 \text{ s}) \frac{I_L}{C_f}$$

Note that the ripple voltage is not given in the peak-to-peak form but is given in the rms form (recall that $V_{pp} = \sqrt{2}\, V_{rms}$). To test the equation out, let's find the ripple voltage for a 5-V supply that has a 4700-μF filter capacitor and a maximum load current of 1.0 A. (Here I am pretending that no voltage regulator is present at this point.) After plugging the numbers into the equation, you get $V_{ripple(rms)} = 510$ mV. But now, recall that a minute ago I said that the amount of variation in the output had to be ±0.25 V% to run digital ICs—510 mV is too big. Now you could keep fiddling around with the capacitance value in the equation and come up with an even better answer, say, letting C equal infinity. In theory, this is fine, but in reality, it is not fine. It is not fine for three basic reasons. The first reason has to do with the simple fact that you cannot find a capacitor at Radio Shack that has an infinite capacitance. If an infinite-capacitance capacitor existed, the universe would not be the same, and you and I probably would not be around to talk about it. The second reason has to do with capacitor tolerances. Unfortunately, the high-capacity electrolytic capacitors used in power supplies have some of the worst tolerances among the capacitor families. It is not uncommon to see a 5 to 20 percent or even larger percentage tolerance for these devices. The mere fact that the tolerances are so bad makes being "nitpicky" about the equation a questionable thing to do. The third and perhaps most important reason for avoiding fiddling around with the equation too much has to do with the inherent ripple-rejection characteristics of voltage regulators. As you will see, the voltage regulator can save us.

Voltage regulators often come with a ripple-rejection parameter given in decibels. For example, the 7805 has a ripple-rejection characteristic of about 60 dB. Using the attenuation expression, you can find the extent of the ripple reduction:

$$-60 \text{ dB} = 20 \log_{10} \frac{V_{out}}{V_{in}}$$

$$-3 = \log_{10} \frac{V_{out}}{V_{in}}$$

$$10^{-3} = \frac{V_{out}}{V_{in}}$$

The last expression says that the output ripple is reduced by a factor of 1000. This means that if you use the regulator with the initial setup, you will only have an output ripple of 0.51 mV—a value well within the safety limits. At this point, it is important to note that the 7805 requires a minimum voltage difference between its input and output of 3 V to function properly. This means that to obtain a 5-V output, the input to the regulator must be at least 8 V. At the same time, it is important to note the voltage drop across the rectifier (typically around 1 to 2 V). The secondary voltage from the transformer therefore should be even larger than 8 V. A transformer with a secondary voltage of 12 V or so would be a suitable choice for the 5-V supply.

Now let's see how well the LM319 adjustable regulator rejects the ripple. Let's say that an LM317 is used in a power supply that has a transformer secondary rms voltage of 12.6 V. The capacitor's peak voltage during a cycle will be 17.8 V (the peak-to-peak voltage of the secondary). An LM317 has a ripple-rejection characteristic of around 65 dB, but this value can be raised to approximately 80 dB by bypassing the LM317's voltage divider with a 10-μF capacitor (see Fig. 10.13).

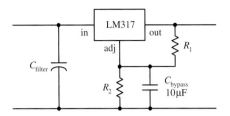

FIGURE 10.13

If you let $C = 4700\ \mu F$ and assume a 1.5-A maximum load current, you get a ripple voltage of

$$V_{r(rms)} = 0.0024\ s(1.5\ A / 4700\ \mu F) = 760\ mV$$

Again, the ripple voltage is too much for sensitive ICs to handle. However, if you consider the LM319's ripple-rejection characteristic (assuming that you use the bypassing capacitor), you get a reduction of

$$-80\ dB = 20\ \log_{10} \frac{V_{out}}{V_{in}}$$

$$-4 = \log_{10} \frac{V_{out}}{V_{in}}$$

$$10^{-4} = \frac{V_{out}}{V_{in}}$$

In other words, the output ripple is reduced by a factor of 10,000, so the final output ripple voltage is only 0.076 mV.

10.7 Loose Ends

Line Filter and Transients Suppressors

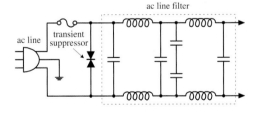

FIGURE 10.14

A line filter is an *LC* filter circuit that is inserted into a supply to filter out unwanted high-frequency interference present in the input line supply. Line filters also can help reduce voltage spikes, as well as help eliminate the emission of radiofrequency interference by the power supply. Line filters are placed before the transformer, as shown in the figure. AC line filters can be purchased in pre-assembled packages. See the electronics catalogs for more info.

A transient suppressor is a device that acts to short out when the terminal voltage exceeds safe limits (e.g., spikes). These devices act like bidirectional high-power zener diodes. They are inexpensive, come in diode-like packages, and come with low-voltage and peak-pulse-voltage ratings.

Overvoltage Protection

crowbar

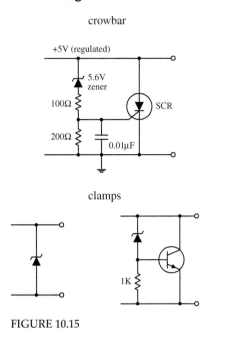

clamps

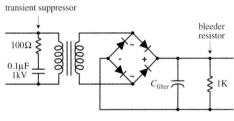

FIGURE 10.15

The crowbar and clamp circuits shown here can be placed across the output of a regulated supply to protect a load against an unregulated voltage that would be present at the output if the voltage regulator failed (shorted internally).

CROWBAR

For the crowbar circuit, when the supply voltage exceeds the zener diode's breakdown voltage by 0.6 V, the zener diode conducts, triggering the SCR into conduction. The SCR then diverts potentially harmful current to ground. The crowbar's SCR will not turn off until the power supply is turned off or the SCR's anode-to-cathode current in interrupted, say, by means of a switch.

CLAMP

A zener diode placed across the output of a supply also can be used for overvoltage protection. However, it may "fry" if the unregulated current is too large. To avoid frying the zener diode, use a high-power transistor to help divert the current. When the zener diode's breakdown voltage is exceeded, some of the current that flows through it will enter the transistor's base, allowing the excessive current to flow toward ground. Using clamps can eliminate false triggering caused by voltage spikes; the crowbar, on the other hand, would need to be reset in such a case.

Bleeder Resistors and Transient Suppressors

transient suppressor

bleeder resistor

FIGURE 10.16

When a resistor is placed across the output of the unregulated supply, it will act to discharge the high-voltage (potentially lethal) filter capacitor when the supply is turned off and the load removed. Such a resistor is referred to as a *bleeder resistor*. A 1-k, 1/2-W resistor is suitable for most applications.

An *RC* network placed across the primary coil of the transformer can prevent large, potentially damaging inductive transients from forming when the supply is turned off. The capacitor must have a high voltage rating. A typical *RC* network consists of a 100-Ω resistor and a 0.1-μF, 1-kV capacitor. Special z-lead transient suppressor devices can also be used, as was mentioned earlier.

10.8 Switching Regulator Supplies (Switchers)

A switching power supply, or *switcher,* is a unique kind of power supply that can achieve power conversion efficiencies far exceeding those of the linear supplies covered earlier in this chapter. With linear regulated supplies, the regulator converts the input voltage that is higher than needed into a desired lower output voltage. To lower the voltage, the extra energy is dissipated as heat from the regulator's control element. The power-conversion efficiency (P_{out}/P_{in}) for these supplies is typically lower than 50 percent. This means that more than half the power is dissipated as heat.

Switchers, on the other hand, can achieve power-conversion efficiencies exceeding 85 percent, meaning that they are much more energy efficient than linearly regulated supplies. Switchers also have a wide current and voltage operating range and can be configured in either step-down (output voltage smaller than input voltage), step-up (output voltage larger than input voltage), or inverting (output is the opposite polarity of the input) configurations. Also, switchers can be designed to run directly off ac line power, without the need for a power transformer. By eliminating the hefty power transformer, the switcher can be made light and small. This makes switchers good supplies for computers and other small devices.

A switching supply resembles a linear supply in many ways. However, two unique features include an energy-storage inductor and a nonlinear regulator network. Unlike a linear supply, which provides regulation by varying the resistance of the regulator's control element, a switcher incorporates a regulation system in which the control element is switched on and off very rapidly. The on/off pulses are controlled by an oscillator/error amplifier/pulse-width modulator network (see Fig. 10.17).

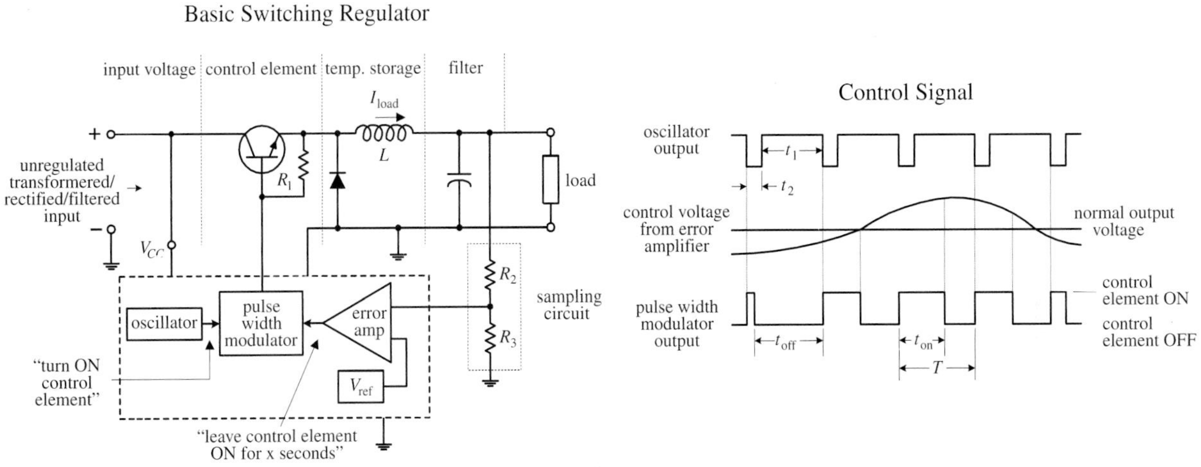

Basic Switching Regulator

FIGURE 10.17

During the on cycle, energy is pumped into the inductor (energy is stored in the magnetic fields around the inductor's coil). When the control element is turned off, the stored energy in the inductor is directed by the diode into the filter and into the load. The sampling circuit (R_2 and R_3) takes a sample of the output voltage and feeds the sample to one of the inputs of the error amplifier. The error amplifier then compares the sample voltage with a reference voltage applied to its other input. If the sample voltage is below the reference voltage, the error amplifier increases its output control voltage. This control voltage is then sent to the pulse-width modulator. (If the sample voltage is above the reference voltage, the error amplifier will decrease the output voltage it sends to the modulator.) While this is going on, the oscillator is supplying a steady series of triggering voltage pulses to the pulse-width modulator. The modulator uses both the oscillator's pulses and the error amplifier's output to produce a modified on/off signal that is sent to the control element's base. The modified signal represents a square wave whose on time is determined by the input error voltage. If the error voltage is low (meaning the sample voltage is higher than it should be), the modulator sends a short-duration on pulse to the control element. However, if the error voltage is high (meaning the sample voltage is lower than it should be), the pulse-width modulator sends a long-duration on pulse to the control element. (The graph in Fig. 10.17 shows how the oscillator, error amplifier, and pulse-width modulator outputs are related.) Using a series of on/off pulses that can be varied in frequency and duration gives the switching regulator its exceptional efficiency; releasing a series of short pulses of energy over time is more efficient than taking excessive supply energy and radiating it off as heat (linear supply).

Figure 10.18 shows a typical switching regulator arrangement. The 556 dual-timer IC houses both the oscillator and pulse-width modulator, while the UA723 voltage-regulator IC acts as the error amplifier. R_2 and R_3 comprise the sampling network, R_6 and R_7 set the reference voltage, and R_4 and R_5 set the final control voltage that is sent to the pulse-width modulator.

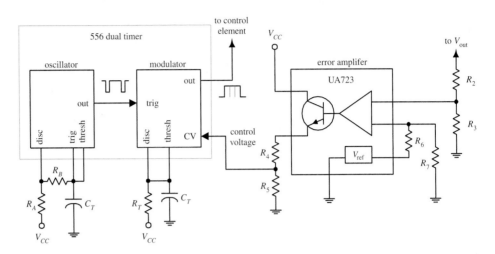

FIGURE 10.18

Step-Up, Step-Down, and Inverting Configurations

The switching regulator in Fig. 10.17 is referred to as a *step-down regulator*. It is used when the regulated output voltage is to be lower than the regulator input voltage. Now, switching regulators also come in step-up and inverting configurations. The step-up version is used when the output is to be higher than the input, whereas the inverting version is used when the output voltage is to be the opposite polarity of the input voltage. Here's an overview of the three configurations.

STEP-DOWN REGULATOR

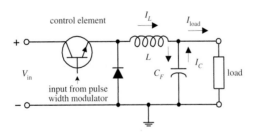

FIGURE 10.19

This is used when output voltage is to be lower than input voltage. When the control element is on, L stores energy, helps supply load current, and supplies current to the filter capacitor. When the control element is off, the energy stored in L helps supply load current but again restores charge on C_F—the charge on C_F is used to supply the load when the control element turns off and L has discharged its energy.

STEP-UP SWITCHER REGULATOR

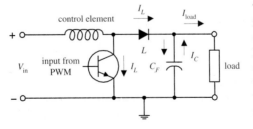

FIGURE 10.20

This is used when output is to be higher than input voltage. When the control element is on, energy is stored in the inductor. The load, isolated by the diode, is supplied by the charge stored in C_F. When the control element is off, the energy stored in L is added to the input voltage. At the same time, L supplies load current, as well as restoring the charge on C_F—the charge on C_F is used to supply the load current when the control element is off and when the energy in L is discharged.

INVERTING SWITCHER REGULATOR

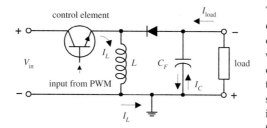

FIGURE 10.21

This is used when output voltage is to be the opposite polarity of the input voltage. When the control element is on, energy is stored in L, while the diode isolates L from load. The load current is supplied by the charge on C_F. When the control element is turned off, the energy stored in L charges C_F to a polarity such that V_{out} is negative. I_L supplies load current and restores the charge on C_F while it is discharging its energy. C_F supplies load current when the control element is off and the inductor is discharged. An inverting switcher regulator can be designed to step up or step down the inverted output.

Eliminating the Need for the Heavy 60-Hz Transformer

By using the unique switching action of the switcher, it is possible to design a supply that does not require the hefty 60-Hz power transformer at the input stage. In other words, you can design a switching power supply to run directly off a 120-V ac line—you still must rectify and filter the line voltage before feeding it to the regulator. However, if you remove the power transformer, you remove the protective isolation that is present between the 120-V ac line and the dc input to the supply. Without the isolation, the dc input voltage will be around 160 V. To avoid this potentially "shocking" situation, the switching regulator must be modified. One method for providing isolation involves replacing the energy-storage inductor with the secondary coil of a high-frequency transformer while using another high-frequency transformer or optoisolator to link the feedback from the error amplifier to the modulating element (see Fig. 10.22).

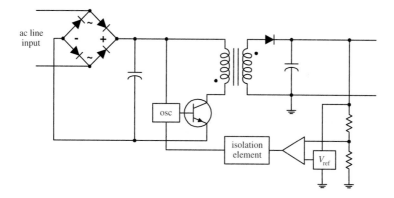

FIGURE 10.22

Now, you may be wondering how removing one transformer and adding another transformer (if not two) makes things smaller and lighter. Well, according to the laws of physics, as the frequency of an alternating signal increases, the need for a large iron core within the transformer decreases. (At high frequencies, the magnetic fields do not need as much help to get from the primary to secondary.) You can use the high-frequency transformer(s) because the switcher's oscillator is beating so fast (e.g., 65 kHz). The difference in size and weight between a switching supply that uses high-frequency transformers and a supply that uses a 60-Hz power transformer is significant. For example, a 500-W switching supply takes up around 640 in³ as compared with 1520 in³ for a linear supply rated at the same power. Also, switching supplies run cooler than linear supplies. In terms of watts per cubic inch, a switching power supply can achieve 0.9 W/in³, while a linear supply usually provides 0.4 W/in³.

There is a slight problem with switchers that should be noted. As a result of the on/off pulsing action of the switching regulator, a switching supply's output will contain a small switching ripple voltage (typically in the tens of millivolts). Usually the ripple voltage does not pose too many problems (e.g., 200-mV noise margins for most digital ICs are not exceeded). However, if a circuit is not responding well to the ripple, an external high-current, low-pass filter can be added.

10.9   Kinds of Commerical Power Supply Packages

To make life easier, you can forget about designing your own supplies and buy one that has been made be the pros. These supplies come in either linear or switcher form and come in a variety of different packages. Here are some of the packages that are available.

Small Modular Units

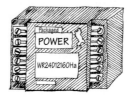

FIGURE 10.23

These are used in low-power applications (e.g., ±5, ±10, ±15 V). Supplies are housed in small modules, usually around 2.5 × 3.5 × 1 in. They often come with pinlike leads that can be mounted directly into circuit boards or come with terminal-strip screw connections along their sides. These supplies may come with single (e.g., +5 V), dual (e.g., ±15 V), or triple (e.g., +5V, ±15V) output terminals. Linear units have power ratings from around 1 to 10 W, while switching units have power ratings from around 10 to 25 W. You must supply the fuses, switches, and filters.

Open Frame

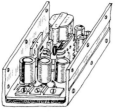
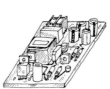

FIGURE 10.24

This supply's circuit board, transformer, etc. are mounted on a metal platform (if it is a low-voltage supply, it may simply be mounted on a circuit board) that is inserted into an instrument. These supplies come in linear and switching types and come with a wide range of voltage, current, and power ratings (around 10 to 200 W for linear supplies, 20 to 400 W for switching supplies). You will probably have to supply the fuses, switches, and filters.

Enclosed

FIGURE 10.25

These supplies are enclosed in a metal box that is especially designed to efficiently radiate off excessive heat. They come in both linear and switching forms. Power rating ranges from around 10 to 800 W for linear supplies and 20 to 1500 W for switching supplies.

Wall Plug-In

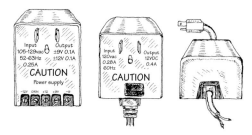

FIGURE 10.26

Wall plug-in power supplies get inserted directly into an ac wall socket. Some of these devices only provide ac transformation, others supply an unregulated dc voltage, and others supply a regulated dc output. The regulated units come in both linear and switching forms. Typical output voltages include +3, +5, +6, +7.5, +9, +12, and +15 V. They also come in dual-polarity form.

10.10   Power Supply Construction

When building a power supply, the following suggestions should help:

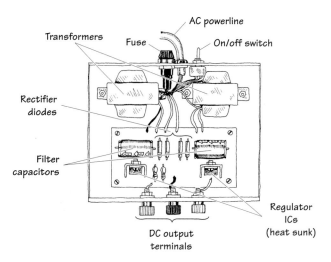

FIGURE 10.27

- Mount the transformer directly to the metal enclosure box, toward the rear.
- Install fuses, power switch, and binding posts at the rear of the box.
- Mount circuit boards on standoffs within the box.
- Place diode or rectifier modules, along with the capacitors and voltage regulators, on the circuit board.
- Make sure to heat-sink voltage regulators.
- Place supply output jacks on the front of the box.
- Drill holes in box to allow cooling.
- Ground the box.
- Place the power-line core through a hole in the rear. Use a rubber grommet for strain relief.
- To avoid shocks, make sure to insulate all exposed 120-V power connections inside the box with heat-shrink tubing.

CHAPTER 11

Audio Electronics

Audio electronics, in part, deals with converting sound signals into electrical signals. This conversion process typically is accomplished by means of a microphone. Once the sound is converted, what is done with the corresponding electrical signal is up to you. For example, you can amplify the signal, filter out certain frequencies from the signal, combine (mix) the signal with other signals, transform the signal into a digitally encoded signal that can be stored in memory, modulate the signal for the purpose of radiowave transmission, use the signal to trigger a switch (e.g., transistor or relay), etc.

Another aspect of audio electronics deals with generating sound signals from electrical signals. To convert electrical signals into sound signals, you can use a speaker. (If you are not interested in retaining frequency response—say, you are only interested in making a warning alarm—you can use an a audible sound device such as a dc buzzer or compression washer.) The electrical signals used to drive a speaker may be sound-generated in origin or may be artificially generated by special oscillator circuits.

11.1   A Little Lecture on Sound

Before you start dealing with audio-related circuits, it is worthwhile reviewing some of the basic concepts of sound. Sound consists of three basic elements: *frequency*, *intensity* (*loudness*), and *timbre* (*overtones*).

The frequency of a sound corresponds to the vibrating frequency of the object that produced the sound. In terms of human physiology, the human ear can perceive frequencies from around 20 to 25,000 Hz; however, the ear is most sensitive to frequencies between 1000 and 2000 Hz.

The intensity of a sound corresponds to the amount of sound energy transported across a unit area per second (or W/m^2) and depends on the amplitude of oscillation of the vibrating object. As you move further away from the vibrating object, the intensity drops in proportion to one over the distance squared. The human ear can perceive an incredible range of intensities, from 10^{-12} to 1 W/m^2. Because this range is so extensive, it is usually more convenient to use a logarithmic scale to describe intensity. For this purpose, decibels are used. When using decibels, sound intensity is defined as dB $= 10 \log_{10} (I/I_0)$, where I is the measured intensity in watts per meter, and $I_0 = 10^{-12}$ W/m^2 is defined as the smallest intensity that is perceived as sound by humans. In terms of decibels, the audio intensity range for humans is between 0 and 120 dB. Figure 11.1 shows a number of sounds, along with their frequency and intensity ranges.

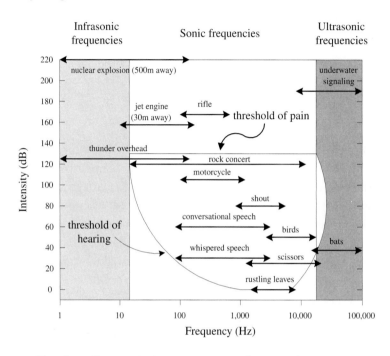

FIGURE 11.1

Tonal quality, or timbre, represents the complex wave pattern that is generated when the overtones of an instrument, voice, etc., are present along with the fundamental frequency. To demonstrate what overtones mean, consider a simple tuning fork that has a resonant frequency of 261.6 Hz (middle C). If you treat the fork as an ideal vibrator, when it is hit, it will vibrate off soundwaves with a frequency of 261.1 Hz. In this case, you have no overtones—you only get one frequency. But now, if you play middle C on a violin, you get an intensely sounding 261.1 Hz, along with a number of other higher, typically less intense frequencies called *overtones* (or *harmonics*). The most intense frequency sounded is typically referred to as the *fundamental frequency*. The overtones of importance have frequencies that are integer multiples of the fundamental frequency (e.g., 2×261.1 Hz is the first harmonic, 3×261.1 Hz is the second harmonic, and $n \times 261.1$ Hz is the nth harmonic). It is the specific intensity of each overtone within the harmonic spectrum of an instrument, voice, etc., that is largely responsible for giving the instrument, voice, etc. its unique tonal quality. (The reason for an instrument's unique set of overtones depends on the construction of the instrument.) Figure 11.2a shows a harmonic spectrum (spectral plot) for an oboe that is tuned to middle C—the fundamental frequency.

In theory, you can create the sound from any type of instrument (e.g., violin, tuba, banjo, etc.) by examining the harmonic spectrum of that instrument. To illus-

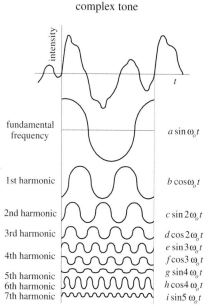

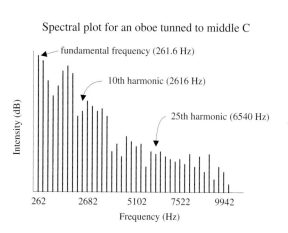

FIGURE 11.2

trate how this can be done, pretend that you have a number of ideal tuning forks. One fork represents the fundamental frequency; the other forks represent the various overtone frequencies. Using the harmonic spectrum of an instrument as a guide, you can mimic the sound of the instrument by varying the intensity of each "overtone fork." (In reality, to accurately mimic an instrument, you also must consider rise and decay times of certain overtones—controlling the intensities of the overtones is not enough.) Mathematically, you can express a complex sound as the sum of all its overtones:

$$\text{Signal} = a \sin \omega_0 t + b \cos \omega_0 t + c \sin 2\omega_0 t + d \cos 2\omega_0 t + e \sin 3\omega_0 t + f \cos \omega_0 t + \cdots$$

The coefficients a, b, c, d, etc., are the intensities of the overtones, and the fundamental frequency $f_0 = \omega_0/2\pi$. This expression is referred to as a *Fourier series*. The coefficients must be calculated from the given waveform or the data from which it is plotted—although an instrument called a *harmonic analyzer* can automatically compute the coefficients. Figure 11.2b shows a complex sound made up of seven of its harmonics.

The art of synthesizing sounds via electric circuits is fairly complex business. To accurately mimic an instrumental sound, train whistle, bird chirp, etc., you must design circuits that can generate complex waveforms that contain all the overtones and decay and rise time information. For this purpose, special oscillator and modulator circuits are needed.

11.2 Microphones

A microphone converts variations in sound pressure into corresponding variations in electric current. The amplitude of the ac voltage generated by a microphone is proportional to the intensity of the sound, while the frequency of the ac voltage corresponds to the frequency of the sound. (Note that if overtones are present within the sound signal, these overtones also will be present in the electrical signal.) Three commonly used microphones are listed below.

Dynamic

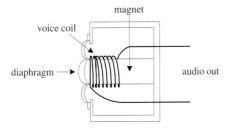

FIGURE 11.3

This type of microphone consists of a plastic diaphragm, voice coil, and a permanent magnet. The diaphragm is connected to one end of the voice coil, while the other end of the coil is loosely supported around (or within) the magnet. When an alternating pressure is applied to the diaphragm, the voice coil alternates in response. Since the voice coil is accelerating through the magnet's magnetic field, an induced voltage is set up across the leads of the voice coil. You can use this voltage to power a very small load, or you can use an amplifier to increase the strength of the signal so as to drive a larger load. Dynamic microphones are extremely rugged, provide smooth and extended frequency response, do not require an external dc source to drive them, perform well over a wide range of temperatures, and have a low impedance output. Some dynamic microphones house internal transformers within their bodies, which give them the ability to have either a high- or low-impedance output—a switch is used to select between the two. Dynamic microphones are widely used in public address, hi-fi, and recording applications.

Condenser

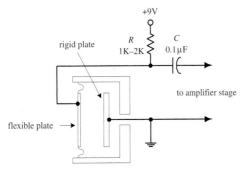

FIGURE 11.4

This type of microphone consists of a pair of charged plates that can be forced closer or further apart by variations in air pressure. In effect, the plates act like a sound-sensitive capacitor. One plate is made of a rigid metal that is fixed in placed and grounded. The other plate is made of a flexible metal or metal-coiled plastic that is positively charged by means of an external voltage source. A very low-noise, high-impedance amplifier is required to operate this type of microphone and to provide low output impedance. Condenser microphones offer crisp, low-noise sound and are used for high-quality sound recording.

Electret

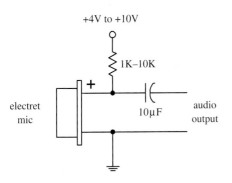

FIGURE 11.5

An electret microphone is a variation of the condenser microphone. Instead of requiring an external voltage source to charge the diaphragm, it uses a permanently charged plastic element (electret) placed in parallel with a conductive metal backplate. Most electret microphones have a small FET amplifier built into their cases. This amplifier requires power to operate—typically a voltage between +4 and +10 V is needed. This voltage is fed into the microphone through a resistor (1–10 K) (see figure). Electret microphones respond best to frequencies within the lower-middle to high-frequency range—they do not respond well to bass frequencies. For this reason, they tend to be restricted to voice communications. Also, the performance of electret microphones decreases over the years; as time passes, the charge on the electret is lost.

11.3   Microphone Specifications

A microphone's *sensitivity* represents the ratio of electrical output (voltage) to the intensity of sound input. Sensitivity is often expressed in decibels with respect to a reference sound pressure of 1 dyn/cm^2.

The *frequency response* of a microphone is a measure of the microphone's ability to convert different acoustical frequencies into ac voltages. For speech, the frequency response of a microphone need only cover a range from around 100 to 3000 Hz. However, for hi-fi applications, the frequency response of the microphone must cover a wider range, from around 20 to 20,000 Hz.

The *directivity characteristic* of a microphone refers to how well the microphone responds to sound coming from different directions. *Omnidirectional* microphones respond equally well in all directions, whereas *directional* microphones respond well only in specific directions.

The *impedance* of a microphone represents how much the microphone resists the flow of an ac signal. Low-impedance microphones are classified as having an impedance of less than 600 Ω. Medium-impedance microphones range from 600 to 10,000 Ω, whereas high-impedance microphones extend above 10,000 Ω. In modern audio systems, it is desirable to connect a lower-impedance microphone to a higher-impedance input device (e.g., 50-Ω microphone to 600-Ω mixer), but it is undesirable to connect a high-impedance microphone to a low-impedance input. In the first case, not much signal loss will occur, whereas in the second case, a significant amount of signal loss may occur. The standard rule of thumb is to allow the load impedance to be 10 times the source impedance. We'll take a closer look at impedance matching later on in this chapter.

11.4 Audio Amplifiers

Electrical signals within audio circuits often require amplification to effectively drive other circuit elements or devices. Perhaps the easiest and most efficient way to amplify a signal is to use an op amp. General-purpose op amps such as the 741 will work fine for many noncritical audio applications, but they may cause distortion and other undesirable effects when audio signals get complex. A better choice for audio applications is to use an audio op amp especially designed to handle audio signals. Audio amplifiers have high slew rates, high gain-bandwidth products, high input impedances, low distortion, high voltage/power operation, and very low input noise. There are a number of good op amps produced by a number of different manufacturers. Some high-quality op amps worth mentioning include the AD842, AD847, AD845, AD797, NE5532, NE5534, NE5535, OP-27, LT1115, LM833, OPA2604, OP249, HA5112, and LT1057.

Inverting Amplifier

The following two circuits act as inverting amplifiers. The gain for both circuits is determined by $-R_2/R_1$ (see Chap. 7 for the theory), while the input impedance is approximately equal to R_1. The first op amp circuit uses a dual power supply, while the second op amp circuit uses a single power supply.

FIGURE 11.6

Inverting amplifier (dual power supply)

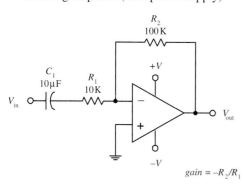

$$gain = -R_2/R_1$$

In both amplifier circuits, C_1 acts as an ac coupling capacitor—it acts to pass ac signals while preventing unwanted dc signals from passing from the previous stage. Without C_1, dc levels would be present at the op amp's output, which in turn could lead to amplifier saturation and distortion as the ac portion of the input signal is amplified. C_1 also helps prevent low-frequency noise from reaching the amplifier's input.

Inverting amplifier (single power supply)

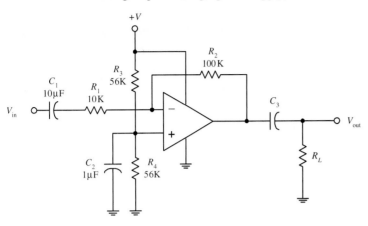

FIGURE 11.6   (*Continued*)

In the single-power-supply circuit, biasing resistors R_3 and R_4 are needed to prevent the amplifier from clipping during negative swings in the audio input signals. They act to give the op amp's output a dc level on which the ac signal can safely fluctuate. Setting $R_3 = R_4$ sets the dc level of the op amp's output to 1/2 (+V). For reliable results, the biasing resistors should have resistance values between 10 and 100 k. Now, to prevent passing the dc level onto the next stage, C_3 (ac coupling capacitor) must be included. Its value should be equal to $1/(2\pi f_C R_L)$, where R_L is the load resistance, and f_C is the cutoff frequency. C_2 acts as a filtering capacitor used to eliminate power-supply noise from reaching the op amp's noninverting input.

Notably, many audio op amps are especially designed for single-supply operation—they do not require biasing resistors.

Noninverting Amplifier

The preceding inverting amplifier works fine for many applications, but its input impedance is not incredibly large. To achieve a larger input impedance (useful when bridging a high-impedance source to the input of an amplifier), you can use one of the following noninverting amplifiers. The left amplifier circuit uses a dual power supply, whereas the right amplifier circuit uses a single power supply. The gain for both circuits is equal to $R_2/R_1 + 1$.

Noninverting amplifier (dual power supply)   Noninverting amplifier (single power supply)

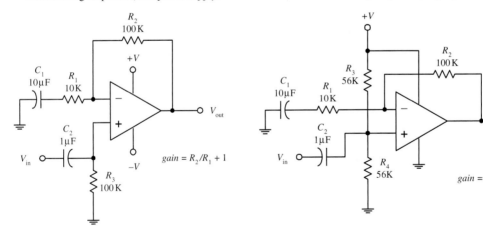

FIGURE 11.7

Components R_1, C_1, R_2, and the biasing resistors serve the same function as was seen in the inverting amplifier circuits. The noninverting input offers an exceptionally high input impedance and can be matched to the source impedance more readily by adjusting C_2 and R_3 (dual-supply circuit) or R_4 (single-supply circuit). The input impedance is approximately equal to R_3 (dual-supply circuit) or R_4 (single-supply circuit).

11.5   Preamplifiers

In most audio applications, the term *preamplifier* refers to a control amplifier that is used to control features such as input selection, level control, gain, and impedance levels. Here are a few simple microphone preamplifier circuits to get you started. (Note that "high Z" refers to a microphone with a high input impedance—one that is greater than ~600 Ω.)

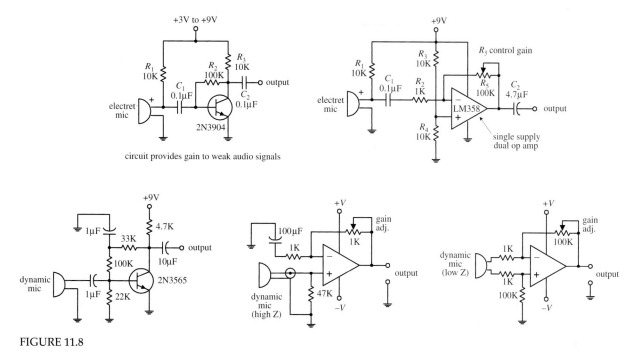

FIGURE 11.8

11.6 Mixer Circuits

Audio mixers are basically summing amplifiers—they add a number of different input signals together to form a single superimposed output signal. The two circuits below are simple audio mixer circuits. The left circuit uses a common-emitter amplifier as the summing element, while the right circuit uses an op amp. The potentiometers are used as independent input volume controls.

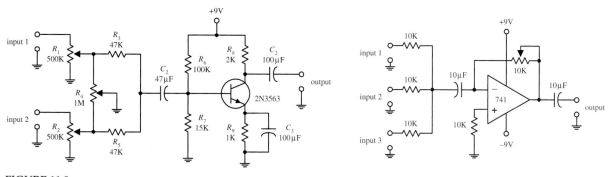

FIGURE 11.9

11.7 A Note on Impedance Matching

Is matching impedances between audio devices necessary? Not any more, at least when it comes to connecting a low-impedance source to a high-impedance load. In the era when vacuum-tube amplifiers were the standard, it was important to match impedances to achieve maximum power transfer between two devices. Impedance matching reduced the number of vacuum-tube amplifiers needed in circuit design (e.g., the number of vacuum-tube amplifiers needed along a telephone transmission line). However, with the advent of the transistor, more efficient amplifiers were created. For these new amplifiers, what was important—and still is important—was maximum voltage transfer, not maximum power transfer. (Think of an op amp with its extremely high input impedance and low output impedance. To initiate a large-output-current response from an op amp, practically no input current is required into its input leads.) For maximum voltage transfer to occur, it was found that the destination device (called the *load*) should have an impedance of at least 10 times that of the sending device (called the *source*). This condition is referred to as *bridging*. (Without applying the bridging rule, if two audio devices with the same impedances are joined, you would see around a 6 dB worth of attenuation loss in the transmitted signal.) Bridging is the most common circuit configuration used when connecting modern audio devices. It is also applied to most other electronic source-load connections, with the exception of certain radiofrequency circuits where matching impedance is usually desired and in cases where the signal being transmitted is a current rather than a voltage. If the transmitted signal is a current, the source impedance should be larger than the load impedance.

Now, if you consider a high-impedance source connected to a low-impedance load (e.g., a high-impedance microphone connected to a low-impedance mixer), voltage transfer can result in significant signal loss. The amount of signal loss in this case would be equal to

$$dB = 20 \log_{10} \frac{R_{\text{load}}}{R_{\text{load}} + R_{\text{source}}}$$

As a rule of thumb, a loss of 6 dB or less is acceptable for most applications.

11.8 Speakers

Speakers convert electrical signals in audible signals. The most popular speaker used today is the dynamic speaker. The dynamic speaker operates on the same basic principle as a dynamic microphone. When a fluctuating current is applied through a moving coil (voice coil) that surrounds a magnet (or that is surrounded by a magnet), the coil is forced back and forth (Faraday's law). A large paper cone attached to the coil responds to the back-and-forth motion by "drumming off" sound waves.

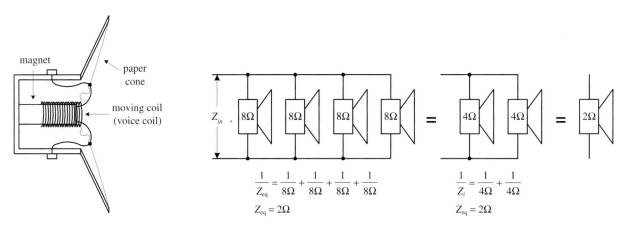

FIGURE 11.10

Every speaker is given a *nominal impedance* Z that represents the average impedance across its leads. (In reality, a speaker's impedance varies slightly with frequency, above and below the nominal level.) In terms of applications, you can treat the speaker like a simple resistive load of impedance Z. For example, if you attach an 8-Ω speaker to an amplifier's output, the amplifier will treat the speaker as an 8-Ω load. The amount of current drawn from the amplifier will be $I = V_{out}/Z_{speaker}$. However, if you replace the 8-Ω speaker with a 4-Ω speaker, the current drawn from the amplifier will double.

Driving two 8-Ω speakers in parallel is equivalent to driving a 4-Ω speaker. Driving two 4-Ω speakers in parallel is equivalent to driving a 2-Ω speaker. By using high-power resistors, it is possible to change the overall impedance sensed by the amplifier. For example, by placing a 4-Ω resistor in series with a 4-Ω speaker, you can create a load impedance of 8 Ω. However, using a series resistor to increase the impedance may hurt the sound quality. There are speaker-matching transformers that can change from 4 to 8 Ω, but a high-quality transformer like this can cost as much as a new speaker and can add slight frequency response and dynamic range errors.

Another important characteristic of a speaker is its frequency response. The *frequency response* represents the range over which a speaker can effectively vibrate off audio signals. Speakers that are designed to respond to low frequencies (typically less than 200 Hz) are referred to as *woofers*. *Midrange speakers* are designed to handle frequencies typically between 500 and around 3000 Hz. A *tweeter* is a special type of speaker (typically dome or horn type) that can handle frequencies above midrange. Some speakers are designed as full-range units that are capable of reproducing frequencies from around 100 to 15,000 Hz. A full-range speaker's sound quality tends to be inferior to a speaker system that incorporates a woofer, midrange, and tweeter speaker all together.

11.9   Crossover Networks

To design a decent speaker system, it would be best to incorporate a woofer, midrange speaker, and tweeter together so that you get good sound response over the entire audio spectrum (20 to 20,000 Hz). However, simply connecting these speakers in parallel will not work because each speaker will be receiving frequencies outside its natural frequency-response range. What you need is a filter network that can divert high-frequency signals to the tweeter, low-frequency signals to the woofer, and midrange-frequency signals to the midrange speaker. The filter network that is used for this sort of application is called a *crossover network*.

There are two types of crossovers: passive or active. Passive crossover networks consist of passive filter elements (e.g., capacitors, resistors, inductors) that are placed between the power amplifier and the speaker—they are placed inside the speaker cabinet. Passive crossover networks are cheap to make, easy to make foolproof, and can be tailored for a specific speaker. However, they are nonadjustable and always use up some amplifier power. Active crossover networks consist of a set of active filters (op amp filters) that are placed before the amplifier section. The fact that active crossover networks come before the power amplifier section makes it easier to manipulate the signal because the signal is still tiny (not amplified). Also, a single active crossover network can be used to control a number of different amplifier-speaker combinations at the same time. Since active crossover networks use active filters, the audio signal will not suffer as much attenuation loss as it would if it were applied through a passive crossover network.

Figure 11.11 shows a simple passive crossover network used to drive a three-speaker system. The graph shows typical frequency-response curves for each speaker. To produce an overall flat response from the system, you use a low-pass, bandpass, and high-pass filter. C_1 and R_t form the low-pass filter, L_1, C_1, and R_m form the bandpass filter, and L_2 and R_w form the low-pass filter (R_t, R_m, and R_w are the nominal impedances of the tweeter, midrange, and woofer speakers).

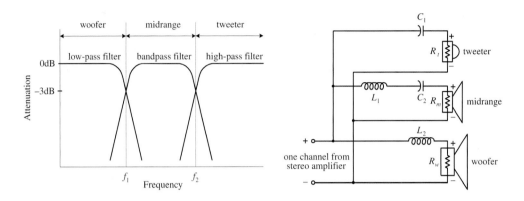

FIGURE 11.11

To determine the component values needed to get the desired response, use the following: $C_1 = 1/(2\pi f_2 R_t)$, $L_1 = R_m/(2\pi f_2)$, $C_2 = 1/(2\pi f_1 R_m)$, and $L_2 = R_w/2\pi f_1$, where f_1 and f_2 represent the 3-dB points shown in the graph. Usually, passive crossover networks are a bit more sophisticated than the one presented here. They often incorporate higher-order filters along with additional elements, such as an impedance compensation network, attenuation network, series notch filter, etc., all of which are used to achieve a flatter overall response.

Here is a more practical passive crossover network used to drive a two-speaker system consisting of an 8-Ω tweeter and an 8-Ω woofer. An 18 × 12 × 8 in fiberboard box acts as a good resonant cavity for this system.

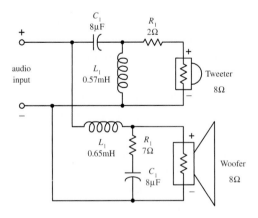

FIGURE 11.12

The following is an active crossover network that is used to drive a two-speaker system that has a crossover frequency (3-dB point) around 500 Hz and 18 dB per octave response. A LF356 high-performance op amp is used as the active element. Remember that for active filters the output signals must be amplified before they are applied to the speaker inputs.

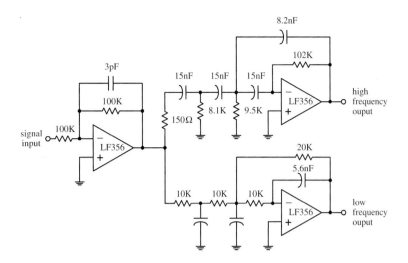

FIGURE 11.13

11.10 Simple ICs Used to Drive Speakers

Audio Amplifier (LM386)

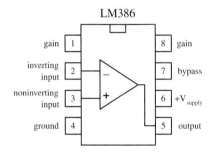

The LM386 audio amplifier is designed primarily for low-power applications. A +4- to +15-V supply voltage is used to power the IC. Unlike a traditional op amp, such as the 741, the 386's gain is internally fixed to 20. However, it is possible to increase the gain to 200 by connecting a resistor-capacitor network between pins 1 and 8. The 386's inputs are referenced to ground, while internal circuitry automatically biases the output signal to one-half the supply voltage. This audio amplifier is designed to drive an 8-Ω speaker.

Audio amplifier (gain of 20)

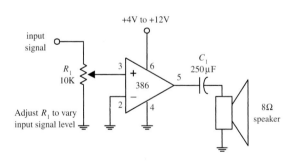

Audio amplifier (gain of 200)

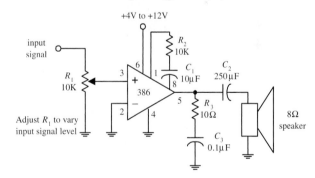

FIGURE 11.14

Audio Amplifier (LM383)

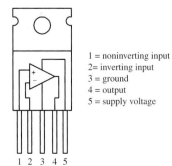

1 = noninverting input
2= inverting input
3 = ground
4 = output
5 = supply voltage

The LM383 is a power amplifier designed to drive a 4-Ω speaker or two 8-Ω speakers in parallel. It contains thermal shutdown circuitry to protect itself from excessive loading. A heat sink is required to avoid meltdown.

FIGURE 11.15

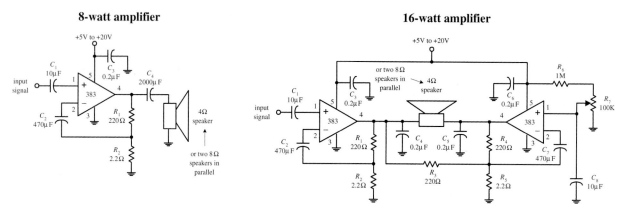

8-watt amplifier

16-watt amplifier

FIGURE 11.15 *(Continued)*

11.11 Audible-Signal Devices

There are a number of unique audible-signal devices that are used as simple warning signal indicators. Some of these devices produce a continuous tone, others produce intermittent tones, and still others are capable of generating a number of different frequency tones, along with various periodic on/off cycling characteristics. Audible-signal devices come in both dc and ac types and in various shapes and sizes. Some of these devices are extremely small—no bigger than a dime. A good electronics catalog will provide a listing of audio-signal devices, along with their size, sound type, dB ratings, voltage ratings, and current-drain specifications.

FIGURE 11.16   Sonalert® audible sound device   Compression washer   DC buzzer

11.12 Miscellaneous Audio Circuits

Simple Tone-Related Circuits

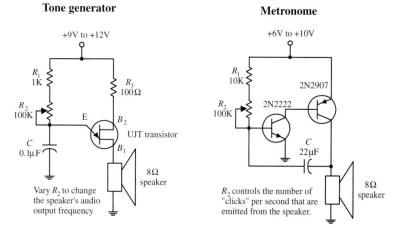

Tone generator

Vary R_2 to change the speaker's audio output frequency

Metronome

R_2 controls the number of "clicks" per second that are emitted from the speaker.

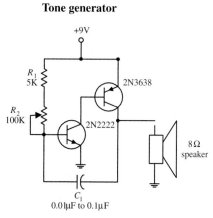

Tone generator

R_2 controls the audio output frequency of the speaker.

FIGURE 11.17

Tone generator   **Warbler siren**

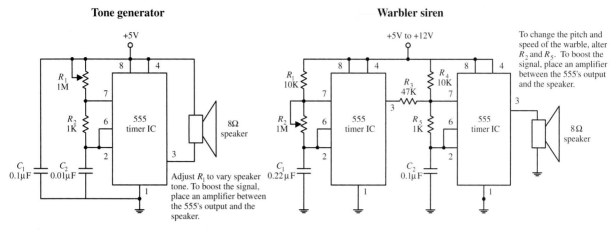

To change the pitch and speed of the warble, alter R_2 and R_5. To boost the signal, place an amplifier between the 555's output and the speaker.

Adjust R_1 to vary speaker tone. To boost the signal, place an amplifier between the 555's output and the speaker.

FIGURE 11.17 (*Continued*)

Simple Buzzer Circuits

Buzzer volume control   **Digitally actuated buzzers**

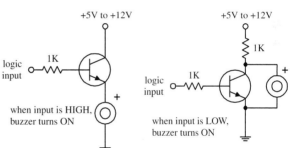

FIGURE 11.18

Megaphone

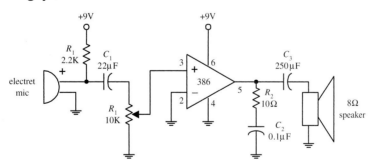

FIGURE 11.19

Sound-Activated Switch

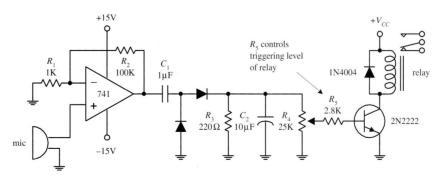

FIGURE 11.20

CHAPTER 12

Digital Electronics

12.1 The Basics of Digital Electronics

Until now I have mainly covered the analog realm of electronics—circuits that accept and respond to voltages that vary continuously over a given range. Such analog circuits included rectifiers, filters, amplifiers, simple *RC* timers, oscillators, simple transistor switches, etc. Although each of these analog circuits is fundamentally important in its own right, these circuits lack an important feature—they cannot store and process bits of information needed to make complex logical decisions. To incorporate logical decision-making processes into a circuit, you need to use digital electronics.

Analog Signal

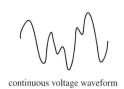

continuous voltage waveform

Digital Signal

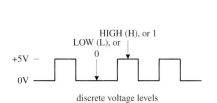

discrete voltage levels

Using a switch to demonstrate logic states

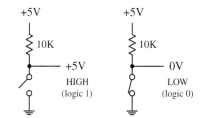

FIGURE 12.1

12.1.1 Digital Logic States

In digital electronics there are only two voltage states present at any point within a circuit. These voltage states are either *high* or *low*. The meaning of a voltage being high or low at a particular location within a circuit can signify a number of things. For example, it may represent the on or off state of a switch or saturated transistor. It may represent one bit of a number, or whether an event has occurred, or whether some action should be taken.

The high and low states can be represented as true and false statements, which are used in Boolean logic. In most cases, high = true and low = false. However, this does not have to be the case—you could make high = false and low = true. The decision to use one convention over the other is a matter left ultimately to the designer. In digital lingo, to avoid people getting confused over which convention is in use, the term

313

positive true logic is used when high = true, while the term *negative true logic* is used when high = false.

In Boolean logic, the symbols 1 and 0 are used to represent true and false, respectively. Now, unfortunately, 1 and 0 are also used in electronics to represent high and low voltage states, where high = 1 and low = 0. As you can see, things can get a bit confusing, especially if you are not sure which type of logic convention is being used, positive true or negative true logic. I will give some examples later on in this chapter that deal with this confusing issue.

The exact voltages assigned to a high or low voltage states depend on the specific logic IC that is used (as it turns out, digital components are entirely IC based). As a general rule of thumb, +5 V is considered high, while 0 V (ground) is considered low. However, as you will see in Section 12.4, this does not have to be the case. For example, some logic ICs may interrupt a voltage from +2.4 to +5 V as high and a voltage from +0.8 to 0 V as low. Other ICs may use an entirely different range. Again, I will discuss these details later.

12.1.2 Number Codes Used in Digital Electronics

Binary

Because digital circuits work with only two voltage states, it is logical to use the binary number system to keep track of information. A binary number is composed of two binary digits, 0 and 1, which are also called *bits* (e.g., 0 = low voltage, 1 = high voltage). By contrast, a decimal number such as 736 is represented by successive powers of 10:

$$736_{10} = 7 \times 10^2 + 3 \times 10^1 + 6 \times 10^0$$

Similarly, a binary number such as 11100 (28_{10}) can be expressed as successive powers of 2:

$$11100_2 = 1 \times 2^4 + 1 \times 2^3 + 1 \times 2^2 + 0 \times 2^1 + 0 \times 2^0$$

The subscript tells what number system is in use (X_{10} = decimal number, X_2 = binary number). The highest-order bit (leftmost bit) is called the *most significant bit* (MSB), while the lowest-order bit (rightmost bit) is called the *least significant bit* (LSB). Methods used to convert from decimal to binary and vice versa are shown below.

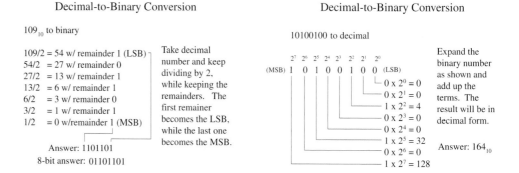

FIGURE 12.2

It should be noted that most digital systems deal with 4, 8, 16, 32, etc., bit strings. In the decimal-to-binary conversion example given here, you had a 7-bit answer. In an 8-bit system, you would have to add an additional 0 in front of the MSB (e.g., 01101101). In a 16-bit system, 9 additional 0s would have to be added (e.g., 0000000001101101).

As a practical note, the easiest way to convert a number from one base to another is to use a calculator. For example, to convert a decimal number into a binary number, type in the decimal number (in base 10 mode) and then change to binary mode (which usually entails a 2d-function key). The number will now be in binary (1s and 0s). To convert a binary number to a decimal number, start out in binary mode, type in the number, and then switch to decimal mode.

Octal and Hexadecimal

Two other number systems used in digital electronics include the octal and hexadecimal systems. In the octal system (base 8), there are 8 allowable digits: 0, 1, 2, 3, 4, 5, 6, 7. In the hexadecimal system, there (base 16) there are 16 allowable digits: 0, 1, 2, 3, 4, 5, 6, 7, 8, 9, A, B, C, D, E, F. Here are example octal and hexadecimal numbers with decimal equivalents:

247_8 (octal) $= 2 \times 8^2 + 4 \times 8^1 + 9 \times 8^0 = 167_{10}$ (decimal)
$2D5_{16}$ (hex) $= 2 \times 16^2 + D\ (=13_{10}) \times 16^1 + 9 \times 16^0 = 725_{10}$ (decimal)

Now, binary numbers are of course the natural choice for digital systems, but since these binary numbers can become long and difficult to interpret by our decimal-based brains (a result of our 10 fingers), it is common to write them out in hexadecimal or octal form. Unlike decimal numbers, octal and hexadecimal numbers can be translated easily to and from binary. This is so because a binary number, no matter how long, can be broken up into 3-bit groupings (for octal) or 4-bit groupings (for hexadecimal)—you simply add zero to the beginning of the binary number if the total numbers of bits is not divisible by 3 or 4. Figure 12.3 should paint the picture better than words.

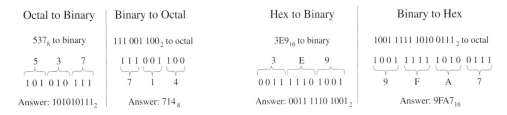

Octal to Binary	Binary to Octal	Hex to Binary	Binary to Hex
537_8 to binary	$111\,001\,100_2$ to octal	$3E9_{16}$ to binary	$1001\,1111\,1010\,0111_2$ to octal
5 3 7	111 001 100	3 E 9	1001 1111 1010 0111
101 010 111	7 1 4	0011 1110 1001	9 F A 7
Answer: 101010111_2	Answer: 714_8	Answer: $0011\,1110\,1001_2$	Answer: $9FA7_{16}$

A 3-digit binary number is replaced for each octal digit, and vise versa. The 3-digit terms are then grouped (or octal terms are grouped).

A 4-digit binary number is replaced for each hex digit, and vise versa. The 4-digit terms are then grouped (or hex terms are grouped).

FIGURE 12.3

Today, the hexadecimal system has essentially replaced the octal system. The octal system was popular at one time, when microprocessor systems used 12-bit and 36-bit words, along with a 6-bit alphanumeric code—all these are divisible by 3-bit units (1 octal digit). Today, microprocessor systems mainly work with 8-bit, 16-bit, 20-bit, 32-bit, or 64-bit words—all these are divisible by 4-bit units (1 hex digit). In other words, an 8-bit word can be broken down into 2 hex digits, a 16-bit word into 4 hex digits, a 20-bit word into 5 hex digits, etc. Hexadecimal representation of binary numbers pops up in many memory and microprocessor applications that use programming codes (e.g., within assembly language) to address memory locations and initiate other specialized tasks that would otherwise require typing in long binary numbers. For example, a 20-bit address code used to identify 1 of 1 million memory locations can be replaced with a hexadecimal code (in the assembly program) that reduces the count to 5 hex digits. [Note that a compiler program later converts the hex numbers within the assembly language program into binary numbers (machine code) which the microprocessor can use.] Table 12.1 gives a conversion table.

TABLE 12.1   Decimal, Binary, Octal, Hex, BCD Conversion Table

DECIMAL	BINARY	OCTAL	HEXADECIMAL	BCD
00	0000 0000	00	00	0000 0000
01	0000 0001	01	01	0000 0001
02	0000 0010	02	02	0000 0010
03	0000 0011	03	03	0000 0011
04	0000 0100	04	04	0000 0100
05	0000 0101	05	05	0000 0101
06	0000 0110	06	06	0000 0110
07	0000 0111	07	07	0000 0111
08	0000 1000	10	08	0001 1000
09	0000 1001	11	09	0000 1001
10	0000 1010	12	0A	0001 0000
11	0000 1011	13	0B	0001 0001
12	0000 1100	14	0C	0001 0010
13	0000 1101	15	0D	0001 0011
14	0000 1110	16	0E	0001 0100
15	0000 1111	17	0F	0001 0101
16	0001 0000	20	10	0001 0110
17	0001 0001	21	11	0001 0111
18	0001 0010	22	12	0001 1000
19	0001 0011	23	13	0001 1001
20	0001 0100	24	14	0010 0000

BCD Code

Binary-coded decimal (BCD) is used to represent each digit of a decimal number as a 4-bit binary number. For example, the number 150_{10} in BCD is expressed as

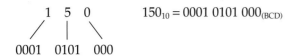

$$150_{10} = 0001\ 0101\ 000_{(BCD)}$$

To convert from BCD to binary is vastly more difficult, as shown in Fig. 12.4. Of course, you could cheat by converting the BCD into decimal first and then convert to binary, but that does not show you the mechanics of how machines do things with 1s and 0s. You will rarely have to do BCD-to-binary conversion, so I will not dwell on this topic—I will leave it to you to figure out how it works (see Fig. 12.4).

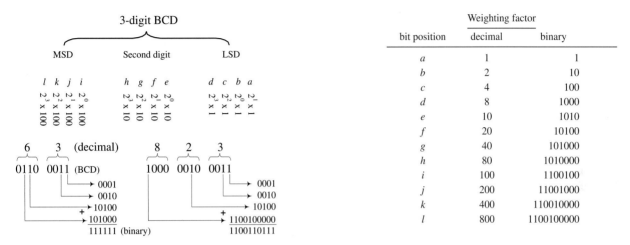

FIGURE 12.4

BCD is commonly used when outputting to decimal (0–9) displays, such as those found in digital clocks and multimeters. BCD will be discussed later in Section 12.3.

Sign-Magnitude and 2's Complement Numbers

Up to now I have not considered negative binary numbers. How do you represent them? A simple method is to use *sign-magnitude representation*. In this method, you simply reserve a bit, usually the MSB, to act as a sign bit. If the sign bit is 0, the number is positive; if the sign bit is 1, the number is negative (see Fig. 12.5). Although the sign-magnitude representation is simple, it is seldom used because adding requires a different procedure than subtracting (as you will see in the next section). Occasionally, you will see sign-magnitude numbers used in display and analog-to-digital applications but hardly ever in circuits that perform arithmetic.

A more popular choice when dealing with negative numbers is to use *2's complement representation*. In 2's complement, the positive numbers are exactly the same as unsigned binary numbers. A negative number, however, is represented by a binary number, which when added to its corresponding positive equivalent results in zero. In this way, you can avoid two separate procedures for doing addition and subtraction. You will see how this works in the next section. A simple procedure outlining how to convert a decimal number into a binary number and then into a 2's complement number, and vice versa, is outlined in Fig. 12.5.

Arithmetic with Binary Numbers

Adding, subtracting, multiplying, and dividing binary numbers, hexadecimal numbers, etc., can be done with a calculator set to that particular base mode. But that's

Decimal, Sign-Magnitude, 2's Complement Conversion Table

DECIMAL	SIGN-MAGNITUDE	2'S COMPLEMENT
+7	0000 0111	0000 0111
+6	0000 0110	0000 0110
+5	0000 0101	0000 0101
+4	0000 0100	0000 0100
+3	0000 0011	0000 0011
+2	0000 0010	0000 0010
+1	0000 0001	0000 0001
0	0000 0000	0000 0000
−1	1000 0001	1111 1111
−2	1000 0010	1111 1110
−3	1000 0011	1111 1101
−4	1000 0100	1111 1100
−5	1000 0101	1111 1011
−6	1000 0110	1111 1010
−7	1000 0111	1111 1001
−8	1000 1000	1111 1000

FIGURE 12.5

Decimal to 2's complement

$+41_{10}$ to 2's complement

true binary = 0010 1001
2's comp   = 0010 1001

-41_{10} to 2's complement

true binary = 0010 1001
1's comp   = 1101 0110
Add 1   =   +1
2's comp   = 0010 1001

If the decimal number is positive, the 2's complement number is equal to the true binary equivalent of the decimal number.

If the decimal number is negative, the 2's complement number is found by:
1) Complementing each bit of the true binary equivalent of the decimal (making 1's into 0's and vise versa). This is called taking the 1's complement.
2) Adding 1 to the 1's complement number to get the magnitude bits. The sign bit will always end up being 1.

2's complement to decimal

1100 1101 (2's comp) to decimal

2's comp   = 1100 1101
Complement = 0011 0010
Add 1   =   +1
True binary = 0011 0011
Decimal eq. = -51_{10}

If the 2's complement number is positive (sign bit = 0), perform a regular binary-to-decimal conversion.

If the 2's complement number is negative (sign bit = 1), the decimal sign will be negative. The decimal is found by:
1) Complementing each bit of the 2's complement number.
2) Adding 1 to get the true binary equivalent.
3) Performing a true binary-to-decimal conversion.

cheating, and it doesn't help you understand the "mechanics" of how it is done. The mechanics become important when designing the actual arithmetical circuits. Here are the basic techniques used to add and subtract binary numbers.

ADDING

FIGURE 12.6

Adding binary numbers is just like adding decimal numbers; whenever the result of adding one column of numbers is greater than one digit, a 1 is carried over to the next column to be added.

SUBTRACTION

FIGURE 12.7

Subtracting decimal numbers is not as easy as it looks. It is similar to decimal subtraction but can be confusing. For example, you might think that if you were to subtract a 1 from a 0, you would borrow a 1 from the column to the left. No! You must borrow a 10 (2_{10}). It becomes a headache if you try to do this by hand. The trick to subtracting binary numbers is to use the 2's complement representation that provides the sign bit and then just add the positive number with the negative number to get the sum. This method is often used by digital circuits because it allows both addition and subtraction, without the headache of having to subtract the smaller number from the larger number.

ASCII

ASCII (American Standard Code for Information Interchange) is an alphanumeric code used to transmit letters, symbols, numbers, and special nonprinting characters between computers and computer peripherals (e.g., printer, keyboard, etc.). ASCII consists of 128 different 7-bit codes. Codes from 000 0000 (or hex 00) to 001 1111 (or hex 1F) are reserved for nonprinting characters or special machine commands like ESC (escape), DEL (delete), CR (carriage return), LF (line feed), etc. Codes from 010 0000 (or hex 20) to 111 1111 (or hex 7F) are reserved for printing characters like a, A, #, &, {, @, 3, etc. See Tables 12.2 and 12.3. In practice, when ASCII code is sent, an additional bit is added to make it compatible with 8-bit systems. This bit may be set to 0 and ignored, it may be used as a parity bit for error detection (I will cover parity bits in Section 12.3), or it may act as a special function bit used to implement an additional set of specialized characters.

TABLE 12.2 **ASCII Nonprinting Characters**

DEC	HEX	7-BIT CODE	CONTROL CHAR	CHAR	MEANING	DEC	HEX	7-BIT	CONTROL CHAR	CHAR	MEANING
00	00	000 0000	ctrl-@	NUL	Null	16	10	001 0000	ctrl-P	DLE	Data line escape
01	01	000 0001	ctrl-A	SOH	Start of heading	17	11	001 0001	ctrl-Q	DC1	Device control 1
02	02	000 0010	ctrl-B	STX	Start of text	18	12	001 0010	ctrl-R	DC2	Device control 2
03	03	000 0011	ctrl-C	ETX	End of text	19	13	001 0011	ctrl-S	DC3	Device control 3
04	04	000 0100	ctrl-D	EOT	End of xmit	20	14	001 0100	ctrl-T	DC4	Device control 4
05	05	000 0101	ctrl-E	ENQ	Enquiry	21	15	001 0101	ctrl-U	NAK	Neg acknowledge
06	06	000 0110	ctrl-F	ACK	Acknowledge	22	16	001 0110	ctrl-V	SYN	Synchronous idle
07	07	000 0111	ctrl-G	BEL	Bell	23	17	001 0111	ctrl-W	ETB	End of xmit block
08	08	000 1000	ctrl-H	BS	Backspace	24	18	001 1000	ctrl-X	CAN	Cancel
09	09	000 1001	ctrl-I	HT	Horizontal tab	25	19	001 1001	ctrl-Y	EM	End of medium
10	0A	000 1010	ctrl-J	LF	Line feed	26	1A	001 1010	ctrl-Z	SUB	Substitute
11	0B	000 1011	ctrl-K	VT	Vertical tab	27	1B	001 1011	ctrl-[	ESC	Escape
12	0C	000 1100	ctrl-L	FF	Form feed	28	1C	001 1100	ctrl-\	FS	File separator
13	0D	000 1101	ctrl-M	CR	Carriage return	29	1D	001 1101	ctrl-]	GS	Group separator
14	0E	000 1110	ctrl-N	SO	Shift out	30	1E	001 1110	ctrl-^	RS	Record separator
15	0F	000 1111	ctrl-O	SI	Shift in	31	1F	001 1111	ctrl-_	US	Unit separator

TABLE 12.3 **ASCII Printing Characters**

DEC	HEX	7-BIT CODE	CHAR	DEC	HEX	7-BIT	CHAR	DEC	HEX	7-BIT CODE	CHAR
32	20	010 0000	SP	64	40	100 0000	@	96	60	110 0000	`
33	21	010 0001	!	65	41	100 0001	A	97	61	110 0001	a
34	22	010 0010	"	66	42	100 0010	B	98	62	110 0010	b
35	23	010 0011	#	67	43	100 0011	C	99	63	110 0011	c
36	24	010 0100	$	68	44	100 0100	D	100	64	110 0100	d
37	25	010 0101	%	69	45	100 0101	E	101	65	110 0101	e
38	26	010 0110	&	70	46	100 0110	F	102	66	110 0110	f
39	27	010 0111	'	71	47	100 0111	G	103	67	110 0111	g
40	28	010 1000	(	72	48	100 1000	H	104	68	110 1000	h
41	29	010 1001	)	73	49	100 1001	I	105	69	110 1001	i
42	2A	010 1010	*	74	4A	100 1010	J	106	6A	110 1010	j
43	2B	010 1011	+	75	4B	100 1011	K	107	6B	110 1011	k
44	2C	010 1100	,	76	4C	100 1100	L	108	6C	110 1100	l
45	2D	010 1101	-	77	4D	100 1101	M	109	6D	110 1101	m
46	2E	010 1110	.	78	4E	100 1110	N	110	6E	110 1110	n
47	2F	010 1111	/	79	4F	100 1111	O	111	6F	110 1111	o
48	30	011 0000	0	80	50	101 0000	P	112	70	111 0000	p
49	31	011 0001	1	81	51	101 0001	Q	113	71	111 0001	q
50	32	011 0010	2	82	52	101 0010	R	114	72	111 0010	r
51	33	011 0011	3	83	53	101 0011	S	115	73	111 0011	s
52	34	011 0100	4	84	54	101 0100	T	116	74	111 0100	t
53	35	011 0101	5	85	55	101 0101	U	117	75	111 0101	u
54	36	011 0110	6	86	56	101 0110	V	118	76	111 0110	v
55	37	011 0111	7	87	57	101 0111	W	119	77	111 0111	w
56	38	011 1000	8	88	58	101 1000	X	120	78	111 1000	x
57	39	011 1001	9	89	59	101 1001	Y	121	79	111 1001	y
58	3A	011 1010	:	90	5A	101 1010	Z	122	7A	111 1010	z
59	3B	011 1011	;	91	5B	101 1011	[	123	7B	111 1011	{
60	3C	011 1100	<	92	5C	101 1100	\	124	7C	111 1100	|
61	3D	011 1101	=	93	5D	101 1101	]	125	7D	111 1101	}
62	3E	011 1110	>	94	5E	101 1110	^	126	7E	111 1110	~
63	3F	011 1111	?	95	5F	101 1111	_	127	7F	111 1111	DEL

12.1.3   Clock Timing and Parallel versus Serial Transmission

Before moving on to the next section, let's take a brief look at three important items: clock timing, parallel transmission, and serial transmission.

Clock Timing

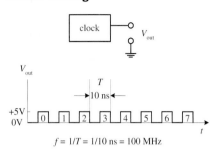

Digital circuits require precise timing to function properly. Usually, a clock circuit that generates a series of high and low pulses at a fixed frequency is used as a reference on which to base all critical actions executed within a system. The clock is also used to push bits of data through the digital circuitry. The period of a clock pulse is related to its frequency by $T = 1/f$. So, if $T = 10$ ns, then $f = 1/(10$ ns$) = 100$ MHz.

FIGURE 12.8

$f = 1/T = 1/10$ ns $= 100$ MHz

Serial versus Parallel Representation

Binary information can be transmitted from one location to another in either a serial or parallel manner. The serial format uses a single electrical conductor (and a common ground) for data transfer. Each bit from the binary number occupies a separate clock period, with the change from one bit to another occurring at each falling or leading clock edge—the type of edge depends on the circuitry used. Figure 12.9 shows an 8-bit (10110010) word that is transmitted from circuit A to circuit B in 8 clock pulses (0–7). In computer systems, serial communications are used to transfer data between keyboard and computer, as well as to transfer data between two computers via a telephone line.

Serial transmission of a 8-bit word (10110010)

Parallel transmission of a 8-bit word (10110010)

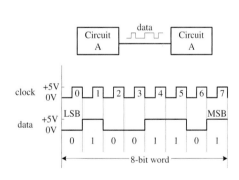

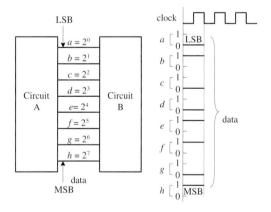

FIGURE 12.9

Parallel transmission uses separate electrical conductors for each bit (and a common ground). In Fig. 12.9, an 8-bit string (01110110) is sent from circuit A to circuit B. As you can see, unlike serial transmission, the entire word is transmitted in only one clock cycle, not 8 clock cycles. In other words, it is 8 times faster. Parallel communications are most frequently found within microprocessor systems that use multiline data and control buses to transmit data and control instructions from the microprocessor to other microprocessor-based devices (e.g., memory, output registers, etc.).

12.2 Logic Gates

Logic gates are the building blocks of digital electronics. The fundamental logic gates include the INVERT (NOT), AND, NAND, OR, NOR, exclusive OR (XOR), and exclusive NOR (XNOR) gates. Each of these gates performs a different logical operation. Figure 12.10 provides a description of what each logic gate does and gives a switch and transistor analogy for each gate.

FIGURE 12.10

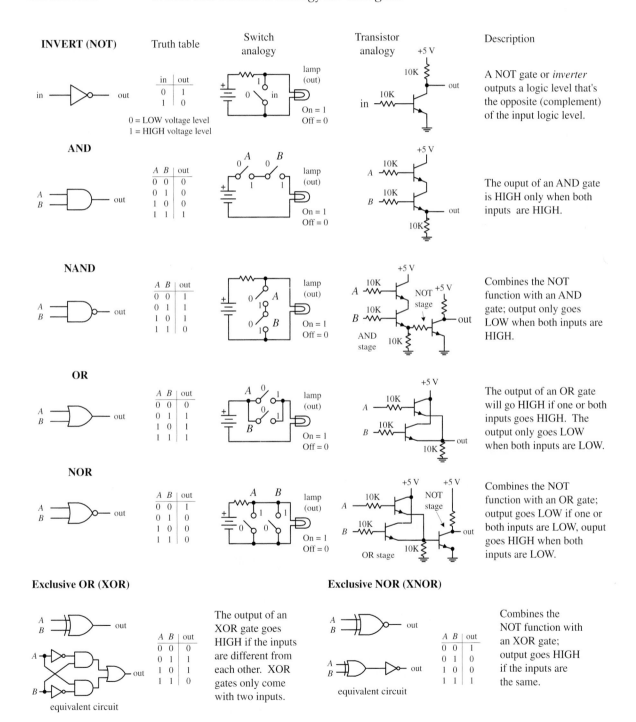

12.2.1   Multiple-Input Logic Gates

AND, NAND, OR, and NOR gates often come with more than two inputs (this is not the case with XOR and XNOR gates, which require two inputs only). Figure 12.11 shows a 4-input AND, an 8-input AND, a 3-input NOR, and an 8-input NOR gate. With the 8-input AND gate, all inputs must be high for the output to be high. With the 8-input OR gate, at least one of the inputs must be high for the output to go high.

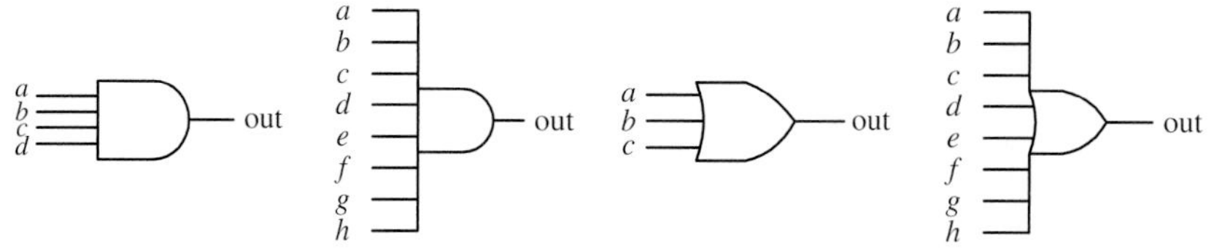

FIGURE 12.11

12.2.2   Digital Logic Gate ICs

The construction of digital gates is best left to the IC manufacturers. In fact, making gates from discrete components is highly impractical in regard to both overall performance (power consumption, speed, drive capacity, etc.) and overall cost and size.

There are a number of technologies used in the fabrication of digital logic. The two most popular technologies include TTL (transistor-transistor logic) and CMOS (complementary MOSFET) logic. TTL incorporates bipolar transistors into its design, while CMOS incorporates MOSFET transistors. Both technologies perform the same basic functions, but certain characteristics (e.g., power consumption, speed, output drive capacity, etc.) differ. There are many subfamilies within both TTL and CMOS. These subfamilies, as well as the various characteristics associated with each subfamily, will be discussed in greater detail in Section 12.4.

A logic IC, be it TTL or CMOS, typically houses more than one logic gate (e.g., quad 2-input NAND, hex inverter, etc.). Each of the gates within the IC shares a common supply voltage that is implemented via two supply pins, a positive supply pin ($+V_{CC}$ or $+V_{DD}$) and a ground pin (GND). The vast majority of TTL and CMOS ICs are designed to run off a +5-V supply. (This does not apply for all the logic families, but I will get to that later.)

Generally speaking, input and output voltage levels are assumed to be 0 V (low) and +5 V (high). However, the actual input voltage required and the actual output voltage provided by the gate are not set in stone. For example, the 74xx TTL series will recognize a high input from 2.0 to 5 V, a low from 0 to 0.8 V, and will guarantee a high output from 2.4 to 5 V and a low output from 0 to 0.4 V. However, for the CMOS 4000B series (V_{CC} = +5 V), recognizable input voltages range from 3.3 to 5 V for high, 0 to 1.7 V for low, while guaranteed high and low output levels range from 4.9 to 5 V and 0 to 0.1 V, respectively. Again, I will discuss specifics later in Section 12.4. For now, let's just get acquainted with what some of these ICs look like—see Figs. 12.12 and 12.13. [CMOS devices listed in the figures include 74HCxx, 4000(B), while TTL devices shown include the 74xx, 74Fxx, 74LS.]

7400 Series

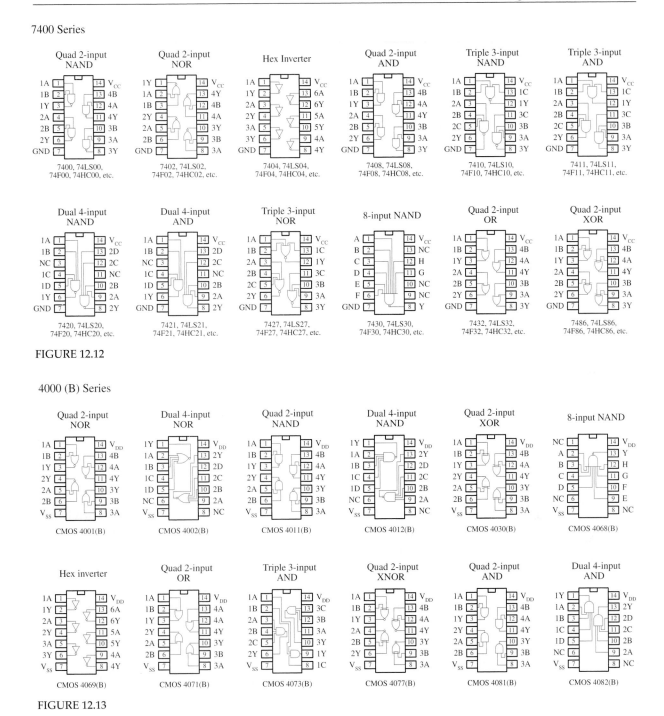

FIGURE 12.12

4000 (B) Series

FIGURE 12.13

12.2.3 Applications for a Single Logic Gate

Before we jump into the heart of logic gate applications that involve combining logic gates to form complex decision-making circuits, let's take a look at a few simple applications that require the use of a single logic gate.

Enable/Disable Control

An enable/disable gate is a logic gate that acts to control the passage of a given waveform. The waveform, say, a clock signal, is applied to one of the gate's inputs, while the other input acts as the enable/disable control lead. Enable/disable gates are used frequently in digital systems to enable and disable control information from reaching various devices. Figure 12.14 shows two enable/disable circuits; the first uses an AND gate, and the second uses an OR gate. NAND and NOR gates are also frequently used as enable gates.

Using an AND as an enable gate

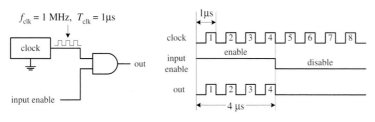

In the upper part of the figure, an AND gate acts as the enable gate. When the input enable lead is made high, the clock signal will pass to the output. In this example, the input enable is held high for 4 µs, allowing 4 clock pulses (where T_{clk} = 1 µs) to pass. When the input enable lead is low, the gate is disabled, and no clock pulses make it through to the output.

Below, an OR gate is used as the enable gate. The output is held high when the input enable lead is high, even as the clock signal is varying. However, when the enable input is low, the clock pulses are passed to the output.

Using an OR as an enable gate

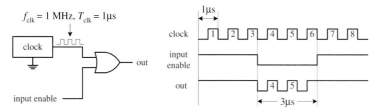

FIGURE 12.14

Waveform Generation

By using the basic enable/disable function of a logic gate, as illustrated in the last example, it is possible, with the help of a repetitive waveform generator circuit, to create specialized waveforms that can be used for the digital control of sequencing circuits. An example waveform generator circuit is the Johnson counter, shown below. The Johnson counter will be discussed in Section 12.8—for now let's simply focus on the outputs. In the figure below, a Johnson counter uses clock pulses to generate different output waveforms, as shown in the timing diagram. Outputs A, B, C, and D go high for 4 µs (four clock periods) and are offset from each other by 1 µs. Outputs $\overline{A}$, $\overline{B}$, $\overline{C}$, and $\overline{D}$ produce waveforms that are complements of outputs A, B, C, and D, respectively.

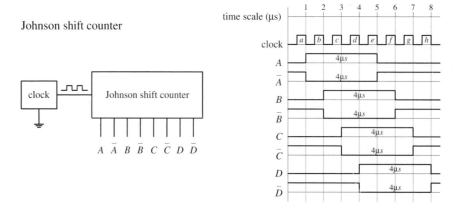

FIGURE 12.15

Now, there may be certain applications that require 4-µs high/low pulses applied at a given time—as the counter provides. However, what would you do if the application required a 3-μs high waveform that begins at 2 μs and ends at 5 μs (relative to the time scale indicated in the figure above)? This is where the logic gates come in handy. For example, if you attach an AND gate's inputs to the counter's *A* and *B* outputs, you will get the desired 2- to 5-μs high waveform at the AND gate's output: from 1 to 2 μs the AND gate outputs a low ($A = 1$, $B = 0$), from 2 to 5 μs the AND gate outputs a high ($A = 1$, $B = 1$), and from 5 to 6 μs the AND gate outputs a low ($A = 0$, $B = 1$). See the leftmost figure below.

Connections for 1μs to 5μs waveform

Other possible connections and waveforms

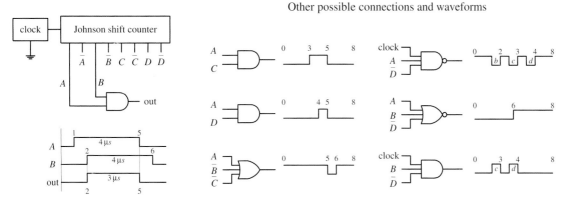

FIGURE 12.16

Various other specialized waveforms can be generated by using different logic gates and tapping different outputs of the Johnson shift counter. In the figure above and to the left, six other possibilities are shown.

12.2.4 Combinational Logic

Combinational logic involves combining logic gates together to form circuits capable of enacting more useful, complex functions. For example, let's design the logic used to instruct a janitor-type robot to recharge itself (seek out a power outlet) only when a specific set of conditions is met. The "recharge itself" condition is specified as follows: when its battery is low (indicated by a high output voltage from a battery-monitor circuit), when the workday is over (indicated by a high output voltage from a timer circuit), when vacuuming is complete (indicated by a high voltage output from a vacuum-completion monitor circuit), and when waxing is complete (indicated by a high output voltage from a wax-completion monitor circuit). Let's also assume that the power-outlet-seeking routine circuit is activated when a high is applied to its input.

Two simple combinational circuits that perform the desired logic function for the robot are shown in Fig. 12.17. The two circuits use a different number of gates but perform the same function. Now, the question remains, how did we come up with these circuits? In either circuit, it is not hard to predict what gates are needed. You simply exchange the word *and* present within the conditional statement with an AND gate within the logic circuit and exchange the word *or* present within the conditional statement with an OR gate within the logic circuit. Common sense takes care of the rest.

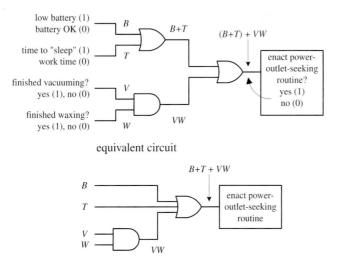

B	T	V	W	B+T	VW	(B+T)+VW
0	0	0	0	0	0	0
0	0	0	1	0	0	0
0	0	1	0	0	0	0
0	0	1	1	0	1	1
0	1	0	0	1	0	1
0	1	0	1	1	0	1
0	1	1	0	1	0	1
0	1	1	1	1	1	1
1	0	0	0	1	0	1
1	0	0	1	1	0	1
1	0	1	0	1	0	1
1	0	1	1	1	1	1
1	1	0	0	1	0	1
1	1	0	1	1	0	1
1	1	1	1	1	1	1

FIGURE 12.17

However, when you begin designing more complex circuits, using intuition to figure out what kind of logic gates to use and how to join them together becomes exceedingly difficult. To make designing combinational circuits easier, a special symbolic language called *Boolean algebra* is used, which only works with true and false variables. A Boolean expression for the robot circuit would appear as follows:

$$E = (B + T) + VW$$

This expression amounts to saying that if B (battery-check circuit's output) *or* T (timer circuit's output) is true *or* V *and* W (vacuum and waxing circuit outputs) are true, than E (enact power-outlet circuit input) is true. Note that the word *or* is replaced by the symbol + and the word *and* is simply expressed in a way similar to multiplying two variables together (placing them side-by-side or using a dot between variables). Also, it is important to note that the term *true* in Boolean algebra is expressed as a 1, while *false* is expressed as a 0. Here we are assuming positive logic, where true = high voltage. Using the Boolean expression for the robot circuit, we can come up with some of the following results (the truth table in Fig. 12.17 provides all possible results):

$E = (B + T) + VW$
$E = (1 + 1) + (1 \cdot 1) = 1 + 1 = 1$ (battery is low, time to sleep, finished with chores = go recharge).
$E = (1 + 0) + (0 \cdot 0) = 1 + 0 = 1$ (battery is low = go recharge).
$E = (0 + 0) + (1 \cdot 0) = 0 + 0 = 0$ (hasn't finished waxing = don't recharge yet).
$E = (0 + 0) + (1 \cdot 1) = 0 + 1 = 1$ (has finished all chores = go recharge).
$E = (0 + 0) + (0 \cdot 0) = 0 + 0 = 0$ (hasn't finished vacuuming and waxing = don't recharge yet).

The robot example showed you how to express AND and OR functions in Boolean algebraic terms. But what about the negation operations (NOT, NAND, NOR) and the exclusive operations (XOR, XNOR)? How do you express these in Boolean terms? To represent a NOT condition, you simply place a line over the NOT'ed variable or variables. For a NAND expression, you simply place a line over an AND expression. For a NOR expression, you simply place a line over an OR expression. For exclusive operations, you use the symbol ⊕. Here's a rundown of all the possible Boolean expressions for the various logic gates.

Boolean expressions for the logic gates

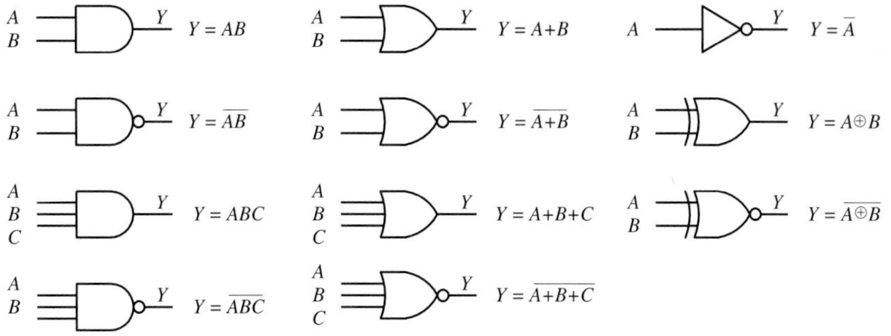

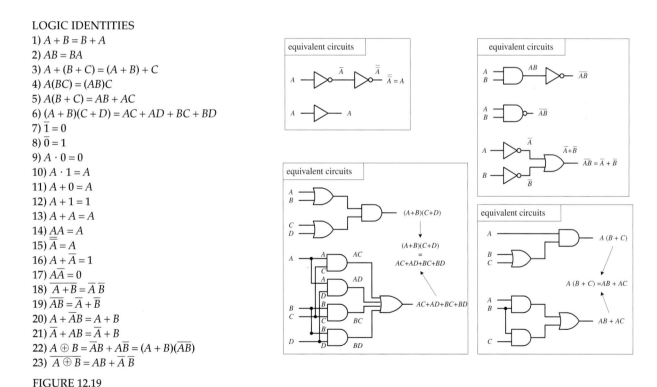

FIGURE 12.18

Like conventional algebra, Boolean algebra has a set of logic identities that can be used to simplify the Boolean expressions and thus make circuits more compact. These identities go by such names as the *commutative law of addition, associate law of addition, distributive law,* etc. Instead of worrying about what the various identities are called, simply make reference to the list of identities provided below and to the left. Most of these identities are self-explanatory, although a few are not so obvious, as you will see in a minute. The various circuits below and to the right show some of the identities in action.

LOGIC IDENTITIES

1) $A + B = B + A$
2) $AB = BA$
3) $A + (B + C) = (A + B) + C$
4) $A(BC) = (AB)C$
5) $A(B + C) = AB + AC$
6) $(A + B)(C + D) = AC + AD + BC + BD$
7) $\overline{1} = 0$
8) $\overline{0} = 1$
9) $A \cdot 0 = 0$
10) $A \cdot 1 = A$
11) $A + 0 = A$
12) $A + 1 = 1$
13) $A + A = A$
14) $AA = A$
15) $\overline{\overline{A}} = A$
16) $A + \overline{A} = 1$
17) $A\overline{A} = 0$
18) $\overline{A + B} = \overline{A}\,\overline{B}$
19) $\overline{AB} = \overline{A} + \overline{B}$
20) $A + \overline{A}B = A + B$
21) $\overline{A} + AB = \overline{A} + B$
22) $A \oplus B = \overline{A}B + A\overline{B} = (A + B)(\overline{AB})$
23) $\overline{A \oplus B} = AB + \overline{A}\,\overline{B}$

FIGURE 12.19

EXAMPLE

Let's find the initial Boolean expression for the circuit in Fig. 12.20 and then use the logic identities to come up with a circuit that requires fewer gates.

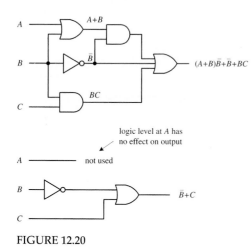

FIGURE 12.20

The circuit shown here is expressed by the following Boolean expression:

$$out = (A + B)\overline{B} + \overline{B} + BC$$

This expression can be simplified by using Identity 5:

$$(A + B)\overline{B} = A\overline{B} + B\overline{B}$$

This makes

$$out = A\overline{B} + B\overline{B} + \overline{B} + BC$$

Using Identities 17 ($B\overline{B} = 0$) and 11 ($B + 0 = B$), you get

$$out = A\overline{B} + 0 + \overline{B} + BC = A\overline{B} + BC$$

Factoring a $\overline{B}$ from the preceding term gives

$$out = \overline{B}(A + 1) + BC$$

Using Identities 12 ($A + 1 = 1$) and 10, you get

$$out = B(1) + BC = \overline{B} + BC$$

Finally, using Identity 21, you get the simplified expression

$$out = \overline{B} + C$$

Notice that A is now missing. This means that the logic input at A has no effect on the output and therefore can omitted. From the reduction, you get the simplified circuit in the bottom part of the figure.

Dealing with Exclusive Gates (Identities 22 and 23)

Now let's take a look at a couple of not so obvious logic identities I mentioned a second ago, namely, those which involve the XOR (Identity 22) and XNOR (Identity 23) gates. The leftmost section below shows equivalent circuits for the XOR gate. In the lower two equivalent circuits, Identity 22 is proved by Boolean reduction. Equivalent circuits for the XNOR gate are show in the rightmost section below. To prove Identity 23, you can simply invert Identity 22.

FIGURE 12.21

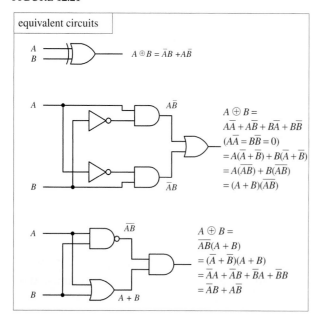

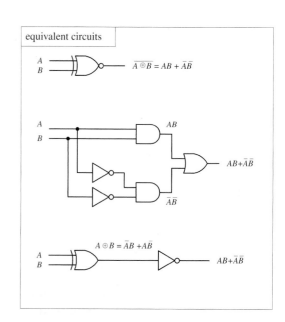

De Morgan's Theorem (Identities 18 and 19)

To simplify circuits containing NANDs and NORs, you can use an incredibly useful theorem known as *De Morgan's theorem*. This theorem allows you to convert an expression having an inversion bar over two or more variables into an expression having inversion bars over single variables only. De Morgan's theorem (Identities 18 and 19) is as follows:

$$\overline{A \cdot B} = \overline{A} + \overline{B} \quad \text{(2 variables)} \qquad \overline{A \cdot B \cdot C} = \overline{A} + \overline{B} + \overline{C} \quad \text{(3 or more variables)}$$
$$\overline{A + B} = \overline{A} \cdot \overline{B} \qquad\qquad\qquad \overline{A + B + C} = \overline{A} \cdot \overline{B} \cdot \overline{C}$$

The easiest way to prove that these identities are correct is to use the figure below, noting that the truth tables for the equivalent circuits are the same. Note the inversion bubbles present on the inputs of the corresponding leftmost gates. The inversion bubbles mean that before inputs A and B are applied to the base gate, they are inverted (negated). In other words, the bubbles are simplified expressions for NOT gates.

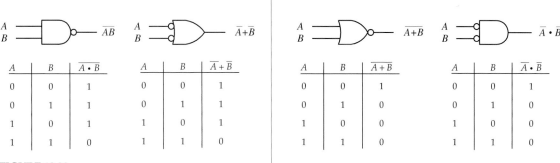

A	B	$\overline{A \cdot B}$
0	0	1
0	1	1
1	0	1
1	1	0

A	B	$\overline{A} + \overline{B}$
0	0	1
0	1	1
1	0	1
1	1	0

A	B	$\overline{A + B}$
0	0	1
0	1	0
1	0	0
1	1	0

A	B	$\overline{A} \cdot \overline{B}$
0	0	1
0	1	0
1	0	0
1	1	0

FIGURE 12.22

Why do you use the inverted-input OR gate symbol instead of a NAND gate symbol? Or why would you use the inverted-input AND gate symbol instead of a NOR gate symbol? This is a choice left up to the designer—whatever choice seems most logical to use. For example, when designing a circuit, it may be easier to think about ORing or ANDing inverted inputs than to think about NANDing or NORing inputs. Similarly, it may be easier to create truth tables or work with Boolean expressions using the inverted-input gate—it is typically easier to create truth tables and Boolean expressions that do not have variables joined under a common inversion bar. Of course, when it comes time to construct the actual working circuit, you probably will want to convert to the NAND and NOR gates because they do not require additional NOT gates at their inputs.

Bubble Pushing

A shortcut method for forming equivalent logic circuits, based on De Morgan's theorem, is to use what's called *bubble pushing*.

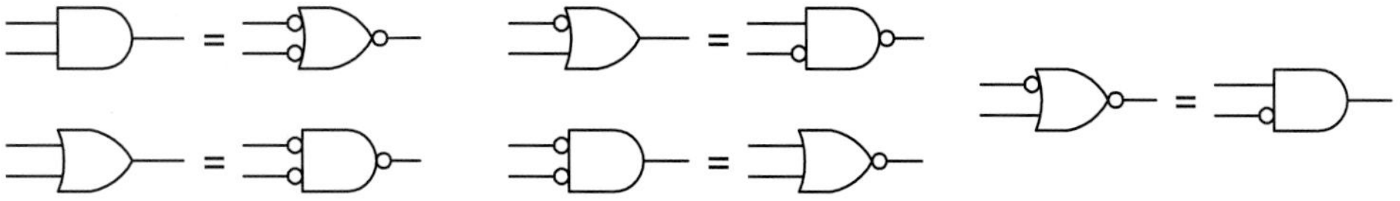

FIGURE 12.23

Bubble pushing involves the flowing tricks: First, change an AND gate to an OR gate or change an OR gate to an AND gate. Second, add inversion bubbles to the inputs and outputs where there were none, while removing the original bubbles. That's it. You can prove to yourself that this works by examining the corresponding truth tables for the original gate and the bubble-pushed gate, or you can work out the Boolean expressions using De Morgan's theorem. Figure 12.23 shows examples of bubble pushing.

Universal Capability of NAND and NOR Gates

NAND and NOR gates are referred to as *universal gates* because each alone can be combined together with itself to form all other possible logic gates. The ability to create any logic gate from NAND or NOR gates is obviously a handy feature. For example, if you do not have an XOR IC handy, you can use a single multigate NAND gate (e.g., 74HC00) instead. The figure below shows how to wire NAND or NOR gates together to create equivalent circuits of the various logic gates.

FIGURE 12.24

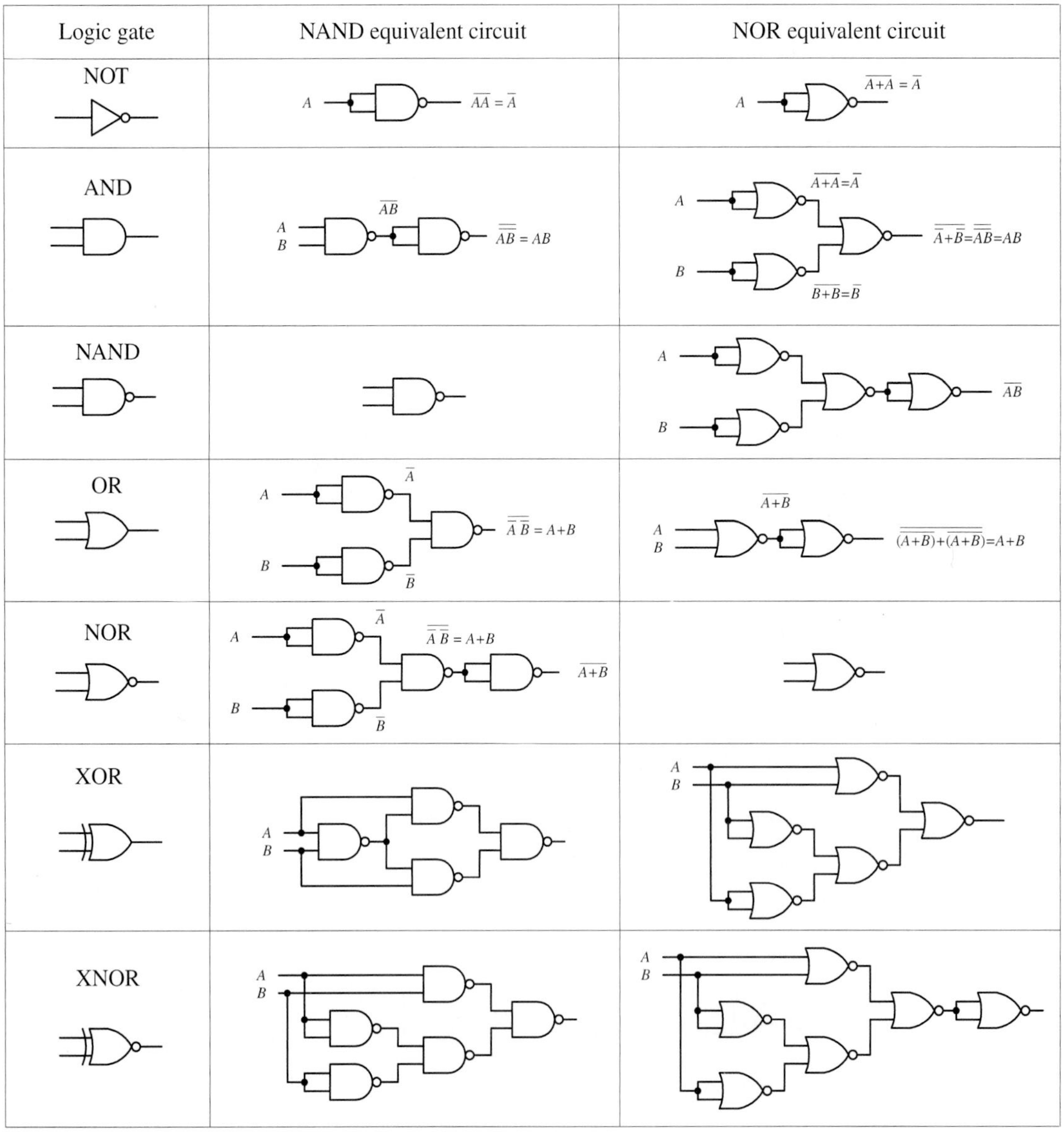

Logic gate	NAND equivalent circuit	NOR equivalent circuit
NOT	$\overline{AA} = \overline{A}$	$\overline{A+A} = \overline{A}$
AND	$\overline{AB}$; $\overline{\overline{AB}} = AB$	$\overline{A+A}=\overline{A}$; $\overline{\overline{A}+\overline{B}}=\overline{\overline{AB}}=AB$; $\overline{B+B}=\overline{B}$
NAND		$\overline{AB}$
OR	$\overline{A}$; $\overline{B}$; $\overline{\overline{A}\,\overline{B}} = A+B$	$\overline{A+B}$; $\overline{\overline{(A+B)}+\overline{(A+B)}}=A+B$
NOR	$\overline{A}$; $\overline{B}$; $\overline{\overline{A}\,\overline{B}} = A+B$; $\overline{A+B}$	
XOR		
XNOR		

AND-OR-INVERT Gates (AOIs)

When a Boolean expression is reduced, the equation that is left over typically will be of one of the following two forms: *product-of-sums* (POS) or *sum-of-products* (SOP). A POS expression appears as two or more ORed variables ANDed together with two or more additional ORed variables. An SOP expression appears as two or more ANDed variables ORed together with additional ANDed variables. The figure below shows two circuits that provide the same logic function (they are equivalent), but the circuit to the left is designed to yield a POS expression, while the circuit to the right is designed to yield a SOP expression.

Table made using SOP expression
(it's easier than POS)

A	B	C	D	$A\bar{C}$	AD	$\bar{B}\bar{C}$	$\bar{B}D$	X
0	0	0	0	0	0	1	0	1
0	0	0	1	0	0	1	1	1
0	0	1	0	0	0	0	0	0
0	0	1	1	0	0	0	1	1
0	1	0	0	0	0	0	0	0
0	1	0	1	0	0	0	0	0
0	1	1	0	0	0	0	0	0
0	1	1	1	0	0	0	0	0
1	0	0	0	1	0	1	0	1
1	0	0	1	1	1	1	1	1
1	0	1	0	0	0	0	0	0
1	0	1	1	0	1	0	1	1
1	1	0	0	1	0	0	0	1
1	1	0	1	1	1	0	0	1
1	1	1	0	0	0	0	0	0
1	1	1	1	0	1	0	0	1

Logic circuit for POS expression

$$X = (A + \bar{B})(\bar{C} + D)$$

Logic circuit for SOP expression

$$X = A\bar{C} + AD + \bar{B}\bar{C} + \bar{B}D$$

FIGURE 12.25

In terms of design, which circuit is best, the one that implements the POS expression or the one that implements SOP expression? The POS design shown here would appear to be the better choice because it requires fewer gates. However, the SOP design is nice because it is easy to work with the Boolean expression. For example, which Boolean expression above (POS or SOP) would you rather use to create a truth table? The SOP expression seems the obvious choice. A more down-to-earth reason for using an SOP design has to do with the fact that special ICs called AND-OR-INVERT (AOI) gates are designed to handle SOP expressions. For example, the 74LS54 AOI IC shown below creates an inverted SOP expression at its output, via two 2-input AND gates and two 3-input AND gates NORed together. A NOT gate can be attached to the output to get rid of the inversion bar, if desired. If specific inputs are not used, they should be held high, as shown in the example circuit below and to the far left. AOI ICs come in many different configurations—check out the catalogs to see what's available.

FIGURE 12.26

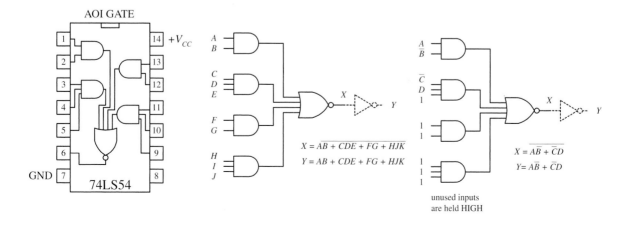

AOI GATE

74LS54

GND

$$X = \overline{AB + CDE + FG + HJK}$$
$$Y = AB + CDE + FG + HJK$$

$$X = \overline{A\bar{B} + \bar{C}D}$$
$$Y = A\bar{B} + \bar{C}D$$

unused inputs
are held HIGH

12.2.5 Keeping Circuits Simple (Karnaugh Maps)

We have just seen how using the logic identities can simplify a Boolean expression. This is important because it reduces the number of gates needed to construct the logic circuit. However, as I am sure you will agree, having to work out Boolean problems in longhand is not easy. It takes time and ingenuity. Now, a simple way to avoid the unpleasant task of using your ingenuity is to get a computer program that accepts a truth table or Boolean expression and then provides you with the simplest expression and perhaps even the circuit schematic. However, let's assume that you do not have such a program to help you out. Are you stuck with the Boolean longhand approach? No. What you do is use a technique referred to as *Karnaugh mapping*. With this technique, you take a given truth table (or Boolean expression that can be converted into a truth table), convert it into a Karnaugh map, apply some simple graphic rules, and come up with the simplest (most of the time) possible Boolean expression for your final circuit. Karnaugh mapping works best for circuits with three to four inputs—below this, things usually do not require much thought anyway; beyond four inputs, things get quite tricky. Here's a basic outline showing how to apply Karnaugh mapping to a three-input system:

1. First, select a desired truth table. Let's choose the one shown in Fig. 12.27. (If you only have a Boolean expression, transform it into an SOP expression and use the SOP expression to create the truth table—refer to Fig. 12.26 to figure out how this is done.)

2. Next, translate the truth table into a Karnaugh map. A Karnaugh map is similar to a truth table but has its variables represented along two axes. Translating the truth

FIGURE 12.27

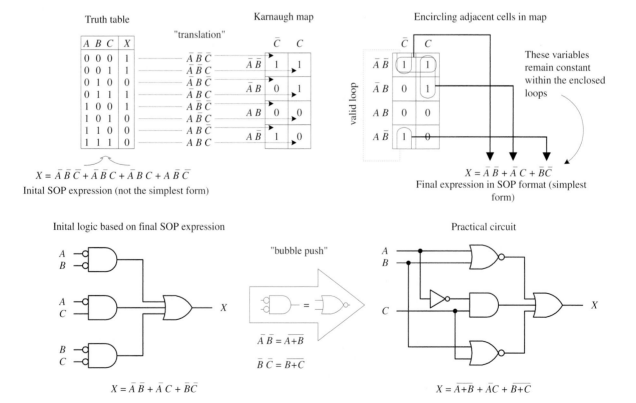

table into a Karnaugh map reduces the number of 1s and 0s needed to present the information. Figure 12.27 shows how the translation is carried out.

3. After you create the Karnaugh map, you proceed to encircle adjacent cells of 1s into groups of 2, 4, or 8. The more groups you can encircle, the simpler the final equation will be. In other words, take all possible loops.

4. Now, identify the variables that remain constant within each loop, and write out an SOP equation by ORing these variables together. Here, *constant* means that a variable and its inverse are not present together within the loop. For example, the top horizontal loop in Fig. 12.27 yields $\overline{A}\,\overline{B}$ (the first term in the SOP expression), since $\overline{A}$'s and $\overline{B}$'s inverses (A and B) are not present. However, the C variable is omitted from this term because C and $\overline{C}$ are both present.

5. The SOP expression you end up with is the simplest possible expression. With it you can create your logic circuit. You may have to apply some bubble pushing to make the final circuit practical, as shown in the figure below.

To apply Karnaugh mapping to four-input circuits, you apply the same basic steps used in the three-input scheme. However, now you use must use a 4 × 4 Karnaugh map to hold all the necessary information. Here is an example of how a four-input truth table (or unsimplified four-variable SOP expression) can be mapped and converted into a simplified SOP expression that can be used to create the final logic circuit:

inputs				out
A	B	C	D	Y
0	0	0	0	0
0	0	0	1	1
0	0	1	0	0
0	0	1	1	1
0	1	0	0	0
0	1	0	1	1
0	1	1	0	1
0	1	1	1	1
1	0	0	0	0
1	0	0	1	1
1	0	1	0	0
1	0	1	1	1
1	1	0	0	0
1	1	0	1	1
1	1	1	0	0
1	1	1	1	1

FIGURE 12.28

Unsimplified SOP expression:

$$\overline{A}\cdot\overline{B}\cdot\overline{C}\cdot D + \overline{A}\cdot\overline{B}\cdot C\cdot D + \overline{A}\cdot B\cdot\overline{C}\cdot D + \overline{A}\cdot B\cdot C\cdot\overline{D}$$
$$+\,\overline{A}\cdot B\cdot C\cdot D + A\cdot\overline{B}\cdot\overline{C}\cdot D + A\cdot\overline{B}\cdot C\cdot D + A\cdot B\cdot\overline{C}\cdot D + A\cdot B\cdot C\cdot D = Y$$

Simplified SOP expression and circuit

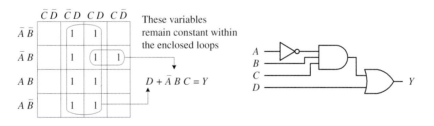

$$D + \overline{A}\,B\,C = Y$$

Here's an example that uses an AOI IC to implement the final SOP expression after mapping. I've thrown in variables other than the traditional *A, B, C,* and *D* just to let you know you are not limited to them alone. The choice of variables is up to you and usually depends on the application.

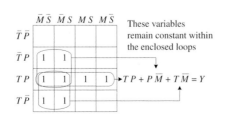

$$T\,P + P\,\overline{M} + T\,\overline{M} = Y$$

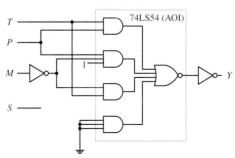

FIGURE 12.29

Other Looping Configurations

Here are examples of other looping arrangements used with 4×4 Karnaugh maps:

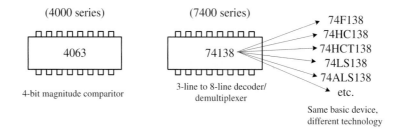

FIGURE 12.30

$$Y = BD + B\bar{C}$$

$$Y = ABD + ABC + CD$$

$$Y = \bar{B} + \bar{A}\,\bar{D}$$

$$Y = \bar{A} + \bar{B}\,\bar{D}$$

12.3   Combinational Devices

Now that you know a little something about how to use logic gates to enact functions represented within truth tables and Boolean expressions, it is time to take a look at some common functions that are used in the real world of digital electronics. As you will see, these functions are usually carried out by an IC that contains all the necessary logic.

A word on IC part numbers before I begin. As with the logic gate ICs, the combinational ICs that follow will be of either the 4000 or 7400 series. It is important to note that an original TTL IC, like the 74138, is essentially the same device (same pinouts and function—usually, but not always) as its newer counterparts, the 74F138, 74HC128 (CMOS), 74LS138, etc. The practical difference resides in the overall performance of the device (speed, power dissipation, voltage level rating, etc.). I will get into these gory details in a bit.

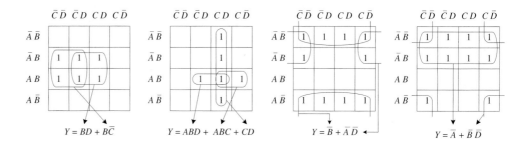

FIGURE 12.31

12.3.1   Multiplexers (Data Selectors) and Bilateral Switches

Multiplexers or data selectors act as digitally controlled switches. (The term *data selector* appears to be the accepted term when the device is designed to act like an SPDT switch, while the term *multiplexer* is used when the throw count of the "switch" exceeds two, e.g., SP8T. I will stick with this convention, although others may not.) A simple 1-of-2 data selector built from logic gates is shown in Fig. 12.32. The data select input of this circuit acts to control which input (*A* or *B*) gets passed to the output: When data select is high, input *A* passes while *B* is blocked. When data select is low, input *B* is passed while *B* is blocked. To understand how this circuit works, think of the AND gates as enable gates.

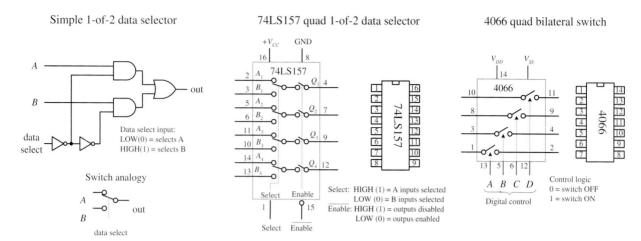

FIGURE 12.32

There are a number of different types of data selectors that come in IC form. For example, the 74LS157 quad 1-of-2 data selector IC, shown in Fig. 12.32, acts like an electrically controlled quad SPDT switch (or if you like, a 4PDT switch). When its select input is set high (1), inputs A_1, A_2, A_3, and A_4 are allowed to pass to outputs Q_1, Q_2, Q_3, and Q_4. When its select input is low (0), inputs B_1, B_2, B_3, and B_4 are allowed to pass to outputs Q_1, Q_2, Q_3, and Q_4. Either of these two conditions, however, ultimately depends on the state of the enable input. When the enable input is low, all data input signals are allowed to pass to the output; however, if the enable is high, the signals are not allowed to pass. This type of enable control is referred to as *active-low* enable, since the active function (passing the data to the output) occurs only with a low-level input voltage. The active-low input is denoted with a bubble (inversion bubble), while the outer label of the active-low input is represented with a line over it. Sometimes people omit the bubble and place a bar over the inner label. Both conventions are used commonly.

Figure 12.33 shows a 4-line-to-1-line multiplexer built with logic gates. This circuit resembles the 2-of-1 data selector shown in Fig. 12.32 but requires an additional select input to provide four address combinations.

FIGURE 12.33

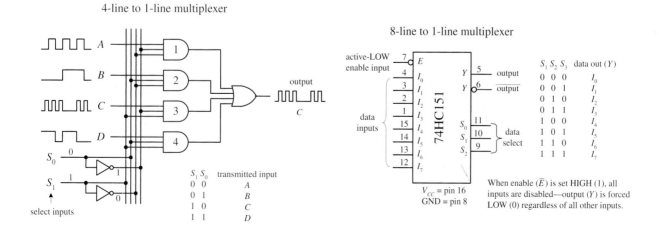

In terms of ICs, there are multiplexers of various input line capacities. For example, the 74151 8-line-to-1-line multiplexer uses three select inputs (S_0, S_1, S_2) to choose among 1 of 8 possible data inputs (I_0 to I_7) to be funneled to the output. Note that this device actually has two outputs, one true (pin 5) and one inverted (pin 6). The active-low enable forces the true output low when set high, regardless of the inputs.

To create a larger multiplexer, you combine two smaller multiplexers together. For example, Fig. 12.34 shows two 8-line-to-1-line 74HC151s combined to create a 16-line-to-1-line multiplexer. Another alternative is to use a 16-line-to-1-line multiplexer IC like the 74HC150 shown below. Check the catalogs to see what other kinds of multiplexers are available.

Combining two 8-line-to-1-line multiplexers to create a 16-line-to-1-line multiplexer

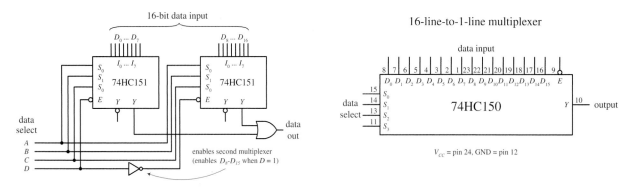

FIGURE 12.34

Finally, let's take a look at a very useful device called a *bilateral switch*. An example bilateral switch IC is the 4066, shown to the far left in Fig. 12.32. Unlike the multiplexer, this device merely acts as a digitally controlled quad SPST switch or quad transmission gate. Using a digital control input, you select which switches are on and which switches are off. To turn on a given switch, apply a high level to the corresponding switch select input; otherwise, keep the select input low.

Later in this chapter you come across analog switches and multiplexers. These devices use digital select inputs to control analog signals. Analog switches and multiplexers become important when you start linking the digital world to the analog world.

12.3.2   Demultiplexers (Data Disctributors) and Decoders

A demultiplexer (or data distributor) is the opposite of a multiplexer. It takes a single data input and routes it to one of several possible outputs. A simple four-line demultiplexer built from logic gates is shown in Fig. 12.35 *left*. To select the output (*A, B, C,* or *D*) to which you want to send the input signal (applied at *E*), you apply logic levels to the data select inputs (S_0, S_1), as shown in the truth table. Notice that the unselected outputs assume a high level, while the selected output varies with the input signal. An IC that contains two functionally separate four-line demultiplexers is the 74HC139, shown in Fig. 12.35 *right*. If you need more outputs, check out the 75*xx*154 16-line demultiplexer. This IC uses four data select inputs to choose from 1 of 16 possible outputs. Check out the catalogs to see what other demultiplexers exist.

4-line demultiplexer logic circuit   74HC139 dual 4-line demultiplexer

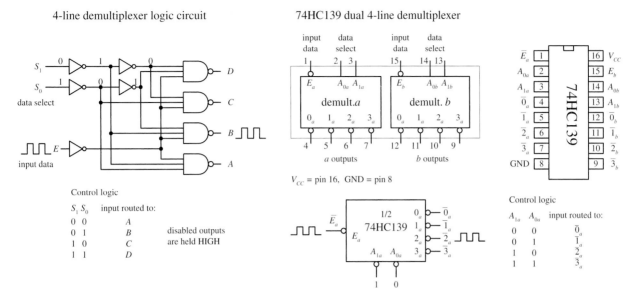

FIGURE 12.35

A decoder is somewhat like a demultiplexer, but it does not route input data to a specific output via data select inputs. Instead, it simply uses the data select inputs to choose which output (or outputs) among many are to be made high or low. The number of address inputs, the number of outputs, and the active state of the selected output vary from decoder to decoder. The variance is of course based on what the decoder is designed to do.

For example, the 74LS138 1-of-8 decoder shown in Fig. 12.36 uses a 3-bit address input to select which of 8 outputs will be made low—all other outputs are held high. Like the demultiplexer in Fig. 12.35, this decoder has active-low outputs.

Logic diagram 74LS138 1-of-8 decoder

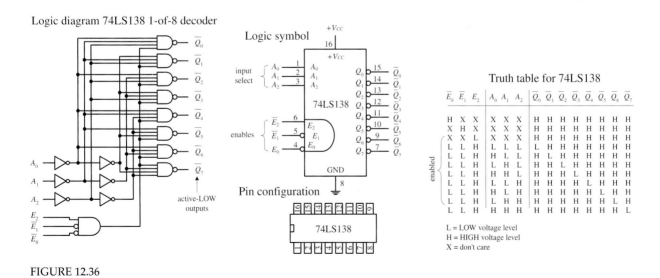

FIGURE 12.36

Now what exactly does it mean to say an output is an active-low output? It simply means that when an active-low output is selected, it is forced to a low logic state; otherwise, it is held high. Active-high outputs behave in the opposite manner. An active-low output is usually indicated with a bubble, although often it is indicated with a bared variable within the IC logic symbol—no bubble included. Active-high outputs have no bubbles. Both active-low and active-high outputs are equally common among ICs. By placing a load (e.g., warning LED) between $+V_{CC}$ and an active-low output, you can sink current through the load and into the active-low output when the output is selected. By placing a load between an active-high output and ground, you can source current from the active-high output and sink it through the load when the output is selected. There are of course limits to how much current an IC can source or sink. I will discuss these limits in Section 12.4, and I will present various schemes used to drive analog loads in Section 12.10.

Now let's get back to the 74LS138 decoder and discuss the remaining enable inputs ($\overline{E}_0$, $\overline{E}_1$, E_2). For the 74LS138 to "decode," you must make the active-low inputs $\overline{E}_0$ and $\overline{E}_1$ low while making the active-high input E_2 high. If any other set of enable inputs is applied, the decoder is disabled, making all active-low outputs high regardless of the selected inputs.

Other common decoders include the 7442 BCD-to-DEC (decimal) decoder, the 74154 1-of-16 (Hex) decoder, and the 7447 BCD-to-seven-segment decoder shown below. Like the preceding decoder, these devices also have active-low outputs. The 7442 uses a binary-coded decimal input to select 1 of 10 (0 through 9) possible outputs. The 74154 uses a 4-bit binary input to address 1 of 16 (or 0 of 15) outputs, making that output low (all others high), provided the enables are both set low.

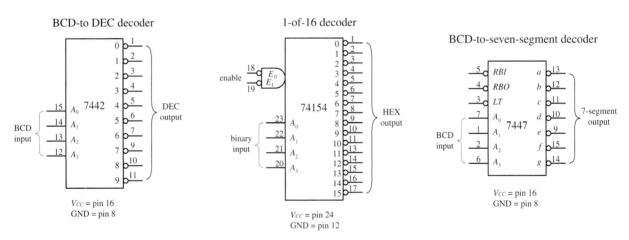

FIGURE 12.37

Now the 7447 is a bit different from the other decoders. With this device, more than one output can be driven low at a time. This is important because it allows the 7447 to drive a seven-segement LED display; to create different numbers requires driving more than one LED segment at a time. For example, in Fig. 12.38, when the BCD number for 5 (0101) is applied to the 7447's inputs, all outputs except $\overline{b}$ and $\overline{c}$ go low. This causes LED segments a, d, e, f, and g to light up—the 7447 sinks current through these LED segments, as indicated by the internal wiring of the display and the truth table.

7447 BCD-to-7-segment decoder/LED driver IC

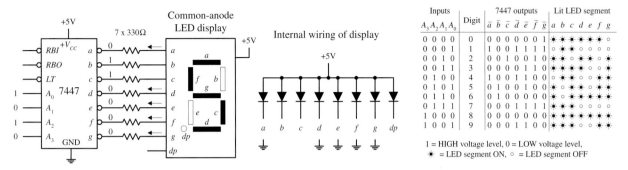

Inputs	Digit	7447 outputs	Lit LED segment
$A_3 A_2 A_1 A_0$		$\bar{a}\ \bar{b}\ \bar{c}\ \bar{d}\ \bar{e}\ \bar{f}\ \bar{g}$	$a\ b\ c\ d\ e\ f\ g$
0 0 0 0	0	0 0 0 0 0 0 1	✹✹✹✹✹✹○
0 0 0 1	1	1 0 0 1 1 1 1	○✹✹○○○○
0 0 1 0	2	0 0 1 0 0 1 0	✹✹○✹✹○✹
0 0 1 1	3	0 0 0 0 1 1 0	✹✹✹✹○○✹
0 1 0 0	4	1 0 0 1 1 0 0	○✹✹○○✹✹
0 1 0 1	5	0 1 0 0 1 0 0	✹○✹✹○✹✹
0 1 1 0	6	0 1 0 0 0 0 0	✹○✹✹✹✹✹
0 1 1 1	7	0 0 0 1 1 1 1	✹✹✹○○○○
1 0 0 0	8	0 0 0 0 0 0 0	✹✹✹✹✹✹✹
1 0 0 1	9	0 0 0 1 1 0 0	✹✹✹○○✹✹

1 = HIGH voltage level, 0 = LOW voltage level,
✹ = LED segment ON, ○ = LED segment OFF

FIGURE 12.38

The 7447 also comes with a lamp test active-low input ($\overline{LT}$) that can be used to drive all LED segments at once to see if any of the segments are faulty. The ripple blanking input ($\overline{RBI}$) and ripple blanking output ($\overline{RBO}$) can be used in multistage display applications to suppress a leading-edge and/or trailing-edge zero in a multi-digit decimal. For example, using the ripple blanking inputs and outputs, it is possible to take an 8-digit expression like 0056.020 and display 56.02, suppressing the two leading zeros and the one trailing zero. Leading-edge zero suppression is obtained by connecting the ripple blanking output of a decoder to the ripple blanking input of the next lower-stage device. The most significant decoder stage should have its ripple blanking input grounded. A similar procedure is used to provide automatic suppression of trailing zeros in the fractional part of the decimal.

12.3.3   Encoders and Code Converters

Encoders are the opposite of decoders. They are used to generate a coded output from a single active numeric input. To illustrate this in a simple manner, let's take a look at the simple decimal-to-BCD encoder circuit shown below.

Simple decimal-to-BCD encoder

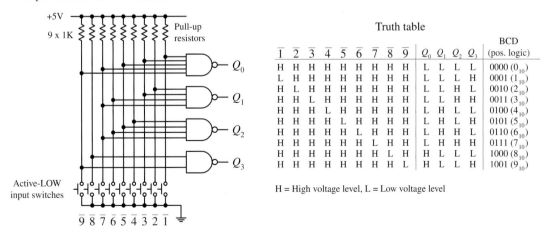

Truth table

$\bar{1}$	$\bar{2}$	$\bar{3}$	$\bar{4}$	$\bar{5}$	$\bar{6}$	$\bar{7}$	$\bar{8}$	$\bar{9}$	Q_0	Q_1	Q_2	Q_3	BCD (pos. logic)
H	H	H	H	H	H	H	H	H	L	L	L	L	0000 (0_{10})
L	H	H	H	H	H	H	H	H	L	L	L	H	0001 (1_{10})
H	L	H	H	H	H	H	H	H	L	L	H	L	0010 (2_{10})
H	H	L	H	H	H	H	H	H	L	L	H	H	0011 (3_{10})
H	H	H	L	H	H	H	H	H	L	H	L	L	0100 (4_{10})
H	H	H	H	L	H	H	H	H	L	H	L	H	0101 (5_{10})
H	H	H	H	H	L	H	H	H	L	H	H	L	0110 (6_{10})
H	H	H	H	H	H	L	H	H	L	H	H	H	0111 (7_{10})
H	H	H	H	H	H	H	L	H	H	L	L	L	1000 (8_{10})
H	H	H	H	H	H	H	H	L	H	L	L	H	1001 (9_{10})

H = High voltage level, L = Low voltage level

FIGURE 12.39

In this circuit, normally all lines are held high by the pull-up resistors connected to +5 V. To generate a BCD output that is equivalent to a single selected decimal input, the switch corresponding to that decimal is closed. (The switch acts as an active-low input.) The truth table in Fig. 12.39 explains the rest.

Figure 12.40 shows a 74LS147 decimal-to-BCD (10-line-to-4-line) priority encoder IC. The 74LS147 provides the same basic function as the circuit shown in Fig. 12.39, but it has active-low outputs. This means that instead of getting an LLHH output when "3" is selected, as in the previous encoder, you get HHLL. The two outputs represent the same thing ("3"); one is expressed in positive true logic, and the other (the 74LS147) is expressed in negative true logic. If you do not like negative true logic, you can slap inverters on the outputs of the 74LS147 to get positive true logic. The choice to use positive or negative true logic really depends on what you are planning to drive. For example, negative true logic is useful when the device that you wish to drive uses active-low inputs.

74LS147 decimal-to-4-bit BCD Priority Encoder IC

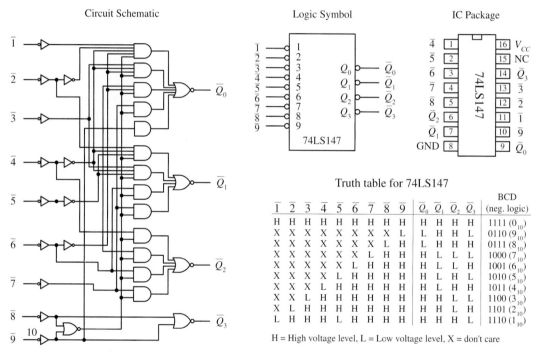

Circuit Schematic   Logic Symbol   IC Package

Truth table for 74LS147

$\overline{1}$	$\overline{2}$	$\overline{3}$	$\overline{4}$	$\overline{5}$	$\overline{6}$	$\overline{7}$	$\overline{8}$	$\overline{9}$	$\overline{Q_0}$	$\overline{Q_1}$	$\overline{Q_2}$	$\overline{Q_3}$	BCD (neg. logic)
H	H	H	H	H	H	H	H	H	H	H	H	H	1111 (0_{10})
X	X	X	X	X	X	X	X	L	L	H	H	L	0110 (9_{10})
X	X	X	X	X	X	X	L	H	L	H	H	H	0111 (8_{10})
X	X	X	X	X	X	L	H	H	H	L	L	L	1000 (7_{10})
X	X	X	X	X	L	H	H	H	H	L	L	H	1001 (6_{10})
X	X	X	X	L	H	H	H	H	H	L	H	L	1010 (5_{10})
X	X	X	L	H	H	H	H	H	H	L	H	H	1011 (4_{10})
X	X	L	H	H	H	H	H	H	H	H	L	L	1100 (3_{10})
X	L	H	H	H	H	H	H	H	H	H	L	H	1101 (2_{10})
L	H	H	H	L	H	H	H	H	H	H	H	L	1110 (1_{10})

H = High voltage level, L = Low voltage level, X = don't care

FIGURE 12.40

Another important difference between the two encoders is the term *priority* that is used with the 74LS147 and not used with the encoder in Fig. 12.39. The term *priority* is applied to the 74LS147 because this encoder is designed so that if two or more inputs are selected at the same time, it will only select the larger-order digit. For example, if 3, 5, and 8 are selected at the same time, only the 8 (negative true BCD LHHH or 0111) will be output. The truth table in Fig. 12.40 demonstrates this—look at the "don't care" or "X" entries. With the nonpriority encoder, if two or more inputs are applied at the same time, the output will be unpredictable.

The circuit shown in Fig. 12.41 provides a simple illustration of how an encoder and a decoder can be used together to drive an LED display via a 0-to-9 keypad. The 74LS147 encodes a keypad's input into BCD (negative logic). A set of inverters then converts the negative true BCD into positive true BCD. The transformed BCD is then fed into a 7447 seven-segment LED display decoder/driver IC.

A decimal-to-BCD encoder being used to convert keypad instructions into BCD instructions used to drive a LED display circuit (7447 decoder plus common-anode display)

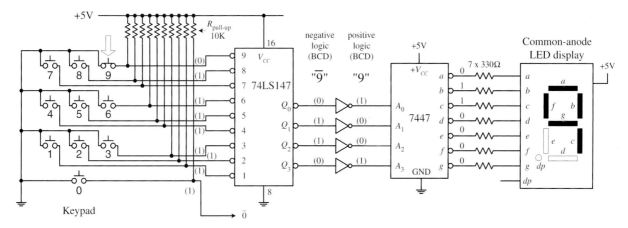

FIGURE 12.41

Figure 12.42 shows a 74148 octal-to-binary priory encoder IC. It is used to transform a specified single octal input into a binary 3-bit output code. As with the 74LS147, the 74148 comes with a priority feature, meaning, again, that if two or more inputs are selected at the same time, only the higher order number is selected.

74148 octal-to-binary priority encoder

74148 truth table

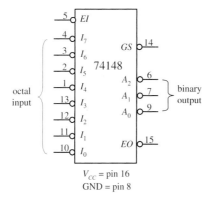

$\overline{EI}$	$\overline{I_0}$	$\overline{I_1}$	$\overline{I_2}$	$\overline{I_3}$	$\overline{I_4}$	$\overline{I_5}$	$\overline{I_6}$	$\overline{I_7}$	$\overline{GS}$	$\overline{A_0}$	$\overline{A_1}$	$\overline{A_2}$	$\overline{EO}$
H	X	X	X	X	X	X	X	X	H	H	H	H	H
L	H	H	H	H	H	H	H	H	H	H	H	H	L
L	X	X	X	X	X	X	X	L	L	L	L	L	H
L	X	X	X	X	X	X	L	H	L	H	L	L	H
L	X	X	X	X	X	L	H	H	L	L	H	L	H
L	X	X	X	X	L	H	H	H	L	H	H	L	H
L	X	X	X	L	H	H	H	H	L	L	L	H	H
L	X	X	L	H	H	H	H	H	L	H	L	H	H
L	X	L	H	H	H	H	H	H	L	L	H	H	H
L	L	H	H	H	H	H	H	H	L	H	H	H	H

FIGURE 12.42

A high applied to the input enable ($\overline{EI}$) forces all outputs to their inactive (high) state and allows new data to settle without producing erroneous information at the outputs. A group signal output ($\overline{GS}$) and an enable output ($\overline{EO}$) are also provided to allow for system expansion. The $\overline{GS}$ output is active level low when any input is low (active). The $\overline{EO}$ output is low (active) when all inputs are high. Using the output enable along with the input enable allows priority coding of N input signals. Both $\overline{EO}$ and $\overline{GS}$ are active high when the input enable is high (device disabled).

Figure 12.43 shows a 74184 BCD-to-binary converter (encoder) IC. This device has eight active-high outputs (Y_1–Y_8). Outputs Y_1 to Y_5 are outputs for regular BCD-to-binary conversion, while outputs Y_6 to Y_8 are used for a special BDC code called *nine's complement* and *ten's complement*. The active-high BCD code is applied to inputs A through E. The $\overline{G}$ input is an active-low enable input.

74185 BCD-to-binary converter

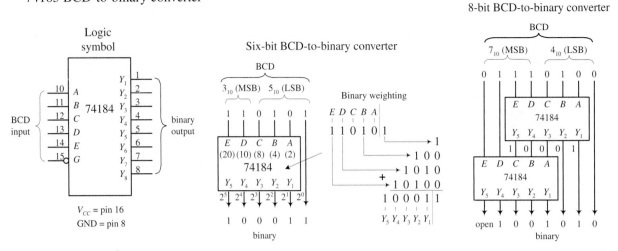

FIGURE 12.43

A sample 6-bit BCD-to-binary converter and a sample 8-bit BCD-to-binary converter that use the 74184 are shown to the right in Fig. 12.43. In the 6-bit circuit, since the LSB of the BCD input is always equal to the LSB of the binary output, the connection is made straight from input to output. The other BCD bits are applied directly to inputs A through E. The binary weighing factors for each input are $A = 2$, $B = 4$, $C = 8$, $D = 10$, and $E = 20$. Because only 2 bits are available for the MSD BCD input, the largest BCD digit in that position is 3 (binary 11). To get a complete 8-bit BCD converter, you connect two 74184s together, as shown to the far right in Fig. 12.43.

Figure 12.44 shows a 74185 binary-to-BCD converter (encoder). It is essentially the same as the 74184 but in reverse. The figure shows 6-bit and 8-bit binary-to-BCD converter arrangements.

74185 binary-to-BCD converter

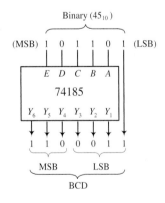

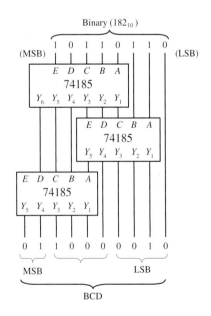

FIGURE 12.44

12.3.4 Binary Adders

With a few logic gates you can create a circuit that adds binary numbers. The mechanics of adding binary numbers is basically the same as that of adding decimal numbers. When the first digit of a two-digit number is added, a 1 is carried and added to the next row whenever the count exceeds binary 2 (e.g., 1 + 1 = 10, or = 0 carry a 1). For numbers with more digits, you have multiple carry bits. To demonstrate how you can use logic gates to perform basic addition, start out by considering the half-adder circuits below. Both half-adders shown are equivalent; one simply uses XOR/AND logic, while the other uses NOR/AND logic. The half-adder adds two single-bit numbers A and B and produces a 2-bit number; the LSB is represented as Σ_0, and the MSB or carry bit is represented as C_{out}.

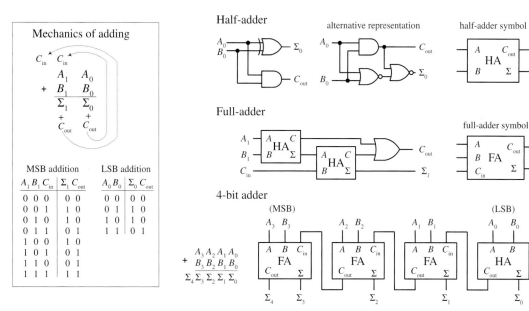

FIGURE 12.45

The most complicated operation the half-adder can do is 1 + 1. To perform addition on a two-digit number, you must attach a full-adder circuit (shown in Fig. 12.45) to the output of the half-adder. The full-adder has three inputs; two are used to input the second digits of the two binary numbers (A_1, B_1), while the third accepts the carry bit from the half-adder (the circuit that added the first digits, A_0 and B_0, of the two numbers). The two outputs of the full-adder will provide the 2d-place digit sum Σ_1 and another carry bit that acts as the 3d-place digit of the final sum. Now, you can keep adding more full-adders to the half-adder/full-adder combination to add larger number, linking the carry bit output of the first full-adder to the next full-adder, and so forth. To illustrate this point, a 4-bit adder is shown in Fig. 12.45.

There are a number of 4-bit full-adder ICs available such as the 74LS283 and 4008. These devices will add two 4-bit binary number and provide an additional input carry bit, as well as an output carry bit, so you can stack them together to get 8-bit, 12-bit, 16-bit, etc. adders. For example, the figure below shows an 8-bit adder made by cascading two 74LS283 4-bit adders.

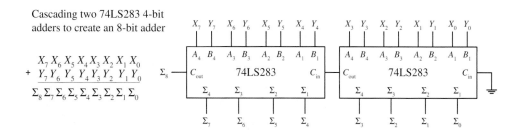

FIGURE 12.46

12.3.5   Binary Adder/Subtractor

Figure 12.47 shows how two 74LS283 4-bit adders can be combined with an XOR array to yield an 8-bit 2's complement adder/subtractor. The first number X is applied to the X_0-X_7 inputs, while the second number Y is applied to the Y_0-Y_7 inputs.

To add X and Y, the add/subtract switch is thrown to the add position, making one input of all XOR gates low. This has the effect of making the XOR gates appear transparent, allowing Y values to pass to the 74LS283s' B inputs (X values are passed to the A inputs). The 8-bit adder then adds the numbers and presents the result to the Σ outputs.

To subtract Y from X, you must first convert Y into 1's complement form; then you must add 1 to get Y into 2's complement form. After that you simply add X to the 2's complemented form of Y to get $X - Y$. When the add/subtract switch is thrown to the subtract position, one input to each XOR gate is set high. This causes the Y bits that are applied to the other XOR inputs to become inverted at the XOR outputs—you have just taken the 1's complement of Y. The 1's complement bits of Y are then presented to the inputs of the 8-bit adder. At the same time, C_{in} of the left 74LS283 is set high via the wire (see figure) so that a 1 is added to the 1's complement number to yield a 2's complement number. The 8-bit adder then adds X and the 2's complement of Y together. The final result is presented at the Σ outputs. In the figure, 76 is subtracted from 28.

8-bit 2's complement adder/subtractor

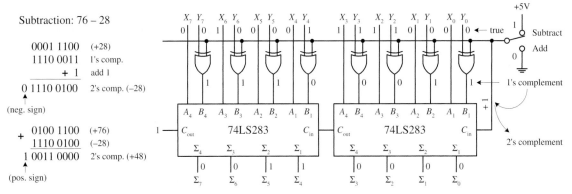

FIGURE 12.47

12.3.6   Arithmetic/Logic Units (ALUs)

An arithmetic/logic unit (ALU) is a multipurpose integrated circuit capable of performing various arithmetic and logic operations. To choose a specific operation to be performed, a binary code is applied to the IC's mode select inputs. The 74181, shown in Fig. 12.48, is a 4-bit ALU that provides 16 arithmetic and 16 logic operations.

To select an arithmetic operation, the 74181's mode control input (M) is set low. To select a logic operation, the mode control input is set high. Once you have decided whether you want to perform a logic or arithmetic operation, you apply a

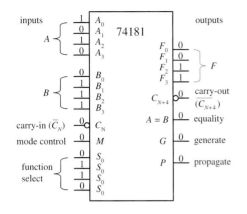

Mode select				Logic functions	Arithmetic operations
S_3	S_2	S_1	S_0	($M = 1$)	($M = 0$, $\overline{C}_n = 1$)
0	0	0	0	$F = \overline{A}$	$F = A$
0	0	0	1	$F = \overline{A + B}$	$F = A + B$
0	0	1	0	$F = \overline{A}B$	$F = A + \overline{B}$
0	0	1	1	$F = 0$	$F =$ minus 1 (2's comp.)
0	1	0	0	$F = \overline{AB}$	$F = A$ plus $A\overline{B}$
0	1	0	1	$F = \overline{B}$	$F = (A + B)$ plus $A\overline{B}$
0	1	1	0	$F = A \oplus B$	$F = A$ minus B minus 1
0	1	1	1	$F = A\overline{B}$	$F = A\overline{B}$ minus 1
1	0	0	0	$F = \overline{A} + B$	$F = A$ plus AB
1	0	0	1	$F = \overline{A \oplus B}$	$F = A$ plus B
1	0	1	0	$F = B$	$F = (A + \overline{B})$ plus AB
1	0	1	1	$F = AB$	$F = AB$ minus 1
1	1	0	0	$F = 1$	$F = A$ plus A
1	1	0	1	$F = A + \overline{B}$	$F = (A + B)$ plus A
1	1	1	0	$F = A + B$	$F = (A + \overline{B})$ plus A
1	1	1	1	$F = A$	$F = A$ minus 1

FIGURE 12.48

4-bit code to the mode select inputs (S_0, S_1, S_2, S_3) to specify which specific operation, as indicated within the truth table, is to be performed. For example, if you select $S_3 = 1$, $S_2 = 1$, $S_1 = 1$, $S_0 = 0$, while $M = 1$, then you get $F_0 = A_0 + B_0$, $F_1 = A_1 + B_1$, $F_2 = A_2 + B_2$, $F_3 = A_3 + B_3$. Note that the + shown in the truth table does not represent addition; it is used to represent the OR function—for addition, you use "plus." Carry-in ($\overline{C}_N$) and carry-out (C_{N+4}) leads are provided for use in arithmetic operations. All arithmetic results generated by this device are in 2's complement notation.

12.3.7 Comparators and Magnitude Comparator ICs

A digital comparator is a circuit that accepts two binary numbers and determines whether the two numbers are equal. For example, the figure below shows a 1-bit and a 4-bit comparator. The 1-bit comparator outputs a high (1) only when the two 1-bit numbers A and B are equal. If A is not equal to B, then the output goes low (0). The 4-bit is basically four 1-bit comparators in one. When all individual digits of each number are equal, all XOR gates output a high, which in turn enables the AND gate, making the output high. If any two corresponding digits of the two numbers are not equal, the output goes low.

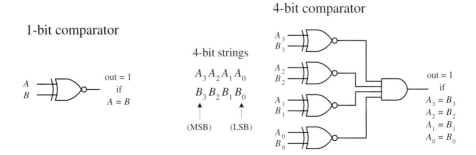

FIGURE 12.49

Now, say you want to know which number, A or B, is larger. The circuits in Fig. 12.49 will not do the trick. What you need instead is a *magnitude comparator* like the 74HC85 shown in Fig. 12.50. This device not only tells you if two numbers are equal; it also tells you which number is larger. For example, if you apply a 1001 (9_{10}) to the $A_3A_2A_1A_0$ inputs and a second number 1100 (12_{10}) to the $B_3B_2B_1B_0$ inputs, the $A < B$ output will go high (the other two outputs, $A > B$ and $A = B$, will remain low). If A and B were equal, the $A = B$ output would have gone high, etc. If you wanted to compare a larger number, say, two 8-bit numbers, you simply cascade two 74HC85's together, as shown to the right in Fig. 12.50. The leftmost 74HC85 compares the lower-order bits, while the rightmost 74HC85 compares the higher-order bits. To link the two devices together, you connect the output of the lower-order device to the expansion inputs of the higher-order device, as shown. The lower-order device's expansion inputs are always set low ($I_A < B$), high ($I_A = B$), low ($I_A > B$).

74HC85 4-bit magnitude comparator   Connecting two 74HC85's together to form an 8-bit magnitude comparator

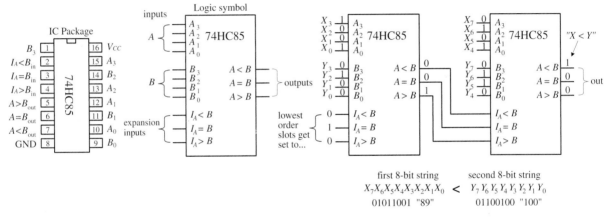

FIGURE 12.50

12.3.8 Parity Generator/Checker

Often, external noise will corrupt binary information (cause a bit to flip from one logic state to the other) as it travels along a conductor from one device to the next. For example, in the 4-bit system shown in Fig. 12.51, a BCD 4 (0100) picks up noise and becomes 0101 (or 5) before reaching its destination. Depending on the application, this type of error could lead to some serious problems.

To avoid problems caused by unwanted data corruption, a parity generator/checker system, like the one shown in Fig. 12.51, can be used. The basic idea is to add an extra bit, called a *parity bit*, to the digital information being transmitted. If the parity bit makes the sum of all transmitted bits (including the parity bit) odd, the transmitted information is of odd parity. If the parity bit makes the sum even, the transmitted information is of even parity. A parity generator circuit creates the parity bit, while the parity checker on the receiving end determines if the information sent is of the proper parity. The type of parity (odd or even) is agreed to beforehand, so the parity checker knows what to look for. The parity bit can be placed next to the MSB or the LSB, provided the device on the receiving end knows which bit is the parity bit and which bits are the data. The arrangement shown in Fig. 12.51 is designed with an even-parity error-detection system.

4-bit even-parity error-detection system

Even vs. odd parity (examples)

4-bit string	parity bit	parity
1 0 0 1	0	even (two 1's)
1 1 0 0	1	odd (three 1's)
0 0 0 0	1	odd (one 1's)
1 1 1 0	1	even (four 1's)

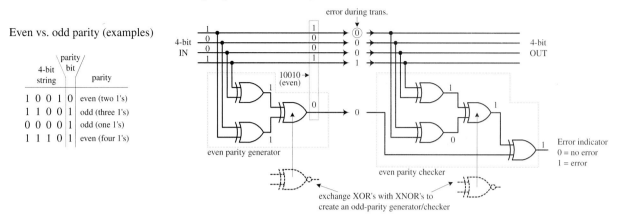

FIGURE 12.51

If you want to avoid building parity generators and checkers from scratch, use a parity generator/checker IC like the 74F280 9-bit odd-even parity generator/checker shown below. To make a complete error-detection system, two 74F280s are used—one acts as the parity generator; the other acts as the parity checker. The generator's inputs A through H are connected to the eight data lines of the transmitting portion of the circuit. The ninth input (I) is grounded when the device is used as a generator. If you want to create an odd-parity generator, you tap the Σ_{odd} output; for even parity, you tap Σ_{even}. The 74F280 checker taps the main line at the receiving end and also accepts the parity bit line at input I. The figure below shows an odd-parity error-detection system used with an 8-bit system. If an error occurs, a high (1) is generated at the Σ_{odd} output.

74F280 9-bit odd-even parity generator/checker

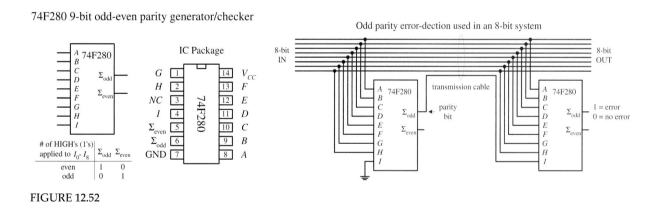

# of HIGH's (1's) applied to I_0- I_8	Σ_{odd}	Σ_{even}
even	1	0
odd	0	1

FIGURE 12.52

12.3.9   A Note on Obsolescence and the Trend Toward Microcontroller Control

You have just covered most of the combinational devices you will find discussed in textbooks and find listed within electronic catalogs. Many of these devices are still used. However, some devices such as the binary adders and code converters are becoming obsolete.

Today, the trend is to use software-controlled devices such as microprocessors and microcontrollers to carry out arithmetic operations and code conversions. Before you attempt to design any logic circuit, I suggest jumping to Section 12.12. In that section, pay close attention to microcontrollers. These devices are quite amazing. They are essentially microprocessors but are significantly easier to program and are easier to interface with other circuits and devices.

Microcontrollers can be used to collect data, store data, and perform logical operations using the input data. They also can generate output signals that can be used to control displays, audio devices, stepper motors, servos, etc. The specific functions a microcontroller is designed to perform depend on the program you store in its internal ROM-type memory. Programming the microcontroller typically involves simply using a special programming unit provided by the manufacturer. The programming unit usually consists of a special prototyping platform that is linked to a PC (via a serial or parallel port) that is running a host program. In the host program, you typically write out a program in a high-level language such as C, or some other specialized language designed for a certain microcontroller, and then, with the press of a key, the program is converted into machine language (1s and 0s) and downloaded into the microcontroller's memory.

In many applications, a single microcontroller can replace entire logic circuits comprised of numerous discrete components. For this reason, it is tempting to skip the rest of the sections of this chapter and go directly to the section on microcontrollers. However, there are three basic problems with this approach. First, if you are a beginner, you will miss out on many important principles behind digital control that are most easily understood by learning how the discrete components work. Second, many digital circuits that you can build simply do not require the amount of sophistication a microcontroller provides. Finally, you may feel intimidated by the electronics catalogs that list every conceivable component available, be it obsolete or not. Knowing what's out there and knowing what to avoid are also important parts of the learning process.

12.4   Logic Families

Before moving on to sequential logic, let's touch on a few practical matters regarding the various logic families available and what kind of operating characteristics these families have. In this section you will also encounter unique logic gates that have open-collector output stages and logic gates that have Schmitt-triggered inputs.

The key ingredient within any integrated logic device, be it a logic gate, a multiplexer, or a microprocessor, is the transistor. The kinds of transistors used within the integrated circuit, to a large extend, specify the type of logic family. The two most popular transistors used in ICs are bipolar and MOSFET transistors. In general, ICs made from MOSFET transistors use less space due to their simpler construction, have very high noise immunity, and consume less power than equivalent bipolar transistor ICs. However, the high input impedance and input capacitance of the MOSFET transistors (due to their insulated gate leads) results in longer time constants for transistor on/off switching speeds when compared with bipolar gates and therefore typically result in a slower device. Over years of development, however, the performance gap between these two technologies has narrowed considerably.

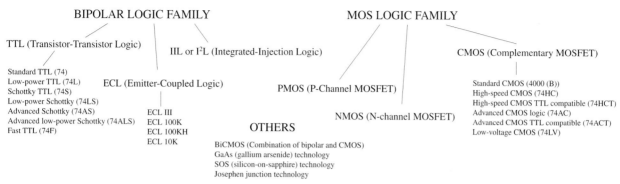

FIGURE 12.53

Both the bipolar and MOSFET logic families can be divided into a number of subclasses. The major subclasses of the bipolar family include TTL (transistor-transistor logic), ECL (emitter-coupled logic), and IIL or I²L (integrated-injection logic). The major subclasses of the MOSFET logic include PMOS (P-channel MOSFET logic), NMOS (N-channel MOSFET logic), and CMOS (complementary MOSFET logic). CMOS uses both NMOS and PMOS technologies (it uses both N-channel and P-channel MOSFETs). The two most popular technologies are TTL and CMOS, while the other technologies are typically used in large-scale integration devices, such as microprocessors and memories. There are new technologies popping up all the time, which yield faster, more energy-efficient devices. Some examples include BiCMOS, GaAS, SOS, and Josephen junction technologies.

As you have already learned, TTL and CMOS devices are grouped into functional categories that get placed into either the 7400 series [74F, 74LS, 74HC (CMOS), etc.] or 4000 CMOS series (or the improved 4000B series). Now, another series you will run into is the 5400 series. This series is essentially equivalent to the 7400 series (same pinouts, same basic logic function), but it is a more expansive chip because it is designed for military applications that require increased supply voltage tolerances and temperature tolerances. For example, a 7400 IC typically has a supply voltage range from 4.75 to 5.25 V with a temperature range from 0 to 70°C, while a 5400 IC typically will have a voltage range between 4.5 and 5.5 V and a temperature range from −55 to 125°C.

12.4.1   TTL Family of ICs

The original TTL series, referred to as the *standard TTL series* (74*xx*), was developed early in the 1960s. This series is still in use, even though its overall performance is inferior to the newer line of TTL devices, such as the 74LS*xx*, 74ALS*xx*, and 74F*xx*. The internal circuitry of a standard TTL 7400 NAND gate, along with a description of how it works, is provided next.

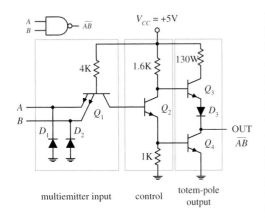

The TTL NAND gate is broken up into three basic sections: multiemitter input, control section, and totem-pole output stage. In the multiemitter input section, a multiemitter bipolar transistor Q_1 acts like a two-input AND gate, while diodes D_1 and D_2 act as negative clamping diodes used to protect the inputs from any short-term negative input voltages that could damage the transistor. Q_2 provides control and current boosting to the totem-pole output stage; when the output is high (1), Q_4 is off (open) and Q_3 is on (short). When the output is low (0), Q_4 is on and Q_3 is off. Because one or the other transistor is always off, the current flow from V_{CC} to ground in that section of the circuit is minimized. The lower figures show both a high and low output state, along with the approximate voltages present at various locations. Notice that the actual output voltages are not exactly 0 or +5V—a result of internal voltage drops across resistor, transistor, and diode. Instead, the outputs are around 3.4V for high and 0.3V for low. As a note, to create, say, an eight-input NAND gate, the multiemitter input transistor would have eight emitters instead of just two as shown.

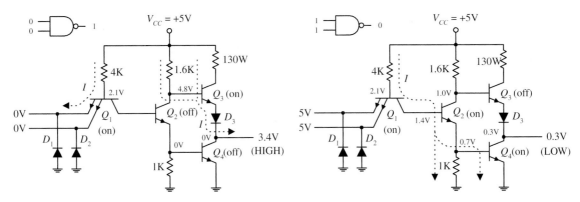

FIGURE 12.54

A simple modification to the standard TTL series was made early on by reducing all the internal resistor values in order to reduce the *RC* time constants and thus increase the speed (reduce propagation delays). This improvement to the original TTL series marked the 74H series. Although the 74H series offered improved speed (about twice as fast) over the 74 series, it had more than double the power consumption. Later, the 74L series emerged. Unlike the 74H, the 74L took the 74 and increased all internal resistances. The net effect lead to a reduction in power but increased propagation delay.

A significant improvement in speed within the TTL line emerged with the development of the 74S*xx* series (Schottky TTL series). The key modifications involved placing Schottky diodes across the base-to-collector junctions of the transistors. These Schottky diodes eliminated capacitive effects caused by charge buildup in the transistor's base region by passing the charge to the collector region. Schottky diodes were the best choice because of their inherent low charge buildup characteristics. The overall effect was an increase in speed by a factor of 5 and only a doubling in power.

Continually over time, by using different integration techniques and increasing the values of the internal resistors, more power-efficient series emerged, like the low-power Schottky 74LS series, with about one-third the power dissipation of the 74S. After the 74LS, the advanced-low-power Schottky 74ALS series emerged, which had even better performance. Another series developed around this time was the 74F series, or FAST logic, which used a new process of integration called *oxide isolation* (also used in the ALS series) that led to reduced propagation delays and decreased the overall size.

Today you will find many of the older series listed in electronics catalogs. Which series you choose ultimately depends on what kind of performance you are looking for.

12.4.2 CMOS Family of ICs

While the TTL series was going through its various transformations, the CMOS series entered the picture. The original CMOS 4000 series (or the improved 4000B series) was developed to offer lower power consumption than the TTL series of devices—a feature made possible by the high input impedance characteristics of its MOSFET transistors. The 4000B series also offered a larger supply voltage range (3 to 18 V), with minimum logic high = $\frac{2}{3}V_{DD}$, and maximum logic low = $\frac{1}{3}V_{DD}$. The 4000B series, though more energy efficient than the TTL series, was significantly slower and more susceptible to damage due to electrostatic discharge. The figure below shows the internal circuitry of CMOS NAND, AND, and NOR gates. To figure out how the gates work, apply high (logic 1) or low (logic 0) levels to the inputs and see which transistor gates turn on and which transistor gates turn off.

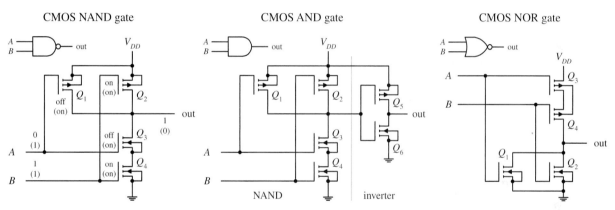

FIGURE 12.55

A further improvement in speed over the original 4000B series came with the introduction of the 40H00 series. Although this series was faster than the 4000B series, it was not quite as fast as the 74LS TTL series. The 74C CMOS series also emerged on the scene, which was designed specifically to be pin-compatible with the TTL line. Another significant improvement in the CMOS family came with the development of the 74HC and the 74HCT series. Both these series, like the 74C series, were pin-compatible with the TTL 74 series. The 74HC (high-speed CMOS) series had the same speed as the 74LS as well as the traditional CMOS low-power consumption. The 74HCT (high-speed CMOS TTL compatible) series was developed to be interchangeable with TTL devices (same input/output voltage level characteristics). The 74HC series is very popular today. Still further improvements in 74HC/74HCT series led to the advanced CMOS logic (74AC/74ACT) series. The 74AC (advanced CMOS) series approached speeds comparable with the 74F TTL series, while the 74ACT (advanced CMOS TTL compatible) series was designed to be TTL compatible.

12.4.3 Input/Output Voltages and Noise Margins

The exact input voltage levels required for a logic IC to perceive a high (logic 1) or low (logic 0) input level differ between the various logic families. At the same time, the high and low output levels provided by a logic IC vary among the logic families. For example, the figure below shows valid input and output voltage levels for both the 74LS (TTL) and 74HC (CMOS) families.

Valid input/output logic levels for the TTL 74LS and the CMOS 74HC

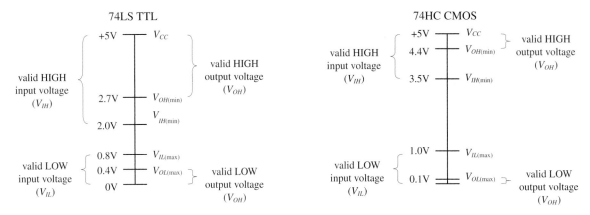

FIGURE 12.56

In Fig. 12.56, V_{IH} represents the valid voltage range that will be interpreted as a high logic input level. V_{IL} represents the valid voltage range that will be interpreted as a low logic input level. V_{OL} represents the valid voltage range that will be guaranteed as a low logic output level, while V_{OH} represents the valid voltage range that will be guaranteed as a high logic output level.

As you can see from Fig. 12.56, if you connect the output of a 74HC device to the input of a 74LS device, there is no problem—the output logic levels of the 74HC are within the valid input range of the 74LS. However, if you turn things around, driving a 74HC device's inputs from a 74LS's output, you have problems—a high output level from the 74LS is too small to be interpreted as a high input level for the 74HC. I will discuss tricks used to interface the various logic families together in a moment.

12.4.4   Current Ratings, Fanout, and Propagation Delays

Logic IC inputs and outputs can only sink or source a given amount of current. I_{IL} is defined as the maximum low-level input current, I_{IH} as the maximum high-level input current, I_{OH} as the maximum high-level output current, and I_{OL} as the maximum low-level output current. As an example, a standard 74xx TTL gate may have an $I_L = -1.6$ mA and $I_{IH} = 40$ μA while having an $I_{OL} = 16$ mA and $I_{OH} = -400$ μA. The negative sign means that current is leaving the gate (the gate is acting as a source), while a positive sign means that current in entering the gate (the gate is acting as sink).

The limit to how much current a device can sink or source determines the size of loads that can be attached. The term *fanout* is used to specify the total number of gates that can be driven by a single gate of the same family without exceeding the current rating of the gate. The fanout is determined by taking the smaller result of I_{OL}/I_{IL} or I_{OH}/I_{IH}. For the standard 74 series, the fanout is 10 (16 mA/1.6 mA). For the 74LS

Standard TTL 74xx maximum input/ouput currents

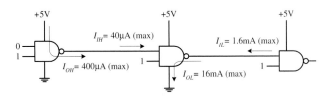

Propagation delays for TTL gates

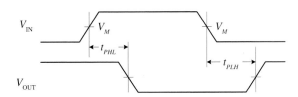

V_M = 1.3V for 74LS; V_M = 1.5V for all other TTL families

FIGURE 12.57

74xx TTL fanout

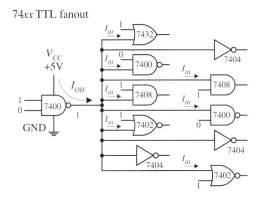

series, the fanout is around 20; for the 74F, it is around 33; and for the 7HC, it is around 50.

If you apply a square pulse to the input of a logic gate, the output signal will experience a sloping rise time and fall time, as shown in the graph in Fig. 12.57. The *rise time* (t_r) is the length of time it takes for a pulse to rise from 10 to 90 percent of its high level (e.g., 5 V = high: 0.5 V = 10%, 4.5 V = 90%). The *fall time* t_f is the length of time it takes for a high level to fall from the 90 to 10 percent. The rise and fall times, however, are not as significant when compared with propagation delays between input transition and output response. Propagation delay results from the limited switching speeds of the internal transistors within the logic device. The low-to-high propagation delay T_{PHL} is the time it takes for the output of a device to switch from low to high after the input transition. The high-to-low propagation delay T_{PLH} is the time it takes for the output to switch from high to low after the input transition. When designing circuits, it is important to take into account these delays, especially when you start dealing with sequential logic, where timing is everything. Figures 12.58 and 12.59 provide typical propagation delays for various TTL and CMOS devices. Manufacturers will provide more accurate propagation information in their data sheets.

12.4.5 A Detailed Look at the Various TTL and CMOS Subfamilies

The following information, shown in Figs. 12.58 and 12.59, especially the data pertaining to propagation delays and current ratings, represents *typical values* for a given logic series. For more accurate data about a specific device, you must consult the manufacturer's literature. In other words, only use the provided information as a rough guide to get a feeling for the overall performance of a given logic series.

TTL Series

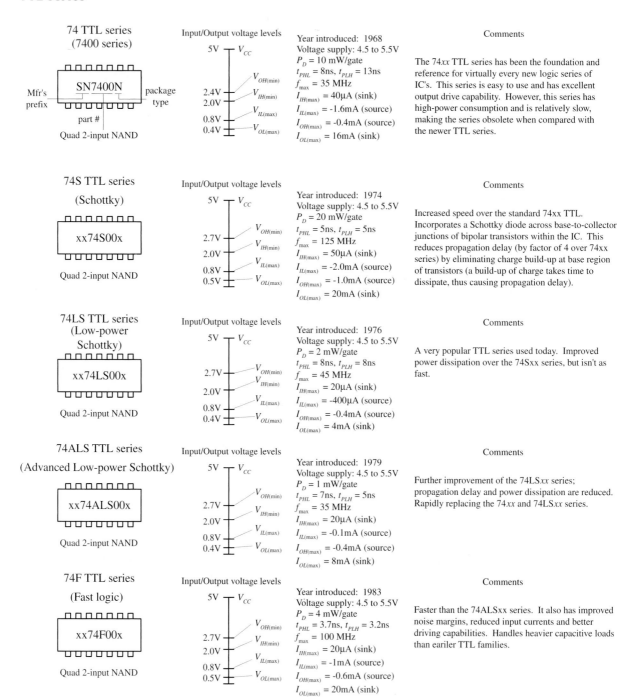

74 TTL series (7400 series)

Mfr's prefix — SN7400N — package type

part #

Quad 2-input NAND

Input/Output voltage levels

5V — V_{CC}

2.4V — $V_{OH(min)}$
2.0V — $V_{IH(min)}$

0.8V — $V_{IL(max)}$
0.4V — $V_{OL(max)}$

Year introduced: 1968
Voltage supply: 4.5 to 5.5V
$P_D = 10$ mW/gate
$t_{PHL} = 8$ns, $t_{PLH} = 13$ns
$f_{max} = 35$ MHz
$I_{IH(max)} = 40\mu A$ (sink)
$I_{IL(max)} = -1.6$mA (source)
$I_{OH(max)} = -0.4$mA (source)
$I_{OL(max)} = 16$mA (sink)

Comments

The 74xx TTL series has been the foundation and reference for virtually every new logic series of IC's. This series is easy to use and has excellent output drive capability. However, this series has high-power consumption and is relatively slow, making the series obsolete when compared with the newer TTL series.

74S TTL series (Schottky)

xx74S00x

Quad 2-input NAND

Input/Output voltage levels

5V — V_{CC}

2.7V — $V_{OH(min)}$
2.0V — $V_{IH(min)}$

0.8V — $V_{IL(max)}$
0.5V — $V_{OL(max)}$

Year introduced: 1974
Voltage supply: 4.5 to 5.5V
$P_D = 20$ mW/gate
$t_{PHL} = 5$ns, $t_{PLH} = 5$ns
$f_{max} = 125$ MHz
$I_{IH(max)} = 50\mu A$ (sink)
$I_{IL(max)} = -2.0$mA (source)
$I_{OH(max)} = -1.0$mA (source)
$I_{OL(max)} = 20$mA (sink)

Comments

Increased speed over the standard 74xx TTL. Incorporates a Schottky diode across base-to-collector junctions of bipolar transistors within the IC. This reduces propagation delay (by factor of 4 over 74xx series) by eliminating charge build-up at base region of transistors (a build-up of charge takes time to dissipate, thus causing propagation delay).

74LS TTL series (Low-power Schottky)

xx74LS00x

Quad 2-input NAND

Input/Output voltage levels

5V — V_{CC}

2.7V — $V_{OH(min)}$
— $V_{IH(min)}$
2.0V

0.8V — $V_{IL(max)}$
0.4V — $V_{OL(max)}$

Year introduced: 1976
Voltage supply: 4.5 to 5.5V
$P_D = 2$ mW/gate
$t_{PHL} = 8$ns, $t_{PLH} = 8$ns
$f_{max} = 45$ MHz
$I_{IH(max)} = 20\mu A$ (sink)
$I_{IL(max)} = -400\mu A$ (source)
$I_{OH(max)} = -0.4$mA (source)
$I_{OL(max)} = 4$mA (sink)

Comments

A very popular TTL series used today. Improved power dissipation over the 74Sxx series, but isn't as fast.

74ALS TTL series (Advanced Low-power Schottky)

xx74ALS00x

Quad 2-input NAND

Input/Output voltage levels

5V — V_{CC}

2.7V — $V_{OH(min)}$
— $V_{IH(min)}$
2.0V

0.8V — $V_{IL(max)}$
0.4V — $V_{OL(max)}$

Year introduced: 1979
Voltage supply: 4.5 to 5.5V
$P_D = 1$ mW/gate
$t_{PHL} = 7$ns, $t_{PLH} = 5$ns
$f_{max} = 35$ MHz
$I_{IH(max)} = 20\mu A$ (sink)
$I_{IL(max)} = -0.1$mA (source)
$I_{OH(max)} = -0.4$mA (source)
$I_{OL(max)} = 8$mA (sink)

Comments

Further improvement of the 74LSxx series; propagation delay and power dissipation are reduced. Rapidly replacing the 74xx and 74LSxx series.

74F TTL series (Fast logic)

xx74F00x

Quad 2-input NAND

Input/Output voltage levels

5V — V_{CC}

2.7V — $V_{OH(min)}$
— $V_{IH(min)}$
2.0V

0.8V — $V_{IL(max)}$
0.5V — $V_{OL(max)}$

Year introduced: 1983
Voltage supply: 4.5 to 5.5V
$P_D = 4$ mW/gate
$t_{PHL} = 3.7$ns, $t_{PLH} = 3.2$ns
$f_{max} = 100$ MHz
$I_{IH(max)} = 20\mu A$ (sink)
$I_{IL(max)} = -1$mA (source)
$I_{OH(max)} = -0.6$mA (source)
$I_{OL(max)} = 20$mA (sink)

Comments

Faster than the 74ALSxx series. It also has improved noise margins, reduced input currents and better driving capabilities. Handles heavier capacitive loads than earlier TTL families.

FIGURE 12.58

CMOS Series

4000 CMOS series
(4000B improved version)

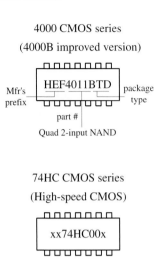

Mfr's prefix — **HEF4011BTD** — package type

part #

Quad 2-input NAND

Input/Output voltage levels

5V — V_{CC}
4.9V — $V_{OH(min)}$
3.3V — $V_{OH(min)}$
$V_{IH(min)}$
1.7V — $V_{IL(max)}$
0.1V — $V_{OL(max)}$

when V_{CC} = +5V

Year introduced: 1970
Voltage supply: 3 to 18V
P_D = 1mW/gate at 1 MHz
t_{PHL} = 50ns, t_{PLH} = 65ns
f_{max} = 6 MHz
$I_{IH(max)}$ = 1µA (sink)
$I_{IL(max)}$ = -1µA (source)
$I_{OH(max)}$ = -3.0mA (source)
$I_{OL(max)}$ = 3.0mA (sink)

Comments

The 4000 series is the orignal CMOS line--the 4000B series is an improvement. Supply voltages range from 3 to 18V, with V_{IH} = 2/3V_{CC} and V_{OH} = 1/3V_{CC} . Was popular due to low power consumption when compared to TTL. Considered obsolete when compared to the newer CMOS line of IC's, and is much slower than any of the TTL series. The 4000 series is also susceptible to damage from electrostatic discharge.

74HC CMOS series
(High-speed CMOS)

xx74HC00x

Quad 2-input NAND

Input/Output voltage levels

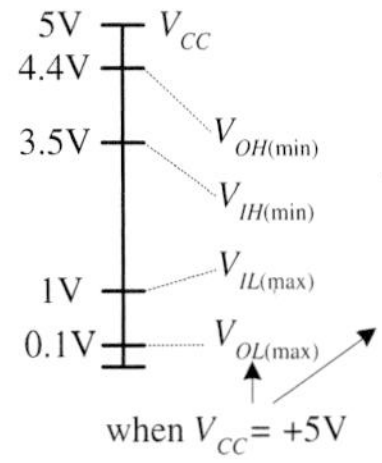

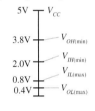

5V — V_{CC}
4.4V
3.5V — $V_{OH(min)}$
$V_{IH(min)}$
1V — $V_{IL(max)}$
0.1V — $V_{OL(max)}$

when V_{CC} = +5V

Year introduced: 1975
Voltage supply: 2 to 6V
P_D = 0.5mW/gate at 1 MHz
t_{PHL} = 20ns, t_{PLH} = 20ns
f_{max} = 20 MHz
$I_{IH(max)}$ = 1µA (sink)
$I_{IL(max)}$ = -1µA (source)
$I_{OH(max)}$ = -4mA (source)
$I_{OL(max)}$ = 4mA (sink)

Comments

Improvement in speed over the 4000 CMOS series; as speedy as the 74LSxx TTL series, depending on the operating frequency. Also provides greater noise immunity and greater voltage and temperature operating ranges than TTL. This is a very popular series used today.

74HCT CMOS series
(High-speed CMOS, TTL compatible)

xx74HCT00x

Quad 2-input NAND

Input/Output voltage levels

5V — V_{CC}
3.8V — $V_{OH(min)}$
2.0V — $V_{IH(min)}$
$V_{IL(max)}$
0.8V
0.4V — $V_{OL(max)}$

Year introduced: 1975
Voltage supply: 4.5 to 5.5V
P_D = 0.5mW/gate at 1 MHz
t_{PHL} = 40ns, t_{PLH} = 40ns
f_{max} = 24 MHz
$I_{IH(max)}$ = 1µA (sink)
$I_{IL(max)}$ = -1µA (source)
$I_{OH(max)}$ = -4mA (source)
$I_{OL(max)}$ = 4mA (sink)

Comments

Like the 74HCxx CMOS series but is also TTL compatable (it's pin compatible, as well as input/output voltage-level compatable).

74AC CMOS series
(Advanced CMOS)

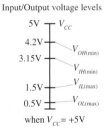

xx74AC00x

Quad 2-input NAND

Input/Output voltage levels

5V — V_{CC}
4.2V
3.15V — $V_{OH(min)}$
$V_{IH(min)}$
1.5V — $V_{IL(max)}$
0.5V — $V_{OL(max)}$

when V_{CC} = +5V

Year introduced: 1985
Voltage supply: 3 to 5.5V
P_D = 0.5mW/gate at 1 MHz
t_{PHL} = 3ns, t_{PLH} = 3ns
f_{max} = 125 MHz
$I_{IH(max)}$ = 1µA (sink)
$I_{IL(max)}$ = -1µA (source)
$I_{OH(max)}$ = -24mA (source)
$I_{OL(max)}$ = 24mA (sink)

Comments

Further improvement of the 74HCxx series; it's faster, has a higher output current drive capacity, and provides shorter propagation delays.

74ACT CMOS series
(Advanced CMOS, TTL compatible)

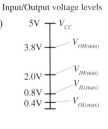

xx74ACT00x

Quad 2-input NAND

Input/Output voltage levels

5V — V_{CC}
3.8V — $V_{OH(min)}$
2.0V — $V_{IH(min)}$
$V_{IL(max)}$
0.8V
0.4V — $V_{OL(max)}$

Year introduced: 1985
Voltage supply: 4.5 to 5.5V
P_D = 0.5mW/gate at 1 MHz
t_{PHL} = 5ns, t_{PLH} = 5ns
f_{max} = 125 MHz
$I_{IH(max)}$ = 1µA (sink)
$I_{IL(max)}$ = -1µA (source)
$I_{OH(max)}$ = -24mA (source)
$I_{OL(max)}$ = 24mA (sink)

Comments

Like the 74ACxx CMOS series but is also TTL compatable (it's pin compatible, as well as input/output voltage-level compatable).

74AHC / 74AHCT CMOS series
(Advanced High-speed/ TTL compatible CMOS)

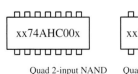

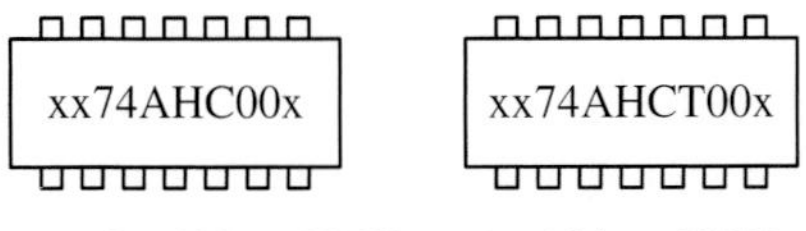

xx74AHC00x   xx74AHCT00x

Quad 2-input NAND   Quad 2-input NAND

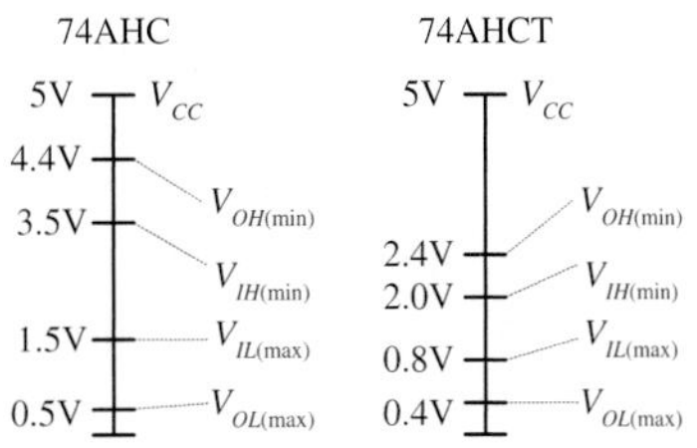

74AHC

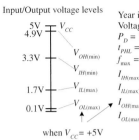

5V — V_{CC}
4.4V
3.5V — $V_{OH(min)}$
$V_{IH(min)}$
1.5V — $V_{IL(max)}$
0.5V — $V_{OL(max)}$

74AHCT

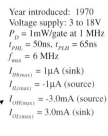

5V — V_{CC}
2.4V — $V_{OH(min)}$
2.0V — $V_{IH(min)}$
0.8V — $V_{IL(max)}$
0.4V — $V_{OL(max)}$

Advanced high-speed CMOS is an enhanced version of the 74HC and 74HCT series; the 74AHC has half the static power consumption, one-third the propagation delay. The 74AHCT is TTL compatible. Introduced in 1996.

FIGURE 12.59

12.4.6   A Look at a Few Other Logic Series

The 74-BiCMOS Series

The 74-BiCMOS series of devices incorporates the best features of bipolar and CMOS technology together in one package. The overall effect is an extremely high-speed, low-power digital logic family. This product line is especially well suited for and is mostly limited to microprocessor bus interface logic. Each manufacturer uses a different suffix to identify its BiCMOS line. For example, Texas Instruments uses 74BCT*xx*, while Signetics (Phillips) uses 74ABT*xx*.

74-Low-Voltage Series

The 74-low-voltage series is a relatively new series that uses a nominal supply voltage of 3.3 V. Members of this series include the 74LV (low-voltage HCMOS), 74LVC (low-voltage CMOS), the 74LVT (low-voltage technology), and the 74ALVC (advanced low-voltage CMOS). See Fig. 12.60.

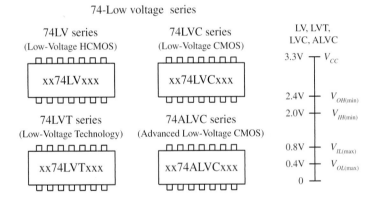

A relatively new series of logic using a nominal supply voltage of 3.3 V which are designed for extremely low-power and low-voltage applications (e.g., battery-powered devices). The switching speed of LV logic is extremely fast, ranging from about 9 ns for LB series down to 2.1 ns for ALVC. Another nice feature of LV logic is high output drive capability. The LVT, for example, can sink up to 64 mA and source up to 32 mA. LVT 1992 BiCMOS, LVC/ALVC 1993 CMOS.

FIGURE 12.60

Emitter-Coupled Logic

Emitter-coupled logic (ECL), a member of the bipolar family, is used for extremely high-speed applications, reaching speeds up to 500 MHz with propagation delays as low as 0.8 ns. There is one problem with ECL—it consumes a considerable amount of power when compared with the TTL and CMOS series. ECL is best suited for use in computer systems, where power consumption is not as big an issue as speed. The trick to getting the bipolar transistors in an ECL device to respond so quickly is to never let the transistors saturate. Instead, high and low levels are determined by which transistor in a differential amplifier is conducting more. Figure 12.61 shows the internal circuitry of an OR/NOR ECL gate. The high and low logic-level voltages (−0.8 and −17 V, respectively) and the supply voltage (−5.2 V/0 V) are somewhat unusual and cause problems when interfacing with TTL and CMOS.

Internal circuitry of ECL OR/NOR gate

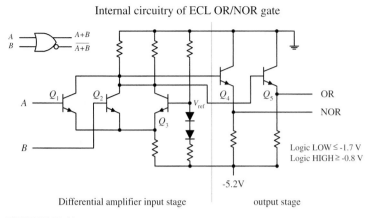

FIGURE 12.61

Differential amplifier input stage   output stage

The OR/NOR gate shown here is composed of a differential amplifier input stage and an output stage. In the differential amplifier stage, a reference voltage is setup at Q_3's base via the voltage divider network (diodes/resistors). This reference voltage determines the threshold between high and low logic levels. When the base of Q_3 is at a more positive potential with respect to the emitter of Q_1 and Q_2, Q_3 conducts. When Q_5 conducts, the OR output goes low. If either input A or B is raised to -0.8 V (high), the base of Q_1 or Q_2 will be at a higher potential than the base of Q_3, and Q_3 will cease conducting, forcing the OR output high. The overall effect of the ECL design prevents transistors from saturating, thereby eliminating charge buildup on the base of the transistors that limits switching speeds.

12.4.7 Logic Gates with Open-Collector Outputs

Among the members of the TTL series there exists a special class of logic gates that have open-collector output stages instead of the traditional totem-pole configuration you saw earlier. (Within the CMOS family, there are similar devices that are said to have open-drain output stages). These devices are not to be confused with the typical logic gates you have seen so far. Logic gates with open-collector outputs have entirely different output characteristics. Figure 12.62 shows a NAND gate with open-collector (OC) output. Notice that the Q_3 transistor is missing in the OC NAND gate. By removing Q_3, the output no longer goes high when A and B logic levels are set to 00, 01, or 10. Instead, the output floats. When A and B logic levels are both high (1), the output is grounded. This means that the OC gate can only sink current, it cannot source current! So how do you get a high output level? You use an external voltage source and a pull-up resistor, as shown in the center figure below. Now, when the output floats, the pullup resistor connected to the external voltage source will "pull" the output to the same level as the external voltage source, which in this case is at +15 V. That's right, you don't have to use +5 V. That is one of the primary benefits of using OC gates—you can drive load-requiring voltage levels different from those of the logic circuitry.

FIGURE 12.62

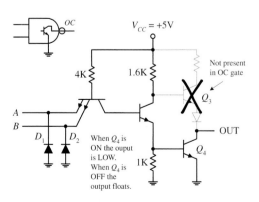

Internal circuitry of an open-collector NAND gate. Note the totem-pole output stage is no longer present.

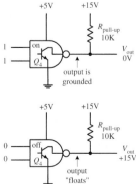

Using a pull-up resistor with open collector logic gates.

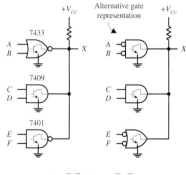

$$X = \overline{A}\,\overline{B} \cdot C\,D \cdot (\overline{E} + \overline{F})$$

Wired-AND logic: outputs of all three gates must float in order to get a HIGH output at X.

Another important feature of OC gates is their ability to sink large amounts of currents. For example, the 7506 OC inverter buffer/driver IC is capable of sinking 40 mA, which is 10 times the amount of current a standard 7404 inverter can sink. (The 7404 OC buffer/driver has the same sinking ability as the 7406 OC but does not provide any logic function—it simply acts as a buffer stage.) The ability for an OC gate to sink a fairly large current makes it useful for driving relays, motors, LED displays, and other high-current loads. Figure 12.63 shows a number of OC logic gate ICs.

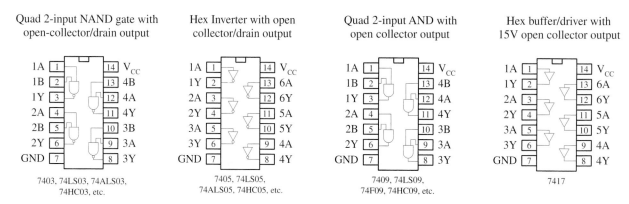

Quad 2-input NAND gate with open-collector/drain output	Hex Inverter with open collector/drain output	Quad 2-input AND with open collector output	Hex buffer/driver with 15V open collector output
7403, 74LS03, 74ALS03, 74HC03, etc.	7405, 74LS05, 74ALS05, 74HC05, etc.	7409, 74LS09, 74F09, 74HC09, etc.	7417

FIGURE 12.63

OC gates are also useful in instances where the output from two or more gates or other devices must be tied together. If you were to use standard gates with totem-pole output stages, if one gate were to output a high (+5 V) while another gate were to output a low (0 V), there would be a direct short circuit created, which could cause either or both gates to burn out. By using OC gates, this problem can be avoided.

When working with OC gates, you cannot apply the same Boolean rules you used earlier with the standard gates. Instead, you must use what is called *wired-AND logic,* which amounts to simply ANDing all gates together, as shown in Fig. 12.62. In other words, the outputs of all the gates must float in order to get a high output level.

12.4.8 Schmitt-Triggered Gates

FIGURE 12.64

There are special-purpose logic gates that come with Schmitt-triggered inputs. Unlike the conventional logic gates, Schmitt-triggered gates have two input threshold voltages. One threshold voltage is called the *positive threshold voltage* (V_T^+), while the other is called the *negative threshold voltage* (V_T^-). Example Schmitt-triggered ICs include the quad 7404 inverter, the quad 2-input NAND gate, and the dual 4-input NAND gate shown below.

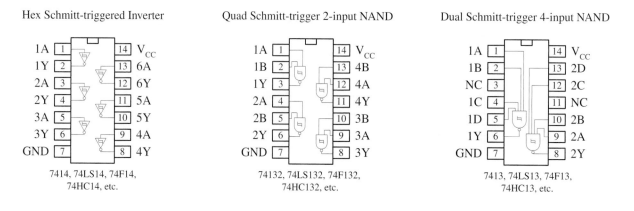

Hex Schmitt-triggered Inverter	Quad Schmitt-trigger 2-input NAND	Dual Schmitt-trigger 4-input NAND
7414, 74LS14, 74F14, 74HC14, etc.	74132, 74LS132, 74F132, 74HC132, etc.	7413, 74LS13, 74F13, 74HC13, etc.

To get a sense of how these devices work, let's compare the Schmitt-triggered 7414 inverter gate with a conventional inverter gate, the 7404. With the 7404, to make the output go from high to low or from low to high, the input voltage must fall above or below the single 2.0-V threshold voltage. However, with the 7414, to make the output go from low to high, the input voltage must dip below V_T- (which is +0.9 V for this particular IC); to make the output go from high to low, the input voltage must pop above V_T+ (which is +1.7 V for this particular IC). The difference in voltage between V_T+ and V_T- is called the *hysteresis voltage* (see Chap. 7 for details). The symbol used to designate a Schmitt trigger is based on the appearance of its transfer function, as shown in the figure below.

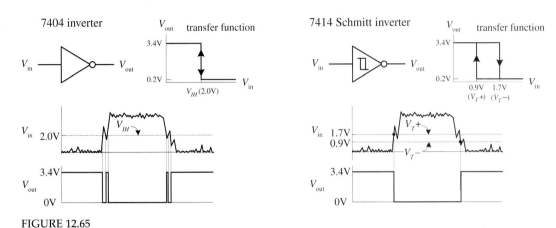

FIGURE 12.65

In terms of applications, Schmitt-triggered devices are quite handy for transforming noisy signals or signals that waver around critical threshold levels into sharply defined, jitter-free output signals. This is illustrated in the lower graphs shown in Fig. 12.65. The conventional 7404 experiences an unwanted output spike resulting from a short-term spike present during low-to-high and high-to-low input voltage transitions. The Schmitt-triggered inverter ignores these spikes because it incorporates hysteresis.

12.4.9 Interfacing Logic Families

Mixing of logic families, in general, should be avoided. Obvious reasons for not mixing include differences in input/output logic levels, supply voltages, and output drive capability that exist among the various families. Another important reason involves differences in speed between the various families; if you mix slow-logic ICs with faster-logic ICs, you can run into timing problems.

There are times, however, when mixing is unavoidable or even desirable. For example, perhaps a desired special-purpose device (e.g., memory, counter, etc.) only exists in CMOS, but the rest of your system consists of TTL. Mixing of families is also common when driving loads. For example, a TTL gate (often with an open-collector output) is frequently used as an interface between a CMOS circuit and an external load, such as a relay or indicator light. A CMOS output, by itself, usually does not provide sufficient drive current to power such loads. I will discuss driving loads in Section 12.10.

Figure 12.66 shows tricks for interfacing various logic families. These tricks take care of input/output incompatibility problems as well as supply voltage incompatibility problems. The tricks, however, do not take care of any timing incompatibility problems that may arise.

Interfacing logic families

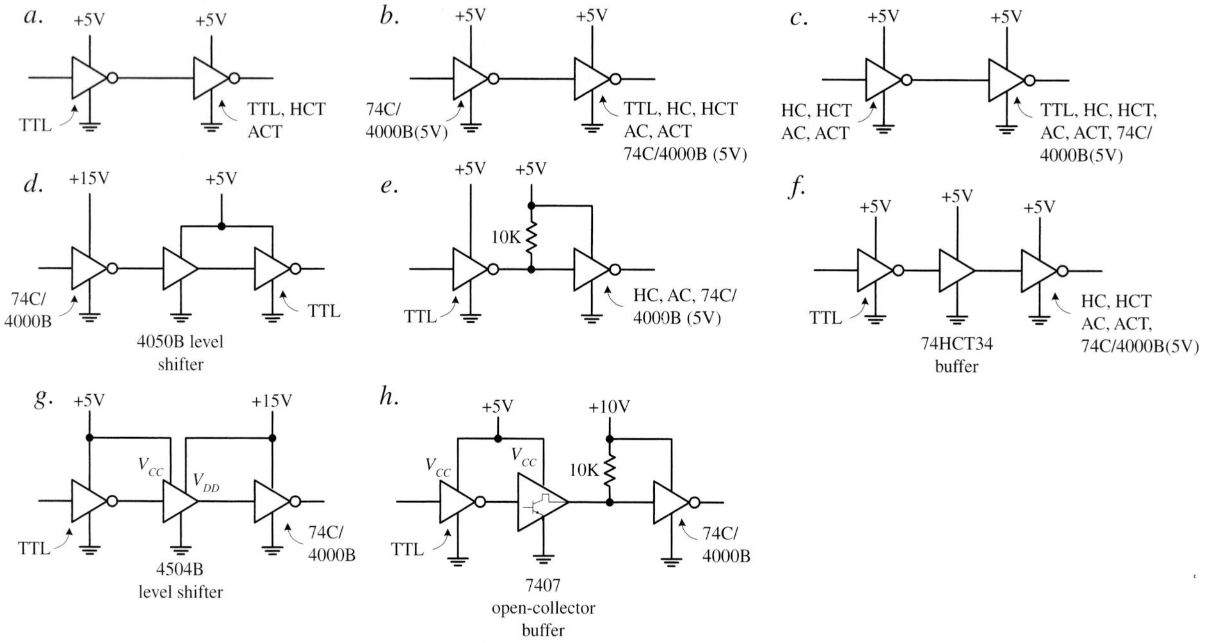

Figure a. TTL can be directly interfaced with itself or with HCT or ACT.

Figure b. CMOS 74C/4000(B) with V_{DD} = +5V can drive TTL, HC, HCT, AC, or ACT.

Figure c. HC, HCT, AC, and ACT can directly drive TTL, HC, HCT, AC, ACT, and 74C/4000B (5V).

Figure d. When 74C/4000(B) uses a supply voltage that is higher than +5V, a level-shifting buffer IC, like the 4050B, can be used. The 4050B is powered by a 5-V supply and can accept 0-V/15-V logic levels at its inputs, while providing corresponding 0-V/5-V logic level outputs. The buffer also provides increased output drive current (4000B has a weak output drive capability when compared to TTL).

Figure e. Recall that the actual high output of a TTL gate is around 3.4 V instead of 5 V. But CMOS (V_{DD} = 5 V) inputs may require from 4.4 (HC) to 4.9V (4000B) for high input levels. If the CMOS device is of the 74C/4000B series, the actual required high input voltage depends on the supply voltage and is equal to $\frac{2}{3}V_{DD}$. To provide enough voltage to match voltage levels, a pullup resistor is used. The pullup resistor pulls the input to the CMOS gate up to the supply voltage to which the pullup resistor is connected.

Figure f. Another trick for interfacing TTL with CMOS is to simply use a CMOS TTL-compatible gate, like the 74HCT or 74ACT.

Figures g, h. These two figures show different methods for interfacing a TTL gate with a CMOS gate set to a higher supply voltage. In Figure g, a 4504B level-shifting buffer is used. The 4504B requires two supply voltages: a TTL supply (for 0/5 V levels) and a CMOS supply (for 0 to 15 V levels). In figure h, an open-collector buffer and 10-k pullup resistor are used to convert the lower-level TTL output voltages into higher-level CMOS input voltages.

FIGURE 12.66

12.5 Powering and Testing Logic ICs and General Rules of Thumb

12.5.1 Powering Logic ICs

Most TTL and CMOS logic devices will work with 5V ± 0.25V (5 percent) supplies like the ones shown in Fig. 12.67. The battery supplies should be avoided when using certain TTL families like the 74xx, 74S, 74AS, and 74F, which dissipate considerably more current than, say, the CMOS 74HC series. Of course, the low-power, low-voltage 74LV, 74LVC, 74LVT, 74ALVC, and 74BCT series, which require from 1.2 to 3.6 V with as low as 2.5 µW/gate power dissipation (for 74BCT), are ideal for small battery-powered applications.

5-V line and battery supplies for digital logic circuits

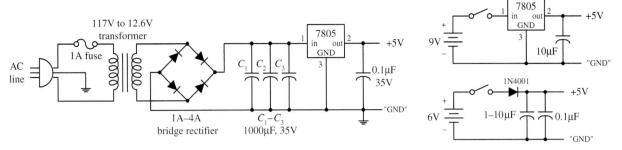

FIGURE 12.67

12.5.2 Power Supply Decoupling

When a TTL device makes a low-to-high or a high-to-low level transition, there is an interval of time that the conduction times in the upper and lower totem-pole output transistors overlap. During this interval, a drastic change in power supply current occurs, which results in a sharp, high-frequency current spike within the supply line. If a number of other devices are linked to the same supply, the unwanted spike can cause false triggering of these devices. The spike also can generate unwanted electromagnetic radiation. To avoid unwanted spikes within TTL systems, decoupling capacitors can be used. A decoupling capacitor, typically tantalum, from 0.01 to 1 µF (>5 V), is placed directly across the V_{CC}-to-ground pins of each IC in the system. The capacitors absorb the spikes and keep the V_{CC} level at each IC constant, thus reducing the likelihood of false triggering and generally electromagnetic radiation. Decoupling capacitors should be placed as close to the ICs as possible to keep current spikes local, instead of allowing them to propagate back toward the power supply. You can usually get by with using one decoupling capacitor for every 5 to 10 gates or one for every 5 counter or register ICs.

12.5.3 Unused Inputs

Unused inputs that affect the logical state of a chip should not be allowed to float. Instead, they should be tied high or low, as necessary (floating inputs are liable to pickup external electrical noise, which leads to erratic output behavior). For example, a four-input NAND TTL gate that only uses two inputs should have its two unused inputs held high to maintain proper logic operation. A three-input NOR gate that only uses two inputs should have its unused input held low to maintain proper logic operation. Likewise, the CLEAR and PRESET inputs of a flip-flop should be grounded or tied high, as appropriate.

If there are unused sections within an IC (e.g., unused logic gates within a multi-gate package), the inputs that link to these sections can be left unconnected for TTL but not for CMOS. When unused inputs are left unconnected in CMOS devices, the inputs may pick up unwanted charge and may reach a voltage level that causes output MOS transistors to conduct simultaneously, resulting in a large internal current spike from the supply (V_{DD}) to ground. The result can lead to excessive supply current drain and IC damage. To avoid this fate, inputs of unused sections of a CMOS IC should be grounded. Figure 12.68 illustrates what to do with unused inputs for TTL and CMOS NAND and NOR ICs.

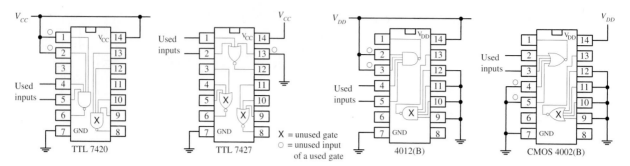

X = unused gate
○ = unused input
      of a used gate

Connect unused inputs of a used NAND gate HIGH to maintain proper logic function. Don't assume they'll be naturally HIGH for TTL. Connect unused input of a NOR gate LOW to maintain proper logic function. Inputs of unused TTL gates can be left unconnected.

Connect unused inputs of a used NAND gate HIGH to maintain proper logic function. Connect unused inputs of a used NOR gate LOW to maintain proper logic function. Inputs of unused CMOS gates should be grounded.

FIGURE 12.68

As a last note of caution, never drive CMOS inputs when the IC's supply voltage is removed. Doing so can damage the IC's input protection diodes.

12.5.4   Logic Probes and Logic Pulsers

Two simple tools used to test logic ICs and circuits include the test probe and logic pulser, as shown below.

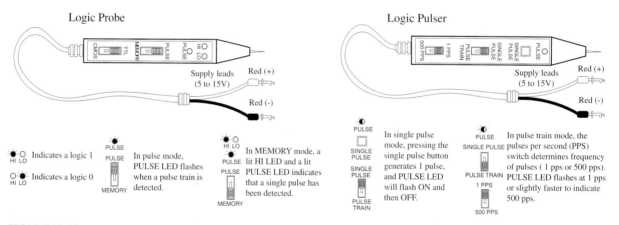

FIGURE 12.69

A typical logic probe comes in a penlike package, with metal probe tip and power supply wires, one red, one black. Red is connected to the positive supply voltage of the digital circuit (V_{CC}), while black is connected to the ground (V_{SS}) of the circuit. To test a logic state within a circuit, the metal tip of the probe is applied. If a high voltage is detected, the probe's high LED lights up; if a low voltage is detected, the probe's low LED turns off. Besides performing simple static tests, logic probes can perform a few simple dynamic tests too, such as detecting a single momentary pulse that is too fast for the human eye to detect or detecting a pulse train, such as a clock signal. To detect a single pulse, the probe's PULSE/MEMORY switch is thrown to the MEMORY position. When a single pulse is detected, the internal memory circuit remembers the single pulse and lights up both the HI LED and

PULSE LED at the same time. To clear the memory to detect a new single-pulse, the PULSE/MEMORY switch is toggled. To detect a pulse train, the PULSE/MEMORY switch is thrown to the PULSE position. When a pulse train is detected, the PULSE LED flashes on and off. Logic probes usually will detect single pulses with widths as narrow as 10 ns and will detect pulse trains with frequencies around 100 MHz. Check the specifications that come with your probe to determine these min/max limits.

A logic pulser allows you to send a single logic pulse or a pulse train through IC and circuits, where the results of the applied pulses can be monitored by a logic probe. Like a logic probe, the pulser comes with similar supply leads. To send a single pulse, the SINGLE-PULSE/PULSE-TRAIN switch is set to SINGLE-PULSE, and then the SINGLE-PULSE button is pressed. To send a pulse train, switch to PULSE-TRAIN mode. With the pulser model shown in Fig. 12.69, you get to select either one pulse per second (1 pps) or 500 pulses per second.

12.6   Sequential Logic

The combinational circuits covered previously (e.g., encoders, decoders, multiplexers, parity generators/checkers, etc.) had the property of input-to-output immediacy. This means that when input data are applied to a combinational circuit, the output responds right away. Now, combinational circuits lack a very important characteristic—they cannot store information. A digital device that cannot store information is not very interesting, practically speaking.

To provide "memory" to circuits, you must create devices that can latch onto data at a desired moment in time. The realm of digital electronics devoted to this subject is referred to as *sequential logic*. This branch of electronics is referred to as *sequential* because for data bits to be stored and retrieved, a series of steps must occur in a particular order. For example, a typical set of steps might involve first sending an enable pulse to a storage device, then loading a group of data bits all at once (parallel load), or perhaps loading a group of data bits in a serial manner—which takes a number of individual steps. At a latter time, the data bits may need to be retrieved by first applying a control pulse to the storage device. A series of other pulses might be required to force the bits out of the storage device.

To push bits through sequential circuits usually requires a clock generator. The clock generator is similar to the human heart. It generates a series of high and low voltages (analogous to a series of high and low pressures as the heart pumps blood) that can set bits into action. The clock also acts as a time base on which all sequential actions can be referenced. Clock generators will be discussed in detail later on. Now, let's take a look at the most elementary of sequential devices, the SR flip-flop.

12.6.1   SR Flip-Flops

The most elementary data-storage circuit is the *SR (set-reset) flip-flop*, also referred to as a *transparent latch*. There are two basic kinds of SR flip-flops, the cross-NOR SR flip-flop and the cross-AND SR flip-flop.

Cross-NOR SR flip-flop

Cross-NAND SR flip-flop

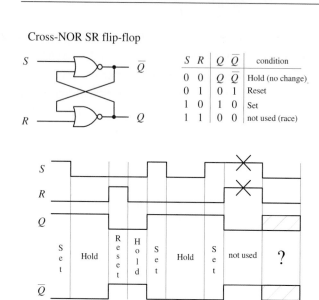

S	R	Q	$\overline{Q}$	condition
0	0	Q	$\overline{Q}$	Hold (no change)
0	1	0	1	Reset
1	0	1	0	Set
1	1	0	0	not used (race)

S	R	Q	$\overline{Q}$	condition
0	0	1	1	not used (race)
0	1	1	0	Set
1	0	0	1	Reset
1	1	Q	$\overline{Q}$	Hold (no change)

Going from $S = 1$, $R = 1$ back to the hold condition ($S = 0$, $R = 0$) leads to an unpredictable output. Therefore $S = 1$, $R = 1$ isn't used.

FIGURE 12.70

First, let's consider the cross-NOR SR flip-flop shown above. At first it appears that figuring out what the cross-NOR SR flip-flop does given only two input voltages is impossible, since each of the NOR gates' inputs depend on the outputs—and what are the outputs anyway? (For now, pretend that Q and $\overline{Q}$ are not complements but separate variables—you could call them X and Y if you like.) Well, first of all, you know that a NOR gate will output a high (logic 1) only if both inputs are low (logic 0). From this you can deduce that if $S = 1$ and $R = 0$, Q must be 1 and $\overline{Q}$ must be 0, regardless of the outputs—this is called the *set* condition. Likewise, by similar argument, we can deduce that if $S = 0$ and $R = 1$, Q must be 0 and $\overline{Q}$ must be 1—this is called the *reset* condition.

But now, what about $R = 0$, $S = 0$? Can you predict the outputs only given input levels? No! It is impossible to predict the outputs because the outputs are essential for predicting the outputs—it is a "catch-22." However, if you know the states of the outputs beforehand, you can figure things out. For example, if you first set the flip-flop ($S = 1$, $R = 0$, $Q = 1$, $\overline{Q} = 0$) and then apply $S = 0$, $R = 0$, the flip-flop would remain set (upper gate: $S = 0$, $Q = 1 \rightarrow \overline{Q} = 0$; lower gate: $R = 0$, $\overline{Q} = 0 \rightarrow Q = 1$). Likewise, if you start out in reset mode ($S = 0$, $R = 1$, $Q = 0$, $\overline{Q} = 0$) and then apply $S = 0$, $R = 0$, the flip-flop remains in reset mode (upper gate: $S = 0$, $Q = 0 \rightarrow \overline{Q} = 1$; lower gate: $R = 0$, $\overline{Q} = 1 \rightarrow Q = 0$). In other words, the flip-flop remembers, or latches onto, the previous output state even when both inputs go low (0)—this is referred to as the *hold* condition.

The last choice you have is $S = 1$, $R = 1$. Here, it is easy to predict what will happen because you know that as long as there is at least one high (1) applied to the input to the NOR gate, the output will always be 0. Therefore, $Q = 0$ and $\overline{Q} = 0$. Now, there are two fundamental problems with the $S = 1$, $R = 1$ state. First, why would you want to set and reset at the same time? Second, when you return to the hold condition from $S = 1$, $R = 1$, you get an unpredictable result, unless you know which input returned low last. Why? When the inputs are brought back to the hold position ($R = 0$, $S = 0$, $Q = 0$,

$\overline{Q} = 0$), both NOR gates will want to be 1 (they want to be held). But let's say one of the NOR gate's outputs changes to 1 a fraction of a second before the other. In this case, the slower flip-flop will not output a 1 as planned but will instead output 0. This is a classic example of a race condition, where the slower gate loses. But which flip-flop is the slower one? This unstable, unpredictable state cannot be avoided and is simply not used.

The cross-NAND SR flip-flop provides the same basic function as the NOR SR flip-flop, but there is a fundamental difference. Its hold and indeterminate states are reversed. This occurs because unlike the NOR gate, which only outputs a low when both its inputs are the same, the NAND gate only outputs a high when both its inputs are the same. This means that the hold condition for the cross-NAND SR flip-flop is $S = 1, R = 1$, while the indeterminate condition is $S = 0, R = 0$.

Here are two simple applications for SR flip-flops.

Switch Debouncer

Say you want to use the far-left switch/pullup resistor circuit (Fig. 12.71) to drive an AND gate's input high or low (the other input is fixed high). When the switch is open, the AND gate should receive a high. When the switch is closed, the gate should receive a low. That's what should happen, but that's not what actually happens. Why? Answer: Switch bounce. When a switch is closed, the metal contacts bounce a number of times before coming to rest due to inherent springlike characteristics of the contacts. Though the bouncing typically lasts no more than 50 ms, the results can lead to unwanted false triggering, as shown in the far left circuit below. A simple way to get rid of switch bounce is to use the switch debouncer circuit, shown at center. This circuit simply uses an SR flip-flop to store the initial switch contact voltage while ignoring all trailing bounces. In this circuit, when the switch is thrown from the B to A position, the flip-flop is set. As the switch bounces alternately high and low, the Q output remains high because when the switch contact bounces away from A, the S input receives a low (R is low too), but that's just a hold condition—the output stays the same. The same debouncing feature occurs when the switch is thrown from position A to B.

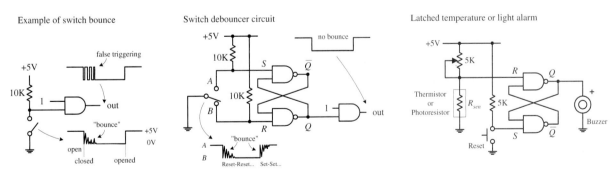

Example of switch bounce   Switch debouncer circuit   Latched temperature or light alarm

FIGURE 12.71

Latched Temperature or Light Alarm

This simple circuit (Fig. 12.71) uses an SR flip-flop to sound a buzzer alarm when the temperature (when using a thermistor) or the light intensity (when using a photoresistor) reaches a critical level. When the temp/light increases, the resistance of the thermistor/photoresistor decreases, and the R input voltage goes down. When the R input voltage goes below the high threshold level of the NAND gate, the flip-flop is set, and the alarm is sounded. The alarm will continue to sound until the RESET switch is pressed and the temp/light level has gone below the critical triggering level. The pot is used to adjust this level.

Level-Triggered SR Flip-Flop (The Beginning of Clocked Flip-Flops)

Now it would be nice to make an SR flip-flop synchronous—meaning making the S and R inputs either enabled or disabled by a control pulse, such as a clock. Only when the clock pulse arrives are the inputs sampled. Flip-flops that respond in this manner are referred to as *synchronous* or *clocked flip-flops* (as opposed to the preceding asynchronous flip-flops). To make the preceding SR flip-flop into a synchronous or clocked device, simply attach enable gates to the inputs of the flip-flop, as shown in Fig. 12.72. (Here, the cross-NAND arrangement is used, though a cross-NOR arrangement also can be used.) Only when the clock is high are the S and R inputs enabled. When the clock is low, the inputs are disabled, and the flip-flop is placed in hold mode. The truth table and timing diagram below help illustrate how this device works.

Clocked level-triggered NAND SR flip-flop

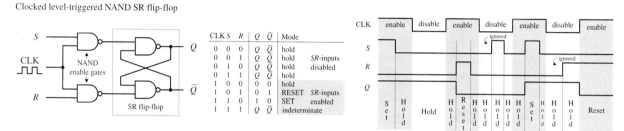

FIGURE 12.72

Edge-Triggered SR Flip-Flops

Now there is an annoying feature with the last level-triggered flip-flop; its S and R inputs have to be held at the desired input condition (set, reset, no change) for the entire time that the clock signal is enabling the flip-flop. With a slight alteration, however, you can make the level-triggered flip-flop more flexible (in terms of timing control) by turning it into an edge-triggered flip-flop. An edge-triggered flip-flop samples the inputs only during either a positive or negative clock edge ($\uparrow$ = positive edge, $\downarrow$ = negative edge). Any changes that occur before or after the clock edge are ignored—the flip-flop will be placed in hold mode. To make an edge-triggered flip-flop, introduce either a positive or a negative level-triggered clock pulse generator network into the previous level-triggered flip-flop, as shown in Fig. 12.73.

In a positive edge-triggered generator circuit, a NOT gate with propagation delay is added. Since the clock signal is delayed through the inverter, the output of the AND gate will not provide a low (as would be the case without a propagation delay) but will provide a pulse that begins at the positive edge of the clock signal and lasts for a duration equal to the propagation delay of the NOT gate. It is this pulse that is used to clock the flip-flop. Within the negative edge-triggered

Clocked edge-triggered NAND SR flip-flop

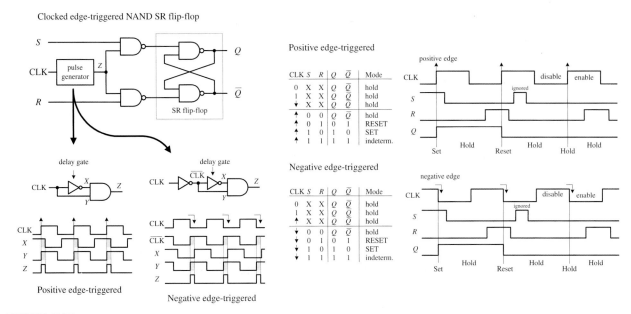

FIGURE 12.73

generator network, the clock signal is first inverted and then applied through the same NOT/AND network. The pulse begins at the negative edge of the clock and lasts for a duration equal to the propagation delay of the NOT gate. The propagation delay is typically so small (in nanoseconds) that the pulse is essentially an "edge."

Pulse-Triggered SR Flip-Flops (Master-Slave Flip-Flops)

A pulse-triggered SR flip-flop is a level-clocked flip-flop; however, for any change in output to occur, both the high and low levels of the clock must rise and fall. Pulse-triggered flip-flops are also called *master-slave flip-flop;* the master accepts the initial inputs and then "whips" the slave with its output when the negative clock edge arrives. Another analogy often used is to say that during the positive edge, the master gets cocked (like a gun), and during the negative clock edge, the slave gets triggered. Figure 12.74 shows a simplified pulse-triggered cross-NAND SR flip-flop.

FIGURE 12.74

Pulse-triggered SR flip-flop (master-slave SR flip-flop)

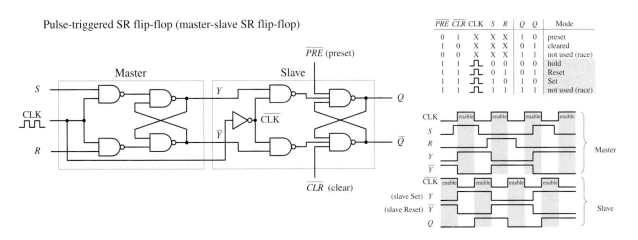

The master is simply a clocked SR flip-flop that is enabled during the high clock pulse and outputs Y and $\overline{Y}$ (either set, reset, or no change). The slave is similar to the master, but it gets enabled only during the negative clock pulse (due to the inverter). The moment the slave is enabled, it uses the Y and $\overline{Y}$ outputs of the master as inputs and then outputs the final result. Notice the preset ($\overline{PRE}$) and clear ($\overline{CLR}$) inputs. These are called *asynchronous inputs.* Unlike the synchronous inputs, S and R, the asynchronous input disregard the clock and either clear (also called *asynchronous reset*) or preset (also called *asynchronous set*) the flip-flop. When $\overline{CLR}$ is high and $\overline{PRE}$ is low, you get asynchronous reset, $Q = 1$, $\overline{Q} = 0$, regardless of the CLK, S, and R inputs. These active-low inputs are therefore normally pulled high to make them inactive. As you will see later when I discuss flip-flop applications, the ability to apply asynchronous set and resets is often used to clear entire registers that consist of an array of flip-flops.

General Rules for Deciphering Flip-Flop Logic Symbols

Now, typically, you do not have to worry about constructing flip-flops from scratch—instead, you buy flip-flop ICs. Likewise, you do not have to worry about complex logic gate schematics—instead, you use symbolic representations like the ones shown below. Although the symbols below apply to SR flip-flops, the basic rules that are outlined can be applied to the D and JK flip-flops, which will be discussed in following sections.

Symbolic representation of level-triggered, edge-triggered and pulse-triggered flip-flops

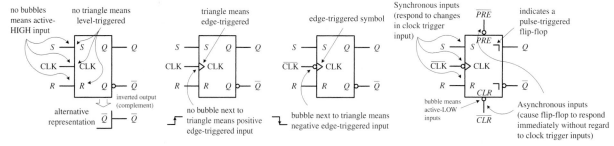

FIGURE 12.75

12.6.2 SR Flip-Flop (Latch) ICs

Figure 12.76 shows a few sample SR flip-flop (latch) ICs. The 74LS279A contains four independent SR latches (note that two of the latches have an extra set input). This IC is commonly used in switch debouncers. The 4043 contains four three-state cross-coupled NOR SR latches. Each latch has individual set and reset inputs, as well as separate Q outputs. The three-state feature is an extra bonus, which allows you to effectively disconnect all Q outputs, making it appear that the outputs are open circuits (high impedance, or high Z). This three-state feature is often used in applications where a number of devices must share a common data bus. When the output data from one latch are applied to the bus, the outputs of other latches (or other devices) are disconnected via the high-Z condition. The 4044 is similar to the 4043 but contains four three-state cross-coupled NAND RS latches.

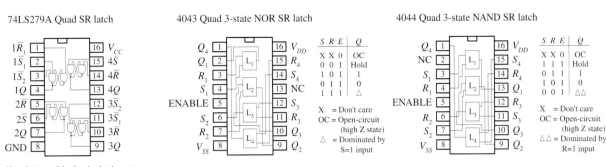

S R E	Q
X X 0	OC
0 0 1	Hold
1 0 1	1
0 1 1	0
1 1 1	△

X = Don't care
OC = Open-circuit (high Z state)
△ = Dominated by S=1 input

S R E	Q
X X 0	OC
1 1 1	Hold
0 1 1	1
1 0 1	0
0 0 1	△△

X = Don't care
OC = Open-circuit (high Z state)
△△ = Dominated by R=1 input

Note that two of the four latches have two S inputs and that inputs are active-LOW

A LOW enable input effectively disconnects the latch states from the Q outputs, resulting in an open-circuit condition or high-impedance (Z) state at the Q outputs.

FIGURE 12.76

12.6.3 D Flip-Flops

A D-type flip-flop (data flip-flop) is a single input device. It is basically an SR flip-flop, where S is replaced with D and R is replaced $\overline{D}$ (inverted D)—the inverted input is tapped from the D input through an inverter to the R input, as shown below. The inverter ensures that the indeterminate condition (race, or not used state, $S=1$, $R=1$) never occurs. At the same time, the inverter eliminates the hold condition so that you are left with only set ($D=1$) and reset ($D=0$) conditions. The circuit below represents a level-triggered D-type flip-flop.

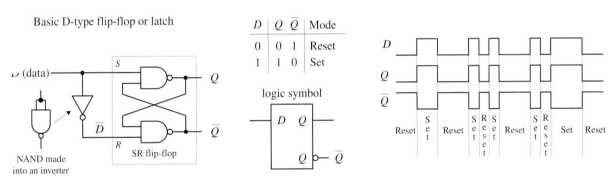

FIGURE 12.77

To create a clocked D-type level-triggered flip-flop, first start with the clocked level-triggered SR flip-flop and throw in the inverter, as shown in Fig. 12.78.

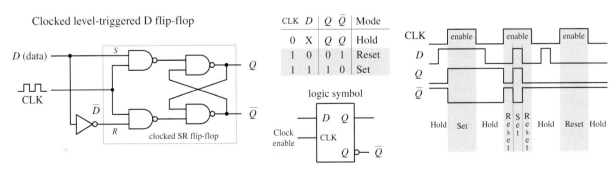

FIGURE 12.78

To create a clocked, edge-triggered D-type flip-flop, take a clocked edge-triggered SR flip-flop and add an inverter, as shown in Fig. 12.79.

Edge-triggered *D* flip-flop

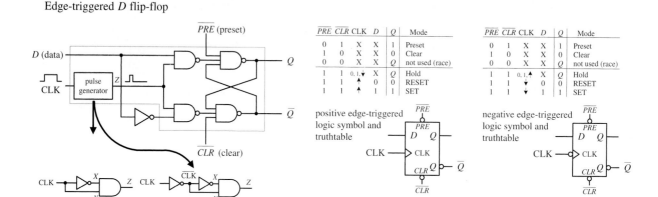

FIGURE 12.79

Here's a popular edge-trigger D-type flip-flop IC, the 7474 (e.g., 74HC74, etc.). It contains two D-type positive edge-triggered flip-flops with asynchronous preset and clear inputs.

74HC74 Dual D-type positive edge-triggered flip-flop with preset and clear

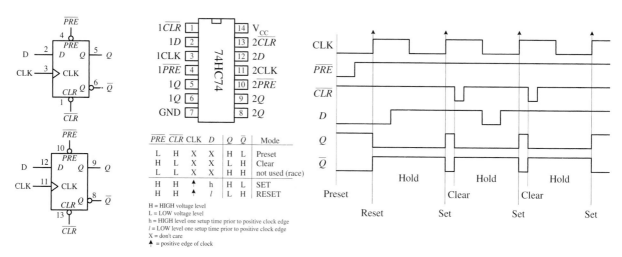

FIGURE 12.80

Note the lowercase letters *l* and *h* in the truth table in this figure. The *h* is similar to the *H* for a high voltage level, and the *l* is similar to the *L* for low voltage level; however, there is an additional condition that must be met for the flip-flop's output to do what the truth table indicates. The additional condition is that the *D* input must be fixed high (or low) in duration for at least one *setup time* (t_s) before the positive clock edge. This condition stems from the real-life propagation delays present in flip-flop ICs; if you try to make the flip-flop switch states too fast (do not give it time to move electrons around), you can end up with inaccurate output readings. For the 7474, the setup time is 20 ns. Therefore, when using this IC, you must not apply input pulses that are within the 20-ns limit. Other flip-flops will have different setup times, so you will have to check the manufacturer's data sheets. I will discuss setup time and some other flip-flop timing parameters in greater detail at the end of this section.

D-type flip-flops are sometimes found in the pulse-triggered (master-slave) variety. Recall that a pulse-triggered flip-flop requires a complete clock pulse before the outputs will reflect what is applied at the input(s) (in this case the D input). The figure below shows the basic structure of a pulse-triggered D flip-flop. It is almost exactly like the pulse-triggered SR flip-flop, except for the inverter addition to the master's input.

Pulse-triggered D-type flip-flop (master-slave D-type flip-flop)

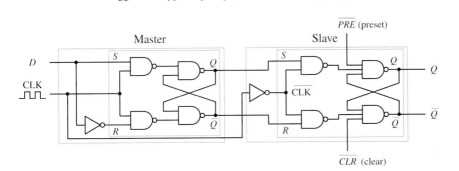

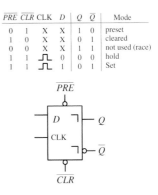

$\overline{PRE}$	$\overline{CLR}$	CLK	D	Q	$\overline{Q}$	Mode
0	1	X	X	1	0	preset
1	0	X	X	0	1	cleared
0	0	X	X	1	1	not used (race)
1	1	⎍	0	0	0	hold
1	1	⎍	1	0	1	Set

FIGURE 12.81

12.6.4 A Few Simple D-Type Flip-Flop Applications

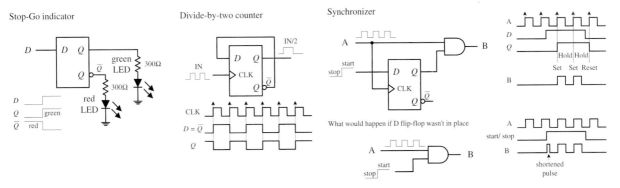

FIGURE 12.82

In the stop-go indicator circuit, a simple level-triggered D-type flip-flop is used to turn on a red LED when its D input is low (reset) and turn on a green LED when the D input is high (set). Only one LED can be turned on at a time.

The divide-by-two counter uses a positive edge-triggered D-type flip-flop to divide an applied signal's frequency by two. The explanation of how this works is simple: The positive edge-triggered feature does not care about negative edges. You can figure out the rest.

A synchronizer is used when you want to use an external asynchronous control signal (perhaps generated by a switch or other input device) to control some action within a synchronous system. The synchronizer provides a means of keeping the phase of the action generated by the control signal in synch with the phase of the synchronous system. For example, say you want an asynchronous control signal to control the number of clock pulses that get from point A to point B within a synchronous system. You might try using a simple enable gate, as shown below the synchronizer circuit in the figure above. However, because the external control signal is not synchronous (in phase) with the clock, when you apply the external control signal, you may shorten the first or last output pulse, as shown in the lower timing diagram. Certain applications do not like shortened clock pulses and will not function properly. To avoid shortened pulses, throw in an edge-triggered D-type flip-flop to create a synchronizer. The flip-flop's CLK input is tapped off the input clock line, its D input receives the external control signal, and its Q output is connected to the AND gate's enable input. With this arrangement, there will never be shortened clock pulses because the Q output of the flip-flop will not supply enable pulses to the AND gate that are out of phase with the input clock signal. This is due to the fact that after the flip-flop's CLK input receives a positive clock edge, the flip-flop ignores any input changes applied to the D input until the next positive clock edge.

12.6.5   Quad and Octal D Flip-Flops

Most frequently you will find a number of D flip-flops or D latches grouped together within a single IC. For example, the 74HC75, shown below, contains four transparent D latches. Latches 0 and 1 share a common active-low enable E_0–E_1, while latches 2 and 3 share a common active-low enable E_2–E_3. From the function table, each Q output follows each D input as long as the corresponding enable line is high. When the enable line goes low, the Q output will become latched to the value that D was one setup time prior to the high-to-low enable transition. The 4042 is another quad D-type latch—an explanation of how it works is provided in the figure below. D-type latches are commonly used as data registers in bus-oriented systems; the figure below explains the details.

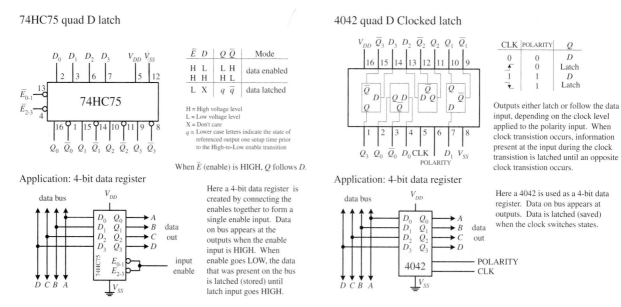

FIGURE 12.83

D flip-flops also come in octal form—eight flip-flops per IC. These devices are frequently used as 8-bit data registers within microprocessor systems, where devices share 8-bit or $2 \times 8 = 16$-bit data or address buses. An example of an octal D-type flip-flop is the 74HCT273 shown in Fig. 12.84. All D flip-flops within the 74HCT273 share a common positive edge-triggered clock input and a common active-low clear input. When the clock input receives a positive edge, data bits applied to D_0 through D_7 are stored in the eight flip-flops and appear at the outputs Q_0 through Q_7. To clear all flip-flops, the clear input is pulsed low. I will talk more about octal flip-flops and other bus-oriented devices later.

74HCT273 octal edge-triggered D-type flip-flop with Clear

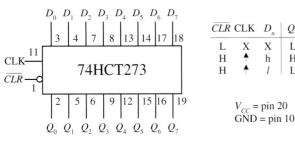

$\overline{CLR}$	CLK	D_n	Q_n	Mode
L	X	X	L	Clear
H	⬆	h	H	Set
H	⬆	l	L	Reset

V_{CC} = pin 20
GND = pin 10

H = High voltage level
L = Low voltage level
h = High voltage level one setup time prior
   to the low-to-high clock transition
l = Low voltage level one setup time prior
   to the low-to-high clock transition
X = Don't care
⬆ = Low-to-high clock transition

FIGURE 12.84

12.6.6 JK Flip-Flops

Finally, we come to the last of the flip-flops, the JK flip-flop. A JK flip-flop resembles an SR flip-flop, where J acts like S and K acts like R. Likewise, it has a set mode ($J = 1$, $K = 0$), a reset mode ($J = 0$, $K = 1$), and a hold mode ($J = 0$, $K = 0$). However, unlike the SR flip-flop, which has an indeterminate mode when $S = 1$, $R = 1$, the JK flip-flop has a *toggle* mode when $J = 1$, $K = 1$. *Toggle* means that the Q and $\overline{Q}$ outputs switch to their opposite states at each active clock edge. To make a JK flip-flop, modify the SR flip-flop's internal logic circuit to include two cross-coupled feedback lines between the output and input. This modification, however, means that the JK flip-flop cannot be level-triggered; it can only be edge-triggered or pulse-triggered. Figure 12.85 shows how you can create edge-triggered flip-flops based on the cross-NAND SR edge-triggered flip-flop.

Edge-triggered JK flip-flops

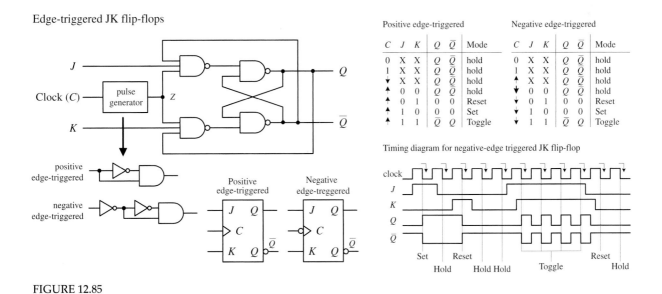

Positive edge-triggered

C	J	K	Q	$\overline{Q}$	Mode
0	X	X	Q	$\overline{Q}$	hold
1	X	X	Q	$\overline{Q}$	hold
⬇	X	X	Q	$\overline{Q}$	hold
⬆	0	0	Q	$\overline{Q}$	hold
⬆	0	1	0	0	Reset
⬆	1	0	0	0	Set
⬆	1	1	$\overline{Q}$	Q	Toggle

Negative edge-triggered

C	J	K	Q	$\overline{Q}$	Mode
0	X	X	Q	$\overline{Q}$	hold
1	X	X	Q	$\overline{Q}$	hold
⬆	X	X	Q	$\overline{Q}$	hold
⬇	0	0	Q	$\overline{Q}$	hold
⬇	0	1	0	0	Reset
⬇	1	0	0	0	Set
⬇	1	1	$\overline{Q}$	Q	Toggle

Timing diagram for negative-edge triggered JK flip-flop

FIGURE 12.85

Edge-triggered JK flip-flops also come with preset (asynchronous set) and clear (asynchronous reset) inputs. See Fig. 12.86.

Edge-triggered JK flip-flops with Preset and Clear

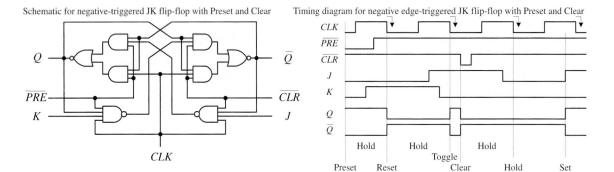

Schematic for negative-triggered JK flip-flop with Preset and Clear

Timing diagram for negative edge-triggered JK flip-flop with Preset and Clear

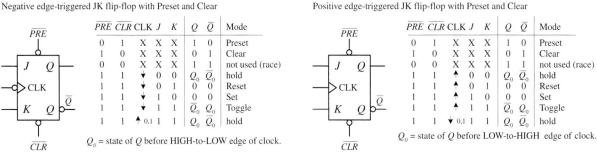

Negative edge-triggered JK flip-flop with Preset and Clear

$\overline{PRE}$	$\overline{CLR}$	CLK	J	K	Q	$\overline{Q}$	Mode
0	1	X	X	X	1	0	Preset
1	0	X	X	X	0	1	Clear
0	0	X	X	X	1	1	not used (race)
1	1	↓	0	0	Q_0	$\overline{Q}_0$	hold
1	1	↓	0	1	0	0	Reset
1	1	↓	1	0	0	0	Set
1	1	↓	1	1	$\overline{Q}_0$	Q_0	Toggle
1	1	↑	0,1	1	1	Q_0	$\overline{Q}_0$	hold

Q_0 = state of Q before HIGH-to-LOW edge of clock.

Positive edge-triggered JK flip-flop with Preset and Clear

$\overline{PRE}$	$\overline{CLR}$	CLK	J	K	Q	$\overline{Q}$	Mode
0	1	X	X	X	1	0	Preset
1	0	X	X	X	0	1	Clear
0	0	X	X	X	1	1	not used (race)
1	1	↑	0	0	Q_0	$\overline{Q}_0$	hold
1	1	↑	0	1	0	0	Reset
1	1	↑	1	0	0	0	Set
1	1	↑	1	1	$\overline{Q}_0$	Q_0	Toggle
1	1	↓	0,1	1	1	Q_0	$\overline{Q}_0$	hold

Q_0 = state of Q before LOW-to-HIGH edge of clock.

FIGURE 12.86

There are pulse-triggered (master-slave) flip-flops too, although they are not as popular as the edge-triggered JK flip-flops for an undesired effect that occurs, which I will talk about in a second. These devices are similar to the pulse-triggered SR flip-flops with the exception of the distinctive JK cross-coupled feedback connections from the slave's Q and $\overline{Q}$ outputs back to the master's input gates. The figure below shows a simple NAND pulse-triggered JK flip-flop.

Pulse-triggered JK flip-flop (master-slave JK flip-flop)

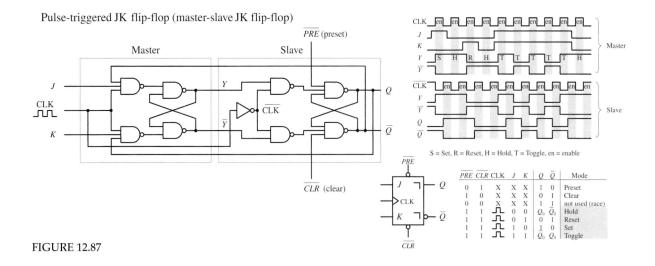

S = Set, R = Reset, H = Hold, T = Toggle, en = enable

$\overline{PRE}$	$\overline{CLR}$	CLK	J	K	Q	$\overline{Q}$	Mode
0	1	X	X	X	1	0	Preset
1	0	X	X	X	0	1	Clear
0	0	X	X	X	1	1	not used (race)
1	1	⊓	0	0	Q_0	$\overline{Q}_0$	Hold
1	1	⊓	0	1	0	1	Reset
1	1	⊓	1	0	1	0	Set
1	1	⊓	1	1	$\overline{Q}_0$	Q_0	Toggle

FIGURE 12.87

Now there is often a problem with pulse-triggered JK flip-flops. They occasionally experience what is called *ones catching*. In ones catching, unwanted pulses or glitches caused by electrostatic noise appear on J and K while the clock is high. The flip-flop remembers these glitches and interprets them as true data. Ones catching normally is not a problem when clock pulses are of short duration; it is when the pulses get long that you must watch out. To avoid ones catching all together, stick with edge-triggered JK flip-flops.

A Few JK Flip-Flop ICs

74LS76 dual negative edge-triggered JK flip-flop with Preset and Clear   74109 dual JK positive edge-triggered flip-flop with Preset and Clear

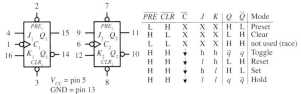

$\overline{PRE}$	$\overline{CLR}$	$\overline{C}$	J	K	Q	$\overline{Q}$	Mode
L	H	X	X	X	H	L	Preset
H	L	X	X	X	L	H	Clear
L	L	X	X	X	H	H	not used (race)
H	H	↓	h	h	$\overline{q}$	q	Toggle
H	H	↓	l	h	L	H	Reset
H	H	↓	h	l	H	L	Set
H	H	↓	l	l	q	$\overline{q}$	Hold

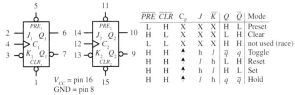

$\overline{PRE}$	$\overline{CLR}$	C_p	J	$\overline{K}$	Q	$\overline{Q}$	Mode
L	H	X	X	X	H	L	Preset
H	L	X	X	X	L	H	Clear
L	L	X	X	X	H	H	not used (race)
H	H	↑	h	l	$\overline{q}$	q	Toggle
H	H	↑	l	h	L	H	Reset
H	H	↑	h	l	H	L	Set
H	H	↑	l	h	q	$\overline{q}$	Hold

7476 dual pulse-triggered JK flip-flop with Preset and Clear   74HC73 dual pulse-triggered JK flip-flop with Clear

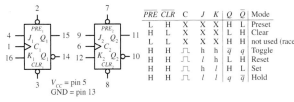

$\overline{PRE}$	$\overline{CLR}$	C	J	K	Q	$\overline{Q}$	Mode
L	H	X	X	X	H	L	Preset
H	L	X	X	X	L	H	Clear
L	L	X	X	X	H	H	not used (race)
H	H	⊓	h	h	$\overline{q}$	q	Toggle
H	H	⊓	l	h	L	H	Reset
H	H	⊓	h	l	H	L	Set
H	H	⊓	l	l	q	$\overline{q}$	Hold

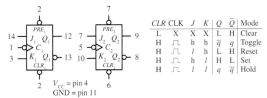

$\overline{CLR}$	CLK	J	K	Q	$\overline{Q}$	Mode
L	X	X	X	L	H	Clear
H	⊓	h	h	$\overline{q}$	q	Toggle
H	⊓	l	h	L	H	Reset
H	⊓	h	l	H	L	Set
H	⊓	l	l	q	$\overline{q}$	Hold

74114 dual pulse-triggered JK flip-flop with common Clock

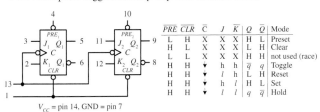

$\overline{PRE}$	$\overline{CLR}$	$\overline{C}$	J	$\overline{K}$	Q	$\overline{Q}$	Mode
L	H	X	X	X	H	L	Preset
H	L	X	X	X	L	H	Clear
L	L	X	X	X	H	H	not used (race)
H	H	↓	h	h	$\overline{q}$	q	Toggle
H	H	↓	l	h	L	H	Reset
H	H	↓	h	l	H	L	Set
H	H	↓	l	l	q	$\overline{q}$	Hold

H = HIGH voltage level steady state
h = HIGH voltage level one setup time prior to the HIGH-to-LOW Clock transition
L = LOW voltage level steady state
l = LOW voltage level one setup time prior to the HIGH-to-LOW Clock transition
q = Lowercase letters indicate the state of the referenced output prior to the
    HIGH-to-LOW Clock transition
X = Don't care
⊓ = Positive Clock pulse
↓ = Negative Clock edge
↑ = Positive Clock edge

FIGURE 12.88

12.6.7   Applications for JK Flip-Flops

Two major applications for JK flip-flops are found within counter and shift register circuits. For now, I will simply introduce a counter application—I will discuss shift registers and additional counter circuits later on in this chapter.

Ripple Counter (Asynchronous Counter)

A simple counter, called a *MOD-16 ripple counter* (or *asynchronous counter*), can be constructed by joining four JK flop-flops together, as shown in Fig. 12.89. (*MOD-16*, or *modulus 16*, means that the counter has 16 binary states.) This means that it can count from 0 to 15—the zero is one of the counts.

MOD-16 ripple counter/divide-by-2,4,8,16 counter

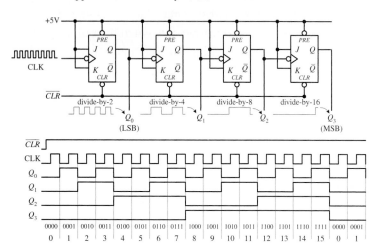

FIGURE 12.89

Each flip-flop in the ripple counter is fixed in toggle mode (J and K are both held high). The clock signal applied to the first flip-flop causes the flip-flop to divide the clock signal's frequency by 2 at its Q_0 output—a result of the toggle. The second flip-flop receives Q_0's output at its clock input and likewise divides by 2. The process continues down the line. What you get in the end is a binary counter with four digits. The least significant bit (LSB) is Q_0, while the most significant bit (MSB) is Q_4. When the count reaches 1111, the counter recycles back to 0000 and continues from there. To reset the counter at any given time, the active-low clear line is pulsed low. To make the counter count backward from 1111 to 0000, you would simply use the $\overline{Q}$ outputs.

The ripple counter above also can be used as a divide-by-2,4,8,16 counter. Here, you simply replace the clock signal with any desired input signal that you wish to divide in frequency. To get a divide-by-2 counter, you only need the first flip-flop; to get a divide-by-8 counter, you need the first three flip-flops.

Ripple counters with higher MOD values can be constructed by slapping on more flip-flops to the MOD-16 counter. But how do you create a ripple counter with a MOD value other than 2, 4, 8, 16, etc.? For example, say you want to create a MOD-10 (0 to 9) ripple counter. Likewise, what do you do if you want to stop the counter after a particular count has been reached and then trigger some device, such as an LED or buzzer. The figure below shows just such a circuit.

Ripple counter that counts from 0 to 9 then stops and activates LED

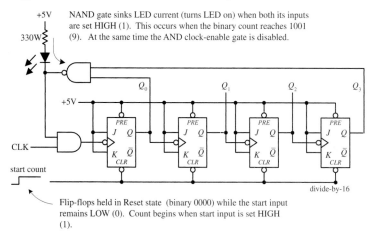

FIGURE 12.90

To make a MOD-10 counter, you simply start with the MOD-16 counter and connect the Q_0 and Q_3 outputs to a NAND gate, as shown in the figure. When the counter reaches 9 (1001), Q_0 and Q_3 will both go high, causing the NAND gate's output to go low. The NAND gate then sinks current, turning the LED on, while at the same time disabling the clock-enable gate and stopping the count. (When the NAND gate is high, there is no potential difference across the LED to light it up.) To start a new count, the active-low clear line is momentarily pulsed low. Now, to make a MOD-15 counter, you would apply the same basic approach used to the left, but you would connect Q_1, Q_2, and Q_3 to a three-input NAND gate.

Synchronous Counter

There is a problem with the ripple counter just discussed. The output stages of the flip-flops further down the line (from the first clocked flip-flop) take time to respond to changes that occur due to the initial clock signal. This is a result of the internal propagation delay that occurs within a given flip-flop. A standard TTL flip-flop may have an internal propagation delay of 30 ns. If you join four flip-flops to create a MOD-16 counter, the accumulative propagation delay at the highest-order output will be 120 ns. When used in high-precision synchronous systems, such large delays can lead to timing problems.

To avoid large delays, you can create what is called a *synchronous counter.* Synchronous counters, unlike ripple (asynchronous) counters, contain flip-flops whose clock inputs are driven at the same time by a common clock line. This means that output transitions for each flip-flop will occur at the same time. Now, unlike the ripple counter, you must use some additional logic circuitry placed between various flip-flop inputs and outputs to give the desired count waveform. For example, to create a 4-bit MOD-16 synchronous counter requires adding two additional AND gates, as shown below. The AND gates act to keep a flip-flop in hold mode (if both input of the gate are low) or toggle mode (if both inputs of the gate are high). So, during the 0–1 count, the first flip-flop is in toggle mode (and always is); all the rest are held in hold mode. When it is time for the 2–4 count, the first and second flip-flops are placed in toggle mode; the last two are held in hold mode. When it is time for the 4–8 count, the first AND gate is enabled, allowing the the third flip-flop to toggle. When it is time for the 8–15 count, the second AND gate is enabled, allowing the last flip-flop to toggle. You can work out the details for yourself by studying the circuit and timing waveforms.

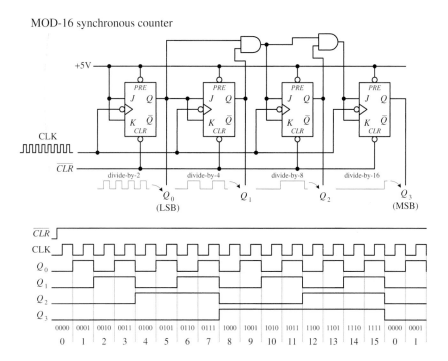

FIGURE 12.91

The ripple (asynchronous) and synchronous counters discussed so far are simple but hardly ever used. In practice, if you need a counter, be it ripple or synchronous, you go out and purchase a counter IC. These ICs are often MOD-16 or MOD-10 counters and usually come with many additional features. For example, many ICs allow you to preset the count to a desired number via parallel input lines. Others allow you to count up or to count down by means of control inputs. I will talk in great detail about counter ICs in a moment.

12.6.8   Practical Timing Considerations with Flip-Flops

When working with flip-flops, it is important to avoid race conditions. For example, a typical race condition would occur if, say, you were to apply an active clock edge at the very moment you apply a high or low pulse to one of the inputs of a JK flip-flop. Since the JK flip-flop uses what is present on the inputs at the moment the clock edge arrives, having a high-to-low input change will cause problems because you cannot determine if the input is high or low at that moment—it is a straight line. To avoid this type of race condition, you must hold the inputs of the flip-flop high or low for at least one setup time t_s before the active clock transition. If the input changes during the t_s to clock edge region, the output levels will be unreliable. To determine the setup time for a given flip-flop, you must look through the manufacturer's data sheets. For example, the minimum setup time for the 74LS76 JK flip-flop is 20 ns. Other timing parameters, such as hold time, propagation delay, etc., are also given by the manufacturers. A description of what these parameters mean is given below.

Flip-Flop Timing Parameters

Clock to output delays, data setup and hold times, clock pulse width

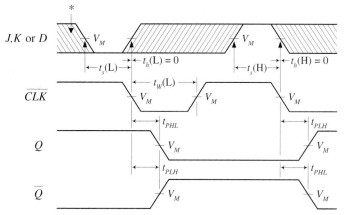

* Shaded region indicates when input is permitted to change for predictable output performance.

Preset and Clear to output delays, Preset and Clear pulse widths

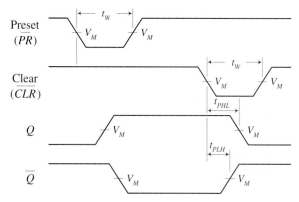

FIGURE 12.92

IMPORTANT TERMS

Setup time t_s—The time required that the input must be held before the active clock edge for proper operation. For a typical flip-flop, t_s is around 20 ns.

Hold time t_h—The time required that the input must be held after the active clock edge for proper operation. For most flip-flops, this is 0 ns—meaning inputs need not be held beyond the active clock signal.

T_{PLH}—Propagation delay from clock trigger point to the low-to-high Q output swing. A typical T_{PLH} for a flip-flop is around 20 ns.

T_{PHL}—Propagation delay from clock trigger point to the high-to-low Q output swing. A typical T_{PLH} for a flip-flop is around 20 ns.

f_{max}—Maximum frequency allowed at the clock input. Any frequency above this limit will result in unreliable performance. Can vary greatly.

$t_W(L)$—Clock pulse width (low), the minimum width (in nanoseconds) that is allowed at the clock input during the low level for reliable operation.

$t_W(H)$—Clock pulse width (high), the minimum width (in nanoseconds) that is allowed at the clock input during the high level for reliable operation.

Preset or clear pulse width—Also given by $t_W(L)$, the minimum width (in nanoseconds) of the low pulse at the preset or clear inputs.

12.6.9   Digital Clock Generator and Single-Pulse Generators

You have already seen the importance of clock and single-pulse control signals. Now let's take a look at some circuits that can generate these signals.

Clocks (Astable Multivibrators)

A clock is simply a squarewave oscillator. In Chap. 9 I discussed ways to generate square waves—so you can refer back there to learn the theory. Here I will simply present some practical circuits. Digital clocks can be constructed from discrete components such as logic gates, capacitors, resistors, and crystals or can be purchased in IC form. Here are some sample clock generators.

FIGURE 12.93

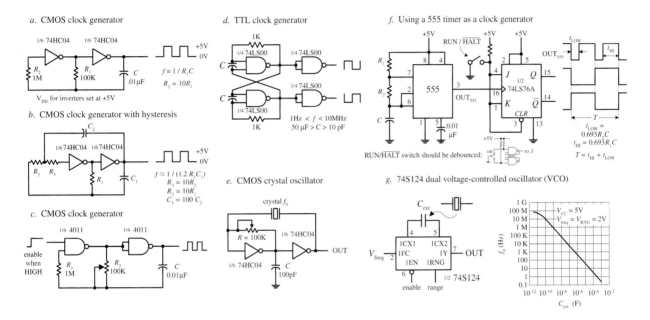

Figure a. Here, two CMOS inverters are connected together to form an RC relaxation oscillator with squarewave output. The output frequency is determined by the RC time constant, as shown in the figure.

Figure b. The previous oscillator has one problem—it may not oscillate if the transition regions of its two gates differ, or it may oscillate at a slightly lower frequency than the equation predicts due to the finite gain of the leftmost gate. The oscillator shown here resolves these problems by adding hysteresis via the additional RC network.

Figure c. This oscillator uses a pair of CMOS NAND gates and RC timing network along with a pot to set the frequency. A squarewave output is generated with a maximum frequency of around 2 MHz. The enable lead could be connected to the other input of the first gate, but here it is brought out to be used as an clock enable input (clock is enabled when this lead is high).

Figure d. Here, a TTL SR flip-flop with dual feedback resistors uses an RC relaxation-type configuration to generate a square wave. The frequency of the clock is determined by the R and C values, as shown in the figure. Changing the C_1-to-C_2 ratio changes the duty cycle.

Figure e. When high stability is required, a crystal oscillator is the best choice for a clock generator. Here, a pair of CMOS inverters and a feedback crystal are used (see Chap. 8 for details). The frequency of operation is determined by the crystal (e.g., 2 MHz, 10 MHz, etc.). Adjustment of the pot may be needed to start oscillations.

Figure f. A 555 timer in astable mode can be used to generate square waves. Here, we slap on a JK flip-flop that is in toggle mode to provide a means of keeping the low and high times the same, as well as providing clock-enable control. The timing diagram and the equations provided within the figure will paint the rest of the picture.

Figure g. The 74S124 dual voltage-controlled oscillator (VCO) outputs square waves at a frequency that is dependent on the value of an external capacitor and the voltage levels applied its frequency-range input (V_{RNG}) and its frequency control input (V_{freq}). The graph in this figure shows how the frequency changes with capacitance, while V_{RNG} and V_{freq} are fixed at 2 V. This device also comes with active-low enable input. Other VCOs that are designed for clock generation include the 74LS624, 4024, and 4046 (PLL). You will find many more listed in the catalogs.

Monostables (One-Shots)

To generate single-pulse signals of a desired width, you can use a discrete device called a *monostable multivibrator,* or *one-shot* for short. A one-shot has only one stable state, high (or low), and can be triggered into its unstable state, low (or high) for a duration of time set by an *RC* network. One-shots can be constructed from simple gates, capacitors, and resistors. These circuits, however, tend to be "finicky" and simply are not worth talking about. If you want a one-shot, you go out and buy a one-shot IC, which is typically around 50 cents.

Two popular one shots, shown below, are the 74121 nonretriggerable monostable multivibrator and the 74123 retriggerable monostable multivibrator.

FIGURE 12.94

74121 non-retriggerable monostable multivibrator (one-shot)

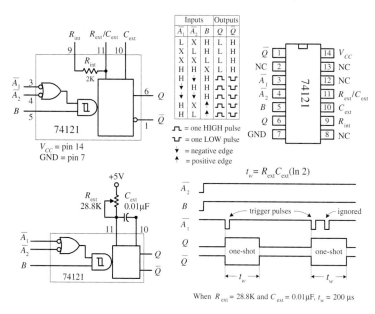

$$t_w = R_{ext} C_{ext} (\ln 2)$$

When $R_{ext} = 28.8K$ and $C_{ext} = 0.01 \mu F$, $t_w = 200 \mu s$

The 74121 has three trigger inputs ($\overline{A_1}, \overline{A_2}, B$), true and complemented outputs ($Q, \overline{Q}$), and timing inputs to which an *RC* network is attached ($R_{ext}/C_{ext}, C_{ext}$). To trigger a pulse from the 74123, you can choose between five possible trigger combinations, as shown in the truth table in the figure. Bringing the input trigger in on B, however, is attractive when dealing with slowly rising or noisy signals, since the signal is directly applied to an internal Schmitt-triggered inverter (recall hysteresis). To set the desired output pulse width (t_w), a resistor/capacitor combination is connected to the R_{ext}/C_{ext} and C_{ext} inputs, as shown. (An internal 2-k resistor is provided, which can be used alone by connecting pin 9 to V_{CC} and placing the capacitor across pins 10 and 11, or which can be used in series with an external resistor attached to pin 9. Here, the internal resistor will not be used.) To determine what values to give to the external resistor and capacitor, use the formula given by the manufacturer, which is shown to the left. The maximum t_w should not exceed 28 s ($R = 40$ k, $C = 1000 \mu F$) for reliable operation. Also, note that with a nonretriggerable one-shot like the 74121, any trigger pulses applied when the device is already in its astable state will be ignored.

74123 retriggerable monostable multivibrator (one-shot)

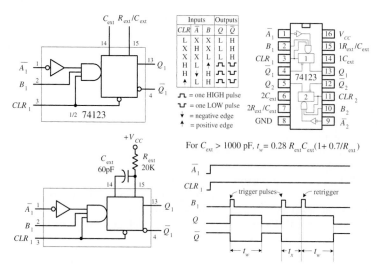

For $C_{ext} > 1000$ pF, $t_w = 0.28 R_{ext} C_{ext} (1 + 0.7/R_{ext})$

The 74123 is a dual, retriggerable one-shot. Unlike nonretriggerable one-shots, this device will not ignore trigger pulses that are applied during the astable state. Instead, when a new trigger pulse arrives during an astable state, the astable state will continue to be astable for a time of t_w. In other words, the device is simply retriggered. The 74123 has two trigger inputs ($\overline{A}, B$) and a clear input (*CLR*). When *CLR* is low, the one-shot is forced back into its stable state ($Q = $ low). To determine t_w, use the formula given to the left, provided $C_{ext} > 1000$ pF. If $C_{ext} < 1000$ pF, use $t_w/C_{ext}/R_{ext}$ graphs provided by the manufacturer to find t_w.

Besides acting as simple pulse generators, one-shots can be combined to make time-delay generators and timing and sequencing circuits. See the figure below.

Time delay circuit

Timing and sequencing circuit

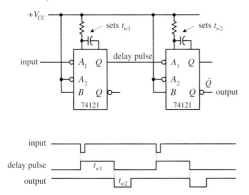

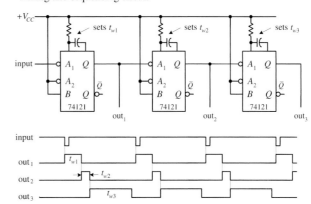

FIGURE 12.95

Now if you do not have a one-shot IC like the 74121, you can use a 555 timer wired in its monostable configuration, as shown below. (I discussed the 555 in Chap. 8—go there if you need the details.)

Using a 555 timer as a one-shot to generate unique output waveforms

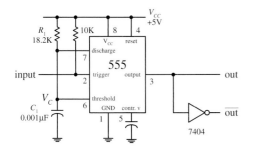

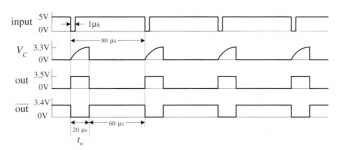

FIGURE 12.96

One-Shot/Continuous-Clock Generator

The circuit below is a handy one-shot/continuous clock generator that is useful when you start experimenting with logic circuits. The details of how this circuit work are explained below.

FIGURE 12.97

One-shot/continuous-clock generator

In this circuit, switch S_2 is used to select whether a single-step or a continuous-clock input is to be presented to the output. When S_2 is in the single-step position, the cross-NAND SR flip-flop (switch debouncer) is set ($Q = 1$, $\bar{Q} = 0$). This disables NAND gate B while enabling NAND gate A, which will allow a single pulse from the one-shot to pass through gate C to the output. To trigger the one-shot, press switch S_1. When S_2 is thrown to the continuous position, the switch debouncer is reset ($Q = 0$, $\bar{Q} = 1$). This disables NAND gate A and enables NAND gate B, allowing the clock signal generated by the 555/flip-flop to pass through gate C and to the output. (Just as a note to avoid confusion, you need gate C to prevent the output from being low and high at the same time.)

12.6.10 Automatic Power-Up Clear (Reset) Circuits

In sequential circuits it is usually a good idea to clear (reset) devices when power is first applied. This ensures that devices, such as flip-flops and other sequential ICs, do not start out in a weird mode (e.g., counter IC does not start counting at, say, 1101 instead of 0000). The following are some techniques used to provide automatic power-up clearing.

Automatic power-up CLEAR circuit

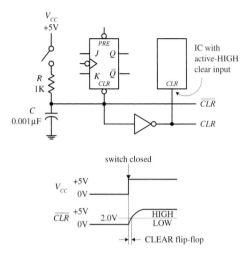

FIGURE 12.98

Improved automatic power-up CLEAR circuit

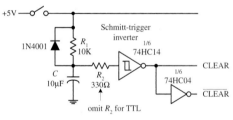

FIGURE 12.99

Let's pretend that one of the devices in a circuit has a JK flip-flop that needs clearing during power-up. In order to clear the flip-flop and then quickly return it to synchronous operations, you would like to apply a low (0) voltage to its active-low clear input; afterwards, you would like the voltage to go high (at least above 2.0V for a 74LS76 JK flip-flop). A simple way to implement this function is to use an *RC* network like the one shown in the figure. When the power is off (switch open), the capacitor is uncharged (0V). This means that the $\overline{CLR}$ line is low (0V). Once the power is turned on (switch closed), the capacitor begins charging up toward V_{CC} (+5V). However, until the capacitor's voltage reaches 2.0V, the $\overline{CLR}$ line is considered low to the active-low clear input. After a duration of $t = RC$, the capacitor's voltage will have reached 63 percent of V_{CC}, or 3.15V; after a duration of $t = 5RC$, its voltage will be nearly equal to +5 V. Since the 74LS76's $\overline{CLR}$ input requires at least 2.0V to be placed back into synchronous operations, you know that $t = RC$ is long enough. Thus, by rough estimate, if you want the $\overline{CLR}$ line to remain low for 1 μs after power-up, you must set $RC = 1$ μs. Setting $R = 1$ k and $C = 0.001$ μF does the trick.

This automatic resetting scheme can be used within circuits that contain a number of resettable ICs. If an IC requires an active-high reset (not common), simply throw in an inverter and create an active-high clear line, as shown in the figure. Depending on the device being reset, the length of time that the clear line is at a low will be about 1 μs. As more devices are placed on the clear line, the low time duration will decrease due to the additional charging paths. To prevent this from occurring, a larger capacitor can be used.

An improved automatic power-up clear circuit is shown in Fig. 12.99. Here a Schmitt-triggered inverter is used to make the clear signal switch off cleanly. With CMOS Schmitt-triggered inverters, a diode and input resistor (R_2) are necessary to protect the CMOS IC when power is removed.

12.6.11 More on Switch Debouncers

The switch debouncer shown to the far left in Fig. 12.100 should look familiar. It is simply a cross-NAND SR-latch-type switch debouncer. Here, a 74LS279A IC is used that contains 4 SR latches—an ideal choice when you need a number of switch debouncers.

Now, a switch debouncer does not have to be constructed from an SR-latch. In fact, most any old flip-flop with preset and clear (reset) inputs can be used. For example, the middle circuit in Fig. 12.100 uses a 74LS74 D-type flip-flop, along with pullup resistor, as a switch debouncer. The *D* input and *CLK* input are tied to ground so that the only two modes that can be enacted are the preset and clear modes. Also, the pullup resistors will always make either the preset input or clear input high, regardless of whether or not the switch is bouncing. From these two facts, you can figure out the rest for yourself, using the truth table for the 74HC74 in Fig. 12.80 as a guide.

Another approach that can be used to debounce an SPST switch is shown in to the far right in Fig. 12.100. This debouncer uses a Schmitt-triggered inverter along with a unique *RC* timing network. When the switch is open, the capacitor is fully charged

(+5 V), and the output is low. When the switch is closed, the capacitor discharges rapidly to ground through the 100-Ω resistor, causing the output to go high. Now, as the switch bounces, the capacitor will repeatedly attempt to charge slowly back to +5 V via the 10-k resistor, and then again will discharge rapidly to zero through the 100-W resistor, making the output high. By making the 10-k pullup resistor larger than the 100-Ω discharge resistor, the voltage across the capacitor or the voltage applied to the inverter's input will not get a chance to exceed the positive threshold voltage (V_T+) of the inverter during a bounce. Therefore, the output remains high, regardless of the bouncing switch. In this example, the charge-up time constant ($R_2C = 10$ k × 0.1 μF) ensures sufficient leeway. When the switch is reopened, the capacitor charges up toward +5 V. When the capacitor's voltage reaches V_T+, the output switches low.

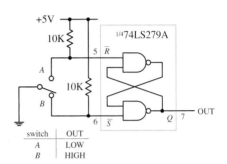

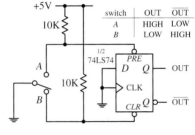

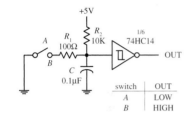

SPDT switch debouncer using 74LS279 quad SR latch.

SPDT switch debouncer using 74LS74 dual D-type flip-flop.

SPST switch debouncer using 74HC14 hex Schmitt-triggered inverter.

FIGURE 12.100

12.6.12 Pullup and Pulldown Resistors

FIGURE 12.101

As you learned when dealing with the switch debouncer circuits, a pullup resistor is used to keep an input high that would otherwise float if left unconnected. If you want to set the "pulled up" input low, you can ground the pin, say, via a switch. Now, it is important to get an idea of the size of pullup resistor to use. The key here is to make

Using a pullup resistor to keep input normally HIGH

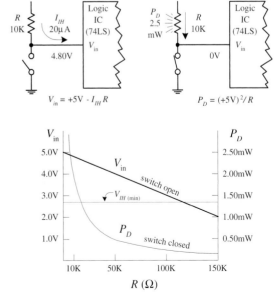

Using a pulldown resistor to keep input normally LOW

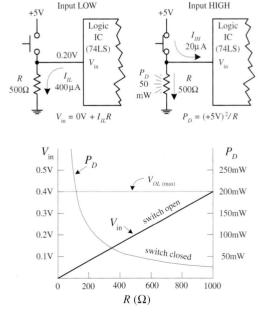

the resistor value small enough so that the voltage drop across it does not weigh down the input voltage below the minimum high threshold voltage ($V_{IH,min}$) of the IC. At the same time, you do not want to make it too small; otherwise, when you ground the pin, excessive current will be dissipated.

In the left figure on p. 383, a 10-k pullup resistor is used to keep a 74LS device's input high. To make the input low, close the switch. To figure out if the resistor is large enough so as not to weigh down the input, use $V_{in} = +5\text{ V} - RI_{IH}$, where I_{IH} is the current drawn into the IC during the high input state, when the switch is open. For a typical 74LS device, I_{IH} is around 20 µA. Thus, by applying the simple formula, you find that $V_{in} = 4.80$ V, which is well above the $V_{IH,min}$ level for a 74LS device. Now, if you close the switch to force the input low, the power dissipated through the resistor ($P_D = V^2/R$) will be $(5\text{ V})^2/10\text{ k} = 25$ mW. The graph shown in the figure provides V_{in} versus R and P_D versus R curves. As you can see, if R becomes too large, V_{in} drops below the $V_{IH,min}$ level, and the output will not go high as planned. As R gets smaller, the power dissipation skyrockets. To determine what value of R to use for a specific logic IC, you look up the $V_{IH,min}$ and $I_{IH,max}$ values within the data sheets and apply the simple formulas. In most applications, a 10-k pullup resistor will work fine.

Now you will run into situations where a pulldown resistor is used to keep a floating terminal low. Unlike a pullup resistor, the pulldown resistor must be smaller because the input low current I_{IL} (sourced by IC) is usually much larger than I_{IH}. Typically, a pulldown resistor is around 100 to 1 kΩ. A lower resistance ensures that V_{in} is low enough to be interpreted as a low by the logic input. To determine if V_{in} is low enough, use $V_{in} = 0\text{ V} + I_{IL}R$. As an example, use a 74LS device with an $I_{IL} = 400$ µA and a 500-Ω pulldown resistor. When the switch is open, the input will be 0.20 V—well below the $V_{IL,max}$ level for the 74LS (~0.8 V). When the switch is closed, the power dissipated by the resistor will be $(5\text{ V})^2/500\ \Omega = 50$ mW. The graph shown in Figure 12.101 provides V_{in} versus R and P_D versus R curves. As you can see by the curves, if R becomes too large, V_{in} surpasses $V_{IL,max}$ and the output will not be low as planned. As R gets small, the power dissipation skyrockets. If you have to use a pulldown resistor/switch arrangement, be wary of the high power dissipation through the resistor when the switch is closed.

12.7 Counter ICs

A few pages back you saw how flip-flops could be combined to make both asynchronous (ripple) and synchronous counters. Now, in practice, using discrete flip-flops is to be avoided. Instead, use a prefab counter IC. These ICs cost a buck or two and come with many additional features, like control enable inputs, parallel loading, etc. A number of different kinds of counter ICs are available. They come in either synchronous (ripple) or asynchronous forms and are usually designed to count in binary or binary-coded decimal (BCD).

12.7.1 Asynchronous Counter (Ripple Counter) ICs

Asynchronous counters work fine for many noncritical applications, but for high-frequency applications that require precise timing, synchronous counters work better. (Recall that unlike an asynchronous counter, a synchronous counter contains flip-flops that get clocked at the same time, and hence the synchronous counter does not accumulate nearly as many propagation delays as is the case with the asynchronous counter.) Here are a few asynchronous counter ICs you will find in the electronics catalogs.

7493 4-Bit Ripple Counter with Separate MOD-2 and MOD-8 Counter Sections

The 7493's internal structure consists of four JK flip-flops connected to provide separate MOD-2 (0-to-1 counter) and MOD-8 (0-to-7 counter) sections. Both the MOD-2 and MOD-8 sections are clocked by separate clock inputs. The MOD-2 section uses C_{p0} as its clock input, while the MOD-8 section uses C_{p1} as its clock input. Likewise, the two sections have separate outputs—MOD-2's output is Q_0, while MOD-8's outputs consist of Q_1, Q_2, and Q_3. The MOD-2 section can be used as a divide-by-2 counter, while the MOD-8 section can be used as either a divide-by-2 counter (output tapped at Q_1), a divide-by-4 counter (output tapped at Q_2), or a divide-by-8 counter (output tapped at Q_3). Now, if you want to create a MOD-16 counter, simply join the MOD-2 and MOD-8 sections by wiring Q_0 to C_{p1}—while using C_{p0} as the single clock input. The MOD-2, MOD-8, or the MOD-16 counter can be cleared by making both AND-gated master reset inputs (MR_1 and MR_2) high. To begin a count, one or both of the master reset inputs must be made low. When the negative edge of a clock pulse arrives, the count advances one step. After the maximum count is reached (1 for MOD-2, 111 for MOD-8, or 1111 for MOD-16), the outputs jump back to zero, and a new count begins.

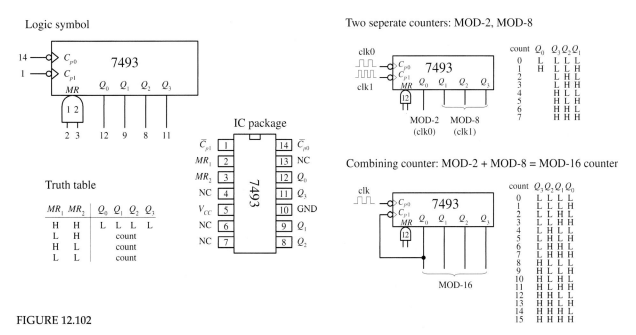

FIGURE 12.102

7490 4-Bit Ripple Counter with MOD-2 and MOD-5 Counter Sections

The 7490, like the 7493, is another 4-bit ripple counter. However, its flip-flops are internally connected to provide MOD-2 (count-to-2) and MOD-5 (count-to-5) counter sections. Again, each section uses a separate clock: C_{p0} for MOD-2 and C_{p1} for MOD-5. By connecting Q_0 to C_{p1} and using C_{p0} as the single clock input, a MOD-10 counter (decade or BCD counter) can be created. When master reset inputs MR_1 and MR_2 are set high, the counter's outputs are reset to 0—provided that master set inputs MS_1 and MS_2 are not both high (the MS inputs override the MR inputs). When MS_1 and MS_2 are high, the outputs are set to $Q_0 = 1$, $Q_1 = 0$, $Q_2 = 0$, and $Q_3 = 1$. In the MOD-10 configuration, this means that the counter is set to 9 (binary 1001). This master set feature comes in handy if you wish to start a count at 0000 after the first clock transition occurs (with master reset, the count starts out at 0001).

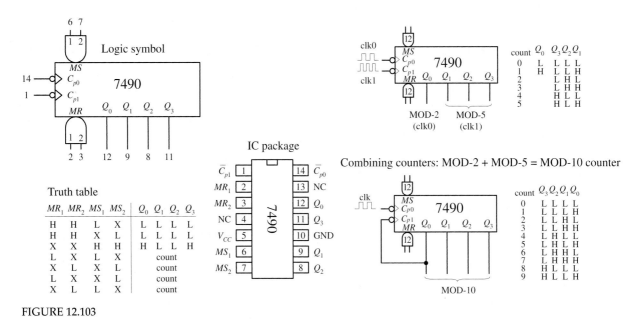

FIGURE 12.103

7492 Divide-by-12 Ripple Counter with MOD-2 and MOD-6 Counter Sections

FIGURE 12.104

The 7492 is another 4-bit ripple counter that is similar to 7490. However, it has a MOD-2 and a MOD-6 section, with corresponding clock inputs C_{p0} (MOD-2) and C_{p1} (MOD-8). By joining Q_0 to C_{p1}, you get a MOD-12 counter, where C_{p0} acts as the single clock input. To clear the counter, high levels are applied to master reset inputs MR_1 and MR_2.

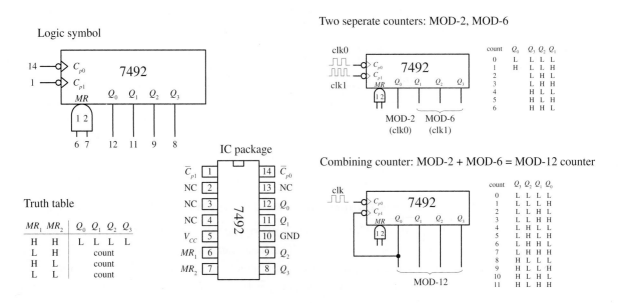

12.7.2 Synchronous Counter ICs

Like the asynchronous counter ICs, synchronous counter ICs come in various MOD arrangements. These devices usually come with extra goodies, such as controls for up or down counting and parallel load inputs used to preset the counter to a desired start count. Synchronous counter ICs are more popular than the asynchronous ICs not only because of these additional features but also because they do not have such long propagation delays as asynchronous counters. Let's take a look at a few popular IC synchronous counters.

74193 Presettable 4-Bit (MOD-16) Synchronous Up/Down Counter

The 74193 is a versatile 4-bit synchronous counter that can count up or count down and can be preset to any count desired—at least a number between 0 and 15. There are two separate clock inputs, C_{pU} and C_{pD}. C_{pU} is used to count up, while C_{pD} is used to count down. One of these clock inputs must be held high in order for the other input to count. The binary output count is taken from Q_0 (2^0), Q_1 (2^1), Q_2 (2^2), and Q_3 (2^3). To preset the counter to any desired count, a corresponding binary number is applied to the parallel inputs D_0 to D_3. When the parallel load input ($\overline{PL}$) is pulsed low, the binary number is loaded into the counter, and the count, either up or down, will start from that number. The terminal count up ($\overline{TC}_U$) and terminal count down ($\overline{TC}_D$) outputs are normally high. The $\overline{TC}_U$ output is used to indicate when the maximum count has been reached and the counter is about to recycle to the minimum count (0000)—carry condition. Specially, this means that $\overline{TC}_U$ goes low when the count reaches 15 (1111) and the input clock (C_{pU}) goes from high to low. $\overline{TC}_U$ remains low until C_{pU} returns high. This low pulse at $\overline{TC}_U$ can be used as an input to the next high-order stage of a multistage counter. The terminal count down ($\overline{TC}_D$) output is used to indicate that the minimum count has been reached (0000) and the counter is about to recycle to the maximum count 15(1111)—borrow condition. Specifically, this means that $\overline{TC}_D$ goes low when the down count reaches 0000 and the input clock (C_{pD}) goes low. The figure below provides a truth table for the 74193, along with a sample load, up-count, and down-count sequence.

FIGURE 12.105

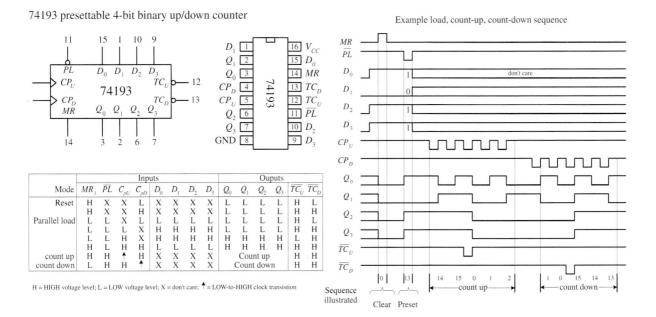

74193 presettable 4-bit binary up/down counter

Example load, count-up, count-down sequence

H = HIGH voltage level; L = LOW voltage level; X = don't care; ↑ = LOW-to-HIGH clock transistion

74192 Presettable Decade (BCD or MOD-10) Synchronous Up/Down Counter

The 74193, shown in Fig. 12.106, is essentially the same device as the 74193, except it counts up from 0 to 9 and repeats or counts down from 9 to 0 and repeats. When counting up, the terminal count up ($\overline{TC_U}$) output goes low to indicate when the maximum count is reached (9 or 1001) and the C_{pU} clock input goes from high to low. $\overline{TC_U}$ remains low until C_{pU} returns high. When counting down, the terminal count down output ($\overline{TC_D}$) goes low when the minimum count is reached (0 or 0000) and the input clock C_{pD} goes low. The truth table and example load, count-up, count-down sequence provided in Fig. 12.106 explains how the 74192 works in greater detail.

74192 presettable decade (BCD) up/down counter

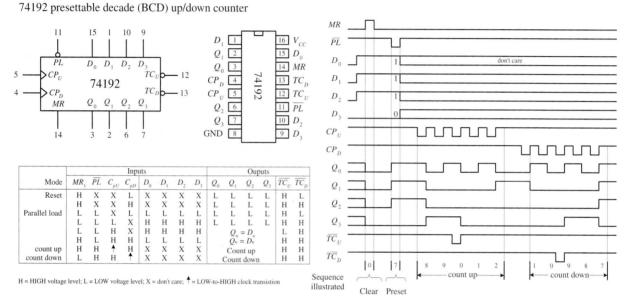

	Inputs								Ouputs					
Mode	MR_1	$\overline{PL}$	C_{pU}	C_{pD}	D_0	D_1	D_2	D_3	Q_0	Q_1	Q_2	Q_3	$\overline{TC_U}$	$\overline{TC_D}$
Reset	H	X	X	L	X	X	X	X	L	L	L	L	H	L
	H	X	X	H	X	X	X	X	L	L	L	L	H	H
Parallel load	L	L	X	L	L	L	L	L	L	L	L	L	H	L
	L	L	L	X	H	H	H	H	L	L	L	L	H	H
	L	L	H	X	H	H	H	H	$Q_n = D_n$				L	H
	H	L	H	H	L	L	L	L	$Q_n = D_n$				H	H
count up	H	H	↑	H	X	X	X	X	Count up				H	H
count down	L	H	H	↑	X	X	X	X	Count down				H	H

H = HIGH voltage level; L = LOW voltage level; X = don't care; ↑ = LOW-to-HIGH clock transistion

FIGURE 12.106

74190 Presettable Decade (BCD or MOD-10) Synchronous Up/Down Counter and the 74191 Presettable 4-Bit (MOD-16) Synchronous Up/Down Counter

The 74190 and the 74191 do basically the same things as the 74192 and 74193, but the input and output pins, as well as the operating modes, are a bit different. (The 74190 and the 74191 have the same pinouts and operating modes—the only difference is the maximum count.) Like the previous synchronous counters, these counters can be preset to any count by using the parallel load ($\overline{PL}$) operation. However, unlike the previous synchronous counters, to count up or down requires using a single input, $\overline{U}/D$. When $\overline{U}/D$ is set low, the counter counts up; when $\overline{U}/D$ is high, the counter counts down. A clock enable input ($\overline{CE}$) acts to enable or disable the counter. When $\overline{CE}$ is low, the counter is enabled. When $\overline{CE}$ is high, counting stops, and the current count is held fixed at the Q_0 to Q_3 outputs. Unlike the previous synchronous counters, the 74190 and the 74191 use a single terminal count output (TC) to indicate when the maximum or minimum count has occurred and the counter is about to recycle. In count-down mode, TC is normally low but goes high when the counter reaches zero (for both the 74190 and 74191). In count-up mode, TC is normally low but goes high when the counter reaches 9 (for the 74190) or reaches 15 (for the 74191). The ripple-clock output ($\overline{RC}$) follows the input clock (CP) whenever TC is high. This means, for example, that in count-down mode, when the count reaches zero, $\overline{RC}$ will go low when CP goes low. The $\overline{RC}$ output can be used as a clock input to the next higher stage of a multistage counter. This, however, leads to a multistage counter that is not truly synchronous because of the small propagation delay from CP to $\overline{RC}$ of each counter. To make a multistage counter that is truly synchronous, you must tie each IC's clock to a common clock input line. You use the TC output to inhibit each successive stage from counting until the previous stage is at its terminal count. The figure below shows various asynchronous (ripple-like) and synchronous multistage counters build from 74191 ICs.

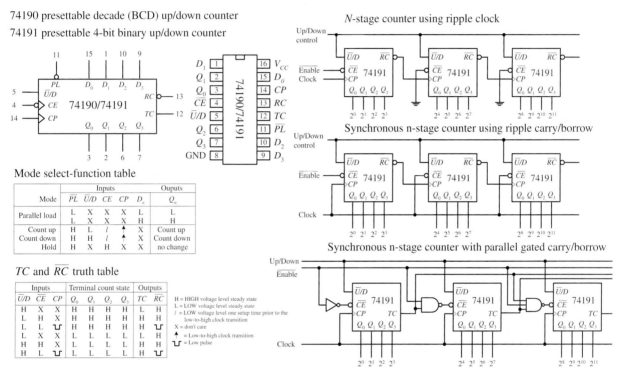

74190 presettable decade (BCD) up/down counter
74191 presettable 4-bit binary up/down counter

Mode select-function table

	Inputs					Ouputs
Mode	PL	$\overline{U/D}$	$\overline{CE}$	CP	D_n	Q_n
Parallel load	L	X	X	X	L	L
	L	X	X	X	H	H
Count up	H	L	l	↑	X	Count up
Count down	H	H	l	↑	X	Count down
Hold	H	X	H	X	X	no change

TC and $\overline{RC}$ truth table

Inputs			Terminal count state				Outputs	
$\overline{U/D}$	$\overline{CE}$	CP	Q_0	Q_1	Q_2	Q_3	TC	$\overline{RC}$
H	X	X	H	H	H	H	L	H
L	H	X	H	H	H	H	H	H
L	L	⊔	H	H	H	H	H	⊔
L	X	X	L	L	L	L	L	H
H	H	X	L	L	L	L	H	H
H	L	⊔	L	L	L	L	H	⊔

H = HIGH voltage level steady state
L = LOW voltage level steady state
l = LOW voltage level one setup time prior to the
  low-to-high clock transition
X = don't care
↑ = Low-to-high clock transition
⊔ = Low pulse

FIGURE 12.107

74163 Presettable 4-Bit (MOD-16) Synchronous Up/Down Counter

The 74160 and 74163 resemble the 74190 and 74191 but require no external gates when used in multistage counter configurations. Instead, you simply cascade counter ICs together, as shown in the last figure below. For both devices, a count can be preset by applying the desired count to the D_0 to D_3 inputs and then applying a low to the parallel enable input ($\overline{PE}$)—the input number is loaded into the counter on the next low-to-high clock transition. The master reset ($\overline{MR}$) is used to force all Q output low, regardless of the other input signals. The two clock enable inputs (CEP and CET) must be high for counting to begin. The terminal count output (TC) is forced high when the maximum count is reached but will be forced low if CET goes low. This is an important feature that makes the multistage configuration synchronous, while avoiding the need for external gating. The truth tables along with the example load, count-up, count-down timing sequences below should help you better understand how these two devices work.

74163 Synchronous 4-bit binary (MOD-16) up counter

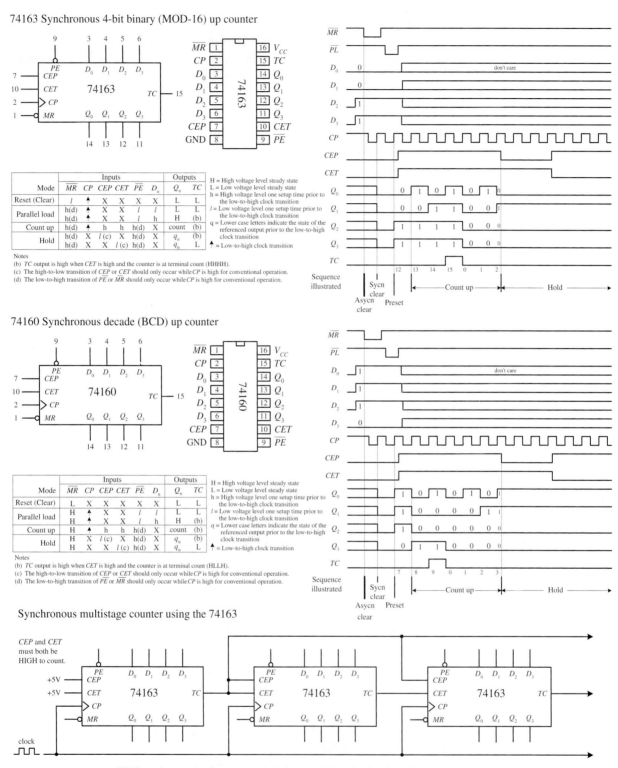

	Inputs						Outputs	
Mode	$\overline{MR}$	CP	CEP	CET	$\overline{PE}$	D_n	Q_n	TC
Reset (Clear)	l	↑	X	X	X	X	L	L
Parallel load	h(d)	↑	X	X	l	l	L	L
	h(d)	↑	X	X	l	h	H	(b)
Count up	h(d)	↑	h	h	h(d)	X	count	(b)
Hold	h(d)	X	l (c)	X	h(d)	X	q_n	(b)
	h(d)	X	X	l (c)	h(d)	X	q_n	L

H = High voltage level steady state
L = Low voltage level steady state
h = High voltage level one setup time prior to the low-to-high clock transition
l = Low voltage level one setup time prior to the low-to-high clock transition
q = Lower case letters indicate the state of the referenced output prior to the low-to-high clock transition
↑ = Low-to-high clock transition

Notes
(b) *TC* output is high when *CET* is high and the counter is at terminal count (HHHH).
(c) The high-to-low transition of *CEP* or *CET* should only occur while *CP* is high for conventional operation.
(d) The low-to-high transition of $\overline{PE}$ or $\overline{MR}$ should only occur while *CP* is high for conventional operation.

74160 Synchronous decade (BCD) up counter

	Inputs						Outputs	
Mode	$\overline{MR}$	CP	CEP	CET	$\overline{PE}$	D_n	Q_n	TC
Reset (Clear)	L	X	X	X	X	X	L	L
Parallel load	H	↑	X	X	l	l	L	L
	H	↑	X	X	l	h	H	(b)
Count up	H	↑	h	h	h(d)	X	count	(b)
Hold	H	X	l (c)	X	h(d)	X	q_n	(b)
	H	X	X	l (c)	h(d)	X	q_n	L

H = High voltage level steady state
L = Low voltage level steady state
h = High voltage level one setup time prior to the low-to-high clock transition
l = Low voltage level one setup time prior to the low-to-high clock transition
q = Lower case letters indicate the state of the referenced output prior to the low-to-high clock transition
↑ = Low-to-high clock transition

Notes
(b) *TC* output is high when *CET* is high and the counter is at terminal count (HLLH).
(c) The high-to-low transition of *CEP* or *CET* should only occur while *CP* is high for conventional operation.
(d) The low-to-high transition of $\overline{PE}$ or $\overline{MR}$ should only occur while *CP* is high for conventional operation.

Synchronous multistage counter using the 74163

74160 synchronous decade counters can also be cascaded together in this multistage configuration.

FIGURE 12.108

Asynchronous Counter Applications

74LS90: Divide-by-n frequency counters

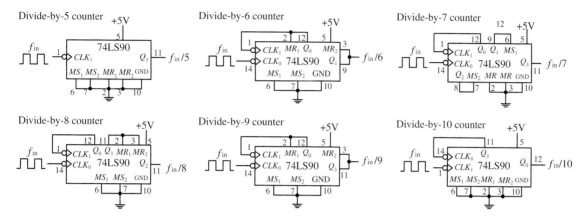

74LS90: 000 to 999 BCD counter

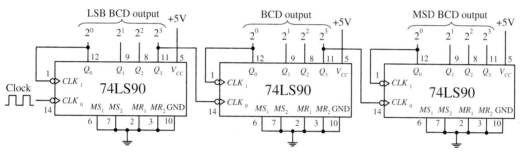

FIGURE 12.109

60-Hz, 10-Hz, 1-Hz Clock-Pulse Generator

FIGURE 12.110

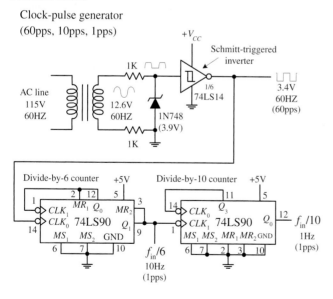

Clock-pulse generator
(60pps, 10pps, 1pps)

This simple clock-pulse generator provides a unique way to generate 60-, 10-, and 1-Hz clock signals that can be used in applications that require real-time counting. The basic idea is to take the characteristic 60-Hz ac line voltage (from the wall socket) and convert it into a lower-voltage square wave of the same frequency. (Note that countries other than the United States typically use 50 Hz instead of 60 Hz. This circuit, therefore, will not operate as planned if used overseas.) First, the ac line voltage is stepped down to 12.6 V by the transformer. The negative-going portion of the 12.6-V ac voltage is removed by the zener diode (acts as a half-wave rectifier). At the same time, the zener diode clips the positive-going signal to a level equal to its reverse breakdown voltage (3.9V). This prevents the Schmitt-triggered inverter from receiving an input level that exceeds its maximum input rating. The Schmitt-triggered inverter takes the rectified/chipped sine wave and converts it into a true square wave. The Schmitt trigger's output goes low (~0.2V) when the input voltage exceeds its positive threshold voltage V_T^+ (~1.7V) and goes high (~3.4V) when its input falls below its negative threshold voltage V_T^- (~0.9V). From the inverter's output, you get a 60-Hz square wave (or a clock signal beating out 60 pulses per second). To get a 10-Hz clock signal, you slap on a divide-by-6 counter. To get a 1-Hz signal, you slap a divide-by-10 counter onto the output of the divide-by-6 counter.

3-digit decimal timer/relay-driver setup to count to 600 and then stop

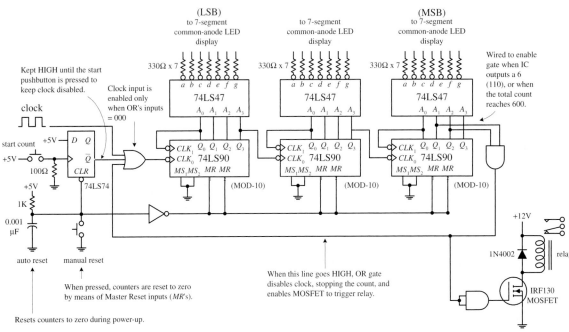

FIGURE 12.111

In the circuit above, three 74LS90 MOD-10 counter ICs are used to create a three-digit (decimal) counter. Features to note here includes an autoreset *RC* network that acts to reset counters during power-up via the master reset inputs. Before the count begins, the D flip-flop's $\overline{Q}$ output is held high, disabling the clock from reaching the first counter's clock input. When the pushbutton switch is closed, the flip-flop's $\overline{Q}$ output goes low, enabling the first counter to count. The BCD outputs of each counter are feed through separate BCD-to-seven-segment decoder/driver ICs, which in turn drive the LED displays. The far-left counter's output represents the count's LSB, while the far-right counter's output represents the count's MSB. As shown, the last counter's output is wired so that when a count of 600 is reached, an AND gate is enabled, causing the three-input OR gate to disable the clock (stop count) while also triggering a relay. To reset the counter, the manual reset switch is momentarily closed.

Synchronous Counter Applications

FIGURE 12.112

74193: Up-down counter/flasher circuit

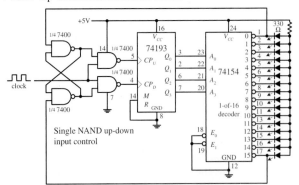

The 74193 is made to count up from 0000 to 1111 and is then swtiched to count down from 1111 to 0000, and then switched to count up again, etc. The NAND gates network provide the count-up and count-down control, while the 74154 1-of-16 decoder accepts the binary count at its address inputs, and then forces an output that corresponds to the address input LOW, thus turning on corresponding LED.

74193: Count up to any number between 0 and 15 and halt

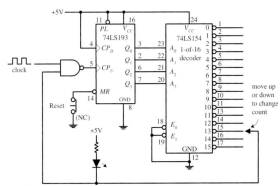

This circuit counts up to a desired number from 0 to 15 and then halts, lighting an LED in the process. A 1-of-16 decoder is used to convert the 4-bit binary output from the counter IC into a single output corresponding to the 4-bit binary number. Here, the circuit is wired to count up to 13 . When the count reaches 13 (all outputs of the decoder are HIGH except the "13" output) the NAND gate is disabled, preventing the clock signal from reaching CP_U.

74193: Divide-by-*n* frequency counter

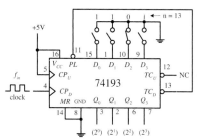

Here the switches set the data input to 1101 (13), and the counter will count down from 13 to 0. This means that the TC_D output will go LOW every 13th CP_D pulse. When TC_D goes LOW, the PL input is triggered and the data input 1101 is read again, and the countdown is repeated.

74193: A larger divide-by-*n* frequency counter

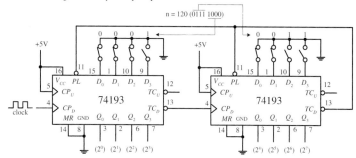

By cascading two 4-bit 74193 IC's together we get an 8-bit down-counter. Here we preload the 8-bit counter with the equivalent of 120 and count down to zero and then repeat. Actually, after the first cycle, the counter counts from 119 down to 0 to give us 120 complete clock pulses between LOW pulses on the second TC_D output.

7493: 4-bit binary counter (MOD-16)

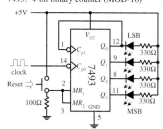

Counts from 0 (LLLL) to 15 (HHHH) and repeats. When a Q output goes LOW, corresponding LED turns ON. Reset counter to zero by pressing Reset switch.

Programmable countdown timer (maximum count 9-to-0)

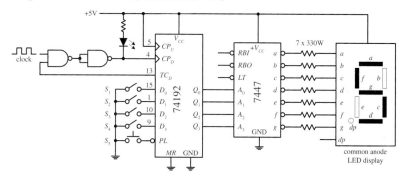

Use switches S_1-S_4 to set D_0-D_4 to the desired count. Press S_5 to load D_0-D_4 and start (or reset) count. When the count is finished (reaches 0000), TC_D goes HIGH, disabling the first NAND gate. This stops the count and causes the LED to light up. The BDC-to-7 segment decoder/driver IC and LED display allow you to see the count in decimal form.

Cascading 74160 BCD counters together to make a 0 to 999 counter

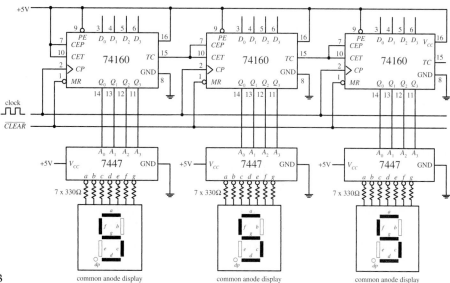

FIGURE 12.113

12.7.3 A Note on Counters with Displays

If you want to build a fairly sophisticated counter that can display many digits, the previous techniques are not worth pursuing—there are simply too many discrete components to work with (e.g., separate seven-segment decoder/driver for each digit). A common alternative approach is to use a microcontroller that functions both as a counter and a display driver. What microcontrollers can do that discrete circuits have a hard time doing is multiplexing a display. In a multiplexed system, corresponding segments of each digit of a multidigit display are linked together, while the common lines for each digit are brought out separately. Right away you can see that the number of lines is significantly reduced; a nonmultiplexed 7-segment 4-digit display has 28 segment lines and 4 common lines, while the 4-digit multiplexed display only has 7 + 4 = 11 lines. The trick to multiplexing involves flashing each digit, one after the other (and recycling), in a fast enough manner to make it appear that the display is continuously lit. In order to multiplex, the microcontroller's program must supply the correct data to the segment lines at the same time that it enables a given digit via a control signal sent to the common lead of that digit. I will talk about multiplexing in greater detail in Appendix H.

Another approach used to create multidigit counters is to use a multidigit counter/display driver IC. One such IC is the ICM7217A four-digit LED display programmable up/down counter made by Intersil. This device is typically used in hardwired applications where thumbwheel switches are used to load data and SPDT switches are used to control the chip. The ICM7217A provides multiplexed seven-segment LED display outputs that are used to drive common-cathode displays.

ICM7217A (Intersil) 4-Digit LED Display, Programmable Up/Down Counter

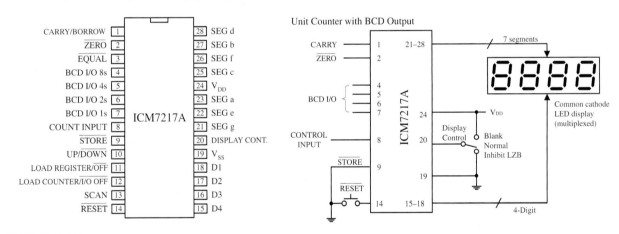

FIGURE 12.114

A simple application of the ICM7217A is a four-digit unit counter shown in the figure above. If you are interested in knowing all the specifics of how this counter works, along with learning about other applications for this device, check out Intersil's data sheets via the Internet (*www.intersil.com*). It is better to learn from the maker in this case. Also, check out the other counter/display driver ICs Intersil has to offer. Other manufacturers produce similar devices, so check out their Web sites as well.

12.8 Shift Registers

Data words traveling through a digital system frequently must be temporarily held, copied, and bit-shifted to the left or to the right. A device that can be used for such applications is the *shift register.* A shift register is constructed from a row of flip-flops connected so that digital data can be shifted down the row either in a left or right direction. Most shift registers can handle parallel movement of data bits as well as serial movement and also can be used to covert from parallel to serial or from serial to parallel. The figure below shows the three types of shift register arrangements: serial-in/serial-out, parallel-in/parallel-out, parallel-in/serial-out, and serial-in/parallel out.

Block diagrams of the serial-in/serial-out, parallel-in/serial-out and serial-in/parallel-out shift registers

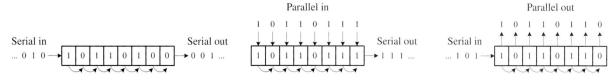

FIGURE 12.115

12.8.1 Serial-In/Serial-Out Shifter Registers

The figure below shows a simple 4-bit serial-in/serial-out shift register made from D flip-flops. Serial data are applied to the D input of flip-flop 0. When the clock line receives a positive clock edge, the serial data are shifted to the right from flip-flop 0 to flip-flop 1. Whatever bits of data were present at flip-flop 2s, 3s, and 4s outputs are shifted to the right during the same clock pulse. To store a 4-bit word into this register requires four clock pulses. The rightmost circuit shows how you can rewire the flip-flops to make a shift-left register. To make larger bit-shift registers, more flip-flops are added (e.g., an 8-bit shift register would require 8 flip-flops cascaded together).

Simple 4-bit serial-in/serial-out shift registers

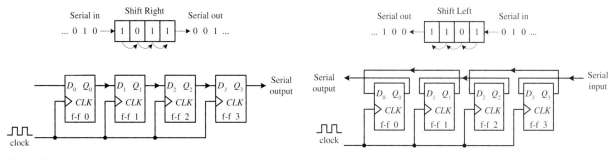

FIGURE 12.116

12.8.2 Serial-In/Parallel-Out Shift Registers

Figure 12.117 shows a 4-bit serial-in/parallel-out shift register constructed from D flip-flops. This circuit is essentially the same as the previous serial-in/serial-out shift register, except now you attach parallel output lines to the outputs of each flip-flop as shown. Note that this shift register circuit also comes with an active-low clear input ($\overline{CLR}$) and a strobe input that acts as a clock enable control. The timing diagram below shows a sample serial-to-parallel shifting sequence.

4-bit serial-in/parallel-out shift register

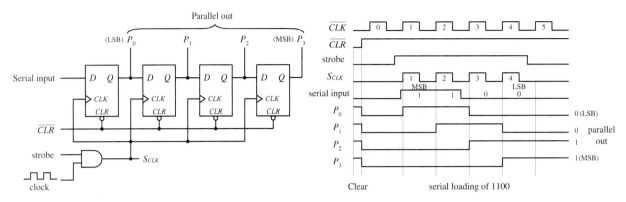

FIGURE 12.117

12.8.3 Parallel-In/Serial-Out Shift Register

Constructing a 4-bit parallel-to-serial shift register from D flip-flops requires some additional control logic, as shown in the circuit below. Parallel data must first be loaded into the D inputs of all four flip-flops. To load data, the SHIFT/$\overline{\text{LOAD}}$ is made low. This enables the AND gates with "X" marks, allowing the 4-bit parallel input word to enter the D_0–D_3 inputs of the flip-flops. When a clock pulse is applied during this load mode, the 4-bit parallel word is latched simultaneously into the four flip-flops and appears at the Q_0–Q_3 outputs. To shift the latched data out through the serial output, the SHIFT/$\overline{\text{LOAD}}$ line is made high. This enables all unmarked AND gates, allowing the latched data bit at the Q output of a flip-flop to pass (shift) to the D input of the flip-flop to the right. In this shift mode, four clock pulses are required to shift the parallel word out the serial output.

Parallel-to-serial shift register

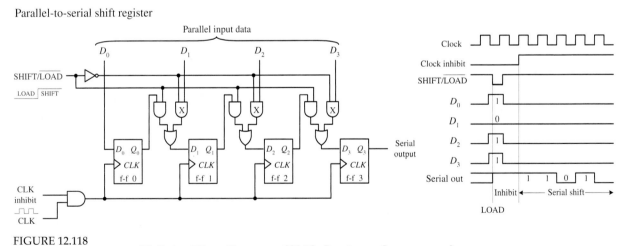

FIGURE 12.118

12.8.4 Ring Counter (Shift Register Sequencer)

The ring counter (shift register sequencer) is a unique type of shift register that incorporates feedback from the output of the last flip-flop to the input of the first flip-flop. Figure 12.119 shows a 4-bit ring counter made from D-type flip-flops. In this circuit, when the $\overline{\text{START}}$ input is set low, Q_0 is forced high by the active-low preset, while Q_1, Q_2, and Q_3 are forced low (cleared) by the active-low clear. This causes the binary word 1000 to be stored within the register. When the $\overline{\text{START}}$ line is brought low, the data bits stored in the flip-flops are shifted right with each positive clock edge. The data bit from the last flip-flop is sent to the D input of the first flip-flop. The shifting cycle will continue to recirculate while the clock is applied. To start fresh cycle, the $\overline{\text{START}}$ line is momentarily brought low.

Ring counter using positive edge-triggered D flip-flops

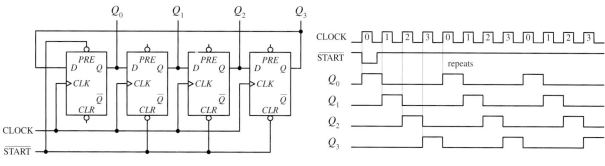

FIGURE 12.119

12.8.5 Johnson Shift Counter

The Johnson shift counter is similar to the ring counter except that its last flip-flop feeds data back to the first flip-flop from its inverted output ($\overline{Q}$). In the simple 4-bit Johnson shift counter shown below, you start out by applying a low to the $\overline{\text{START}}$ line, which sets presets Q_0 high and Q_1, Q_2, and Q_3 low—$\overline{Q}_3$ high. In other words, you load the register with the binary word 1000, as you did with the ring counter. Now, when you bring $\overline{\text{START}}$ line low, data will shift through the register. However, unlike the ring counter, the first bit sent back to the D_0 input of the first flip-flop will be high because feedback is from $\overline{Q}_3$ not Q_3. At the next clock edge, another high is fed back to D_0; at the next clock edge, another high is fed back; at the next edge, another high is fed back. Only after the fourth clock edge does a low get fed back (the 1 has shifted down to the last flip-flop and $\overline{Q}_3$ goes high). At this point, the shift register is full of 1s. As more clock pulses arrive, the feedback loop supplies lows to D_0 for the next four clock pulses. After that, the Q outputs of all the flip-flops are low while $\overline{Q}_3$ goes high. This high from $\overline{Q}_3$ is fed back to $\overline{D}_0$ during the next positive clock edge, and the cycle repeats. As you can see, the 4-bit Johnson shift counter has 8 output stages (which require 8 clock pulses to recycle), not 4, as was the case with the ring counter.

FIGURE 12.120

Johnson counter using positive edge-triggered D flip-flops

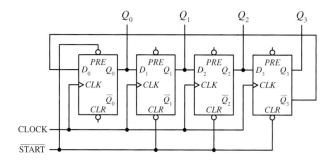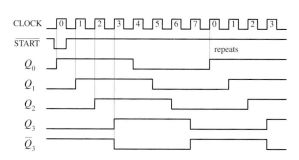

12.8.6 Shift Register ICs

I have just covered the basic theory of shift registers. Now let's take a look at practical shift register ICs that contain all the necessary logic circuitry inside.

7491A 8-Bit Serial-In/Serial-Out Shift Register

The 7491A is an 8-bit serial-in/serial-out shift register that consists of eight internally linked SR flip-flops. This device has positive edge-triggered inputs and a pair of data inputs (A and B) that are internally ANDed together, as shown in the logic diagram below. This type of data input means that for a binary 1 to be shifted into the register, both data inputs must be high. For a binary 0 to be shifted into the register, either input can be low. Data are shifted to the right at each positive clock edge.

7491A 8-bit serial-in/serial-out shift register IC

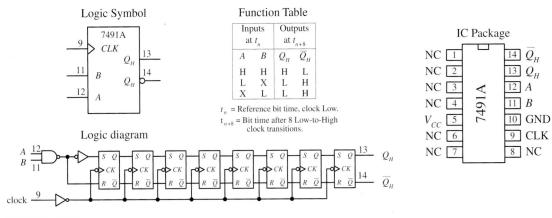

FIGURE 12.121

74164 8-Bit Serial-In/Parallel-Out Shift Register IC

The 74164 is an 8-bit serial-in/parallel-out shift register. It contains eight internally linked flip-flops and has two serial inputs D_{sa} and D_{sb} that are ANDed together. Like the 7491A, the unused serial input acts as an enable/disable control for the other serial input. For example, if you use D_{sa} as the serial input, you must keep D_{sb} high to allow data to enter the register, or you can keep it low to prevent data from entering

FIGURE 12.122

The 74164 8-bit serial-in/parallel-out shift register IC

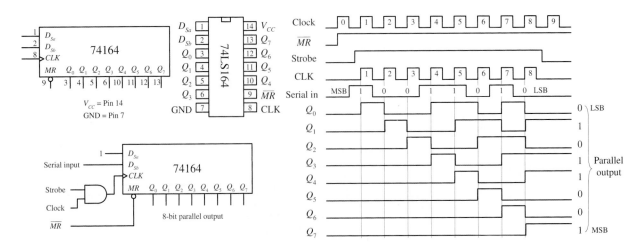

the register. Data bits are shifted one position to the right at each positive clock edge. The first data bit entered will end up at the Q_7 parallel output after the eighth clock pulse. The master reset ($\overline{MR}$) resets all internal flip-flops and forces the Q outputs low when it is pulsed low. In the sample circuit shown below, a serial binary number 10011010 (154_{10}) is converted into its parallel counterpart. Note the AND gate and strobe input used in this circuit. The strobe input acts as a clock enable input; when it is set high, the clock is enabled. The timing diagram paints the rest of the picture.

75165 8-Bit (Serial-In or Parallel-In)/Serial-Out Shift Register

The 75165 is a unique 8-bit device that can act as either a serial-to-serial shift register or as a parallel-to-serial shift register. When used as a parallel-to-serial shift register, parallel data are applied to the D_0–D_7 inputs and then loaded into the register when the parallel load input ($\overline{PL}$) is pulsed low. To begin shifting the loaded data out the serial output Q_7 (or $\overline{Q}_7$ if you want inverted bits), the clock enable input ($\overline{CE}$) must be set low to allow the clock signal to reach the clock inputs of the internal D-type flip-flops. When used as a serial-to-serial shift register, serial data are applied to the serial data input DS. A sample shift, load, and inhibit timing sequence is shown below.

74165 8-Bit (serial-in or parallel-in)/serial-out shift register

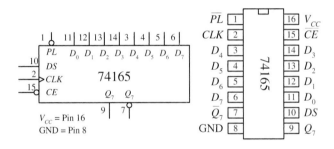

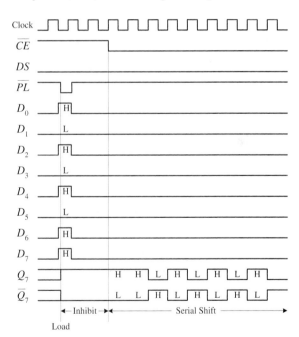

Sample shift, load, and inhibit sequence for parallel load case

Operating Modes	Inputs					Q_n Register			Outputs	
	$\overline{PL}$	$\overline{CE}$	CLK	DS	D_0 - D_7	Q_0	Q_1 - Q_6	Q_7	$\overline{Q}_7$
Parallel load	L	X	X	X	L	L	L - L	L	H
	L	X	X	X	H	H	H - H	H	L
Serial shift	H	L	↑	l	X	L	q_0 - q_5	q_6	$\overline{q}_6$
	H	L	↑	h	X	H	q_0 - q_5	q_6	$\overline{q}_6$
Hold ("do nothing")	H	H	X	X	X	q_0	q_1 - q_6	q_7	$\overline{q}_7$

H = High voltage level; h = High voltage level one setup time prior to the low-to-high clock transition; L = Low voltage level; l = Low voltage level one setup time prior to the low-to-high clock transiton; q_n = Lower case letters indicate the state of the referenced output one setup time prior to the low-to-high clock transition; X = Don't care; ↑ = Low-to-high clock transition.

FIGURE 12.123

74194 Universal Shift Register IC

Figure 12.124 shows the 74194 4-bit bidirectional universal shift register. This device can accept either serial or parallel inputs, provide serial or parallel outputs, and shift left or shift right based on input signals applied to select controls S_0 and S_1. Serial data can be entered into either the serial shift-right input (D_{SR}) or the serial shift-left input (D_{SL}). Select control S_0 and S_1 are used to initiate either a hold (S_0 = low, S_1 = low), shift left (S_0 = low, S_1 = high), shift-right (S_0 = high, S_1 = low), or to parallel load (S_0 = high, S_1 = high) mode—a clock pulse must then be applied to shift or parallel load the data.

In parallel load mode (S_0 and S_1 are high), parallel input data are entered via the D_0 through D_3 inputs and are transferred to the Q_0 to Q_3 outputs following the next low-to-high clock transition. The 74194 also has an asynchronous master reset ($\overline{MR}$) input that forces all Q outputs low when pulsed low. To make a shift-right recirculating register, the Q_3 output is wired back to the D_{SR} input, while making S_0 = high and S_1 = low. To make a shift-left recirculating register, the Q_0 output is connected back to the D_{SL} input, while making S_0 = low and S_1 = high. The timing diagram below shows a typical parallel load and shifting sequence.

74194 4-bit bidirectional universal shift register

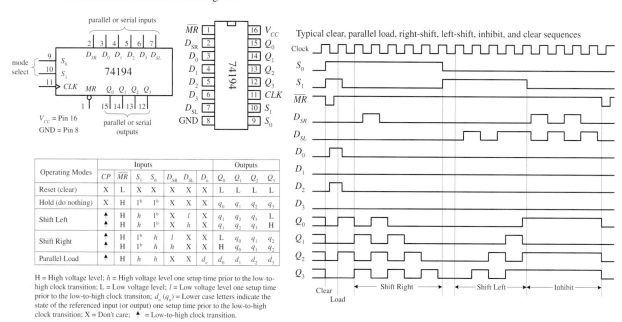

Operating Modes	Inputs							Outputs			
	CP	$\overline{MR}$	S_1	S_0	D_{SR}	D_{SL}	D_n	Q_0	Q_1	Q_2	Q_3
Reset (clear)	X	L	X	X	X	X	X	L	L	L	L
Hold (do nothing)	X	H	1[b]	1[b]	X	X	X	q_0	q_1	q_2	q_3
Shift Left	▲	H	h	1[b]	X	l	X	q_1	q_2	q_3	L
	▲	H	h	1[b]	X	h	X	q_1	q_2	q_3	H
Shift Right	▲	H	1[b]	h	l	X	X	L	q_0	q_1	q_2
	▲	H	1[b]	h	h	X	X	H	q_0	q_1	q_2
Parallel Load	▲	H	h	h	X	X	d_n	d_0	d_1	d_2	d_3

H = High voltage level; h = High voltage level one setup time prior to the low-to-high clock transition; L = Low voltage level; l = Low voltage level one setup time prior to the low-to-high clock transiton; d_n (q_n) = Lower case letters indicate the state of the referenced input (or output) one setup time prior to the low-to-high clock transition; X = Don't care; ▲ = Low-to-high clock transition.

b = The high-to-low transition of S0 and S1 input should only take place while the clock is HIGH for convential operation.

FIGURE 12.124

74299 8-Bit Universal Shift/Storage Register with Three-State Interface

There are a number of shift registers that have three-state outputs—outputs that can assume either a high, low, or high impedance state (open-circuit or float state). These devices are commonly used as storage registers in three-state bus interface applications. An example 8-bit universal shift/storage register with three-state outputs is the 74299, shown in Fig. 12.125. This device has four synchronous operating modes that are selected via two select inputs, S_0 and S_1. Like the previous 74194 universal shift register, the 74299's select modes include shifting right, shifting left, holding, and parallel loading (see function table in Fig. 12.125). The mode-select inputs, serial data inputs (D_{S0} and D_{S7}) and the parallel-data inputs (I/O_0 through I/O_7) inputs are positive edge triggered. The master reset ($\overline{MR}$) input is an asynchronous active-low input that clears the register when pulsed low. The three-state bidirectional I/O port has three modes of operation: read register, load register, and disable I/O. The read-register mode allows data within the register to be available at the I/O outputs. This mode is selected by making both output-enable inputs ($\overline{OE}_1$ and $\overline{OE}_2$) low and making one or both select inputs low. The load-register mode sets up the register for a parallel load during the next low-to-high clock transition. This mode is selected by setting both select inputs high. Finally, the disable-I/O mode acts to disable the outputs (set to a high impedance state) when a high is applied to one or both of the output-enable inputs. This effectively isolates the register from the bus to which it is attached.

74299 8-bit universal shift/storage register with 3-state outputs

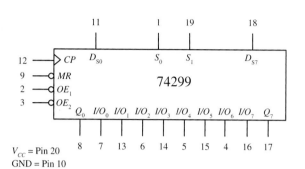

Operating Modes	Inputs							Outputs			
	$\overline{MR}$	CP	S_0	S_1	D_{S0}	D_{S7}	I/O_n	Q_0	$Q_1 - Q_6$		Q_7
Reset (clear)	L	H	X	X	X	X	X	L	L - L		L
Shift right	H	▲	h	l	l	X	X	L	$q_0 - q_5$		q_6
	H	▲	h	l	h	X	X	H	$q_0 - q_5$		q_6
Shift left	H	▲	l	h	X	l	X	q_1	$q_0 - q_5$		L
	H	▲	l	h	X	h	X	q_1	$q_0 - q_5$		H
Hold ("do nothing")	H	▲	l	l	X	X	X	q_0	$q_1 - q_6$		q_7
Parallel load	H	▲	h	h	X	X	l	L	L - L		L
	H	▲	h	h	X	X	h	H	H - H		H

3-state I/O port operating mode	Inputs					Inputs/Outputs
	$\overline{OE_1}$	$\overline{OE_2}$	S_0	S_1	Q_n (register)	$I/O_0 -- I/O_7$
Read register	L	L	L	X	L	L
	L	L	L	X	H	H
	L	L	X	L	L	L
	L	L	X	L	H	H
Load register	X	X	H	H	$Q_n = I/O_n$	I/O_n = inputs
Disable I/O	H	X	X	X	X	High Z
	X	H	X	X	X	High Z

V_{CC} = Pin 20
GND = Pin 10

H = High voltage level; h = High voltage level one setup time prior to the low-to-high clock transition; L = Low voltage level; l = Low voltage level one setup time prior to the low-to-high clock transiton; q_n = Lowercase letters indicate the state of the referenced output one setup time prior to the low-to-high clock transition; X = Don't care; ▲ = Low-to-high clock transition.

FIGURE 12.125

12.8.7 Simple Shift Register Applications

16-Bit Serial-to-Parallel Converter

A simple way to create a 16-bit serial-to-parallel converter is to join two 74164 8-bit serial-in/parallel-out shift registers, as shown below. To join the two ICs, simply wire the Q_7 output from the first register to one of the serial inputs of the second register. (Recall that the serial input that is not used for serial input data acts as an active-high enable control for the other serial input.) In terms of operation, when data are shifted out Q_7 of the first register (or data output D_7), they enter the serial input of the second (I have chosen D_{Sa} as the serial input) and will be presented to the Q_0 output of the second register (or data output D_8). For an input data bit to reach the Q_7 output of the second register (or data output D_{15}), 16 clock pulses must be applied.

Using two 74164s to create a 16-bit serial-to-parallel converter

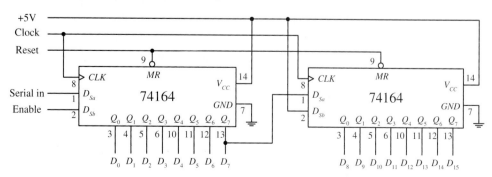

FIGURE 12.126

8-Bit Serial-to-Parallel Converter with Simultaneous Data Transfer

Figure 12.127 shows a circuit that acts as a serial-to-parallel converter that only outputs the converted 8-bit word when all 8-bits have been entered into the register. Here, a 74164 8-bit serial-in/parallel-out shift register is used, along with a 74HCT273 octal D-type flip-flop and a divide-by-8 counter. At each positive clock edge, the serial data are loaded into the 74164. After eight clock pulses, the first serial bit entered is shifted down to the 74164's Q_7 output, while the last serial bit entered resides at the 74164's Q_0 output. At the negative edge of the eighth clock pulse, the negative-edge triggered divide-by-8 circuit's output goes high. During this high transition, the data present on the inputs of the 74HCT273 (which hold the same data present at the 74164's Q outputs) are passed to the 74HCT273's outputs at the same time. (Think of the 74HCT273 as a temporary storage register that dumps its contents after every eighth clock pulse.)

8-Bit Serial-to-Parallel Data Converter

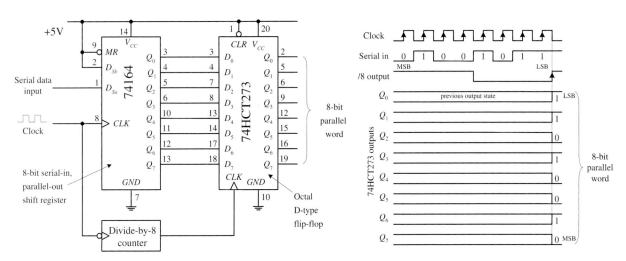

FIGURE 12.127

8-Bit Parallel-to-Serial Interface

Here, a 74165 8-bit parallel-to-serial shift register is used to accept a parallel ASCII word and convert it into a serial ASCII word that can be sent to a serial device. Recall that ASCII codes are only 7 bits long (e.g., the binary code for & is 010 0110). How do you account for the missing bit? As it turns out, most 8-bit devices communicating via serial ASCII will use an additional eighth bit for a special purpose, perhaps to act as a parity bit, or as a special function bit to enact a special set of characters. Often the extra bit is simply set low and ignored by the serial device receiving it. To keep things simple, let's set the extra bit low and assume that that is how the serial device likes things done. This means that you will set the D_0 input of the 74165 low. The MSB of the ASCII code will be applied to the D_1 input, while the LSB of the ASCII code will be applied to the D_7 input. Now, with the parallel ASCII word applied to the inputs of the register, when you pulse the parallel load line ($\overline{PL}$) low, the ASCII word, along with the "ignored bit," is loaded into the register. Next, you must enable the clock to allow the loaded data to be shifted out, serially, by setting the clock enable input ($\overline{CE}$) low for the duration it takes for the clock pulses to shift out the parallel word. After the eighth clock pulse (0 to 7), the serial device will have received all 8 serial data bits. Practically speaking, a microprocessor or microcontroller is necessary to provide the $\overline{CE}$ and $\overline{PL}$ lines with the necessary control signals to ensure that the register and serial device communicate properly.

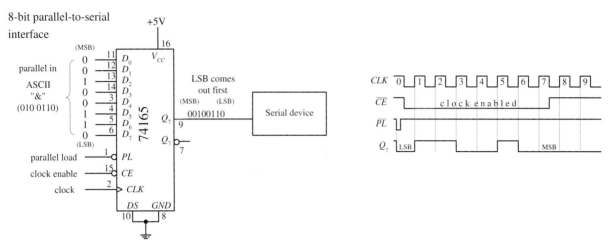

FIGURE 12.128

Recirculating Memory Registers

A *recirculating memory register* is a shift register that is preloaded with a binary word that is serially recirculated through the register via a feedback connection from the output to the input. Recirculating registers can be used for a number of things, from supplying a specific repetitive waveform used to drive IC inputs to driving output drivers used to control stepper motors.

In the leftmost circuit in Fig. 12.129, a parallel 4-bit binary word is applied to the D_0 to D_3 inputs of a 74194 universal shift register. When the S_1 select input is brought high (switch opened), the 4-bit word is loaded into the register. When the S_1 input is then brought low (switch closed), the 4-bit word is shifted in a serial fashion through the register, out Q_3, and back to Q_0 via the D_{SR} input (serial shift-right input) as positive clock edges arrive. Here, the shift register is loaded with 0111. As you begin shifting the bits through the register, a single low output will propagate down through high outputs, which in turn causes the LED attached to the corresponding low output to turn on. In other words, you have made a simple Christmas tree flasher.

FIGURE 12.129

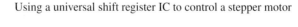

Simple shift register sequence generator

Using a universal shift register IC to control a stepper motor

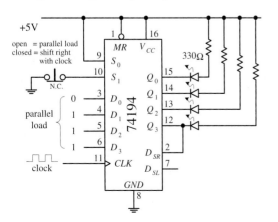

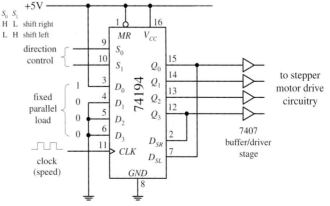

The rightmost circuit in Fig. 12.129 is basically the same thing as the last circuit. However, now the circuit is used to drive a stepper motor. Typically, a stepper motor has four stator coils that must be energized in sequence to make the motor turn a given angle. For example, to make a simple stepper motor turn clockwise, you must energize its stator coils 1, 2, 3, and 4 in the following sequence: 1000, 0100, 0010, 0001, 1000, etc. To make the motor go counterclockwise, apply the following sequence: 1000, 0001, 0010, 0100, 1000, etc. You can generate these simple firing sequences with the 74194 by parallel loading the D_0 to D_3 inputs with the binary word 1000. To output the clockwise firing sequence, simply shift bits to the right by setting S_0 = high and S_1 = low. As clock pulses arrive, the 1000 present at the outputs will then become 0100, then 0010, 0001, 1000, etc. The speed of rotation of the motor is determined by the clock frequency. To output the counterclockwise firing sequence, simply shift bits to the left by setting S_0 = low and S_1 = high. To drive steppers, it is typically necessary to use a buffer/driver interface like the 7407 shown below, as well a number of output transistors, not shown. Also, different types of stepper motors may require different firing sequences than the one shown here. Stepper motors and the various circuits used to drive them are discussed in detail in Chap. 13.

12.9   Three-State Buffers, Latches, and Transceivers

As you will see in a moment, digital systems that use microprocessors require that a number of different devices (e.g., RAM, ROM, I/O devices, etc.) share a common bus of some sort. For simple microprocessor systems, the data bus is often 8 bits wide (eight separate conductors). In order for devices to share the bus, only one device can be transmitting data at a time—the microprocessor decides which devices get access to the bus and which devices do not. In order for the microprocessor to control the flow of data, it needs help from an external register-type device. This device accepts a control signal issued by the microprocessor and responds by either allowing parallel data to pass or prohibiting parallel data to pass. Three popular devices used for such applications are the octal three-state buffer, octal latch/flip-flop, and the transceiver.

12.9.1   Three-State Octal Buffers

A three-state octal buffer is a device that when enabled, passes data present on its eight inputs to its outputs unchanged. When disabled, input data are prevented from reaching the outputs—the outputs are placed in a high-impedance state. This high-impedance state makes data bus sharing between various devices possible. The octal buffer also can provide additional sink or source current required to drive output devices. Three popular three-state octal buffers are shown below. The 74xx240 is a three-state inverting octal buffer, the 74xx241 is a three-state Schmitt-triggered inverting octal buffer, and the 74244 is a conventional three-state octal buffer. The enable/disable control for all three devices is the same. To enable all eight outputs (allow data to pass from I inputs to Y outputs), both output enable inputs, $\overline{OE_a}$ and $\overline{OE_b}$ must be set low. If you wish to enable only four outputs, make one output enable high while setting the other low (refer to ICs below to see which output enable controls which group of inverters). To disable all eight outputs, both output enables are set high. To disable only four outputs, set only one output enable input high.

74*xx*240 Inverting octal buffer with three-state outputs

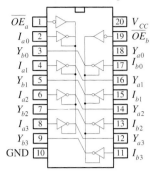

74*xx*241 Inverting octal buffer with three-state outputs

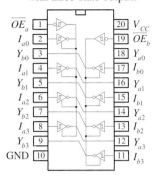

74*xx*244 Octal buffer with three-state outputs

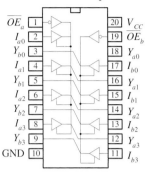

74240, 74LS240, 74F240, 74HC240, etc.   74241, 74LS241, 74F241, 74HC241, etc.   74244, 74LS244, 74F244, 74HC244, etc.

FIGURE 12.130

Here's an example of how three-state inverting octal buffers can be used in an 8-bit microprocessor system. The upper buffer links one bus to a common data bus. The two lower buffers are used to link input devices to the common data bus. With programming and the help of an additional control bus, the microprocessor can select which buffer gets enabled and which buffers get disabled.

8-bit bus buffer system using three-state inverting octal buffer ICs

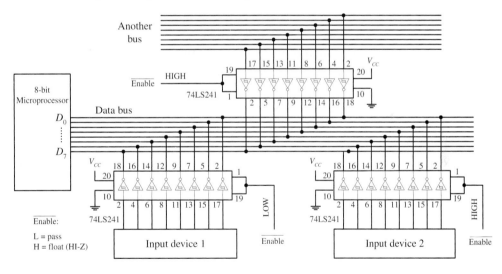

FIGURE 12.131

Data from an input device or another bus pass to the data bus only when the corresponding octal buffer is enabled (input enables made low). Only one input device or bus is allowed to pass data to the data bus at a time. Here, input device 1 is the only device allowed to pass data to the data bus because its enable inputs are set low. Note that data are inverted when passed through the inverting octal buffer.

12.9.2   Three-State Octal Latches and Flip-Flops

A three-state octal latch or octal flip-flop, unlike a three-state octal buffer, has the ability to hold onto data present at its data inputs before transmitting the data to its outputs. In microprocessor applications, where a number of devices share a common data bus, this memory feature is handy because it allows the processor to store data, go onto other operations that require the data bus, and come back to the stored data if necessary. This feature also allows output devices to sample held bus data at leisure while the current state of the data bus is changing. To understand how three-state octal latches and flip-flops work, let's first consider the $73xx373$ three-state octal latch and the $74xx374$ three-state octal flip-flop shown in Fig. 12.132.

74xx373 Octal latch with three-state outputs

74xx374 Octal D flip-flop with three-state outputs

$\overline{OE}$	E	D_n	Q_n
L	H	H	H
L	H	L	L
L	L	X	Q_n
H	X	X	HI-Z

$\overline{OE}$	CP	D_n	Q_n
L	↑	H	H
L	↑	L	L
L	H	X	Q_n
H	X	X	HI-Z

74132, 74LS132,
74F132, 74HC132, etc.

74132, 74LS132,
74F132, 74HC132, etc.

FIGURE 12.132

The $73xx373$ octal latch contains eight D-type "transparent" latches. When its enable input (E) is high, the outputs (Q_0–Q_7) follow the inputs (D_0–D_7). When E is low, data present at the inputs are loaded into the latch. To place the output in a high-impedance state, the output enable input ($\overline{OE}$) is set high. Figure 12.133 shows a simple bus-oriented system that uses two 73HC373s to communicate with an input device and output device. Again, like the octal buffers, control signals are typically supplied by a microprocessor.

Using three-state octal latch IC's as data bus registers

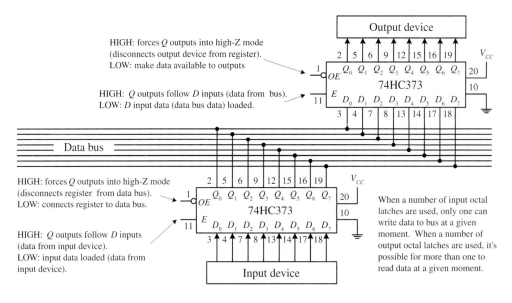

HIGH: forces Q outputs into high-Z mode (disconnects output device from register).
LOW: make data available to outputs

HIGH: Q outputs follow D inputs (data from bus).
LOW: D input data (data bus data) loaded.

HIGH: forces Q outputs into high-Z mode (disconnects register from data bus).
LOW: connects register to data bus.

HIGH: Q outputs follow D inputs (data from input device).
LOW: input data loaded (data from input device).

When a number of input octal latches are used, only one can write data to bus at a given moment. When a number of output octal latches are used, it's possible for more than one to read data at a given moment.

FIGURE 12.133

The $74xx374$ octal flip-flop comes with eight edge-triggered flip-flops. Unlike the octal latch, the $74xx374$'s outputs are not "transparent"—they do not follow the inputs. Instead, a positive clock edge at clock input CP must be applied to load the device before data are presented at the Q outputs. To place the output in a high-impedance state, the output enable ($\overline{OE}$) input is set high. Figure 12.134 shows a simple bus-oriented system that uses two 73HC374s to communicate with two output devices.

Using octal D flip-flops as clocked, three-state data bus registers

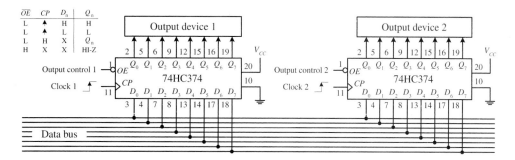

FIGURE 12.134

12.9.3 Transceivers

Another method for connecting devices that share a common bus is to use a transceiver. Unlike the three-state octal buffer, octal latch, and octal flip-flop, the transceiver is a bidirectional device. This means that when used in a bus-oriented system, external devices can read or write from the data bus. The figure below shows the 74xx245 octal transceiver, along with sample application. In the application circuit, a 74LS245 is used as a bidirectional interface between two data buses. To send data from bus A to bus B, the 74LS245's transmit/receive input ($T/\overline{R}$) is set high, while the output enable input ($\overline{OE}$) is set low. To send data from bus B to bus A, $T/\overline{R}$ is set low. To disable the transceiver's outputs (place output in a high impedance state) a high is applied to $\overline{OE}$.

FIGURE 12.135

74xx245 octal transceiver

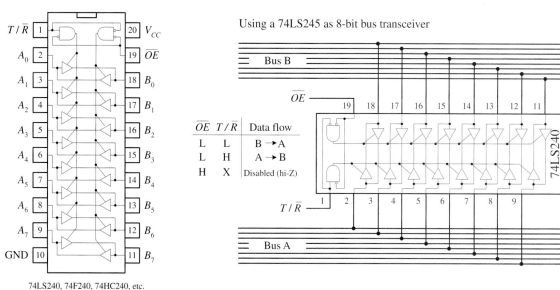

74LS240, 74F240, 74HC240, etc.

12.10   Additional Digital Topics

Appendices H to K contain important digital topics regarding digital-to-analog and analog-to-digital conversion, digital display, memory, and microprocessors and microcontrollers.

DC Motors, RC Servos, and Stepper Motors

Perhaps one of the most entertaining things to do with electronics is make some mechanical device move. Three very popular devices used to "make things move" include dc motors, RC servos, and stepper motors.

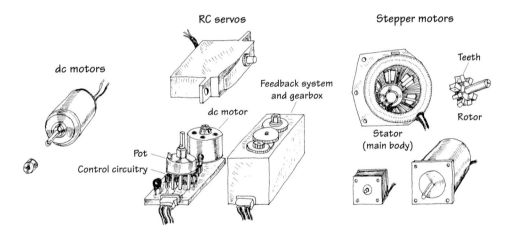

FIGURE 13.1

13.1 DC Continuous Motors

DC motors are simple two-lead, electrically controlled devices that come with a rotary shaft on which wheels, gears, propellers, etc., can be mounted. DC motors generate a considerable amount of revolutions per minute (rpm's) for their size and can be made to rotate clockwise or counterclockwise by reversing the polarity applied to the leads. At low speeds, dc motors provide little torque and minimal position control, making them obsolete for pointlike position-control applications.

DC motors are available in many different shapes and sizes. Most dc motors provide rotational speeds anywhere between 3000 and 8000 rpm at a specific operating voltage typically set between 1.5 and 24 V. The operating voltage provided by the manufacturer tells you at what voltage the motor runs most efficiently. Now, the actual voltage applied to a motor can be made slightly lower to make the motor slower or can be elevated to make the motor faster. However, when the applied voltage drops to below around 50 percent of the specified operating voltage, the motor usually will cease to rotate. Conversely, if the applied voltage exceeds the operating voltage by around 30 percent, there is a chance that the motor will overheat and become damaged. In practice, as you will see in a second, the speed of a dc motor is most efficiently controlled by means of pulse-width modulation, whereby the motor is rapidly turned on and off. The width of the applied pulse, as well as the period between pulses, controls the speed of the motor. Also, it is worth noting that a freely running motor (no load) may draw little current (power). However, if a load is applied, the amount of current drawn by the motor's inner coils goes up immensely (up to 1000 percent or more). Manufacturers usually will provide what is called a *stall current* rating for their motors. This rating specifies the amount of current drawn at the moment the motor stalls. If your motor's stall current rating is not listed, it is possible to determine it by using an ammeter; slowly apply a force to the motor's shaft, and note the current level at the point when the motor stalls. Another specification given to dc motors is a torque rating. This rating represents the amount of force the motor can exert on a load. A motor with a high torque rating will exert a larger force on a load placed at a tangent to its rotational arm than a motor with a lower torque rating. The torque rating of a motor is usually given in lb/ft, g/cm, or oz/in.

13.2 Speed Control of DC Motors

Bad Designs

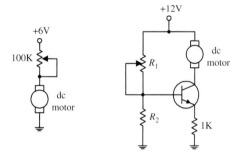

FIGURE 13.2

A seemingly obvious approach to control the speed of a dc motor would be simply to limit the current flow by using a potentiometer, as shown in the circuit to the left in the figure. According to Ohm's law, as the resistance of the pot increases, the current decreases, and the motor will slow down. However, using a pot to control the current flow is inefficient. As the pot's resistance increases, the amount of current energy that must be converted into heat increases. Producing heat in order to slow a motor down is not good—it consumes supply power and may lead to potentiometer meltdown. Another seemingly good but inefficient approach to control the speed of a motor is to use a transistor amplifier arrangement like the one shown to the right in the figure. However, again, there is a problem. As the collector-to-emitter resistance increases with varying base voltage/current, the transistor must dissipate a considerable amount of heat. This can lead to transistor meltdown.

Better Designs

UJT/SCR Control Circuit

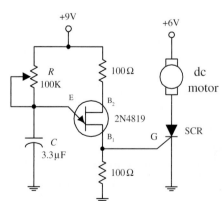

FIGURE 13.3

In order to conserve energy and prevent component meltdown, an approach similar to what was used in switching power supplied is used to control the speed of the motor. This approach involves sending the motor short pulses of current. By varying the width and frequency of the applied pulses, the speed of the motor can be controlled. Controlling a motor's speed in this manner prevents any components from experiencing continuous current stress. Figure 13.3 shows three simple circuits used to provide the desired motor-control pulses.

In the first circuit, a UJT relaxation oscillator generates a series of pulses that drives an SCR on and off. To vary the speed of the motor, the UJT's oscillatory frequency is adjusted by changing the *RC* time constant.

CMOS/MOSFET Control Circuit

In the second circuit, a pair of NAND gates make up the relaxation oscillator section, while an enhancement-type power MOSFET is used to drive the motor. Like the preceding circuit, the speed of the motor is controlled by the oscillator's RC time constant. Notice that if one of the input leads of the left NAND gate is pulled out, it is possible to create an extra terminal that can be used to provide on/off control that can be interfaced with CMOS logic circuits.

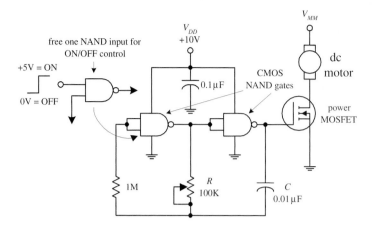

555 Timer/MOSFET Control Circuit

The third circuit is a 555 timer that is used to generate pulses that drive a power MOSFET. By inserting a diode between pins 7 and 6, as shown, the 555 is placed into low-duty cycle operation. R_1, R_2, and C set the frequency and on/off duration of the output pulses. The formulas accompanying the diagram provide the details.

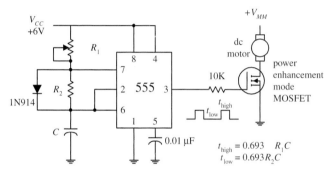

FIGURE 13.3   (*Continued*)

13.3   Directional Control of DC Motors

FIGURE 13.4

To control the direction of a motor, the polarity applied to the motor's leads must be reversed. A simple manual-control approach is to use a DPDT switch (see leftmost circuit in Fig. 13.4). Alternately, a transistor-driven DPDT relay can be used (see middle circuit). If you do not like relays, you can use a push-pull transistor circuit (see leftmost circuit). This circuit uses a complementary pair of transistors (similar betas and power rating)—one is an *npn* power Darlington, and the other is a *pnp* power Darlington. When a high voltage (e.g., +5 V) is applied to the input, the upper transistor (*npn*) conducts, allowing current to pass from the positive supply through the motor and into ground. If a low voltage (0 V) is applied to the input, the lower transistor (*pnp*) conducts, allowing current to pass through the motor from ground into the negative supply terminal.

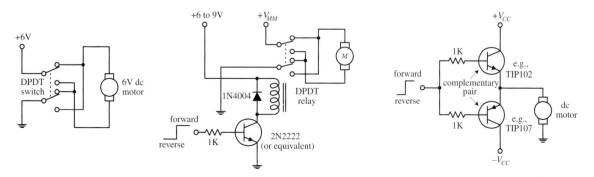

Another very popular circuit used to control the direction of a motor (as well as the speed) is the H-bridge. Figure 13.5 shows two simple versions of the H-bridge circuit. The left H-bridge circuit is constructed with bipolar transistors, whereas the right H-bridge circuit is constructed from MOSFETs. To make the motor rotate in the forward direction, a high (+5-V) signal is applied to the forward input, while no signal is applied to the reverse input (applying a voltage to both inputs at the same time is not allowed). The speed of the motor is controlled by pulse-width modulating the input signal. Here is a description of how the bipolar H-bridge works: When a high voltage is applied to Q_3's base, Q_3 conducts, which in turn allows the *pnp* transistor Q_2 to conduct. Current then flows from the positive supply terminal through the motor in the right-to-left direction (call it the *forward direction* if you like). To reverse the motor's direction, the high voltage signal is removed from Q_3's base and placed on Q_4's base. This sets Q_4 and Q_1 into conduction, allowing current to pass through the motor in the opposite direction. The MOSFET H-bridge works in a similar manner. The diodes within the H-bridge circuits help dampen transient spikes that are generated by the motor's coils so that they do not damage the other components within the circuit. All transistors (except the bipolar within the MOSFET circuit) should have high power ratings.

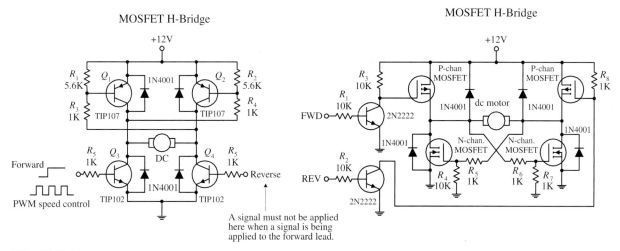

FIGURE 13.5

Now, it is possible to construct these H-bridge circuits from scratch, but it is far easier and usually cheaper to buy a motor-driven IC. For example, National Semiconductor's LMD18200 motor-driver IC is a high-current, easy-to-use H-bridge chip that has a rating of 3 A and 12 to 55 V. This chip is TTL and CMOS compatible and includes clamping diodes, shorted load protection, and a thermal warning interrupt output lead. The L293D (Unitrode) is another popular motor-driver IC. This chip is very easy to use and is cheaper than the LMD18200, but it cannot handle as much current and does not provide as many additional features. There are many other motor-driver ICs out there, as well as a number of prefab motor-diver boards that are capable of driving a number of motors. Check the electronics catalogs and Internet to see what is available.

13.4   RC Servos

Remote control (RC) servos, unlike dc motors, are motorlike devices designed specifically for pointerlike position-control applications. An RC servo uses an external pulse-width-modulated (PWM) signal to control the position of its shaft to within a small fraction of its maximum range of rotation. To alter the position of the shaft, the pulse width of the modulated signal is varied. The amount of angular rotation of an RC servo's shaft is limited to around 180 or 210° depending on the specific brand of servo. These devices can provide a significant amount of low-speed torque (due to an internal gearing system) and provide moderate full-swing displacement switching speeds. RC servos frequently are used to control steering in model cars, boats, and airplanes. They are also used commonly in robotics as well as in many sensor-positioning applications.

The standard RC servo looks like a simple box with a drive shaft and three wires coming out of it. The three wires consist of a power supply wire (usually black), a ground wire (usually red), and the shaft-positioning control wire (color varies based on manufacturer). Within the box there is a dc motor, a feedback device, and a control circuit. The feedback device usually consists of a potentiometer whose control dial is mechanically linked to the motor through a series of gears. When the motor is rotated, the potentiometer's control dial is rotated. The shaft of the motor is usually limited to a rotation of 180° (or 210°)—a result of the pot not being able to rotate indefinitely. The potentiometer acts as a position-monitoring device that tells the control circuit (by means of its resistance) exactly how far the shaft has been rotated. The control circuit uses this resistance, along with a pulse-width-modulated input control signal, to drive the motor a specific number of degrees and then hold. (The amount of holding torque varies from servo to servo.) The width of the input signal determines how far the servo's shaft will be rotated.

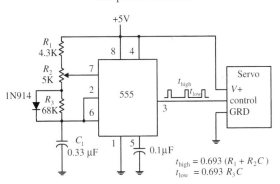

FIGURE 13.6

By convention, when the pulse width is set to 1.5 ms, the servo rotates its shaft to neutral position (e.g., 90° if the servo is constrained within a 0 to 180° range). Now, to rotate the shaft a certain number of degrees from neutral position, the pulse width of the control signal is varied. In order to make the shaft go counterclockwise from neutral, a pulse wider than 1.5 ms is applied to the control input. Conversely, to make the shaft go clockwise from neutral, a pulse narrower than 1.5 ms is applied (see figure). Knowing exactly how much wider or how much narrower to make the pulse to achieve exact angular displacements depends largely on what brand of servo you are using. For example, one brand of servo may provide maximum counterclockwise rotation at 1 ms and maximum clockwise rotation at 2 ms, whereas another brand of servo may provide maximum counterclockwise rotation at 1.25 ms and maximum clockwise rotation at 1.75 ms. The supply voltage used to power servos is commonly

4.8 V but may be 6.0 V or so depending on the specific brand of servo. Unlike the supply voltage, the supply current drawn by a servo varies greatly, depending of the servo's power output.

A simple 555 timer circuit like the one shown in Fig. 13.6 can be used to generate the servo control signal. In this circuit, R_2 acts as the pulse-width control. Servos also can be controlled by a microprocessor or microcontroller.

Now, when controlling servos within model airplanes, an initial control signal (generated by varying position-control potentiometers) is first sent to a radiowave modulator circuit that encodes the control signal within a carrier wave. This carrier wave is then radiated off as a radiowave by an antenna. The radiowave, in turn, is then transmitted to the model's receiver circuit. The receiver circuit recovers the initial control signal by demodulating the carrier. After that, the control signal is sent to the designated servo within the model. If there is more than one servo per model, more channels are required. For example, most RC airplanes require a four-channel radio set; one channel is used to control the ailerons, another channel controls the elevator, another controls the rudder, and another controls the throttle. More complex models may use five or six channels to control additional features such as flaps and retractable landing gear. The FCC sets aside 50 frequencies in the 72-MHz band (channels 11–60) dedicated to aircraft use only. No license is needed to operate these radios. However, with an amateur (ham) radio operator's license, it is possible to use a radio within the 50-MHz band. Also, there are frequencies set aside within the 27-MHz band that are legal for any kind of model use (surface or air). If you are interested in radio-controlled RC servos, a good starting point is to check out an RC model hobby shop. These shops carry a number of transmitter and receiver sets, along with the servos.

As a final note, with a bit of rewiring, a servo can be converted into a drive motor with unconstrained rotation. A simple way to modify the servo is to break the feedback loop. This involves removing the three-lead potentiometer (and unlinking the gear system so that it can rotate 360°) and replacing it with a pair of voltage-divider resistors (the output of the voltage divider replaces the variable terminal of the potentiometer). The voltage divider is used to convince the servo control circuit that the servo is in neutral position. The exact values of the resistors needed to set the servo in neutral position can be determined by using the old potentiometer and an ohmmeter. Now, to turn the motor clockwise, a pulse wider than 1.5 ms is applied to the control input. As long as the control signal is in place, the motor will keep turning and not stop—you have removed to feedback system. To turn the motor counterclockwise, a pulse narrower than 1.5 ms is applied to the control input.

13.5   Stepper Motors

Stepper motors, or *steppers,* are digitally controlled brushless motors that rotate a specific number of degrees (a step) every time a clock pulse is applied to a special translator circuit that is used to control the stepper. The number of degrees per step (resolution) for a given stepper motor can be as small as 0.72° per step or as large as 90° per step. Common general-purpose stepper resolutions are 15 and 30° per step. Unlike RC servos, steppers can rotate a full 360° and can be made to rotate in a continuous manner like a dc motor (but with a lower maximum speed) with the help of proper digital control circuitry. Unlike dc motors, steppers provide a large amount of torque at low speeds, making them suitable in applications where low-speed and high-precision position control is needed. For example, they are used in printers to control paper feed and are used to help a telescope track stars. Steppers are also found in plotter- and sensor-positioning applications. The list goes on. To give you a basic idea of how a stepper works, take a look at the Fig. 13.7.

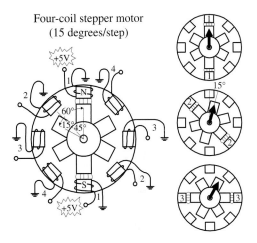

Four-coil stepper motor
(15 degrees/step)

Here is a simple model depicting a 15° per step variable-reluctance stepper. The stationary section of the motor, called the *stator*, has eight poles that are spaced 45° apart. The moving section of the motor, called the *rotor*, is made from a ferromagnetic material (a material that is attracted to magnetic fields) that has six teeth spaced 60° apart. To make the rotor turn one step, current is applied, at the same time, through two opposing pole pairs, or coil pairs. The applied current causes the opposing pair of poles to become magnetized. This in turn causes the rotor's teeth to align with the poles, as shown in the figure. To make the rotor rotate 15° clockwise from this position, the current through coil pair 1 is removed and sent through coil pair 2. To make the rotor rotate another 15° clockwise from this position, the current is removed from coil pair 2 and sent through coil pair 3. The process continues in this way. To make the rotor spin counterclockwise, the coil-pair firing sequence is reversed.

FIGURE 13.7

13.6 Kinds of Stepper Motors

The model used in the last example was based on a variable-reluctance stepper. As it turns out, this model is incomplete—it does not show how a real variable-reluctance stepper is wired internally. Also, the model does not apply to a class of steppers referred to as *permanent-magnet steppers*. To make things more realistic, let's take a look at some real-life steppers.

FIGURE 13.8

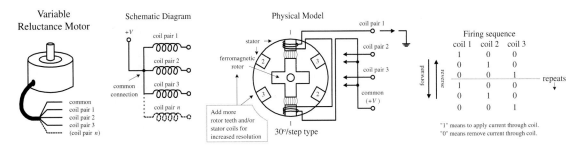

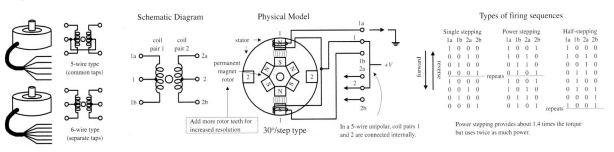

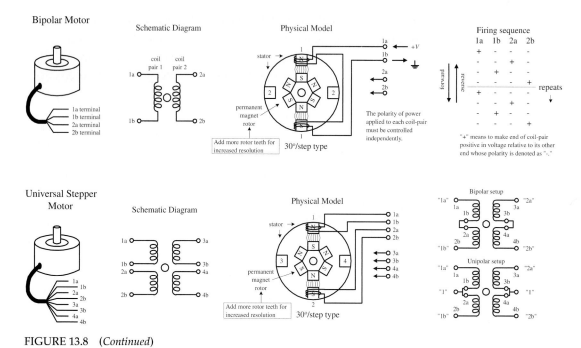

FIGURE 13.8 (*Continued*)

Variable-Reluctance Steppers

Figure 13.8 shows a physical model and schematic diagram of a 30° per step variable-reluctance stepper. This stepper consists of a six-pole (or three-coil pair) stator and a four-toothed ferromagnetic rotor. Variable-reluctance steppers with higher angular resolutions are constructed with more coil pairs and/or more rotor teeth. Notice that in both the physical model and the schematic, the ends of all the coil pairs are joined together at a common point. (This joining of the coil ends occurs internally within the motor's case.) The common and the coil pair free ends are brought out as wires from the motor's case. These wires are referred to as the *phase wires.* The common wire is connected to the supply voltage, whereas the phase wires are grounded in sequence according to the table shown in Fig. 13.8.

Permanent-Magnet Steppers (Unipolar, Bipolar, Universal)

UNIPOLAR STEPPERS

These steppers have a similar stator arrangement as the variable-reluctance steppers, but they use a permanent-magnet rotor and different internal wiring arrangements. Figure 13.8 shows a 30° per step unipolar stepper. Its consists of a four-pole (or two-coil pair) stator with center taps between coil pairs and a six-toothed permanent-magnetic rotor. The center taps may be wired internally and brought out as one wire or may be brought out separately as two wires. The center taps typically are wired to the positive supply voltage, whereas the two free ends of a coil pair are alternately grounded to reverse the direction of the field provided by that winding. As shown in the figure, when current flows from the center tap of winding 1 out terminal 1a, the top stator pole "goes north," while the bottom stator pole "goes south." This causes the rotor to snap into position. If the current through winding 1 is removed, sent through winding 2, and out terminal 2a, the horizontal poles will become energized, causing the rotor will turn 30°, or one step. In Fig. 13.8, three firing sequences are shown. The first sequence provides full stepping action (what I just discussed). The second sequence, referred to as the *power stepping sequence,* provides full stepping action with 1.4 times the torque but twice the power consumption. The third sequence provides half stepping (e.g., 15° instead of the rated 30°). Half stepping is made possible by energizing adjacent poles at the same time. This pulls the rotor in-between the poles, thus resulting in one-half the stepping angle. As a final note, unipolar steppers with higher angular resolutions are constructed with more rotor teeth. Also, unipolars come in either five- or six-wire types. The five-wire type has the center taps joined internally, while the six-wire type does not.

BIPOLAR STEPPERS

These steppers resemble unipolar steppers, but their coil pairs do not have center taps. This means that instead of simply supplying a fixed supply voltage to a lead, as was the case in unipolar steppers (supply voltage was fixed to center taps), the supply voltage must be alternately applied to different coil ends. At the same time, the opposite end of a coil pair must be set to the opposite polarity (ground). For example, in Fig. 13.8, a 30° per step bipolar stepper is made to rotate by applying the polarities shown in the firing sequence table to the leads of the stepper. Notice that the firing sequence uses the same basic drive pattern as the unipolar stepper, but the "0" and "1" signals are replaced with "+" and "−" symbols to show that the polarity matters. As you will see in the next section, the circuitry used to drive a bipolar stepper requires an H-bridge network for every coil pair. Bipolar steppers are more difficult to control than both unipolar steppers and variable-reluctance steppers, but their unique polarity-shifting feature gives them a better size-to-torque ratio. As a final note, bipolar steppers with higher angular resolutions are constructed with more rotor teeth.

UNIVERSAL STEPPERS

These steppers represent a type of unipolar-bipolar hybrid. A universal stepper comes with four independent windings and eight leads. By connecting the coil windings in parallel, as shown in Fig. 13.8, the universal stepper can be converted into a unipolar stepper. If the coil windings are connected in series, the stepper can be converted into a bipolar stepper.

13.7   Driving Stepper Motors

Every stepper motor needs a driver circuit that can control the current flow sent through the coils within the stepper's stator. The driver, in turn, must be controlled by a logic circuit referred to as a *translator*. I will discuss translator circuits after I have covered the driver circuits.

Figure 13.9 shows driver networks for a variable-reluctance stepper and for a unipolar stepper. Both drivers use transistors to control current flow through the motor's individual windings. In both driver networks, input buffer stages are added to protect the translator circuit from the motor's supply voltage in the event of transistor collector-to-base breakdown. Diodes are added to both drivers to protect the transistors and power supply from inductive kickback generated by the motor's coils. (Notice that the unipolar driver uses extra diodes because inductive kickback can leak out on either side of the center tap. As you will see in a moment, a pair of diodes within this driver can be replaced with a single diode, keeping the diode count to four.) The single driver section shown in Fig. 13.9 provides a general idea of what kinds of components can be used within the driver networks. This circuit uses a high-power Darlington transistor, a TTL buffer, and a reasonably fast protection diode (the extra diode should be included in the unipolar circuit). If you do not want to bother with discrete components, transistor-array ICs, such as the ULN200x series by Allegro Microsystems or the DS200x series by National Semiconductor, can be used to construct the driver section. The ULN2003, shown in Fig. 13.9, is a TTL-compatible chip that contains seven Darlington transistors with protection diodes included. The 7407 buffer IC can be used with the ULN2003 to construct a full-stepper driver. Other ICs, such as Motorola's MC1413 Darlington array IC, can drive multiple motor winding directly from logic inputs.

Single Driver Section

Transistor and Buffer Arrays

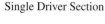

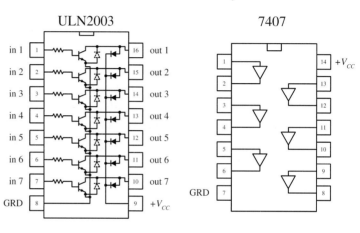

Variable-Reluctance Driver Network

Unipolar Driver Network

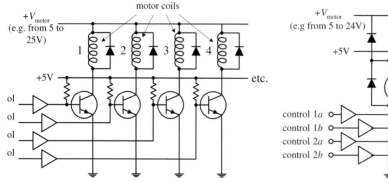

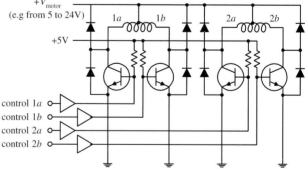

FIGURE 13.9

The circuitry used to drive a bipolar stepper requires the use of an H-bridge circuit. The H-bridge circuit acts to reverse the polarity applied across a given coil pair within the stepper. (Refer back to the section on dc direction control for details on how H-bridges work.) For each coil pair within a stepper, a separate H-bridge is needed. The H-bridge circuit shown in Fig. 13.10 uses four power Darlington transistors that are protected from the coil's inductive kickback by diodes. An XOR logic circuit is added to the input to prevent two high (1's) signals from being applied to the inputs at the same time. [If two high signals are placed at both inputs (assuming that there is no logic circuit present), the supply will short to ground. This is not good for the supply.] The table in Fig. 13.10 provides the proper firing sequence needed to create the desired polarities.

H-bridge used with Bipolar Stepper

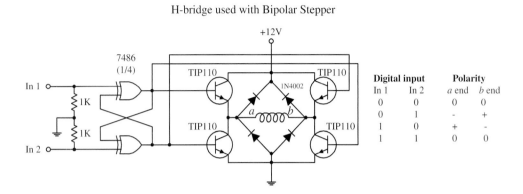

Digital input		Polarity	
In 1	In 2	a end	b end
0	0	0	0
0	1	-	+
1	0	+	-
1	1	0	0

FIGURE 13.10

As mentioned in the dc motor section of this chapter, H-bridges can be purchased in IC form. SGS Thompson's L293 dual H-bridge IC is a popular choice for driving small bipolar steppers drawing up to 1 A per motor winding at up to 36 V. The L298 dual H-bridge is similar to the L293 but can handle up to 2 A per winding. National Semiconductor's LMD18200 H-bridge IC can handle up to 3 A, and unlike the L293 and L298, it has protection diodes built in. More H-bridge ICs are available, so check the catalogs.

13.8   Controlling the Driver with a Translator

A translator is a circuit that enacts the sequencing pulses used to drive a driver. In some instances, the transistor may simply be a computer or programmable interface controller, with software directly generating the outputs needed to control the driver leads. In most cases, the translator is a special IC that is designed to provide the proper firing sequences from its output leads when a clock signal is applied to one of its input leads; another input signal may control the direction of the firing sequence (the direction of the motor). There are a number of steppers translator ICs available that are easy to use and fairly inexpensive. Let's take a look at one of these devices in a second. First, let's take a look at some simple translator circuits that can be built from simple digital components.

A simple way to generate a four-phase drive pattern is to use a CMOS 4017 decade counter/divider IC (or a 74194 TTL version). This device sequentially makes 1 of 10 possible output high (others stay low) in response to clock pulses. Tying the fifth output (Q_4) to ground makes the decade counter into a quad counter. To enact the drive sequence, a clock signal is applied to the clock input (see Fig. 13.11). Another four-phase translator circuit that provides power stepping control as well as direction control can be constructed with a CMOS 4027 dual JK flip-flop IC (or a 7476 TTL version). The CMOS 4070 XOR logic (or 7486 TTL XOR logic) is used to set up directional control.

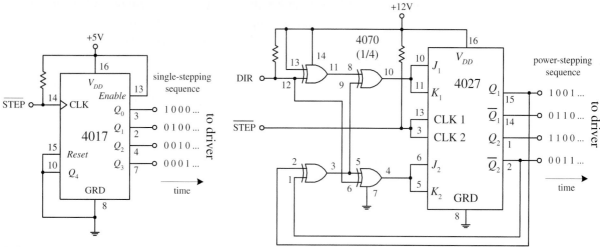

FIGURE 13.11

Figure 13.12 shows a circuit that contains the translator, driver, and stepper all in one. The motor, in this case, is a unipolar stepper, while the translator is a TTL 74194 shift counter. The 555 timer provides clock signals to the 74194, while the DPDT switch acts to control the direction of the motor. The speed of the motor is dependent on the frequency of the clock, which in turn is dependent on R_1's resistance. The translator in this circuit also can be used to control a variable reluctance stepper. Simply use the variable-reluctance diver from Fig. 13.9 and the firing sequence shown in Fig. 13.8 as your guides.

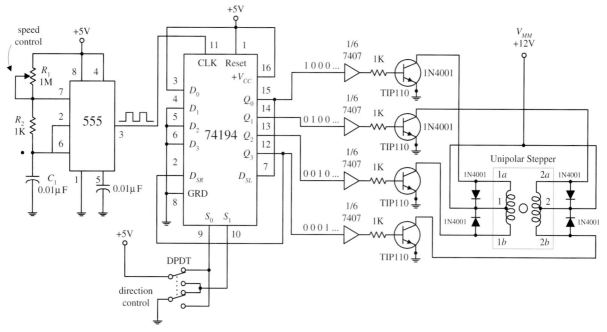

FIGURE 13.12

Perhaps the best translator circuits you can hope for come in integrated packages. A number of manufacturers produce stepper motor controller ICs that house both the translator and driver sections. These chips are fairly simple to use and inexpensive. A classic stepper controller chip is the Philips SAA1027. The SAA1027 is a bipolar IC that is designed to drive four-phase steppers. It consists of a bidirectional four-state counter and a code converter that are used to drive four outputs in sequence. This chip has high-noise-immunity inputs, clockwise and counterclockwise capability, a reset control input, high output current, and output voltage protection. Its supply voltage runs from 9.5 to 18 V, and it accepts input voltages of 7.5 V minimum for high (1) and 4.5 V maximum for low (0). It has a maximum output current of 500 mA. Figure 13.13 will paint the rest of the picture.

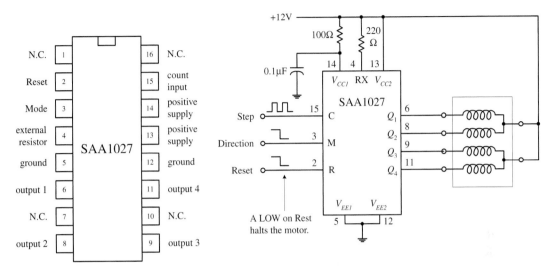

FIGURE 13.13

Q_1	Q_2	Q_3	Q_4	Q_1	Q_2	Q_3	Q_4
	M = 0				M = 1		
0	1	0	1	0	1	0	1
1	0	0	1	0	1	1	0
1	0	1	0	1	0	1	0
0	1	1	0	1	0	0	1
0	1	0	1	0	1	0	1

Count input C (pin 15)—A low-to-high transition at this pin causes the outputs to change states.

Mode input M (pin 3)—Controls the direction of the motor. See table to the left.

Reset input R (pin 2)—A low (0) at the R input resets the counter to zero. The outputs take on the levels shown in the upper and lower line of the table to the left.

External resistor RX (pin 4)—An external resistor connected to the RX terminal sets the base current of the transistor drivers. Its value is based on the required output current.

Outputs Q_1 through Q_4 (pins 6, 8, 9, 11)—Output terminals that are connected to the stepper motor.

As mentioned, the SAA1027 is a classic chip (old chip). Newer and better stepper control ICs are available from a number of different manufacturers. If you are interested in learning more about these chips, try searching the Internet. You will find a number of useful Web sites that discuss stepper controller ICs in detail. Also, these Web sites often will provide links to manufacturers and distributors of stepper motors and controller ICs.

13.9   A Final Word on Identifying Stepper Motors

When it comes to identifying the characteristics of an unknown stepper, the following suggestions should help. The vast majority of the steppers on the market today are unipolar, bipolar, or universal types. Based on this, you can guess that if your stepper has four leads, it is most likely a bipolar stepper. If the stepper has five leads, then the motor is most likely a unipolar with common center taps. If the stepper has six leads, it is probably a unipolar with separate center taps. A motor with eight leads would most likely be a universal stepper. (If you think your motor might be a variable-reluctance stepper, try spinning the shaft. If the shaft spins freely, the motor is most likely a variable-reluctance stepper. A coglike resistance indicates that the stepper is a permanent-magnet type.)

Once you have determined what kind of stepper you have, the next step is to determine which leads are which. A simple way to figure this out is to test the resistance between various leads with an ohmmeter.

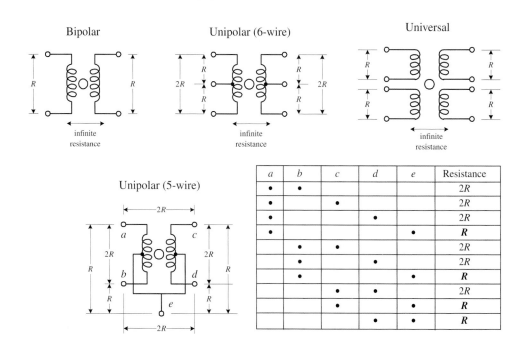

a	b	c	d	e	Resistance
•	•				2R
•		•			2R
•			•		2R
•				•	R
	•	•			2R
	•		•		2R
	•			•	R
		•	•		2R
		•		•	R
			•	•	R

FIGURE 13.14

Decoding the leads of a bipolar stepper is easy. Simply use an ohmmeter to determine which wire pair yields a low resistance value. A low resistance indicates that the two wires are ends of the same winding. If the two wires are not part of the same winding, the resistance will be infinite. A universal stepper can be decoded using a similar approach. Decoding a six-wire unipolar stepper requires isolating two three-wire pairs. From there, you figure out which wire is the common by noticing which measured pair among the isolated three wires gives a unit R worth of resistance and which pair gives a unit of $2R$ worth of resistance (see Fig. 13.14). Now, decoding a five-wire unipolar (with common center tap) is a bit more tricky than the others because of the common, but hidden, center tap. To help decode this stepper, you can use the diagram and table shown in Fig. 13.14. (The dots within the table represent where the ohmmeter's two probes are placed within the diagram.) With the table you isolate e (common tap wire) by noting when the ohmmeter gives a resistance of R units. Next, you determine which of the two wires in your hand is actually e by testing one of the two with the rest of the wires. If you always get R, then you are holding e, but if you get $2R$, you are not holding e. Once the e wire is determined, any more ohmmeter deducing does not work—at least in theory—because you will always get $2R$. The best bet now is to connect the motor to the driver circuitry and see if the stepper steps. If it does not step, fiddle around with the wires until it does.

CHAPTER 14

Hands-On Electronics

14.1   Safety

Your body is a complex system that is controlled by electrochemical signals that are sent to and from your brain. If you screw these signals up by introducing an external flow of electrons, vital organs may cease to function properly, which ultimately may lead to death.

Applying a current of about 1 mA through your body will have the effect of providing a tingling feeling or mild sensation. A current of about 10 mA will result in a shock of sufficient intensity to cause involuntary loss of muscle control. A current of 100 mA lasting for more than 1 second can result in crippling effects and may result in death. Beyond 100 mA, extreme shock occurs, which may result in ventricular fibrillations (irregular heartbeat) that could easily lead to death.

The resistance of a human body to current flow is between 1 MΩ when it is dry and a few hundred ohms when it is wet. To figure out the amount of current flow through a body that is connected to an ideal voltage source, simply use Ohm's law ($I = V/R_{body}$). For example, say you attach wires from your hand and foot to the terminals of a 6-V battery. If your pretend that your internal resistance is 300,000 Ω (you are perspiring that day), the amount of current that would flow from one arm to the other would be 6 V/300,000 Ω = 20 nA. This amount of current is well within the safety limits, and you probably would not feel a thing. However, if you were unfortunate enough to drop a 120-V ac-powered dryer into the bathtub (when you are in it), the amount of current that would flow through your body—assuming for now that a wet body has a resistance of 1000 Ω—would be 120 V/1000 Ω = 0.12 A. This amount of current most likely would be fatal.

Now, you have probably heard the saying, "It's not the voltage that will kill you, it's the current." But according to Ohm's law, it appears the voltage term (*V*) sets the current term (*I*), so the two would appear to have an equal part to play. What's going on? Is this statement true, or what? Well, this is where it is important to understand what exactly is meant by *voltage,* at least in terms of using it in Ohm's law, since this is how we are determining the currents. When using Ohm's law, it is always assumed that voltage sources are ideal. As you learned in the theory section, an ideal voltage source is a device that maintains a fixed voltage no matter what size load is attached to it. For an ideal voltage source to maintain its voltage no matter the resistance placed between it, it must be able to supply varying amounts of current. If you were to treat all sources of power as ideal voltage sources, then the saying, "It's not the voltage, it's the current that kills you" does not make sense. The problem with this approach is that in the real world you are often dealing with less than ideal voltage sources—ones that can only produce a limited output current. For such cases, blindly applying Ohm's law does not work. A good example to demonstrate this point is to consider the static electric charge that accumulates on a brush when you are combing your hair. During this simple operation, it is possible that as many as 10^{10} electrons will be stripped from your hair and deposited onto the brush, resulting in a voltage (relative to ground) of 2000 V. If you plug this voltage into Ohm's law, taking R_{body} to be, say, 10,000 Ω, you would get a result of 0.2 A—a potentially lethal current. But, wait a minute! How many people do you know who have been killed by statically charged combs? There has got to be something wrong. The problem lies in the fact that you are dealing with a classic nonideal voltage source. Unlike an ideal voltage source, the number of charges needed to make up a current runs out very quickly. You can make a very rough calculation of the time required for these charges to run out by finding the initial charge on the comb. If there are 10^{10} electrons, each of which have a charge of 1.6×10^{-19} C, then you get a net charge of 1.6×10^{-9} C. Next, using the definition of current $I = \Delta Q / \Delta t$, set $\Delta Q = 1.6 \times 10^{-9}$ C and $I = 0.2$ A, and then solve for the time Δt. Doing this calculation, you get $\Delta t = 8 \times 10^{-9}$ s, or 8 ns. Being exposed to such a short pulse of current is not going to do you any harm, unlike the case where you stick a fork into an ac outlet. In the fork in the ac outlet case, you would be receiving a continuous flow of current, frying your body tissues in the process.

A more intuitive explanation as to what the statement, "it's not the voltage that will kill you, it's the current," refers to is given by the following analogy: Suppose that you drop a grain of sand from a second-story window onto a pedestrian. As long as the grain of sand does not hit the individual in the eye, there is little chance that any permanent damage will be done. If you drop the same grain from a tenth-story window onto the same person, the sand may prick the top of the individuals head with a little more intensity but still will not kill the individual. However, if you repeat this experiment but this time replace the single grain of sand with a hundred-pound bag of sand, the results, as you can image, will be drastically different. The sand in this analogy represents electrons (current), the height from which the sand is drop represents the voltage, and the pedestrian represents the internal organs and tissue of a human being.

14.1.1   Some Safety Tips

The following tips can help you stay alive, or at least prevent you from getting a shock.

- Make sure all components that are connected to ac power lines meet required power ratings.

- Use one hand while taking measurements, and keep your other hand at your side or in your pocket. If you do get shocked, it will be less likely that current will pass through your heart.

- When building power supplies or other instruments, make sure that all wires and components are enclosed in a metal box or insulative plastic container. If you use a metal box, it is important that you ground the conductive shell (attach a wire from the inner surface of the box to the ground wire of the power cable). Grounding a metal box eliminates shocks incurred as a result of a hot wire coming lose and falling on the box, thereby making the whole box "hot."

- When making holes in metal boxes through which to insert an ac line cable, place a rubber grommet around the inner edges of the hole to eliminate the chance of fraying the cable.

- Do not attempt repairs on line-powered circuits with the power on. Always turn off the main power first.

- Watch out for large filters, voltage-multiplier circuits, and energy-storage capacitors in general. These devices can store a lethal amount of charge and may retain the charge for a number of days. Even capacitors rated at voltages as low as 5 or 10 V can be dangerous. Never touch both terminals of a capacitor at the same time. When working with capacitors, disarm them by shorting their two leads together with a screwdriver (with insulated handle).

- When working on ac line circuits, wearing rubber soled shoes or standing on a sheet of rubber or wood can reduce possible shocks.

- Avoid standing in a position that could be dangerous if you were to lose muscle control due to shock. Often the fall alone is more dangerous than the initial shock.

- When working on high-power circuits, bring someone else along who can assist you if something goes wrong. If you see that someone cannot let go of a "hot" object, do not grab onto him or her. Instead, use a stick or insulated object to push him or her away from the source. It is not a bad idea to know CPR too.

- All high-voltage test instruments (e.g., power supplies, signal generators, oscilloscopes) that are operated on a 120-V ac supply should use three-wire polarized line cable. To minimize the chance of getting shocked, run equipment through an isolation transformer.

- Use only shielded (insulated) lead probes when testing circuits. Never allow your fingers to slip down to the metal probe tip when testing a "hot" circuit. Also, make sure to remove power from a circuit when making connections with wires and cables.

14.1.2 Damaging Components with Electrostatic Discharge

Scuffling across a carpet while wearing sneakers on a dry day can result in a transfer of electrons from the carpet to your body. In such a case, it is entirely possible that you will assume a potential of 1000 V relative to ground. Handling a polyethylene bag can result in static voltages of 300 V or more, whereas combing your hair can result in voltages as high as 2500 V. The drier the conditions (lower the humidity), the greater the change is for these large voltages to form. Now, the amount of electrostatic discharge that can result from a electrostatically charged body coming in contact with a grounded object is not of much concern, in terms of human standards. However, the situation is entirely different when subjecting certain types of semiconductive devices to similar discharges.

Devices that are particularly vulnerable to damage include field-effect transistors, such as MOSFETs and JFETs. For example, a MOSFET, with its delicate gate-channel oxide insulator, can be destroyed easily if an electrostatically charged individual touches its gate; the gate-channel breakdown voltage will be exceeded, and a hole will be blown through the insulator that will destroy the transistor. Here is a rundown of the vulnerable devices out there.

Extremely vulnerable: MOS transistors, MOS ICs, junction FETs, laser diodes, microwave transistors, metal film resistors

Moderately vulnerable: CMOS ICs, LS TTL ICs, Schottky TTL ICs, Schottky diodes, linear ICs

Somewhat vulnerable: TTL ICs, small signal diodes and transistors, piezoelectric crystals

Not vulnerable: Capacitors, carbon-composite resistors, inductors, and many other analog devices

Devices that are highly vulnerable to damage are often marked with "Caution, components subject to damage by static electricity." If you see a label like this, use the following precautions.

14.1.3 Handling Precautions

- Store components in their original packages, in electrically conductive containers (e.g., metal sheet, aluminum foil), or in conductive foam packages.

- Do not touch leads of ESD-sensitive components.

- Discharge the static electricity on your body before touching components by touching a grounded metal surface such as a water pipe or large appliance.

- Never allow your clothing to make contact with components.

- Ground tabletops and soldering irons (or use a battery-powered soldering iron). You also should ground yourself with a conductive wrist guard that's connected with a wire to ground.

- Never install or remove an ESD-sensitive component into or from a circuit when power is applied. Once the component is installed, the chances for damage are greatly reduced.

14.2 Constructing Circuits

In this section I will briefly discuss the steps involved in building a working circuit, namely, drawing a circuit schematic, building a prototype, making a permanent circuit, finding a suitable enclosure for the circuit, and applying a sequence of troubleshooting steps to fix improperly functioning circuits.

14.2.1 Drawing a Circuit Schematic

The *circuit schematic,* or *circuit diagram,* is a blueprint of a circuit. For a schematic to be effective, it must include all the information necessary so that you or anyone else reading it can figure out what parts to buys, how to assemble the parts, and possibly what kind of output behavior to expect. To make an easy-to-read, unambiguous schematic, the following guidelines should be followed:

- The standard convention used when drawing a schematic is to place inputs on the left, outputs on the right, positive supply terminals on top, and negative supply or ground terminals on the bottom of the drawing.

- Keep functional groups, such as amplifiers, input stages, filters, etc., separated within the schematic. This will make it easier to isolate problems during the testing phase.

- Give all circuit components symbol designations (e.g., R_1, C_3, Q_1, $IC4$), and provide the exact size or type of component information (e.g., 100 k, 0.1 μF, 2N2222, 741). It is also important to include power rating for certain devices, such as resistors, capacitors, relays, speakers, etc.

- Use abbreviations for large-valued components (e.g., 100 kΩ instead of 100,000 Ω, 100 pF instead of 100×10^{-12} F). Common unit prefixes include $p = 10^{-12}$, $n = 10^{-9}$, $\mu = 10^{-6}$, $k = 10^3$, $M = 10^6$.

- When labeling ICs, place lead designations (e.g., pin numbers) on the outside of the device symbol, and place the name of the device on the inside.

- In certain circuits where the exact shape of a waveform is of importance (e.g., logic circuits, inverting circuits, etc.), it is helpful to place a picture of the expected waveforms at locations of interest on your circuit diagram. This will help isolate problems later during the testing phase.

- Power-supply connections to op amps and digital ICs are usually assumed—they are typically left out of the schematic. However, if you anticipate confusion later on, include these supply voltages in you drawing.

- To indicate joined wires, place a dot at the junction point. Unjoined wires simply cross (do not include a dot in such cases).

- Include a title area near the bottom of the page that contains the circuit name, the designer's name, and the date. It is also useful to leave room for a list of revisions.

FIGURE 14.1   Figure 14.1 shows a sample schematic.

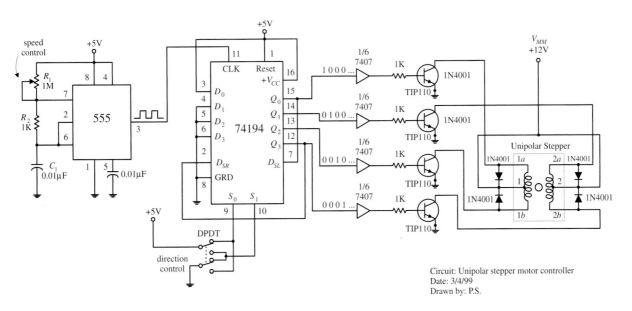

Circuit: Unipolar stepper motor controller
Date: 3/4/99
Drawn by: P.S.

Once you have completed the circuit schematic, see if anything looks fishy. Are there missing connections or missing component values? Are component polarities indicated? Have you considered the power ratings of the components? Have you made connections as simple as possible? It is better to check things now; finding an error when you are soldering things together is much more annoying than erasing a few lines on a drawing.

14.2.2 A Note on Circuit Simulator Programs

Now, before you build a circuit, or even before you finish the schematic, you might consider using a circuit simulator program to test your idea to see if it works. Circuit simulator programs allow you to construct a computer model of your circuit and then test it (measure voltages, currents, wave patterns, logic states, etc.) without ever having to touch a real component. A typical simulator program contains a library of analog and digital devices, both discrete and integrated in form. If you wish to model an oscillator circuit—one that is built from a few bipolar transistors, some resistors, a capacitor, and a dc power supply—all you do is select the parts from the library, set the values of these parts, and then arrange the parts to form an oscillator circuit. To test the circuit, simply choose one of the simulator's test instruments, and then attach the test instrument's probes to the desired test points within the circuit. For example, if you are interested in what the output waveform of the oscillator looks like, choose the simulator's oscilloscope, and then attach the test leads and measure the output. The computer screen will then display a voltage versus time graph of the output. Other instruments found in simulator programs include multimeters, logic analyzers, function generators, and bode plotters, to name a few items.

Why make a computer simulation of a circuit before building a real one? For one thing, when you make a computer simulation, you do not have to worry about working with faulty components. Second, there is no need to worry about destroying components with excessive current—the computer program does not "care about the numbers." And finally, a simulator program does all the mathematical work for you. The simulator allows you to fiddle around with component values until the circuit is working as desired. Using a simulator program can make learning electronics an intuitive process and saves time spent at the workbench.

Some popular simulator programs include Electronic Workbench, CircuitMaker, and MicroSim/Pspice. Electronic Workbench and CircuitMaker are relatively easy to use, while Pspice is a bit more technical. You can learn more about these simulator programs by looking them up on the Internet.

14.2.3 Making a Prototype of Your Circuit

Once you are satisfied with the schematic, the next step is to make a prototype of your circuit. The most common tool used during the prototype phase is a solderless modular breadboard. A breadboard acts as a temporary assembly board on which electrical parts such as resistors, transistors, and ICs are placed and joined together by wires or built-in conductive pathways hidden underneath the surface of the breadboard (see Fig. 14.2).

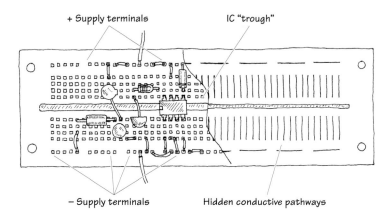

FIGURE 14.2

Breadboards come with an array of small square sockets spaced about 0.100 in from center to center. When a wire or component lead is inserted into one of these sockets, a springlike metal sleeve built into the socket acts to hold the wire or lead in place. Breadboard sockets are designed to accept 22-gauge wire but can expand to fit wire diameters between 0.015 and 0.032 in (0.38 and 0.81 mm). The upper and lower rows of sockets of a breadboard typically are reserved for power-supply connections, while the sockets between the central gap region are reserved for DIP ICs.

14.2.4 The Final Circuit

Once you are finished making a successful prototype, the next step is to construct a more permanent circuit. At this point, you must choose the type of mounting board on which to place your circuit. Your choices include a perforated board, a wire-wrap board, a preetched board, or a custom-etched PC board. Let's take a closer look at each of these boards.

Perforated Board

A *perforated board* is an insulated board with an array of holes drilled into it (see Fig. 14.3). To join a lead from one electrical component to another, each of the components' leads are placed through neighboring holes. The lead ends sticking out the backside of the board are then twisted together (and possibly soldered).

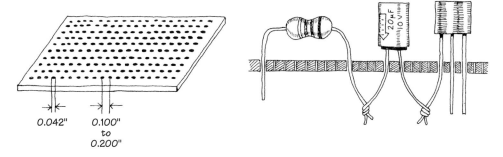

FIGURE 14.3

Constructing a circuit on a perforated board is easy. Few supplies are needed, and making connections does not require much skill. However, what you get in the end is a fairly bulky circuit that is liable to fall apart over time and which may pick up noise inadvertently (jumper wires will act like little antennas). In general, perforated boards are used for constructing simple, noncritical types of circuits.

Wire-Wrap

Using a *wire-wrap board* is perhaps the fastest way to assemble moderately complex circuits containing ICs. Every wire-wrap board is made up of an array of sockets, each of which has a corresponding pinlike extension sticking out the opposite side of the board (see Fig. 14.4).

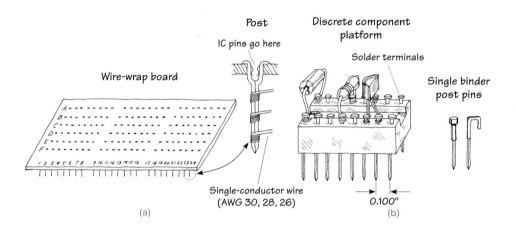

FIGURE 14.4

(a)   (b)

IC leads are inserted directly into the wire-wrap's sockets, while discrete components, such as resistors, capacitors, and transistors, must be mounted on special platforms like blocks or single postlike pins (see Fig. 14.4*b*). Each of these platforms contains a number of nail-like heads on which discrete component leads are attached, either by coiling the leads around the nail head or by fastening the leads to the nail head with solder. The nail-like tips of the platforms are then inserted into the board's sockets. To connect components together, the pins on the backside of the wire-wrap board are joined together with wires (typically 30-, 28-, or 26-gauge single-conductor wires). In order to fasten the joining wires to the pins, a special wire-wrap tool is used (see Fig. 14.5). This tool wraps the joining wire around the pin by means of a bit section that rotates around a hollow core. The wire is inserted into the bit, the hollow core is placed over the pin, then the wire-wrap tool is twisted several times (usually around seven times) and removed.

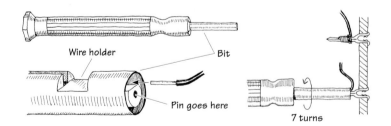

FIGURE 14.5

In practice, to save time and avoid making mistakes, it is desirable to do all the wrapping in a single pass. To follow this approach, some bookkeeping is required. Notice in Figure 14.4*a* that each socket/pin is given a row/column designation. For example, a pin located three rows down and two rows in from the left is given a C2 designation, whereas a pin that is located five rows down and seven rows in is given an E7 designation. Now, to figure out how to arrange electrical components on the board, making a simple sketch of the wire-wrap board is helpful. On the sketch, draw in all the components, fixing each component lead to a specific row/column coordinate. Once the sketch is complete, simply grab your wire-wrap tool and start making wire connections between the pins, using the sketch as a guide.

Wire-wrap boards are suitable platforms for circuits that contain a number of ICs, such as logic circuits. However, because the sockets of these boards are not designed for linear component leads (you must use platforms in such cases), it may be easier to use a preetched, or custom-etched PC board for building analog circuits.

Preetched Perforated Boards

A *preetched perforated board* is made of an insulating material that is coated with a preetched copper pattern and which has a number of holes drilled into it. To join electrical components, simply place the leads of the components in the appropriate holes that are joined by a copper etched strip and then apply solder. Preetched boards come with a variety of different etched patterns. Figure 14.6 shows samples of the kinds of patterns available.

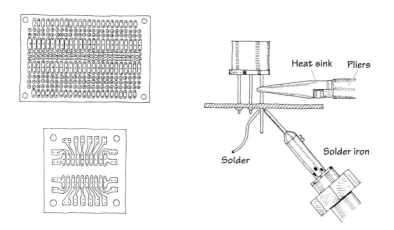

FIGURE 14.6

Custom-Etched Boards (PC Boards)

If you are looking to build a circuit with a professional appearance, designing a *custom-etched circuit board*, or *PC board* (printed circuit board), is the ticket. Custom etching involves using graphic and chemical techniques to convert a copper-covered board into a custom-etched one. By doing your own custom etching, you can construct highly reliable, tightly compact circuits that require few jumper wires (see Fig. 14.7).

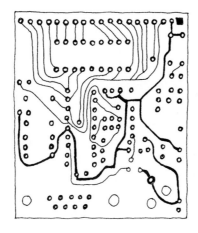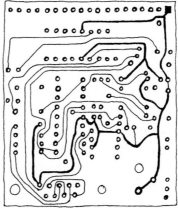

FIGURE 14.7

Designing custom-etched circuit boards takes a bit longer, but the time spent is often well worth the effort. There are times when making a custom-etched board is essential to ensure proper circuit operation, especially when dealing with circuits that contain components whose properties are greatly influenced by the length of the leads. For example, emitter-coupled logic circuits require the unique microstrip line geometries and precise placement of components to achieve fast rise times and at the same time avoid crosstalk among circuit elements. Sensitive low-level amplifier circuits also benefit from well-placed microstrip interconnections—the short and direct interconnections help eliminate noise pickup.

To design a custom-etched PC board, you first need an insulated board (usually $\frac{1}{16}$ in thick and made from a fire-resistant epoxy-bonded fiberglass) that is completely covered on one or both sides with a very thin copper coating. Next, you must transform your circuit schematic into a hardwired sketch. This involves rearranging components in such ways as to make all conductive pathways short and direct. The hardwired sketch also should eliminate any wire crossing—if possible.

Once you feel that your hardwired sketch is complete, the next step is to transfer it onto the copper-coated board. Afterward, the trick is to etch out all the undesired copper-coated sections while leaving the conductive pathways intact. At this point, there are a number of different transferring/etching techniques from which you can choose. Perhaps the simplest technique involves using a PC board kit—which you can buy for a few bucks from a store such as Radio Shack. A typical kit comes with a single or dual copper-coated board, a bottle of etching solvent, a magic marker, a bottle of rubbing alcohol, and a drill bit. To make your custom-etched board, you first transfer your hardwired sketch onto the surface of the board with, say, a pencil. Next, you drill in the appropriate holes where component leads are to go. Now, with the magic marker, you trace over the pencil sketch, making sure to encircle the drilled-out holes. After that, you place the board in a tub of etching solvent (typically ferric chloride) and wait until the copper dissolves away from the sections of the board that are not coated with magic-marker ink (the ink does not dissolve in the solvent—it acts to protect the underlying copper). After the board is removed from the solvent bath, it is washed off with water, and the magic-marker ink is then removed with a rag doused in rubbing alcohol. Using one of these kits is great for simple, single-run productions. They are easy to use, inexpensive, and require practically no special equipment—other than what is provided in the kit itself. However, one problem with these kits is that you can only construct one circuit board at a time. Another problem with these kits is the limited precision you get by using a magic marker to create conductive pathways. If you are interested in making multiple copies of a circuit, and if you are looking for greater line precision, a more sophisticated technique that involves photochemical processes is required.

One such technique involves transferring a hardwired sketch onto a Mylar sheet, replacing pencil lines with cutout tap patterns. Once the Mylar template is completed, it is photographed and reduced to a desired size. Next, a copper-coated circuit board, covered with a thin layer of light-sensitive material (called *resist*), is placed under the negative. When the negative is placed over the photoresist-coated board, the photoresist underneath the clear areas of the negative undergoes chemical and physical changes when light is shined on the combination. These chemical and physical changes make the resist impervious to solvents. When the board is placed in an etching solution, all unexposed regions dissolve away, while the regions with the photo-treated resist remain. To remove the resist, a separate solvent is used. Afterwards, holes can be drilled into the board at the desired lead insertion points. This whole process is fairly time-consuming. Perhaps more important, this technique is outdated—at least the drafting-stage part of it. Today, computer programs are used to make the hardwired sketches, and special machines are used to transform these sketches into negatives.

If you require a high-quality PC board or need a large number of similar PC boards made, your best bet is to get some professional help from a company that specializes in PC board production. Now, the nice thing about letting specialists design and construct your PC board is that they have all the computer programs and manufacturing equipment needed to build a high-quality PC board. There are hundreds of such companies out there—some of which go by snappy names like Circuit Board Express. These companies can take a circuit schematic and transform it into a working PC board within a surprisingly short period of time. Some companies even guarantee a 24-hour turnaround time. These specialists also can design and construct multilayer boards (typically between 1 and 18 layers). (Multilayer boards increase the number of interconnections and help eliminate jumpers). If you are interested in using one of these companies, try searching the Web or Yellow Pages using *printed circuit boards* as your keyword.

14.2.5   A Note About Board Layout

When arranging components on a circuit board, ICs and resistors should be placed in rows and should all be pointing in the same direction. Also, make sure to leave about a quarter-inch border around the circuit board to allow room for card lifters, guides, and standoffs. Bring power supply leads or other input/output leads to the edge of the board, connecting them through either an edge connector, D-connector, barrier-strip connector, or single-binder posts fixed to the edge of the board. Avoid mounting heavy components on circuit boards to prevent damage in case of a fall. It is also a good idea to place polarity marking on the board next to devices such as diodes and electrolytic capacitors. Placing labels next to IC pins is also helpful. Consider labeling test points, trimmer functions (e.g., zero adjustment), inputs and outputs, indicator light functions, and power-supply terminals as well.

14.2.6   Special Pieces of Hardware Used in Circuit Construction

Three pieces of hardware are used frequently during the construction phase. These items include prototyping boards with I/O interfacing gold-plated fingers, IC and transistor socket holders, and heat sinks.

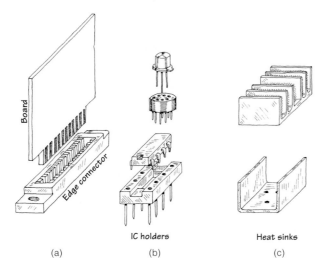

FIGURE 14.8   (a)   (b)   (c)

Prototyping boards with gold-plated fingers typically are inserted into a card cage along with a number of other boards. Each board is inserted through a plastic guide and into an edge connector. Separate boards can be linked by means of a flat multiple-conductor cable (see Fig. 14.8*a*). The nice thing about these boards is that you can easily remove them from a cage to work on them, without making a mess of things in the process. When designing multiple-board devices, it is wise to use a separate board for each functional group of a circuit (e.g., amplifier sections, memory-chip sections, etc.). This will make it easy to find and fix problems later on.

IC and transistor sockets holders are used in situations where there is a good chance that the device they house will need replacing (see Fig. 14.8*b*). It is tempting to use such holders everywhere within a circuit. However, placing too many IC and transistor socket holders in a circuit board can lead to headaches latter on. Often the socket sections of these holders are poorly designed and may prove unreliable over time.

Heat sinks are metal devices with large surface areas that are connected to heat-generating devices (e.g., power diodes and transistors) to help dissipate heat energy. A heat sink is usually connected to a component by means of a screw and washer fastener (see Fig. 14.8*c*). Silicon grease placed between the washer and heat sink is often used to enhance the thermal conductivity between the electrical component and heat sink.

14.2.7   Soldering

Solder is a tin-lead alloy that is used to join component leads together. Electrical solder often comes with a rosin flux mixed in, which is used to dissolve oxides that are present on the metal surfaces to be joined. Before solder can be applied with a soldering iron, all metal surfaces must be cleansed of oils, silicones, waxes, or grease. Use a solvent, steel wool, or fine sandpaper to remove undesirable residue.

When soldering PC boards, use a low-wattage iron (25 to 40 W). To ensure good soldering connections, a thin bright coating of molten solder should be present on the tip of the iron. With time, this coat becomes contaminated with oxides and should be renewed by wiping its surface with a sponge and then reapplying fresh solder. (Applying a fresh coat of solder to the tip of an iron is referred to as *tinning* an iron.)

The trick to making good soldering connections is to first heat the two metal pieces to be joined. Do not melt the solder first; you will never be able to control the placement of the molten solder. Solder likes to flow toward hot spots.

Also when working on a circuit, make sure not to splatter solder on your board. If a small piece of solder lands between two separate conductive lines, you will end up shorting them. When you are done soldering, inspect your work carefully for stray solder spatters.

To protect sensitive components from the heat of the soldering iron's tip, heat-sink components by gripping the component lead with a needle-nose pliers. Special heat-sink clips are also available for this purpose.

14.2.8   Desoldering

If you make a bad connection or have to replace a component, you must melt the solder and start over again. Simply melting the solder and then attempting to yank the part, while the solder is still wet, however, is not always easy. This is especially true when it comes to dealing with ICs. The trick to freeing a component from the solder's hold is to first melt the solder and then to remove the solder with an aspiration tool, or "sucker." An aspiration tool typically resembles a turkey baster or a large syringe-like device. Another method that can be used to remove solder is to use a solder wick. This device resembles a braided copper wire and acts to draw solder away from a connection by means of capillary action.

14.2.9   Enclosing the Circuit

Circuits typically are enclosed in either an aluminum or plastic box. Aluminum enclosures are often used when designing high-voltage devices, whereas plastic containers typically are used for lower-voltage applications. If you design a high-voltage circuit housed in an aluminum box, make sure to ground the box so as to avoid getting shocked.

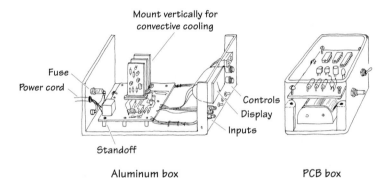

FIGURE 14.9

Circuit boards that are placed within an aluminum box should be supported off the ground with standoffs. If the circuit is ac powered, drill a hole through the backside of the box, and insert a strain relief (e.g., rubber grommet) around the edges of the hole before feeding through the cable. Place frequently used switches, dials, and indicators on the front panel, and place seldom-used switches and fuses on the back panel. If you expect that your circuit will be generating a lot of heat (running more than 10 W or so), consider installing a blower fan. For circuits running on moderately low power, simple perforated holes placed on the top and bottom of the box will aid in conductive cooling. Place major heat-producing components, such as power resistors and transistors, toward the back of the box, connecting them through the back panel to heat sinks. Make sure to orient heat sinks with their fins in the vertical direction. Also, if you are building a multiple-board cage device, align all boards vertically to allow for efficient ventilation.

Plastic boxes usually come with built-in standoffs on which to rest the circuit board. Typically these boxes allow for extra room underneath the board for such items as batteries and speakers.

14.2.10   Useful Items to Keep Handy

The following items are worth having at your workbench: needle-nose pliers, snippers, solder, soldering iron, solder sucker, IC inserter, lead bender, solvent, clip-on heat sinks, circuit-board holder, screws (flathead and roundhead), nuts, flat and lock washers (4-40, 6-32, 10-24), binding posts, grommets, standoffs, cable clamps, line cord, hookup wire, shrink tubing in assorted sizes, eyelets, and fuse holders.

14.2.11 Troubleshooting the Circuits You Build

If your circuit is malfunctioning, see if you have overlooked any suggestion in the troubleshooting flowchart in Fig. 14.10.

FIGURE 14.10

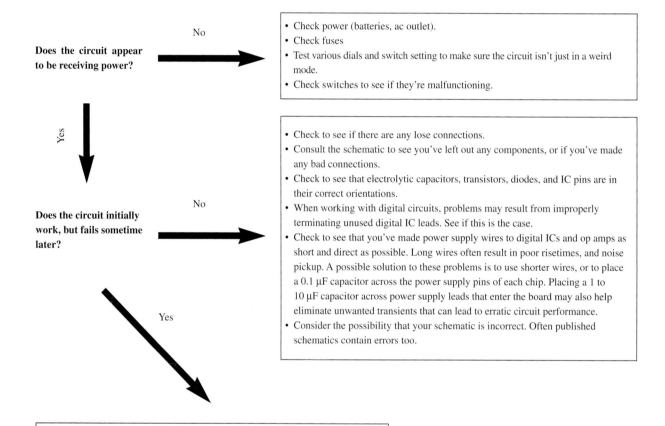

STARTING POINT

Does the circuit appear to be receiving power?

No →

- Check power (batteries, ac outlet).
- Check fuses
- Test various dials and switch setting to make sure the circuit isn't just in a weird mode.
- Check switches to see if they're malfunctioning.

Yes ↓

Does the circuit initially work, but fails sometime later?

No →

- Check to see if there are any lose connections.
- Consult the schematic to see you've left out any components, or if you've made any bad connections.
- Check to see that electrolytic capacitors, transistors, diodes, and IC pins are in their correct orientations.
- When working with digital circuits, problems may result from improperly terminating unused digital IC leads. See if this is the case.
- Check to see that you've made power supply wires to digital ICs and op amps as short and direct as possible. Long wires often result in poor risetimes, and noise pickup. A possible solution to these problems is to use shorter wires, or to place a 0.1 µF capacitor across the power supply pins of each chip. Placing a 1 to 10 µF capacitor across power supply leads that enter the board may also help eliminate unwanted transients that can lead to erratic circuit performance.
- Consider the possibility that your schematic is incorrect. Often published schematics contain errors too.

Yes ↓

Check to see if any components are becoming excessively hot to the touch. Do you smell anything burning? If so, the problem may be that you're overheating a component. In such a case, simply replacing the component with another that has a higher power rating may be the answer. Or perhaps, a heat-sink is all that's needed.

If all else fails . . . →

Attempt to isolate the area of the circuit were you think the problem lays. Perhaps, simply replacing a suspect part may be the answer. If you're still in doubt, consult literature that talks about circuits similar to yours. These sources often mention possible sources of error, and then give you a solution.

14.3 Multimeters

A *multimeter,* or *VOM* (volt-ohm-milliammeter), is an instrument that is used to measure current, voltage, and resistance. The two most common types of multimeters include the analog VOM and the digital VOM, as shown in Fig. 14.11.

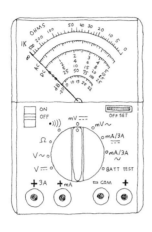

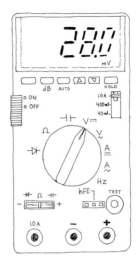

FIGURE 14.11   Analog multimeter   Digital multimeter

The obvious difference between the two types of VOMs is that an analog VOM uses a moving-pointer mechanism that swings along a calibrated scale, while the digital VOM uses some complex digital circuitry to convert input measurements into a digitally displayed reading. Technically speaking, analog VOMs are somewhat less accurate than digital VOMs (they typically have a 3 percent higher error in reading than a digital VOM), and they are a harder to read. Also, the resolution (displayable accuracy) for an analog VOM is roughly 1 part in 100, as compared with a 1 part in 1000 resolution for a digital VOM. Despite these limitations, analog VOMs are superior to digital VOM when it comes to testing circuits that contain considerable electrical noise. Unlike digital VOMs, which may go blank when noise is present, analog VOMs are relatively immune to such disturbances.

14.3.1 Basic Operation

Measuring Voltages

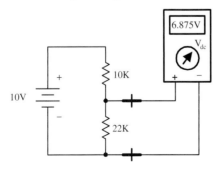

FIGURE 14.12

The trick to measuring voltages with a VOM is to turn the selector knob to the voltage setting. If you want to measure a dc voltage, the knob is turned to the appropriate dc voltage-level setting. If you wish to measure an ac voltage, the knob is turned to the ac voltage setting (V_{ac}, or V_{rms}). Note that the displayed voltage in the V_{ac} setting is the rms voltage ($V_{rms} = 0.707$ $V_{peak-to-peak}$). Once the VOM is set correctly, the voltage between two points in a circuit can be measured by touching the VOM's probes to these points (the VOM is placed in parallel). For example, Fig. 14.12 shows the procedure used to measure the voltage drop across a resistor.

Measuring Currents

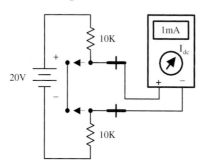

FIGURE 14.13

Measuring currents with a VOM is almost as easy as measuring voltages. The only difference (besides changing the setting) is that you must break the test circuit at the location were you wish to make a current reading. Once the circuit is open, the two probes of the VOM are placed across the break to complete the circuit (VOM is placed in series). Figure 14.13 shows how this is done. When measuring ac currents, the VOM must be set to the rms current setting.

Measuring Resistances

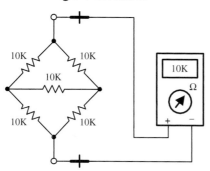

FIGURE 14.14

Measuring resistances with a VOM is simple enough—remove the power to the resistive section of interest, and then place the VOM's probes across this section. Of course, make sure to turn the VOM selector knob to the ohms setting beforehand.

14.3.2 How Analog VOMs Work

An analog VOM contains an ammeter, voltmeter, and ohmmeter all in one. In principle, understanding how each one of these meters works individually will help to explain how an analog VOM works as a whole.

Ammeter

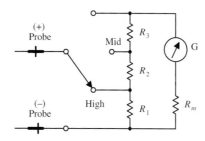

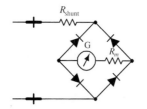

FIGURE 14.15

An ammeter uses a D-Arsonval galvanometer that consists of a current-controlled electromagnetic that imparts a torque on a spring-loaded rotatable needle. The amount of deflection of the needle is proportional to the current flow through the electromagnet. The electromagnetic coil has some resistance built in, which means you have to throw R_m into the circuit, as shown in Fig. 14.15. (R_m is typically around 2 k or so). Now, a galvanometer alone could be used to measure currents directly; however, if the input current is excessively large, it will force the needle beyond the viewable scale. To avoid this effect, a number of *shunt resistors* placed in parallel with the galvanometer make up a current divider capable of diverting some of the "needle bending" current away from the galvanometer. The current value read from the display must be read from the appropriate ruler marking on the display that correspond to the shunt resistance chosen. To make this device capable of measuring ac currents, a bridge rectifier can be incorporated into the design so as to provide a dc current to the galvanometer (see lower circuit). The dc current will produce a needle swing that is proportional to the alternating voltage measured. A typical ammeter has about a 2-k input resistance. Ideally, an ammeter should have zero input resistance.

Voltmeter

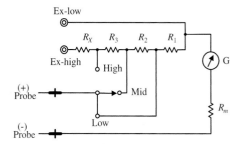

FIGURE 14.16

An analog voltmeter, like the ammeter, also uses a D-Arsonval galvanometer. Again, the galvanometer has some internal resistance (R_m). When the voltmeter's leads are placed across a voltage difference, a current will flow from the higher potential to lower potential, going through the galvanometer in the process. The current flow and the needle deflection are proportional to the voltage difference. Again, like the ammeter, shunt resistors are used to calibrate and control the amount of needle deflection. To make ac voltage measurements, a bridge rectifier, like the one shown in the last example, can be incorporated into the meter's design. A typical voltmeter has an input resistance of 100 k. An ideal voltmeter should have infinite input resistance.

Ohmmeter

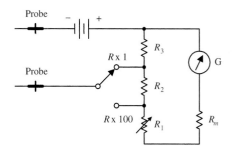

FIGURE 14.17

To measure resistance, an ohmmeter uses an internal battery to supply a current through the measured load and through a galvanometer (the load and galvanometer are in series). If the tested load is small, a large current will flow through the galvanometer, and a large deflection will occur. However, if the tested load resistance is large, the current flow and the deflection will be small. (In a VOM, the ohmmeter calibration markings are backwards—0 Ω is set to the right of the scale.) The amount of current flow through the galvanometer is proportional to the load resistance. The ohmmeter is first calibrated by shorting the probe leads together and then zeroing the needle. Like the other meters, an ohmmeter uses a number of shunt resistors to control and calibrate the needle deflections. A typical ohmmeter has an input resistance of about 50 Ω. An ideal ohmmeter should have zero input resistance.

14.3.3 How Digital Multimeters Work

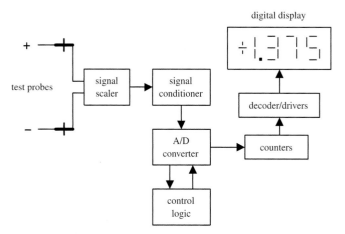

FIGURE 14.18

A digital multimeter is composed of a number of functional groups, as shown in the block diagram. The signal scaler is an attenuator amplifier that acts as a range selector. The signal conditioner converts the scaled input signal to a dc voltage within the range of the analog-to-digital converter (A/D converter). In the case of ac voltage measurements, the ac voltage is converted into a dc voltage via a precision rectifier-filter combination. The gain of the active filter is set to provide a dc level equal to the rms value of the ac input voltage or current. The signal conditioner also contains circuits to convert current and/or resistance into proportional dc voltages. The A/D converter converts the dc analog input voltage into a digital output voltage. The digital display provides a digital readout of the measured input. Control logic is used to synchronize the operation of the A/D converter and digital display.

14.3.4   A Note on Measurement Errors

When measuring the current through (or voltage/resistance across) a load, the reading obtained from the VOM will always be different when compared with the true value present before the meter was connected. This error comes from the internal resistance of the VOM. For each setting (ammeter, voltmeter, ohmmeter), there will be a different internal resistance. A real ammeter typically will have a small input resistance of around 2 k, while a voltmeter may have an internal resistance of 100 k or more. For an ohmmeter, the internal resistance is usually around 50 Ω. It is crucial to know these internal resistances in order to make accurate measurements. The following examples show how large the percentage error in readings can be for meters with corresponding input resistances.

Current-Measurement Error

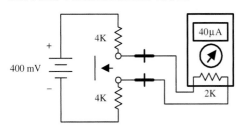

FIGURE 14.19

If an ammeter has an input resistance of 2 k, calculate the percentage error in reading for the circuit shown here.

$$I_{\text{true}} = \frac{400\text{ mV}}{4\text{ k} + 4\text{ k}} = 50\ \mu\text{A}$$

$$I_{\text{measured}} = \frac{400\text{ mV}}{4\text{ k} + 4\text{ k} + 2\text{ k}} = 40\ \mu\text{A}$$

$$\%\text{ error} = \frac{50\ \mu\text{A} - 40\ \mu\text{A}}{50\ \mu\text{A}} \times 100\% = 20\%$$

Voltage-Measurement Error

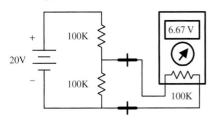

FIGURE 14.20

If a voltmeter has an input resistance of 100 k, calculate the percentage error in reading for the circuit shown here.

$$V_{\text{true}} = \frac{100\text{ k}}{100\text{ k} + 100\text{ k}}(20\,\text{V}) = 10\,\text{V}$$

$$V_{\text{measured}} = \frac{100\text{ k}}{100\text{ k} + (100\text{ k} \times 100\text{ k})/(100\text{ k} + 100\text{ k})} = 6.67\,\text{V}$$

$$\%\text{ error} = \frac{10\,\text{V} - 6.67\,\text{V}}{10\,\text{V}} \times 100\% = 33\%$$

Resistance-Measurement Error

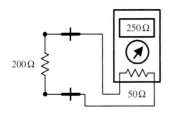

FIGURE 14.21

If an ohmmeter has an input resistance of 50 Ω, calculate the percentage error in reading for the circuit shown here.

$$R_{\text{true}} = 200\ \Omega$$

$$R_{\text{measured}} = 200\ \Omega + 50\ \Omega = 250\ \Omega$$

$$\%\text{ error} = \left|\frac{200\ \Omega - 250\ \Omega}{200\ \Omega}\right| \times 100\% = 25\%$$

To minimize the percentage error, an ammeter's input resistance should be less than the Thevenin resistance of the original circuit by 20 times or more. Conversely, a voltmeter should have an input resistance that is larger than the Thevenin resistance of the original circuit by 20 times or more. The same goes for the ohmmeter; it should have an input resistance that is at least 20 times the Thevenin resistance of the original circuit. By following these simple rules, it is possible to reduce the error to below 5 percent. Another approach (perhaps a bit more tedious) is to look up the internal resistances of your VOM, make your measurements, and then add or subtract the internal resistances afterwards.

14.4 Oscilloscopes

Oscilloscopes measure voltages, not currents and not resistances, just voltages—an important point to get straight from the start. An oscilloscope is an extremely fast *xy* plotter capable of plotting an input signal versus time or versus another input signal. The screen of an oscilloscope looks a lot like the screen of a television. As a signal is supplied to the input of a scope, a luminous spot appears on the screen. As changes in the input voltage occur, the luminous spot responds by moving up or down, left or right. In most applications, the oscilloscope's vertical-axis (*y*-axis) input receives the voltage part of an incoming signal and then moves the spot up or down depending on the value of the voltage at a particular instant in time. The horizontal axis (*x* axis) is usually used as a time axis, where an internally generated linear ramp voltage is used to move the spot across the screen from left to right at a rate that can be controlled by the operator. If the signal is repetitive, such as a sinusoidal wave, the oscilloscope can make the sinusoidal pattern appears to stand still. This makes a scope a useful tool for analyzing time-varying voltages.

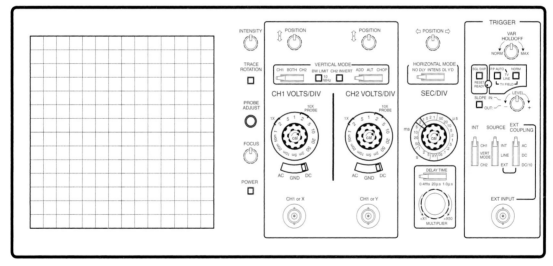

FIGURE 14.22

Even though oscilloscopes only measure voltage, it is possible to convert quantities such as current, strain, acceleration, pressure, etc., into voltages that the scope can use. To convert a current into a voltage, a resistor of known resistance is used; the current is measured indirectly by measuring the voltage drop across the resistor and then applying Ohm's law. To convert strain, movement, etc. into voltage requires the use of transducers (electromechanical devices). By applying some calibrating tricks, the magnitude of, say, a pressure applied to a pressure transducer can be measured accurately.

14.4.1 How Scopes Work

An oscilloscope is built around a cathode-ray tube. All the circuits inside the scope are designed to take an input signal and modify it into a set of electrical instructions that supply the tube's electron gun with aiming instructions (location where to focus the beam). Most of the knobs and switches on the face of a scope that are connected to the interior circuitry are designed to help modify the instructions sent to the cathode-ray tube. For example, these controls set voltage scale, time scale, intensity of beam, focus of beam, channels selection, triggering, etc.

Cathode-Ray Tube

A cathode-ray tube consists of an electron gun (filament, cathode, control grid, and anode), a second anode, vertical deflection plates, horizontal deflection plates, and a phosphor-coated screen. When current flows through the filament, the filament heats the cathode to a point where electrons are emitted. The control grid controls the amount of electrons that flow through the electron gun and thus controls the intensity of the beam. If this grid is made more negative in voltage, more electrons will be repelled away from the grid, thus reducing the electron flow. The electron beam is focused into a sharp pointlike beam by applying a controlling voltage or focus voltage to anode 1. The second anode is supplied with a large voltage that is used to give the electrons within the beam the additional momentum needed to cause a photon emission once they collide with the phosphor screen. The beam-focusing section of the tube is referred to as the *electron gun*. There are two sets of electrostatic deflection plates (vertical and horizontal) that are set between the second anode and the inner face of the phosphor screen. One set of plates is used to deflect the electron beam vertically; the other set is used to deflect the beam horizontally. For example, when one of the plates of a pair of plates is made more negative in charge than the other, the electron beam will bend away from the negative plate and veer toward the more positive plate. (The electrons in the beam are usually moving with sufficient forward velocity that they never actually come in contact with the plates.) When a sawtooth voltage is applied to the horizontal plates, the gradually rising potential across the plates pulls the electron beam from the negative plate to the positive plate, causing the beam to scan across the phosphor screen. The vertical plates cause the electron beam to move up and down.

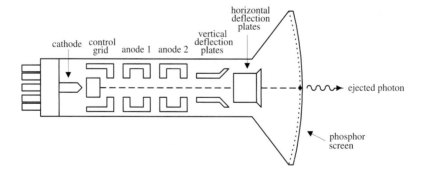

FIGURE 14.23

The next step in understanding how an oscilloscope works is understanding how an incoming signal is converted into a set of electrical signals or applied voltages that control the beam-aiming mechanisms of the cathode-ray tube. This is where the interior circuitry comes in.

Interior Circuitry of a Scope

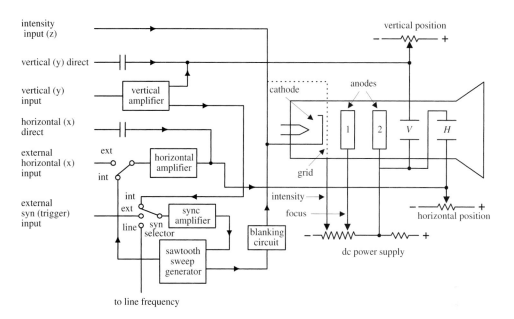

FIGURE 14.24

Let's take a sinusoidal signal and see how the interior circuits of a scope convert it into something you can see on a display. The first thing you do is apply the sinusoidal signal to the vertical input. From the vertical input, the sinusoidal signal is sent to a vertical amplifier, where it is amplified so that it can supply enough voltage to deflect the electron beam. The vertical amplifier then sends a signal to the sweep selector. When the sweep selector is switched to the internal position (the other positions will be explained later), the signal from the vertical amplifier will enter the sync amplifier. The sync amplifier is used to synchronize the horizontal sweep (sawtooth in this case) with the signal under test. Without the sync amplifier, the display pattern would drift across the screen in a random fashion. The sync amplifier then sends a signal to the sawtooth sweep generator, telling it to start a cycle. The sawtooth sweep generator then sends a sawtooth signal to a horizontal amplifier (when horizontal input is set to internal). At the same time, a signal is sent from the sawtooth sweep generator to the blanking circuit. The blanking circuit creates a high negative voltage on the control grid (or high positive voltage on the cathode-ray tube cathode), which turns off the beam as it snaps back to the starting point. Finally, voltages from the vertical and horizontal amplifiers (sawtooth) are sent to the vertical and horizontal plates in a synchronized fashion; the final result is a sinusoidal pattern displayed on the scope's screen.

The other features—such as vertical direct and horizontal direct inputs, external horizontal input, external trigger, line frequency, xy mode, etc.—are described later in this chapter. It is important to note that the scope does not always use a sawtooth voltage applied to the horizontal plates. You can change the knobs and inputs and use another input signal for the horizontal axis. Controls such as intensity, focus, and horizontal and vertical position of the beam can be understood by looking at the oscilloscope circuit diagram.

14.4.3   Aiming the Beam

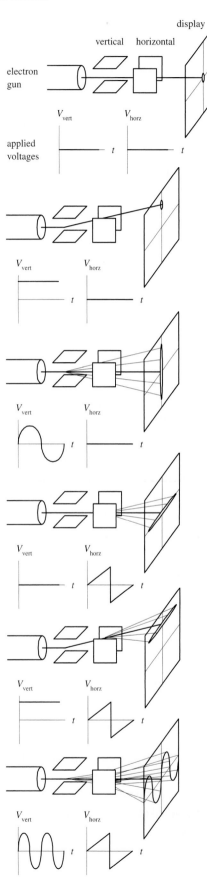

When no voltages are applied to the horizontal and vertical plates, the electron beam is focused at the center of the scope's display.

When a dc voltage is applied to the vertical plates, while no voltage is applied to the horizontal plates, the electron beam shifts up or down depending on the sign of the applied voltage.

When a sinusoidal voltage is applied to the vertical plates, while no voltage is applied to the horizontal plates, a vertical line is traced on the y axis.

When a sawtooth voltage is applied to the horizontal plates, while no voltage is applied to the vertical plate, the electron beam traces a horizontal line from left to right. After each sawtooth, the beam jumps back to the left and repeats its left-to-right sweep.

When a dc voltage is applied to the vertical plates, while a sawtooth voltage is applied to the horizontal plates, a horizontal line is created that is shifted up or down depending on the sign of the dc vertical plate voltage (+ or −).

When a sinusoidal voltage is applied to the vertical plates and a sawtooth voltage is applied to the horizontal plates, the electron beam moves up as the signal voltage increases and at the same time moves to the left as the sawtooth voltage is applied to the horizontal plates. The display gives a sinusoidal graph. If the applied sinusoidal frequency is twice that of the sawtooth frequency, two cycles appear on the display.

FIGURE 14.25

14.4.4 Scope Applications

DC Voltmeter

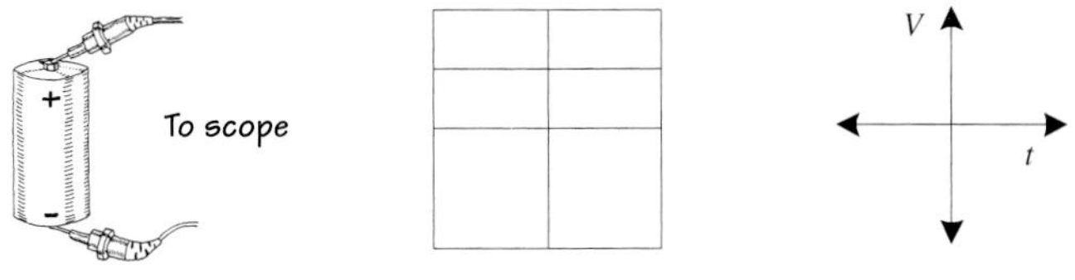

**AC Voltmeter/
AC Frequency
Meter**

T = period

f = frequency

$f = 1/T$

$V_{rms} = \dfrac{1}{\sqrt{2}}\,V_{max}$

**Phase Relationships
Between Two Signals**

source 1 → chan 1
source 2 → chan 2

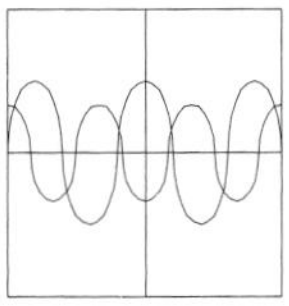

The scope can be used to compare two source signals (e.g., measure phase shifts, voltage and frequency differences, etc.).

Digital Measurements

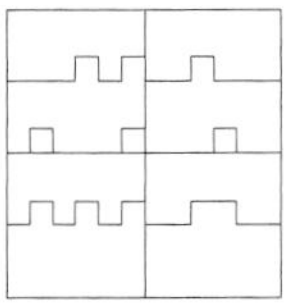

A scope can be used to create timing diagrams for digital circuits.

xy Graphics (xy Mode)

chan 1 input → x-axis
chan 2 input → y-axis

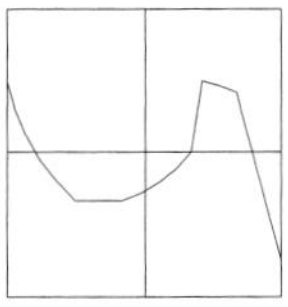

The scope no longer uses the x axis as the time axis but uses signal voltage from another external source.

Measurements Using Transducers

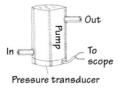

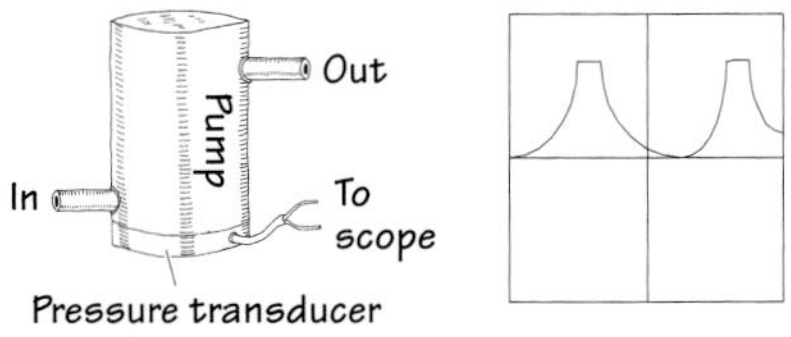

By using transducers to convert an input quantity, such as pressure, into a voltage, the scope can be transformed into a pressure meter.

y axis → pressure

x axis → time

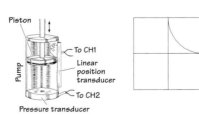

Here, a scope is used in xy mode, where

y axis → pressure

x axis → piston

FIGURE 14.26

14.4.5 What All the Little Knobs and Switches Do

Figure 14.27 shows a typical layout of an oscilloscope control panel. The control panel of the scope you use may look slightly different (knob positions, digital display, number of input channels, etc.), but the basic ingredients are the same. If you do not find what you need in this section, refer to the oscilloscope user manual that comes with the scope.

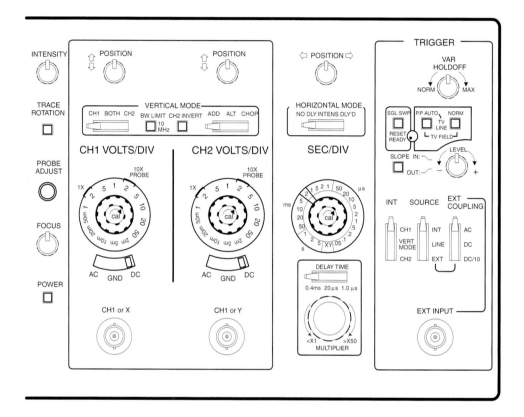

FIGURE 14.27

The control panel of an oscilloscope is broken down into the following sections:

Vertical mode. This section of the scope contains all the knobs, buttons, etc., that are used to control the vertical graphics of the scope, most of which are associated with the voltage amplitude of an incoming signal.

Horizontal mode. This section of the scope contains all the knobs, buttons, etc. that control the horizontal portion of the graphics display, which usually are associated with the time base for the scope.

Trigger mode. This section of the scope contains all the knobs, buttons, etc. that control the way in which the scope "reads" an incoming signal. This section of the scope is probably the most technical. To understand triggering, read the next section.

Figuring Out What the Knobs Do

VERTICAL MODE

- *CH1, CH2 coaxial inputs.* Where input signals enter the scope.
- *AC, GRD, DC switches:*

    AC. Blocks dc component of the signal, passing only the ac part of the signal.

    DC. Measures direct input of both ac and dc components of the input signal.

GRD. Grounds input, causing vertical plates in cathode-ray tube to become uncharged, thus eliminating electron beam deflection. Used to recalibrate vertical component of the electron beam to a reference position on the display after altering the vertical position of the knob.

- *CH1 VOLTS/DIV, CH2 VOLTS/DIV knobs.* Used to set the voltage scale on the display. For example, 5 volts/div means that each division (1 cm) on the display is 5 V high.

- *MODE switches:*

  CH1, BOTH (DUAL), CH2 switch. This switch allows you to pick between displaying a signal from channel 1 or channel 2 or allows you to display both channels at the same time.

  NORM, INVERT. This switch lets you choose to display a signal normally or inverted.

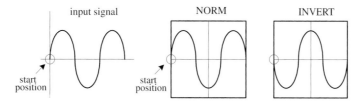

FIGURE 14.28

  ADD, ALT, CHOP:

  - *ADD.* Adds signals from channel 1 and channel 2 together arithmetically.

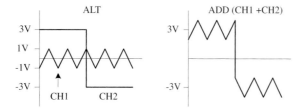

FIGURE 14.29

  - *ALT.* Alternate sweep is selected regardless of sweep time, and the NORM-CHOP switch has no effect.
  - *CHOP.* Operates the triggering SOURCE switch, providing automatic or manual selection of alternate or chop method of dual-trace sweep generation.

- *POSITION knob.* This knob allows you to move the displayed image up or down on the screen.

- *XY mode.* When selected, the sweep rate (time base) supplied by the scope is switched off, and an external signal voltage applied to the channel 2 input replaces it.

HORIZONTAL MODE

- *SEC/DIV knob.* This knob sets the sweep speed or the scale for the horizontal time display. For example, 0.5 ms/DIV means each division (1 cm) on the display is 0.5 ms wide.

- *MODE switches:*

  NO DLY. This setting takes the horizontal signal and presents it to display immediately.

  DLY'D. This setting delays the horizontal signal for a time you specify on the delay time section of scope. Use this to set the delay time of a signal.

- *SWEEP-TIME variable control.* Sometimes known as *sweep frequency control, fine frequency control,* or *frequency vernier.* Used as a fine sweep-time adjustment. In the extreme clockwise (CAL) position, the sweep-time is calibrated by using the SWEEP TIME/CM switch. In the other positions, the variable control provides a continuously variable sweep rate.

- *POSITION knob.* Moves the horizontal display left or right. This feature is useful when comparing two input signals; it allows you to align the wave patterns for comparison.

TRIGGER MODE

- *EXT TRIG jack.* Input terminal for external trigger signals.

- *CAL terminal.* Provides a calibrated 1-kHz, 0.1-V peak-to-peak squarewave signal. The signal can be used to calibrate the vertical-amplifier attenuators and to check frequency compensation of probes used with the scope.

- *HOLDOFF control.* Used to adjust holdoff time (trigger inhibit period beyond sweep duration).

- *TRIGGERING mode switch:*

  SINGLE. When a signal is nonrepetitive, or if it varies in amplitude, shape, or time, a conventional repetitive display can produce an unstable presentation. SINGLE enables the RESET switch for triggered single sweep operation. The signal sweep can be used to photograph a nonrepetitive signal. Pushing the RESET button initiates a single sweep that begins when the next sync trigger occurs.

  NORM. Used for triggered sweep operation. The triggering threshold is adjustable by means of the triggering LEVEL control. No sweep is generated in the absence of the triggering signal or if the LEVEL control is set in such a way as to allow the threshold to exceed the amplitude of the triggering signal (see Fig. 14.31).

  AUTO. Selects automatic sweep operation, where the sweep generator free-runs and generates a sweep without a trigger signal (this is often referred to as *recurrent sweep operation*). In AUTO, the sweep generator automatically switches to triggered sweep operation if an acceptable trigger signal is present. The AUTO position is useful when first setting up the scope in order to observe a waveform; it provides sweep for waveform operation until other controls can be set properly. DC AUTO sweep must be used for dc measurements and for signals of such low amplitude that the sweep is not triggered.

  FIX. This mode is the same as the AUTO mode, except that triggering always occurs at the center of the sync trigger waveform regardless of the LEVEL control setting.

• *SLOPE button.* Selects the point at which a scope will trigger. When positive slope is selected, the scope will only begin a sweep when the signal voltage crosses the LEVEL voltage during a positive sloping rise. (LEVEL voltage is explained a few lines down.) A negative slope setting initiates a sweep to occur when the signal crosses the LEVEL voltage during a negative sloping fall (see Fig. 14.30).

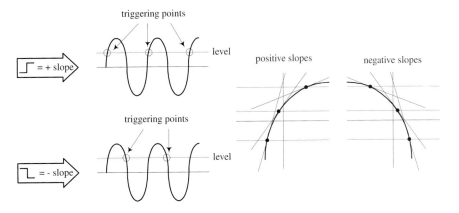

FIGURE 14.30

• *RESET button.* When triggering MODE switch is set to SINGLE, pushing the RESET button initiates a single sweep that begins when the next sync trigger occurs.

• *READY/TRIGGER indicator.* In SINGLE trigger mode, an indicator light turns on when the RESET button is pushed in, indicating that a sweep is beginning. The light turns off when the sweep is completed. In the NORM, AUTO, and FIX triggering modes, the indicator turns on for the duration of the triggered sweep. The indicator also shows when the LEVEL control is set properly to obtain triggering.

• *LEVEL knob.* Used to trigger a sweep. LEVEL sets the point when scope will trigger based on the amplitude of the applied signal. The level can be shifted up or down. The READY/TRIGGER indicator turns on when the sweep is triggered, indicating that the triggering LEVEL control is within the proper range (see Fig. 14.31).

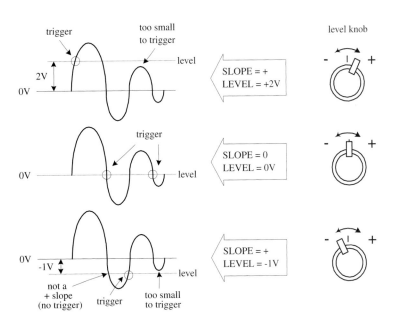

FIGURE 14.31

- *COUPLING switch.* Used to select input coupling for sync trigger signal.

    AC. This is the most commonly used position. AC position permits triggering from 10 Hz to over 35 MHz and blocks any dc component of the sync trigger signal.

    LF REJ. DC signals are rejected, and signals below 10 kHz are attenuated; sweep is triggered only by the higher-frequency components of the signal. Useful for providing stable triggering when the trigger signal contains line-frequency components, such as 60-Hz hum.

without LF REJ   with LF REJ

FIGURE 14.32
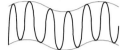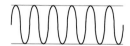

    HF REJ. Attenuates signals above 100 kHz. Used to reduce high-frequency noise or to initiate a trigger from the amplitude of a modulated envelope rather than the carrier frequency.

without HF REJ   with HF REJ

FIGURE 14.33
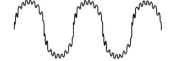

    VIDEO. Used to view composite video signals. Use in television and VCR service.

    DC. Permits triggering from dc to over 35 MHz. DC can be used to provide stable triggering for low-frequency signals that would otherwise be attenuated if measured in the ac setting. The LEVEL control can be adjusted to provide triggering at the desired dc level on the waveform.

14.4.6 Measuring Things with Scopes

A scope's buttons and knobs must be set to a particular set of positions for accurate measurements to occur. If just one of these buttons or switches is set wrong, things can go haywire. You must make sure every button is set correctly.

In the next few pages, a number of applications for oscilloscopes will be covered. When I start to discuss a new application, such as making phase measurements between two signals, I frequently will tell you to set the scope to the initial settings listed below. Afterwards, I will tell you which buttons and knobs to adjust to put the scope in the proper configuration needed for that particular application.

Initial Scope Settings

STEP ONE

1. *Power switch:* off
2. *Internal recurrent sweep (TRIGGER MODE switch):* off (NORM or AUTO position)
3. *Focus:* lowest setting
4. *Gain:* lowest setting

5. *Intensity:* lowest setting

6. *Sync controls* (LEVEL, HOLDOFF): lowest settings

7. *Sweep selector:* external (EXT)

8. *Vertical position control:* midpoint

9. *Horizontal position control:* midpoint

STEP TWO

1. *Power switch:* on.

2. *Focus:* until beam is in focus.

3. *Intensity:* desired luminosity.

4. *Sweep selector:* internal (use the linear internal sweep if more than one sweep is available).

5. *Vertical position control:* until beam is centered on display.

6. *Horizontal position control:* until beam is centered on display.

7. *Internal recurrent sweep:* on. Set the sweep frequency to any frequency above 100 Hz.

8. *Horizontal gain control:* Check that luminous spot has expanded into a horizontal trace or line. Return horizontal gain control to zero or lowest setting.

9. *Internal recurrent sweep:* off.

10. *Vertical gain control:* to midpoint. Touch the vertical input with a finger. The stray signal pickup should cause the spot to be deflected vertically into a trace or line. Check that the line length is controllable by adjusting the vertical-gain control. Return vertical-gain control to zero or lowest setting.

11. *Internal recurrent sweep:* on. Advance the horizontal-gain control to expand the spot into a horizontal line.

Measuring a Sinusodial Voltage Signal

1. Connect the equipment as shown in Fig. 14.34.

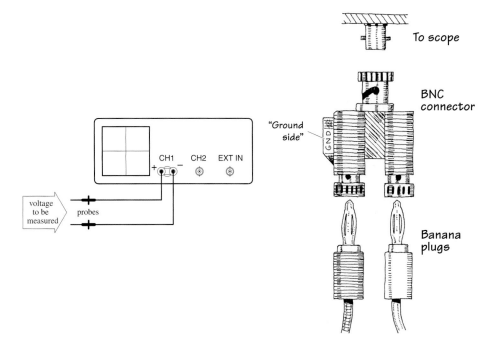

FIGURE 14.34

2. Set scope to the *initial settings* (see pp. 450–451).

3. Fiddle with vertical VOLT/DIV knob until signal comes into view.

4. Set input selector (AC/GRD/DC) to ground (GRD).

5. Switch scope to internal recurrent sweep. Fiddle with the SEC/DIV knob until electron beam is tracing out a desired path of the screen.

6. Now you should have a horizontal line in view. Next, center this line on the *x* axis or some desired reference position by fiddling around with the vertical position knob. (Make sure you do not fiddle around with the vertical position knob after it has been set to the desired reference point. If you do, your measurements will be offset. If you think you have accidentally moved the vertical position line, set input selector to GRD and recalibrate.)

7. Set input selector switch (AC/GRD/DC) to dc. Connect the probe to the signal being measured.

8. Fiddle with vertical and horizontal VOLT/DIV knob and SEC/DIV knob to get signal into view.

9. Once you have an image of your signal on the screen, take a look at your VOLT/DIV and SEC/DIV knobs and record these settings. Now visually measure the period, peak-to-peak voltage, etc. of the displayed image, using the centimeter gridlines on the scope's screen as a ruler. To find the actual voltages and times, multiply the measurement made in centimeters by the VOLT/DIV (or VOLTS/cm) and SEC/DIV (SEC/cm) recorded set values. The example in Fig. 14.35 shows how to calculate the peak-to-peak voltage, root-mean-voltage, period, and frequency of a sinusoidal waveform.

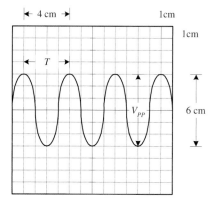

FIGURE 14.35

Peak-to-peak voltage (V_{pp}):

$$V_{pp} = (6 \text{ cm}) \frac{(2 \text{ V})}{(1 \text{ cm})} = 12 \text{ V}$$

Root-mean-square voltage (V_{rms})

$$V_{rms} = \frac{1}{\sqrt{2}} V_{pp} = 8.5 \text{ V}$$

Period (T):

$$T = (4 \text{ cm}) \frac{(10 \text{ ms})}{(1 \text{ cm})} = 40 \text{ ms}$$

Frequency (f):

$$f = \frac{1}{T} = \frac{1}{40 \text{ ms}} = 25 \text{ Hz}$$

Measuring Current

As I mentioned earlier, oscilloscopes can only measure voltages; they do not directly measure currents. However, with the help of a resistor and Ohm's law, you can trick the scope into making current measurements. You simply measure the voltage drop across a resistor of known resistance and let $I = V/R$ do the rest. Typically, the resistor's resistance must be small to avoid disturbing the operating conditions within the circuit that is being measured. A high-precision 1-Ω resistor is often used for such instances.

Let's now take a look at the specifics as to how to measure currents with a scope.

1. Set up equipment as shown in Fig. 14.36.

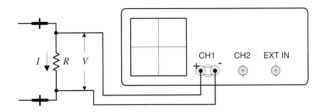

FIGURE 14.36

2. Set scope to *initial settings* (see pp. 450–451).

3. Apply a dc current to be measured through the resistor. Here, we'll use a 1-Ω resistor to make the calculations simple and to avoid altering the dynamics of the circuit under test. The wattage of the resistor must be at least 2Ω times the square of the maximum current (expressed in amps). For example, if the maximum anticipated current is 0.5 A, the minimum wattage of the resistor should be 2Ω × $(0.5\text{ A})^2 = \frac{1}{2}$ W.

4. Measure the voltage drop across the resistor using the scope. The unknown current will equal the magnitude of the voltage measured, provided you stick with the 1-Ω resistor. Figure 14.37 shows some example measurements, two of which describe how to measure rms and total (dc + ac) effective currents.

FIGURE 14.37

DC CURRENT	AC CURRENT	DC + AC CURRENT

20 mV/DIV, $R = 1\ \Omega$

$$I = \frac{V}{R} = \frac{3\text{ cm}(20\text{ mV})/\text{DIV}}{1\ \Omega} = 60\text{ mA}$$

2 mV/DIV, $R = 1\ \Omega$

$$I_{\text{rms}} = \frac{V_{\text{rms}}}{R} = \frac{8\text{ cm}(2\text{ mV})/\text{DIV}}{\sqrt{2}\ (1\ \Omega)}$$
$$= 11.3\text{ mA}$$

1 mV/DIV, $R = 1\ \Omega$

$$I_{\text{tot}} = \sqrt{I_{\text{rms}}^2 + I_{\text{dc}}^2}$$
$$= \sqrt{\left(\frac{1}{\sqrt{2}}\ \frac{4\text{ cm}(1\text{ mV/cm})}{1\ \Omega}\right)^2 + \left(\frac{3\text{ cm}(1\text{ mV/cm})}{1\ \Omega}\right)^2}$$
$$= 4.6\text{ mA}$$

Phase Measurements Between Two Signals

Say you wish to compare the phase relationship between two voltage signals. To do so, apply one of the signals to CH1 and the other to CH2. Then, using the DUAL setting (or BOTH setting), you can display both signals at the same time and then align them side by side to compare the phased difference between them. The specifics of how to do this are explained below.

1. Step up equipment as shown in Fig. 14.38.

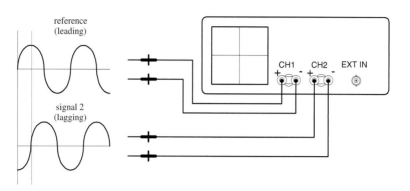

FIGURE 14.38

2. Set the scope to the *initial settings* (see pp. 450–451). *Note:* Cables should be short, of the same length, and should have similar electrical characteristics. At high frequencies, a difference in cable length or a difference in electrical characteristics between cables can introduce improper phase shifts.

3. Switch the scope's internal recurrent sweep to on.

4. Set scope to dual trace (DUAL) mode.

5. Fiddle with the CH1 and CH2 VOLT/DIV settings until both signals are of similar amplitudes. This makes measuring phase differences easier.

6. Determine the phase factor of the reference signal. If one period (360°) of a signal is 8 cm, then 1 cm equals one-eighth of 360°, or 45°. The 45° value represents the phase factor (see Fig. 14.39).

7. Measure the horizontal distance between corresponding points (e.g., corresponding peaks or troughs) of the two waveforms. Multiply this measured distance by the phase factor to get the phase difference (see Fig. 14.39). For example, if the measured difference between the two signals is 2 cm, then the phase difference is $2 \times 45°$, or 90°.

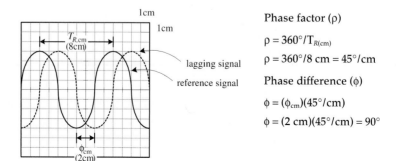

FIGURE 14.39

14.4.7 Scope Applications

The ability for an oscilloscope to "freeze" a high-frequency waveform makes it an incredibly useful instrument for testing electronic components and circuits whose response curves, transient characteristics, phase relationships, and timing relationships are of fundamental importance. For example, scopes are used to study the shape of particular waveforms (e.g., squarewave, sawtooth, etc.). They are used to measure static noise (current variation caused by poor connections between components), pulse delays, impedances, digital signals, etc. The list goes on. Here are a few example scope applications.

Checking Potentiometers for Static Noise

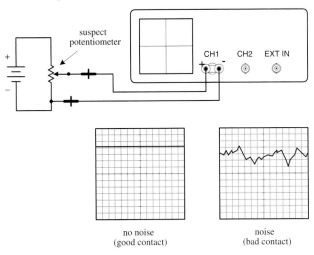

Here, a scope is used to determine if the sliding contact of a potentiometer is faulty. A good potentiometer will present a solid voltage line on the scope's screen, whereas a bad potentiometer will present a noisy pattern on the display. Before concluding that a potentiometer is bad, make sure that noise was not present beforehand. For example, the cables used in this test may have been at fault.

FIGURE 14.40

Pulse Measurements

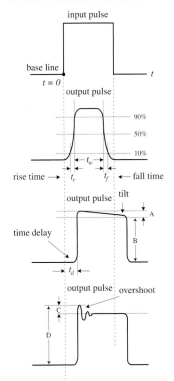

FIGURE 14.41

Scopes are often used to study how square pulses change as they pass through a circuit. This figure, along with the definitions below, shows some of the pulse alterations that can occur.

PULSE DESCRIPTIONS

Rise time (t_r). The time interval during which the amplitude of the output pulse changes from 10 to 90 percent of the maximum value.

Fall time (t_f). The time interval during which the amplitude of the output pulse changes from 90 to 10 percent of maximum value.

Pulse width (t_w). The time interval between the two 50 percent maximum values of the output pulse.

Time delay (t_d). The time interval between the beginning of the output pulse ($t = 0$) and the 10 percent maximum value of the output pulse.

Tilt. A measure of the fall of the upper portion of the output pulse.

$$\text{Percent tilt} = \frac{A}{B} \times 100\%$$

Overshoot. A measure of the how much of the output pulse exceeds the upper portion of the input pulse.

$$\text{Percent overshoot} = \frac{C}{D} \times 100\%$$

Measuring Impedances

The following method of measuring impedance makes use of comparing the reflected pulse with the output pulse. When the output signal travels down a transmission line, part of the signal will be reflected and sent back along the line to the source whenever the signal encounters a mismatch or difference in impedance. The line has a characteristic impedance. If the line impedance is greater than the source impedance (thing being measured), the reflected signal will be inverted. If the line impedance is lower than the source impedance, the reflected signal will not be inverted.

1. Set up the equipment as shown in Fig. 14.42.

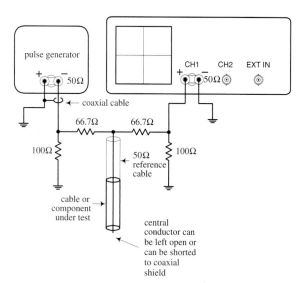

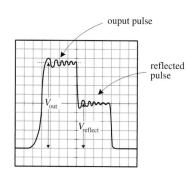

FIGURE 14.42

2. Set knobs and switches to the *initial settings* (see above).

3. Switch internal recurrent sweep on.

4. Set sweep selector to INTERNAL.

5. Set sync selector to INTERNAL.

6. Switch on pulse generator.

7. Fiddle with the VOLT/DIV, SEC/DIV knobs until output pulse is displayed.

8. Observe the output and reflected pulses on the scope. Measure the output voltage (V_{out}) and the reflected voltage ($V_{reflect}$).

9. To find the unknown impedance, use the following equation:

$$Z = \frac{50\ \Omega}{2V_{out}/V_{reflect}} - 1$$

The 50-Ω value represents the characteristic impedance of the coaxial reference cable.

Digital Applications

INPUT/OUTPUT RELATIONSHIPS

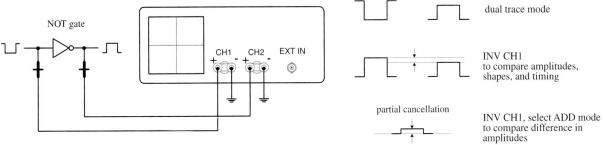

dual trace mode

INV CH1
to compare amplitudes,
shapes, and timing

partial cancellation

INV CH1, select ADD mode
to compare difference in
amplitudes

FIGURE 14.43

CLOCK TIMING RELATIONSHIPS

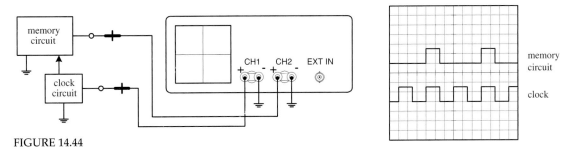

memory
circuit

clock

FIGURE 14.44

FREQUENCY-DIVISION RELATIONSHIPS

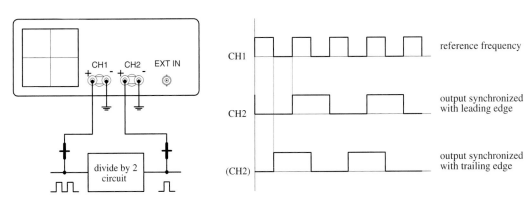

CH1 — reference frequency

CH2 — output synchronized
with leading edge

(CH2) — output synchronized
with trailing edge

FIGURE 14.45

CHECKING PROPAGATION TIME DELAY

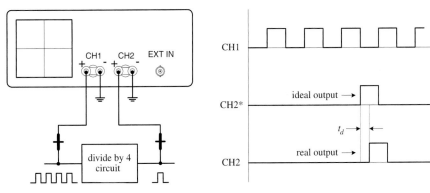

CH1

CH2* — ideal output →

t_d →

CH2 — real output →

FIGURE 14.46

CHECKING LOGIC STATES

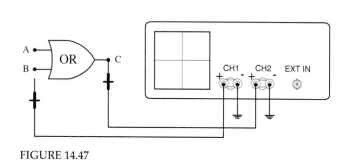

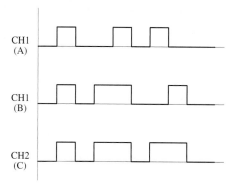

FIGURE 14.47

Power Distribution
and Home Wiring

A.1 Power Distribution

Figure A.1 shows a typical power-distribution system found in the United States (region in California). The voltages listed are sinusoidal and are represented by their rms values. Note that the system in your area may look a bit different from the system shown here. Contact your local utility to learn about how things are set up in your region.

As a note, ac is used in electrical distribution instead of dc because ac can be stepped up or down easily by using a transformer. Also, over long distances, it is more efficient to send electricity via high-voltage/low-current transmission lines. By reducing the current, less power is lost to resistive heating during transmission (lowering I within $P = I^2 R$ lowers P). Once electricity reaches a substation, it must be stepped down to a safe level before entering homes and businesses.

Notice that industry typically uses three-phase electricity. The natural sequencing of the three phases is particularly useful for devices that perform rhythmic tasks. For example, three-phase electric motors (found in grinders, lathes, welders, air conditioners, and other high-power devices) often turn in near synchrony with the rising and falling voltages of the phases. Also, with three-phase electric power, there is never a time when all three phases are at the same voltage. With single-phase power, whenever the two phases have the same voltage, there is temporarily no electric power available. This is why single-phase electric devices must store energy to carry them over these dry spells. In three-phase power, a device can always obtain power from at least one pair of phases.

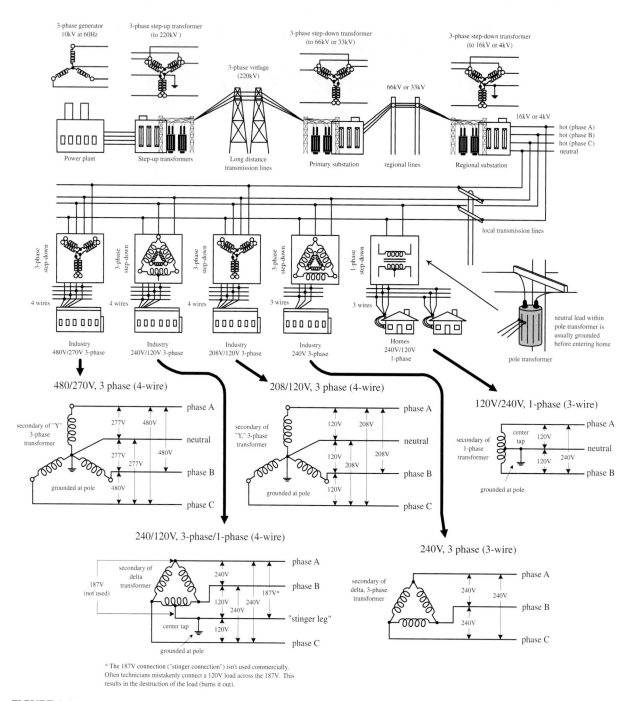

FIGURE A.1

A.2   A Closer Look at Three-Phase Electricity

The simple generator shown in Fig. A.2 can be used to generate single-phase voltage. As the magnet is rotated by a mechanical force, a voltage is induced within the two coils (spaced 180° apart) that yields a single sinusoidal voltage. The output voltage is usually expressed as an rms voltage ($V_{\mathrm{rms}} = 1/\sqrt{2}\ V_0$).

In the three-phase generator, three separate voltages are generated by using three different coils spaced 120° apart. As the magnet rotates, a voltage is induced across each of the generator's coils. All the coil voltages are equal in magnitude but are 120° out of phase with each other. With this generator, you could power three separate

Single-phase generator

Three-phase generator

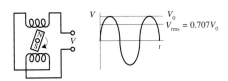
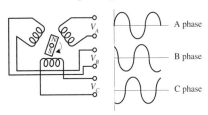

FIGURE A.2

loads of equal resistance, or you could drive a three-phase motor with a similar coil configuration; however, this requires using six separate wires. To reduce the number of wires, there are two tricks that can be used. The first trick involves rearranging the three-phase coil connections into what is called a *delta connection*—which yields three wires. The second trick involves rearranging the coil connection into what is called a *Y connection*—which yields four wires. Here are the details.

Y Connection

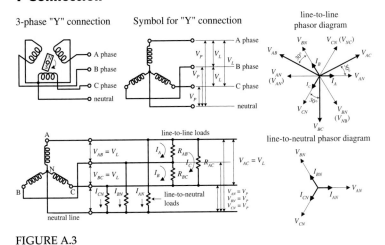

FIGURE A.3

A Y configuration is made by connecting one end of each of the generator's coils together to form what is called the *neutral* lead. The remaining three coil ends are brought out separately and are considered the "hot" leads. The voltage between the neutral and any one of the hot lines is called the *phase voltage* V_p. The total voltage, or *line voltage* V_L, is the voltage across any two hot leads. The line voltage is the vector sum of the individual phase voltages. In a Y-loaded circuit, each load has two phases in series. This means that the current and voltage through and across a load must be determined by superimposing the phase currents and phase voltages. One way to do this is to make a phasor diagram, as shown in the figure. For practical purposes, what is important to note is the line voltage is about $\sqrt{3}$ times the phase voltage. Also, the line currents are 30° out of phase with the line voltages. When a neutral is used, the line currents and line voltages are equal to the individual phases from the generator. No current flows in the neutral when the loads are equal across each phase (balanced load).

Delta Connection

3-phase delta connection

Symbol for delta connection

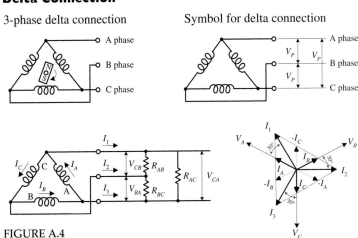

FIGURE A.4

A delta connection is made by placing the three-phase generator's coils end to end, as shown in the figure. Because the delta configuration has no neutral lead, the phase voltages are equal to the line voltages. Again, like the Y configuration, all line voltages are 120° out of phase with each other. However, unlike the Y configuration, the line currents (I_1, I_2, I_3) are equal to the vector sum of the phase currents (I_A, I_B, I_C). When each of the phases are equally loaded, the line currents are all equal but 120° apart. The line currents are 30° out of phase with the line voltages, and they are $\sqrt{3}$ times the phase currents.

A.3   Home Wiring

In the United States, three wires run from the pole transformer (or "green box" trans-former) to the main service panel at one's home. One wire is the A-phase wire (black in color), another is the B-phase wire (black in color), and the third is the neutral wire (white in color). (Figure A.5 shows where these three wires originate within the pole/green box transformer.) The voltage between the A-phase and B-phase wires, or the "hot to hot" voltage, is 240 V, while the voltage between the neutral wire and either A-phase or B-phase wire, or the "neutral to hot" voltage, is 120 V. (These volt-ages are nominal and may vary from region to region.)

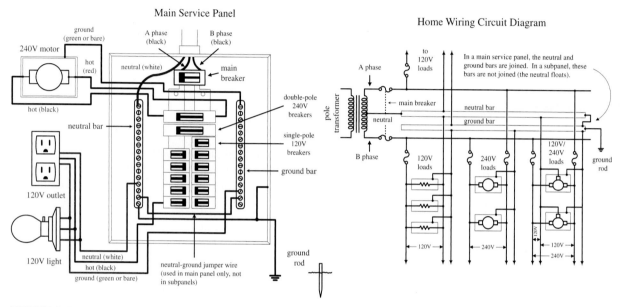

FIGURE A.5

At the home, the three wires from the pole/green box transformer are connected through a wattmeter and then enter a main service panel that is grounded to a long copper rod driven into the ground or to the steel in a home's foundation. The A-phase and B-phase wires that enter the main panel are connected through a main disconnect breaker, while the neutral wire is connected to a terminal referred to as the *neutral bar* or *neutral bus*. A *ground bar* also may be present within the main service panel. The ground bar is connected to the grounding rod or to the foundation's steel supports.

Within main service panels, the neutral bar and the ground bar are connected together (they act as one). However, within subpanels (service panels that get their power from the main service panel but which are located some distance from the main service panel), the neutral and ground bars are not joined together. Instead, the subpanel's ground bar receives a ground wire from the main service panel. Often the metal conduit that is used to transport the wires from the main service panel to the subpanel is used as the "ground wire." However, for certain critical applications (e.g., computer and life-support systems), the ground wire probably will be included within the conduit. Also, if a subpanel is not located in the same building as the main panel, a new ground rod typically is used to ground the subpanel. Note that different regions within the United States may use different wiring protocols. Therefore, do not assume that what I am telling you is standard practice where you live. Contact your local electrical inspector.

Within the main service panel, there are typically two bus bars into which circuit breaker modules are inserted. One of these bus bars is connected to the A-phase wire; the other bus bar is connected to the B-phase wire. To power a group of 120-V loads (e.g., upstairs lights and 120-V outlets), you throw the main breaker to the off position and then insert a single-pole breaker into one of the bus bars. (You can choose either the A-phase bus bar or the B-phase bus bar. The choice of which bus bar you use only becomes important when it comes to balancing the overall load—more on that in a second.) Next, you take a 120-V three-wire cable and connect the cable's black (hot) wire to the breaker, connect the cable's white (neutral) wire to the neutral bar, and connect the cable's ground wire (green or bare) to the ground bar. You then run the cable to where the 120-V loads are located, connect the hot and neutral wires across the load, and fasten the ground wire to the case of the load (typically a ground screw is supplied on an outlet mounting or light figure for this purpose). To power other 120-V loads that use their own breakers, you basically do the same thing you did in the last setup. However, to maximize the capacity of the main panel (or subpanel) to supply as much current as possible without overloading the main circuit breaker in the process, it is important to balance the number of loads connected to the A-phase breakers with the number of loads connected to the B-phase breakers. This is referred to as *balancing the load.*

Now, if you want to supply power to 240-V appliances (e.g., ovens, washers, etc.), you insert a double-pole breaker between the A-phase and B-phase bus bars in the main panel (or subpanel). Next, you take a 240-V three-wire cable and attach one of its hot wires to the A-phase terminal of the breaker and attach its other hot wire to the B-phase terminal of the breaker. The ground wire (green or bare) is connected to the ground bar. You then run the cable to where the 240-V loads are located and attach the wires to the corresponding terminals of the load (typically within a 240-V outlet). Also, 120-V/240-V appliances are wired in a similar manner, except you use a four-wire cable that contains an additional neutral (white) wire that is joined at the neutral bar within the main panel (or subpanel). (As a practical note, you could use a four-wire 120-V/240-V cable instead of a 240-V three-wire cable for 240-V applications—you would just leave the neutral wire alone in this case.)

As a note of caution, do not attempt home wiring unless you are sure of your abilities. If you feel that you are capable, just make sure to flip the main breaker off before you start work within the main service panel. When working on light fixtures, switches, and outlets that are connected to an individual breaker, tag that breaker with tape so that you do not mistakenly flip the wrong breaker when you go back to test your connections.

A.4   Electricity in Other Countries

In the United States, homes receive a 60-Hz, 120-V single-phase voltage, whereas industry typically receives a 60-Hz, 208-V/120-V three-phase voltage. Most other countries, on the other hand, work with a 50-Hz, 230-V single-phase voltage and a 415-V three-phase voltage. Now, if you were to take a U.S.-built 120-V, 60-Hz device over to Norway—where 230 V is used—and plug that device directly into the outlet (you would need an adapter to do this—their outlets look different), you run a risk of damaging the device. Some devices may not "care" about the voltage and frequency differences, but others will. You could use a converter (transformer plug-in device) to step down the voltage from the outlet, but you would still be stuck with 50 Hz. The 10-Hz difference will not affect most devices, but others devices, such as TVs and VCRs, may not function properly.

Here's a listing of single-phase voltages found in some countries. Note the plug types.

FIGURE A.6

Country	Voltage V	Frequency Hz	Plug Type
Australia	240	50	I
Belgium	230	50	C, E
Brazil	110/220	60	A, B, C, D, G
Canada	120	60	A, B
Chile	220	50	C, L
China	220	50	I
Congo	230	50	C, E
Costa Rica	120	60	A, B
Egypt	220	60	C
France	230	50	C, E, F
Germany	230	50	F
Hong Kong	230	50	D, G
India	230	50	C, D
Iraq	220	50	C, D, G
Italy	127/220	50	F, L
Japan	100	50/60	A, B
Korea	110/220	60	A, B, D, G, I, K
Mexico	127	60	A
Netherlands	230	50	C, E
Norway	230	50	C, F
Philippines	110/220	60	A, B, C, E, F, I
Russia & former Soviet Republics)	220	50	C, F
Spain	127/220	50	C, E
Switzerland	220	50	C, E, J
Taiwan	110	60	A, B, I
US	120	60	A, B
United Kingdom	230	50	G

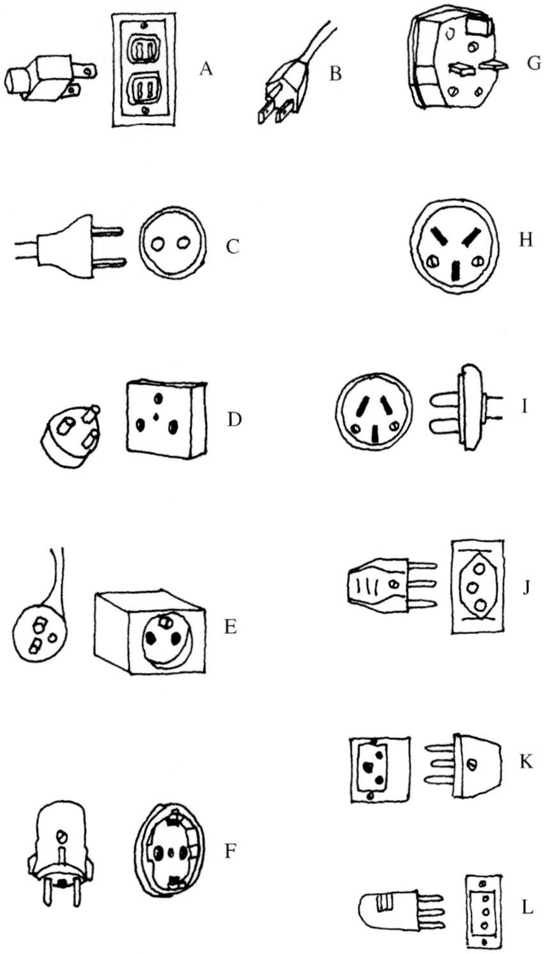

APPENDIX B:

Electronic Symbols

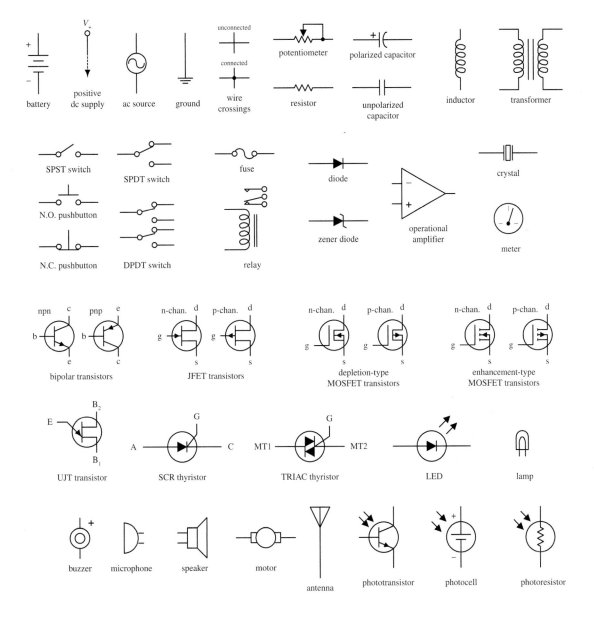

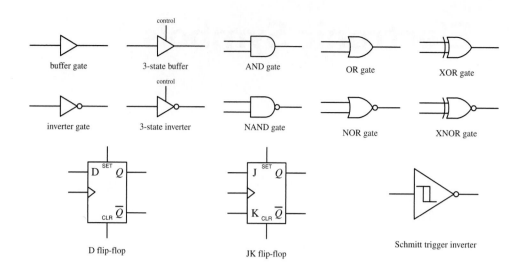

buffer gate   3-state buffer   AND gate   OR gate   XOR gate

inverter gate   3-state inverter   NAND gate   NOR gate   XNOR gate

D flip-flop   JK flip-flop   Schmitt trigger inverter

APPENDIX C:

Useful Facts and Formulas

C.1 Greek Alphabet

Alpha	A	α	Eta	H	η	Nu	N	ν	Tau	T	τ
Beta	B	β	Theta	Θ	θ	Xi	Ξ	ξ	Upsilon	Y	υ
Gamma	Γ	γ	Iota	I	ι	Omicron	O	o	Phi	Φ	ϕ
Delta	Δ	δ	Kappa	K	κ	Pi	Π	π	Chi	X	χ
Epsilon	E	ε	Lambda	Λ	λ	Rho	P	ρ	Psi	Ψ	ψ
Zeta	Z	ζ	Mu	M	μ	Sigma	Σ	σ	Omega	Ω	ω

C.2 Powers of 10 Unit Prefixes

PREFIX	SYMBOL	MULTIPLYING FACTOR
tera	T	$\times 10^{12}$
giga	G	$\times 10^{9}$
mega	M	$\times 10^{6}$
kilo	K	$\times 10^{3}$
centi	c	$\times 10^{-2}$
milli	m	$\times 10^{-3}$
micro	μ	$\times 10^{-6}$
nano	n	$\times 10^{-9}$
pico	p	$\times 10^{-12}$

C.3 Linear Functions ($y = mx + b$)

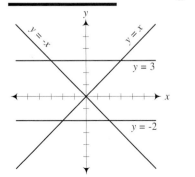

The equation $y = mx + b$ represents the equation of a line. The slope of the line ($\Delta y/\Delta x$) is equal to m, while the vertical shift, or point were the line crosses the y axis, is equal to b.

FIGURE C.1

467

C.4 Quadratic Equation ($y = ax^2 + bx + c$)

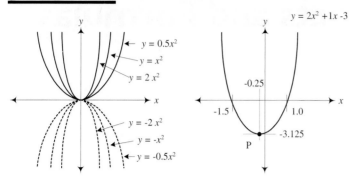

FIGURE C.2

The equation $y = ax^2 + bx + c$ traces out a parabola in the xy plane. The narrowness of the parabola is influenced by a, the horizontal shift is given by $-b/2a$, and the vertical shift is given by $-b^2/a + c$. To determine the roots of the equation (points where the parabola crosses the x axis), use

$$x = \frac{-b \pm \sqrt{b^2 - 4ac}}{2a}$$

C.5 Exponents and Logarithms

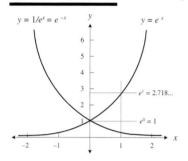

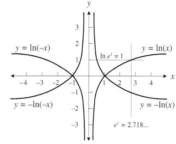

FIGURE C.3

EXPONENTS	LOGARITHMS
$x^0 = 1$	Base 10: if $10^n = x$, then $\log_{10} x = n$
$1/x^n = x^{-n}$	Base e: if $e^m = y$, then $\ln y = m$
$x^{1/n} = \sqrt[n]{x}$	($\log_{10} 100 = 2$, since $10^2 = 100$,
$x^m \cdot x^n = x^{m+n}$	$\ln e = 1$, since $e^1 = e = 2.718\ldots$)
$(xy)^n = x^n \cdot y^n$	Properties of any logarithm to the base b:
$(x^n)^m = x^{n \cdot m}$	$\log_b 1 = 0$
	$\log_b b = 1$
	$\log_b 0 = \begin{cases} +\infty & b < 1 \\ -\infty & b > 1 \end{cases}$
	$\log_b (x \cdot y) = \log_b x - \log_b y$
	$\log_b (x/y) = \log_b x - \log_b y$
	$\log_b (x^y) = y \log_b x$

C.6 Trigonometry

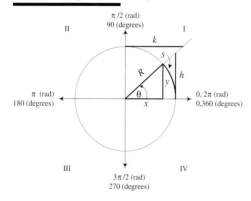

FIGURE C.4

The angle θ subtended by an arc S of a circle of radius R is equal to the ratio $\theta = S/R$, where θ is in radians. 1 radian $= 180°/\pi = 57.296°$, while $1° = \pi/180° = 0.17453$ radians. If R is rotated counterclockwise from the positive x axis, θ is positive in sign. If R is rotated clockwise from the positive x axis, θ is negative in sign. The trigonometric functions of the angle θ are defined as specific ratios between the sides of the triangles shown in the figure and are expressed as

$$\sin \theta = \frac{y}{R} \qquad \text{if } R = 1 \rightarrow y = \sin \theta$$

$$\cos \theta = \frac{x}{R} \qquad \text{if } R = 1 \rightarrow x = \cos \theta$$

$$\tan \theta = \frac{y}{x} \qquad \text{if } R = 1 \rightarrow h = \tan \theta$$

$$\cot \theta = \frac{x}{y} = \frac{1}{\tan \theta} \qquad \text{if } R = 1 \rightarrow k = \cot \theta$$

$$\sec \theta = \frac{R}{x} = \frac{1}{\cos \theta} \qquad \text{if } R = 1 \rightarrow \frac{1}{x} = \sec \theta$$

$$\csc \theta = \frac{R}{y} = \frac{1}{\sin \theta} \qquad \text{if } R = 1 \rightarrow \frac{1}{y} = \csc \theta$$

Sine and Cosine Functions

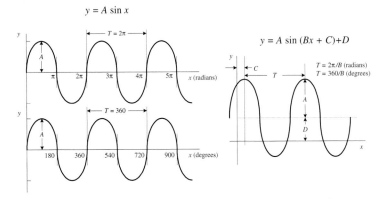

$y = A \sin x$

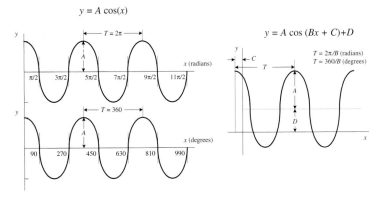

$y = A \cos(x)$

$y = A \sin (Bx + C) + D$

$y = A \cos (Bx + C) + D$

The graph of $y = A \sin \theta$ is shown to the far left. To alter the vertical, horizontal, period, and phase of this function, alter the equation to the form $y = A \sin (Bx + C) + D$. A is amplitude, $2\pi/B$ is the period (T), C is the phase shift, and D is the vertical shift. In electronics, a voltage can be expressed as

$$V(t) = V_0 \sin (\omega t + \phi) + V_{dc}$$

V_0 is the peak voltage, V_{dc} is the dc offset, ϕ is the phase shift, and ω is the angular frequency (rad/s), which is related to the conventional frequency (cycles/s) by

$$f = \frac{1}{T} = \frac{\omega}{2\pi}$$

The graph of $y = A \cos x$ is shifted in phase by $\pi/2$ radians (or 90°) with respect to the graph $y = A \sin x$. The following relations show how the sine and cosine functions are related:

$$\sin\left(\frac{\pi}{2} \pm x\right) = +\cos x \quad \text{or} \quad \sin (90° \pm x) = +\cos x$$

$$\sin\left(\frac{3\pi}{2} \pm x\right) = -\cos x \quad \text{or} \quad \sin (270° \pm x) = -\cos x$$

$$\cos\left(\frac{\pi}{2} \pm x\right) = \pm\sin x \quad \text{or} \quad \cos (90° \pm x) = \pm\sin x$$

$$\cos\left(\frac{3\pi}{2} \pm x\right) = \pm\sin x \quad \text{or} \quad \cos (270° \pm x) = \pm\sin x$$

FIGURE C.5

C.7   Complex Numbers

Complex numbers are covered in detail in Chap. 2.

C.8   Differential Calculus

Say you have a function $f(x)$. This function may represent a line, parabola, exponential curve, trigonometric curve, etc. Now pretend that you take a point and move it along the curve of $f(x)$. At the same time, you envision a tangent line touching the curve at the point. As the point moves along the curve, the slope of the tangent line changes ("teeter-totters")—provided the curve is not a line. Now, the slope of the tangent line has great significance in real-life situations. For example, if you graph a curve of the position of an object versus time, the slope at a particular time along the curve represents the instaneous velocity of the object. Likewise, if you have an electrical charge versus time graph, the slope at time t represents the instanteous current-flow. Now, the trick to finding the slope of a tangent line for any point along a curve involves using differential calculus. What differential calculus does is this: If you have a function, say, $y = x^2$—through the tricks of differential calculus—you can find another function, which is called the *derivative of y* (usually expressed as y' or dy/dx), that tells you the slope at every point along the curve of y. For $y = x^2$, the derivative is $dy/dx = 2x$. If you are interested in the slope of y at $x = 2$, you plug 2 into dy/dx to give you a slope of 4. But how do you find the derivative of $y = x^2$? Better yet, how do you find the derivative of any given function? The following provides the basic theory.

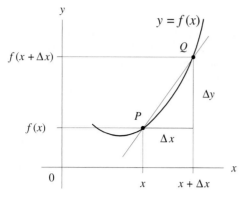

FIGURE C.6

To find the derivative of a function $y = f(x)$, you let $P(x,y)$ be a point of the graph of $y = f(x)$ and let $Q(x + \Delta x, y + \Delta y)$ be another point of the graph. The slope of the line between P and Q is then simply

$$\frac{f(x + \Delta x) - f(x)}{\Delta x}$$

Now you substitute the function into the preceding equation. For example, if $f(x) = x^2$, then $f(x + \Delta x) = (x + \Delta x)^2$, and the whole expression equals $[(x + \Delta x)^2 - x^2]/\Delta x$. Next, you hold x fixed and let Δx approach zero. If the slope approaches a value that depends only on x, you call this the slope of the curve at point P. The slope of the curve at P is itself a function of x, defined at every value of x at which the limit exists. You denote the slope by $f'(x)$ ("f prime"), or dy/dx ("$dydx$"), or df/dx ("$dfdx$"), and call any one of these terms (you choice) the derivative of $f(x)$:

$$f'(x) = \frac{dy}{dx} = \lim_{\Delta x \to 0} \frac{f(x + \Delta x) - f(x)}{\Delta x}$$

For the function $f(x) = x^2$, after carrying out the limit, you would get $f'(x) = dy/dx = 2x$ as the derivative.

Now, in practice, if you need to find the derivative of a function, you do not bother using the preceding equation. To do so would be very time-consuming and might require a number of nasty mathematical tricks, especially if you are trying to find the derivative of a complex function such as $2e^x \sin(3x + 2)$. Instead, what you do is memorize a few simple rules and memorize a few simple derivatives. The table below shows some of the rules and simple derivatives that will come in handy for many applications. In the table, a and n are constants, while u and v are functions.

DERIVATIVE	EXAMPLES
$\dfrac{d}{dx} a = 0$	$\dfrac{d}{dx} 4 = 0$
$\dfrac{d}{dx} x^n = nx^{n-1}$ (Note: $\dfrac{1}{x^n} = x^{-n}$)	$\dfrac{d}{dx} x = 1, \dfrac{d}{dx} x^2 = 2x, \dfrac{d}{dx} x^5 = 5x^4, \dfrac{d}{dx} x^{-1/2} = -\tfrac{1}{2}x^{-3/2}$
$\dfrac{d}{dx} e^x = e^x$	—
$\dfrac{d}{dx} \ln x = \dfrac{1}{x}$	—
$\dfrac{d}{dx} \sin x = \cos x$	—
$\dfrac{d}{dx} \cos x = -\sin x$	—
$\dfrac{d}{dx} au(x) = a\dfrac{d}{dx} u(x)$	$\dfrac{d}{dx} 3x^2 = 3\dfrac{d}{dx} x^2 = 6x, \dfrac{d}{dx} 3e^x = 3e^x, \dfrac{d}{dx} 7 \sin x = 7 \cos x$
$\dfrac{d}{dx}(u + v) = \dfrac{du}{dx} + \dfrac{dv}{dx}$	$\dfrac{d}{dx}(2x + x^2) = \dfrac{d}{dx}(2x) + \dfrac{d}{dx}(x^2) = 2 + 2x$
$\dfrac{d}{dx}\left(\dfrac{u}{v}\right) = \dfrac{v\, du/dx - u\, dv/dx}{v^2}$	$\dfrac{d}{dx}\left(\dfrac{x^2 + 1}{x^2 - 1}\right) = \dfrac{(x^2 - 1) \cdot 2x - (x^2 + 1) \cdot 2x}{(x^2 - 1)^2} = \dfrac{-4x}{(x^2 - 1)^2}$
Chain rule: If u is a function of y, and y is in turn a function of x, then $\dfrac{d}{dx}\{u[v(x)]\} = \dfrac{du}{dx} \cdot \dfrac{dv}{dx}$	$\dfrac{d}{dx} \sin(ax) = a \cos(ax), \dfrac{d}{dx} e^{2x} = 2e^{2x}$

Finding Components

A great way to find a wide variety of inexpensive electrical components is through mail-order catalogs. There are a number of suppliers to choose from. The following list provides the Web-site addresses and telephone numbers for some of the major suppliers. Most of these companies have on-line catalogs to browse through. You also can request a hard copy of their catalog (usually free) by either e-mailing them (once you are in their Web site) or contacting them by phone. Prices may vary from supplier to supplier, so compare before you buy.

COMPANY	WEB SITE	TELEPHONE NUMBER
All Electronics	http://www.allelectronics.com	(888) 826-5432
B.G Micro	http://www.bgmicro.com	(800) 276-2206
Circuit Specialists, Inc.	http://www.cir.com	(800) 528-1417
Debco Electronics	http://www.debco.com	(800) 423-4499
Digi-Key	http://www.digi-key.com	(800) 344-4539
Electronix Express	http://www.elexp.com	(800) 972-2225
Electronic Expediter, Inc.	http://www.expediters.com	(805) 987-7171
Gateway Electronics, Inc.	http://www.gatewayelex.com	(800) 669-5810
Hosfelt	http://www.hosfelt.com	(800) 524-6464
Jameco	http://www.jameco.com	(800) 831-4242
Marlin P. Jones & Assoc., Inc.	http://www.mpja.com	(800) 652-6733
Mouser Electronics	http://www.mouser.com	(800) 346-6873
Newark Electronics	http://www.newark.com	(800) 463-9275
RadioShack	http://www.radioshack.com	(800) 843-7422
Surplus Sales of Nebraska	http://www.surplussales.com	(800) 244-4567

Miscellaneous Items

There are times when you will undoubtedly need some kind of part, tool, device, etc. that is not necessarily electrical in nature. If you are looking for something that fits your needs but are not exactly sure where to find it or what to call it, a good place to start your search is to page through the *Thomas Register* or the *Grainger Catalog*.

The *Thomas Register* is a publication (a series of extremely large green books) that lists a wide variety of American manufacturers and services. Within this publication you will find companies that specialize in such things as injection molding, metal hardware, metal casting, machine tooling, electronic components, PC board design, fabrics, test equipment, motors, custom engineering, lasers, packaging, product marketing, chemicals, ceramic parts, computers, rubber parts, etc. The register lists company addresses, telephone numbers, and Web sites and provides you with pictures of company products. Often the pictures provide you with new insights into how to construct your invention. The *Thomas Register* is found in most local libraries or can be accessed via the Internet by entering the *Thomas Register*'s Web site (*http://www.thomasregister.com*).

The *Grainger Catalog*, unlike the *Thomas Register*, is geared toward actually selling the various companies' items they present. The catalog lists almost anything an inventor would find useful. You will find motors, welders, electrical adapters, tools, hydraulic pumps, home wiring service panels, adhesives, batteries, ball bearings, metals, plastics, electrical test equipment, various gauges (thickness, temperature, wire, etc.), grinders, etc. There are pictures, prices, and specifications for the various products presented. You can order a free catalog from Grainger, provided you are affiliated with a company or institution of some sort. Otherwise, you can page through the catalog at your local library or access an on-line catalog by entering Grainger's Web site (*http://www.grainger.com*).

Injection Molding and Patents

Injection Molding

Injection molding is a production process where plastic pellets are melted under great pressure and forced into a mold or die cut into aluminum or steel. Injection molding is used to create a phenomenal number of everyday items such as computer monitor cases and keyboards, cellular radio cases, plastic spoons, toys, and almost every plastic gadget. If you are planning to make a special holder for your circuit, or if you are interested in making some nonelectrical invention out of plastic, injection molding is the ticket. Before you get to the stage were you need to use injection molding, you will want to make a wood or clay prototype of your invention. Typically, the next step is to show your prototype to an injection molder. He or she will take a look at your prototype, assess the design, and give you a quote on the molding and manufacturing costs. In order for these companies to make a mold for you, you will need to have a computer model of your design made by an injection mold design engineer. Such engineers typically use a program called SolidWorks or some equivalent to make the computer model. The computer model is fed into a special computer attached to a milling machine that cuts out a mold for either aluminum or steel (aluminum is cheaper than steel but does not last as long during the injection process). Injection molding is not cheap. A simple mold of, say, something the size of a telephone receiver may cost you from $3000 to $10,000, depending on if the model is aluminum or steel and if the surface will need special etching, etc.

Patents

A *patent* is a legal document issued by the U.S. government granting property rights to individuals or corporations. It allows the holder to exclude others from making, using, or selling the invention "claimed" in the patent deed. The term of the patent is 20 years from the date of filing. If another party attempts to develop or use a patented invention, the owner of the patent may file a patent infringement lawsuit against the individual in federal court. If you believe you have an invention that is both unique and commercially viable, you definitely should look into getting a patent. If you attempt to market an unpatented "winner," sneaky individuals (or most likely, sneaky corporations) may steal the idea and price you out of the market. Without a patent, you have extremely limited legal rights to your invention. For more information about the United States Patent and Trademark Office (USPTO), check out their Web site at *www.uspto.gov.*

To determine if getting a patent is necessary, there are a few things you must consider. First, you must find out if your invention is unique—you need assurance that someone else has not already patented your idea. The best way to find out is to do a patent search. Patent lawyers and professional patent searchers can do this for you for a hefty fee. However, if you have the time, there are a couple of do-it-yourself avenues for independent searching. Every state in the United States has at least one library that is designated as a patent and trademark depository library. In these libraries, you can sit down at a computer, type in keywords, and take down reference numbers. You can then go to the stacks and search through monstrously large books filled with patents. Another, faster and easier approach is to do a search via the Internet. Currently, the most extensive patent search engine is the IBM site, *www.patents .ibm.com,* which allows you to do a free search of all patents issued since 1971. You only need to type in some keywords, and boom, a list of patented inventions will appear. Using a lot of cross-checking between similar patents, information about the uniqueness of your patent should turn up.

If it appears that your patent is unique, the next thing to consider before you file for a patent is whether your invention is commercially viable. Also, you must decide if you want to do all the marketing, manufacturing, and distributing yourself, or if you want to sell your invention to a manufacturer. Things can get a bit tricky at this point because in order to determine if a product will sell or if a manufacturer will be interested in your invention, you will need to show your invention to people. This is an absolute must. Without feedback from people in the business world, you will never know if it is worth your effort to push forward. Should you risk it and show your invention to people without a patent in order to see if a patent is worth the effort ("catch 22")? Or should you spend the (approximately) $1000 dollars for filing fees and patent certificate plus attorney fees (if you wisely choose to use one for the patent filing) and then show it to people? Another "salt in the wound" thing to consider is that perhaps your invention is not quite perfected; that is, you are still testing it, making revisions, etc. Filing a traditional patent application would be a waste—you would have to refile later once you are satisfied that your invention is complete.

The best suggestion I have for dealing with these problems is to consider getting what is called a *provisional patent.* A provisional patent costs only $75 (as of 1999) and provides a "patent pending" status for your invention. Unlike regular patents, provisional patents only last a year—after which time you must file for a regular patent. The main benefit of a provisional patent is an early filing date, which only goes into effect when the regular patent is granted. If a court battle ensues, it can be used to prove that you came up with the invention before another party. A neat thing about the provisional patent is that the invention listed in the application does not have to be fully polished. With a provisional patent, you need not keep meticulous records of product development and testing. After you have shown your invention around, you are free to make some tough decisions about whether or not you want to sell the invention to a manufacturer or develop it yourself. In sum, you can assess the viability of the product and weigh the benefits of shelling out big bucks for a patent.

An incredibly useful book detailing the patent process is *Patent It Yourself,* by David Pressman, published by Nolo Press. This book discusses the different kinds of patents available (utility, design, plant), provides important resources (such as Website addresses, phone numbers, and addresses), talks about how to do patent searches and how to write your own patent application, discusses the legal aspects of patents, the costs, marketing, and much more. If you are an inventor considering marketing an invention, you must get this book!

History of Electronics Timeline

DATE	INVENTION/DISCOVERY	DISCOVERER(S)
1745	Capacitor	Leyden
1780	Galvanic action	Galvani
1800	Dry cell	Volta
1808	Atomic theory	Dalton
1812	Cable insulation	Sommering and Schilling
1820	Electromagnetism	Oersted
1821	Thermoelectricity	Seebeck
1826	Ohm's law	Ohm
1831	Electromagnetic induction	Faraday
1831	Transformer	Faraday
1832	Self-induction	Henry
1834	Electrolysis	Faraday
1837	Relays	Cooke, Wheatstone, and Davy
1839	Photovoltaic effect	Becquerel
1843	Wheatstone bridge	Wheatstone
1845	Kirchhoff's laws	Kirchhoff
1850	Thermistor	Faraday
1860	Microphone diaphragm	Reis
1865	Radiowave propagation	Maxwell
1866	Transatlantic telegraph cable	T.C. & M. Co.

(continued)

DATE	INVENTION/DISCOVERY	DISCOVERER(S)
1874	Capacitors, mica	Bauer
1876	Rolled-paper capacitor	Fitzgerald
1876	Telephone	Bell
1877	Phonograph	Edison
1877	Microphone, carbon	Edison
1877	Loudspeaker moving coil	Siemens
1878	Cathode rays	Crookes
1878	Carbon-filament incandescent lamp	Swan, Stearn, Topham, and Cross
1879	Hall effect	Hall
1880	Piezoelectricity	Curie
1887	Gramophone	Berliner
1887	Aerials, radiowave	Hertz
1888	Induction motor	Tesla
1893	Waveguides	Thomson
1895	X-rays	Roentgen
1896	Wireless telegraphy	Marconi
1900	Quantum theory	Planck
1901	Fluorescent lamp	Cooper and Hewitt
1904	Two-electrode tube	Fleming
1905	Theory of relativity	Einstein
1906	Radio broadcasting	Fessenden
1908	Television	Campell, Swinton
1911	Superconductivity	Onnes
1915	Sonar	Langevin
1918	Multivibrator circuit	Abraham & Bloch
1918	Atomic transmutation	Rutherford
1919	Flip-flop circuits	Eccles and Jordan
1921	Crystal control of frequency	Cady
1924	Radar	Appleton, Briet, Watson, and Watt
1927	Negative-feedback amplifier	Black
1932	Neutron	Chadwick
1932	Particle accelerator	Crockcroft and Walton

DATE	INVENTION/DISCOVERY	DISCOVERER(S)
1934	Liquid crystals	Dreyer
1935	Transistor field effect	Hieil
1935	Scanning electron microscope	Knoll
1937	Xerography	Carlson
1937	Oscillograph	Van Ardenne, Dowling, and Bullen
1938	Nuclear fission	Fristsch and Meitner
1939	Early digital computer	Aitken and IBM
1943	First general-purpose computer (ENIAC: 10 ft tall, 11,000 sq ft, 30 tons, 70,000 resistors, 10,000 capacitors, 6000 switches, 18,000 vacuum tubes, 150 kW power, programmed with knobs and switches)	Mauchly and Eckert
1943	Printed wiring	Eisler
1945	First commercial computer (UNIVAC I)	—
1948	Transistor (bipolar)	Bardeen, Bratlain, and Shockley
1948	Holography	Gabor and Shockley
1950	Modem	MIT & Bell Labs
1950	Karnaugh mapping technique (digital logic)	Karnaugh
1952	Digital voltmeter	Kay
1953	Unijunction transistor	GEC
1954	Transistor radioset	Regency
1954	Solar battery	Chapin, Fuller, and Pearson
1956	Transatlantic telephone cable	U.K. & U.S.A.
1957	Sputnik I satellite	U.S.S.R.
1957	FORTRAN programming language	Watson Scientific
1958	Video tape recorder	U.S.A.
1958	Laser	Schalow and Townes
1959	Planar manufacturing process for transistors	Fairchild Semicondutor
1959	First integrated circuits	Fairchild Semiconductor and Texas Instruments
1960	Light-emitting diodes	Allen and Gibbons
1961	Electronic clock	Vogel and Cie
1962	MOSFET transistors	Hofstein, Heiman, and RCA
1963	Electronic calculator	Bell Punch Co.

(continued)

DATE	INVENTION/DISCOVERY	DISCOVERER(S)
1964	BASIC programming language	Kemeny and Kurtz
1966	Optical fiber communications	Kao and Hockham
1969	UNIX operating system	AT&T's Bell Labs
1970	Floppy disk recorder	IBM
1970	First microprocessor (4004, 60,000 oper/s)	Intel
1971	EPROM	—
1971	PASCAL programming language	Wirth
1971	First microcomputer-on-a-chip	Texas Instruments
1972	8008 processor (200 kHz, 16 kB)	Intel
1972	Ping Pong (early video game)	Atari
1972	First programmable word processor	Automatic Electronic Systems
1972	5.25-in diskette	—
1973	Josephson junction	IBM
1973	Tunable continuous-wave laser	Bell Labs
1973	Ethernet	Metcalfe
1974	C programming language	Kernighan, Ritchie
1974	Programmable pocket calculator	Hewlett Packard
1975	BASIC for personal computers	Allen
1975	Liquid-crystal display	United Kingdom
1975	First personal computer (Altair 8800)	Roberts
1975	Integrated optical circuits	Reinhart and Logan
1975	Microsoft founded	Gates and Allen
1976	Apple I computer	Wozniak, Jobs
1977	Commodore PET (14 K ROM, 4 K RAM)	Commodore Business Machines
1978	Space Invaders video game	Taito
1978	WordPerfect 1.0	Satellite Software
1980	3.5-in floppy (2-sided, 875 kB)	Sony Electronics
1980	Commodore 64	Commodore Business Machines
1980	Macintosh computer	Apple Computer
1981	IBM PC (8088 processor)	IBM
1981	MS-DOS 1.0	Microsoft
1982	Laser printer	IBM

DATE	INVENTION/DISCOVERY	DISCOVERER(S)
1983	Satellite television	U.S. Satellite Communications, Inc.
1983	"Wet" solar cell	Germany/U.S.A.
1983	First built-in hard drive (IBM PC-XT)	IBM
1983	Microsoft Word	Microsoft
1983	C++ programming language	Stroostrup
1984	CD-ROM player for personal computers	Philips
1985	300,000 simultaneous telephone conversations over single optical fiber	AT&T, Bell Laboratory
1987	Warm superconductivity	Karl Alex Mueller
1987	80386 microprocessor (25 MHz)	Intel
1990	486 microprocessor (33 MHz)	Intel
1994	Pentium processor (60/90 MHz 166.2 mips)	Intel
1996	Alpha 21164 processor (550 MHz)	Digital Equipment
1996	P2SC processor (15 million transistors)	IBM
1997	Deep Blue (IBM RS/6000SP supercomputer) defeats world chess champ Garry Kasparov	IBM

Component Data, List of Logic ICs, Foreign Semiconductor Codes

Standard Resistance Values for 5% Carbon-Film Resistors

1.0 Ω	8.2 Ω	33 Ω	120 Ω	470 Ω	1.8 kΩ	6.8 kΩ	27 kΩ	100 kΩ	390 kΩ	1.5 MΩ	6.2 MΩ
1.1 Ω	9.1 Ω	36 Ω	130 Ω	510 Ω	2.0 kΩ	7.5 kΩ	30 kΩ	110 kΩ	430 kΩ	1.6 MΩ	6.8 MΩ
1.2 Ω	10 Ω	39 Ω	150 Ω	560 Ω	2.2 kΩ	8.2 kΩ	33 kΩ	120 kΩ	470 kΩ	1.8 MΩ	7.5 MΩ
1.3 Ω	11 Ω	43 Ω	160 Ω	620 Ω	2.4 kΩ	9.1 kΩ	36 kΩ	130 kΩ	510 kΩ	2.0 MΩ	8.2 MΩ
1.5 Ω	12 Ω	47 Ω	180 Ω	680 Ω	2.7 kΩ	10 kΩ	39 kΩ	150 kΩ	560 kΩ	2.2 MΩ	9.1 MΩ
1.6 Ω	13 Ω	51 Ω	200 Ω	750 Ω	3.0 kΩ	11 kΩ	43 kΩ	160 kΩ	620 kΩ	2.4 MΩ	10 MΩ
1.8 Ω	15 Ω	56 Ω	220 Ω	820 Ω	3.3 kΩ	12 kΩ	47 kΩ	180 kΩ	680 kΩ	2.7 MΩ	

Selection of Diodes

DEVICE	TYPE	MATERIAL	PEAK INVERSE VOLTAGE (V)	AVERAGE FORWARD CURRENT (mA)	SURGE CURRENT (A)	FORWARD VOLTAGE DROP (V)
1N34	Signal	Ge	60	8.5	—	1.0
1N34A	Signal	Ge	80	—	—	1.0
1N60	Signal	Ge	80	0	—	1.0
1N67A	Signal	Ge	100	4.0	—	1.0
1N191	Signal	Ge	90	5.0	—	1.0
1N914	Fast switch	Si	75	75	0.05	1.0
1N4001	Rectifier	Si	50	1000	30	1.1
1N4002	Rectifier	Si	100	1000	30	1.1

continues

Selection of Diodes *(Continued)*

DEVICE	TYPE	MATERIAL	PEAK INVERSE VOLTAGE (V)	AVERAGE FORWARD CURRENT (mA)	SURGE CURRENT (A)	FORWARD VOLTAGE DROP (V)
1N4003	Rectifier	Si	200	1000	30	1.1
1N4004	Rectifier	Si	400	1000	30	1.1
1N4005	Rectifier	Si	600	1000	30	1.1
1N4006	Rectifier	Si	800	1000	30	1.1
1N4007	Rectifier	Si	1000	1000	30	1.1
1N4148	Signal	Si	75	10	—	1.0
1N4152	Fast switch	Si	40	20	—	1.0
1N5400	Rectifier	Si	50	1000	200	—
1N5401	Rectifier	Si	100	3000	200	—
1N5402	Rectifier	Si	200	3000	200	—
1N5404	Rectifier	Si	400	3000	200	—
1N5406	Rectifier	Si	600	3000	200	—
1N5408	Rectifier	Si	1000	3000	200	—
1N4448	Signal	Si	75	—	0.72	0.72
1N5817	Schottky	Si	20	1000	25	0.45
1N5818	Schottky	Si	30	1000	25	0.55
1N5819	Schottky	Si	40	1000	25	0.60
SB1100	Schottky	Si	100	1000	30	0.85
1N5820	Schottky	Si	20	3000	80	0.475
1N5821	Schottky	Si	30	3000	80	0.400
1N5822	Schottky	Si	40	3000	150	0.525
P600A	Rectifier	Si	50	6000	400	—
P600B	Rectifier	Si	100	6000	400	—
P600D	Rectifier	Si	200	6000	400	—
P600M	Rectifier	Si	1000	6000	400	—

Selection of Zener Diodes

VOLTS	Power (W)							
	0.25	0.4	0.5	1.0	1.5	5.0	10.0	50.0
1.8	IN4614							
2.0	IN4615							
2.2	IN4616							
2.4	IN4617	IN4370,A						
3.0	IN4619	IN4372,A	IN5987					
3.3	IN4620	IN5518	IN5988	IN3821	IN5913	IN5333,B		
3.6	IN4621	IN5519	IN5989	IN3822	IN5914	IN5334,B		
3.9	IN4622	IN5520	IN5844	IN3823	IN5915	IN5335,B	IN3993,A	IN4549,B
4.7	IN4624	IN5522	IN5846	IN3825	IN5917	IN5337,B	IN3995,A	IN4551,B
5.6	IN4626	IN5524	IN5848	IN3827	IN5919	IN5339,B	IN3997,A	IN4553,B
7.5	IN4100	IN5527	IN5997	IN3830	IN3786	IN5343,B	IN4000,A	IN4556
10.0	IN4104	IN5531	IN6000	IN4740	IN3789	IN5347,B	IN2974,B	IN2808,B
12.0		IN5532	IN6002	IN3022,B	IN3791	IN5349,B	IN2976,B	IN2810,B
14.0	IN4108	IN5534	IN5860			IN5351,B	IN2978,B	IN2812,B
16.0	IN4110	IN5536	IN5862	IN3025,B	IN3794	IN5353,B	IN2980,B	IN2814,B
20	IN4114	IN5540	IN5866	IN3027,B	IN3796	IN5357,B	IN2984,B	IN2818,B
24	IN4116	IN5542	IN6009	IN3929,B	IN3798	IN5359,B	IN2986,B	IN2820,B
28	IN4119	IN5544	IN5871			IN5362,B		
60	IN4128		IN5264,A,B			IN5371,B		
100	IN4135	IN985	16024	IN3044,A,B	IN3813	IN5378,B	IN3005,B	12838,B
120		IN987	IN6026	IN3046,B	IN5951	IN5380,B	IN3008A,B	IN2841,B

General-Purpose Bipolar Transistors

DEVICE	TYPE	V_{CEO} MAX (V)	V_{CBO} MAX (V)	V_{EBO} MAX (V)	I_C MAX (mA)	P_D (W)	H_{FE} $I_C =$ 0.1 mA	$I_C =$ 150 mA	F_T
2N918	NPN	15	30	3.0	50	0.2 (3 mA)	20	—	600
2N2102	NPN	65	120	7.0	1000	1.0	20	40	60
2N2219	NPN	30	60	5.0	800	3.0	35	100	250
2N2219A	NPN	40	—	—	150	—		100 (min)	300
2N2222	NPN	30	60	5.0	800	1.2	35	100	250
2N2222A	NPN	40	—	—	150	—		100 (min)	250
2N2484	NPN	60	—	—	50	—		100 (min)	60
2N2857	NPN	15	—	—	40	—		30 (min)	1200
2N2905	PNP	40	60	5.0	600	0.6	35	—	200
2N2907	PNP	40	60	5.0	600	0.4	35	—	200
2N3019	NPN	80	—	—	1000	—		100 (min)	100
2N3053	NPN	40	60	5.0	700	5.0	0	50	100
2N2904	NPN	40	60	6.0	200	0.625	40	—	300
2N2907A	PNP	40	—	—	600	—		75 (min)	200
2N3439	NPN	350	—	—	1000	—		40 (min)	15
2N3467	PNP	40	—	—	1000	—		40 (min)	175
2N3704	NPN	30	—	—	600	—		100 (min)	250
2N3904	NPN	40	—	—	10	—		100 (min)	200
2N3906	PNP	20	—	—	10	—		100 (min)	250
2N3906	PNP	40	40	5.0	200	1.5	60	—	250
2N4037	PNP	40	60	7.0	1000	5.0	—	50	—
2N4125	PNP	30	—	—	200	—		50 (min)	200
2N4126	PNP	25	—	—	200	—		120 (min)	200
2N4401	NPN	40	60	6.0	600	0.625	20	100	250
2N4403	PNP	40	40	5.0	600	0.625	30	100	200
2N4400	NPN	40	—	—	600	—		20 (min)	200
2N5087	PNP	50	—	—	50	—		40 (min)	40
2N5088	NPN	30	—	—	50	—		50 (min)	50
2N5415	PNP	200	200	4.0	1000	10.000	—	30	15

General-Purpose Power Bipolar Transistors

DEVICE	TYPE	V_{CEO} MAX (V)	I_c MAX (A)	P_D (W)	H_{FE} (MIN)	F_T (MHz)
2N5172	NPN	25	10	360	100	100
2N5210	NPN	50	50	625	250	30
2N5307	NPN-D	40	0.3	400	2000	60
2N5308	NPN-D	40	0.3	400	7000	60
2N5400	PNP	120	0.6	625	30	100
2N5401	PNP	150	0.6	625	40	100
2N5415	PNP	260	50	5	30	15
2N6036	PNP-D	80	4	40	750	25
2N6038	NPN-D	60	4	40	750	25
2N6043	NPN-D	60	8	75	1000	4
2N6052	PNP-D	100	12	150	750	4
2N6284	NPN-D	100	20	160	750	4
2N6287	PNP-D	100	20	160	750	4
2N6388	NPN-D	80	10	65	1000	20
2N6668	PNP-D	80	10	65	1000	20
TIP29C	NPN	100	1.0	30	40	—
TIP30C	PNP	100	1.0	30	40	—
TIP31C	NPN	100	3	40	25	—
TIP32C	PNP	100	3	40	25	—
TIP35C	NPN	100	25	125	25	—
TIP36C	PNP	100	25	125	25	—
TIP102	NPN-D	100	8	80	2500	—
TIP107	PNP-D	100	8	80	2500	—
TIP110	NPN-D	60	4	60	500	—
TIP112	NPN-D	100	4	50	500	—
TIP117	PNP-D	100	4	50	500	—
TIP120	NPN	60	5	65	1000	>5
TIP142	NPN-D	100	10	125	500	—
TIP147	PNP-D	100	10	125	500	—
D4H11	NPN	80	10	50	40	50
D44H10	PNP	80	15	83	40	50
MJ11015	PNP-D	120	30	200	1000	4
MJ11016	NPN-D	120	30	200	1000	4
MJ11032	NPN-D	120	50	300	1000	—
MJ11033	PNP-D	120	50	300	1000	—
MJE13007	NPN	400	8	80	8	4
MJE200	NPN	25	5	15	70	65
MJE210	PNP	20	5	15	70	65

Selection of RF Transistors

DEVICE	TYPE	MAX V_{CEO} (V)	MAX I_c (mA)	GAIN (dB)	FREQ. (MHz)	PACKAGE TYPE
MPS5175	NPN	12	50	—	2000	TO-92
MPSH10	NPN	25	—	—	650	TO-92
MPSH17	NPN	12	—	—	800	TO-92
MPSH81	PNP	20	50	—	—	TO-92
MPS918	NPN	30	600	—	—	TO-92
MRF531	NPN	100	100	—	—	TO-39
MRF544	NPN	70	400	16.5	250	TO-39
MRF545	PNP	70	400	—	—	TO-39
MRF586	NPN	17	200	9.0	500	TO-39
MRF904	NPN	15	30	16	450	MICRO-X
MRF571	NPN	10	70	12	1000	MICRO-X
MRF901	NPN	15	30	12	1000	MICRO-X
BRF90	NPN	15	30	—	—	MICRO-T
BRF91	NPN	12	35	—	—	MICRO-T
BRF90	NPN	15	30	—	—	MICRO-T

Selection of Small-Signal JFETs

DEVICE	TYPE	BV_{GS} (V)	MAX $V_{GS,OFF}$ (V)	INPUT C (pF)	MAX I_D (mA)	APPLICATION
2N4338	N-JFET	50	−1	6	0.6	—
2N4416	N-JFET	30	−6	4	15	VHF/UHF amp, mix, osc.
2N5114	P-JFET	30	10	25	90	switch: $R_{on} = 75\ \Omega$ (max)
2N5265- 2N5270	N-JFET	60 60	−3 −8	7 7	1 14	series of 6, 2N5358-64 P-JFET complement
2N5432	N-JFET	25	−10	30	—	switch: $R_{on} = 5\ \Omega$ (max)
2N5358- 2N5364	P-JFET	40 40	3 8	6 6	1 18.6	2N5265-70; N-JFET compliment
2N5457- 2N5459	N-JFET	25 25	−6 −8	7 7	5 16	general purpose; 2N5460-2 P-JFET complement
2N5460- 2N5462	P-JFET	40 40	6 9	7 7	5 16	2N5457-9; N-JFET complement

DEVICE	TYPE	BV_{GS} (V)	MAX $V_{GS,OFF}$ (V)	INPUT C (pF)	MAX I_D (mA)	APPLICATION
2N5484 MPF106	N-JFET	25	−6	5	30	HF/VHF/UHF amp, mix, osc.
2N5486	N-JFET	25	−2	5	15	VHF/UHF amp, mix, osc.
U304	P-JFET	30	10	27	50	analog switch chopper common-gate VHF/UHF
U310	N-JFET	30	−6	2.5	60	amp
U350	Quad N-JFET	25	−6	5	60	matched JFETs

Selection of Power FETs

DEVICE	TYPE	BV_{DS} (V)	$R_{DS(ON)}$ MAX (Ω)	$V_{GS(TH)}$ MAX (V)	I_D MAX (A)	P_D	CASE
IRFZ30	N-channel	50	0.050	4	30	75	TO-220
IRFZ42	N-channel	50	0.035	4	50	150	TO-220
VN0610L	N-channel	60	6	2.5	0.27	—	TO-92
VN10KM	N-channel	60	6	2.5	0.3	1	TO-237
IR511	N-channel	60	0.6	4	2.5	20	TO-220AB
MTP2955E	P-channel	60	0.12	—	11.5	125	TO-220AB
ZVN2110B	N-channel	100	4	—	0.85	5	TO-39
ZVN3310B	P-channel	100	20	—	0.3	5	TO-39
IRF510	N-channel	100	0.6	4	2	20	TO-220AB
IRF520	N-channel	100	0.27	4	5	40	TO-220AB
IRF150	N-channel	100	0.055	4	40	150	TO-204AE
ZVN0120B	N-channel	200	16	—	0.42	5	TO-39
ZVN1320B	P-channel	200	80	—	0.1	5	TO-39
IRF620	N-channel	200	0.8	4	5	40	TO-220AB
IRF220	N-channel	200	0.4	4	8	75	TO-220AB
IRF640	N-channel	200	0.18	4	10	125	TO-220AB
VP1320N3	P-channel	200	0.6	3.5	0.15	—	TO-92
IR9640	P-channel	200	0.5	4	11	—	TO-220
IR820	N-channel	500	3	4	2.5		TO-220
VP0650N3	P-channel	500	25	4	0.1	—	TO-92

Selection of Op Amps

TYPE	OP AMPS PER PACKAGE	SUPPLY VOLTAGE		INPUT OFFSET CURRENT, MAX (nA)	INPUT OFFSET VOLTAGE, TYPICAL (μV)	SLEW RATE (V/μS)	FREQUENCY, TYPICAL (MHz)	MAX OUTPUT CURRENT (mA)	COMMENTS
		MIN (V)	MAX (V)						
324A	4	3	32	30	2	0.5	1	20	Popular, single-supply bipolar
349	4	10	36	50	1	2	4	15	
355B	1	10	36	0.2	3	5	2.5	20	JFET, popular
741	1	10	36	200	2	0.5	1.2	20	Classic chip; 1458 (quad), 348 (dual)
1436	1	10	80	10	10	2	1	10	High-voltage
1463		30	80	—	—	165	17	1000	High-voltage
4558	2	8	36	200	2	1	2.5	15	Bipolar
AD841K	1	10	36	200	0.5	300	40	50	Bipolar, high-speed
AD848J	1	9	36	15	0.5	300	250	25	Bipolar, high-speed
AD744C	1	9	36	0.02	0.1	75	13	20	High-speed, low-distortion JFET
CA3410A	4	4	36	0.01	3	10	5.4	6	High-speed MOSFET
HA5141A	1	2	40	10	0.5	1.5	0.4	1	low-power, single-supply bipolar
HA2541	1	10	35	7μA	—	280	40	10	High-speed, low distortion
HA2542	1	10	35	7μA	—	375	120	100	High-power
HA2544	1	10	33	2μA	6	150	33	35	High-speed
HA5151	1	2	40	30	2	4.5	1.3	3	Bipolar
ICL7641B	4	1	18	0.03	—	1.6	1.4	5	MOSFET, low voltage, general purpose
LF351	1	10	36	0.1	5	13	4	10	JFET, 353 = ual, 347 = quad
LF411	1	10	36	0.1	0.8	15	4	15	General purpose, low-noise JFET
LM10	1	1	45	0.7	0.3	0.12	0.1	20	Low voltage, precision
LM11	1	5	40	10 pA	0.1	0.3	0.5	2	Precision, low bias
LM12	1	20	80	—	—	9	0.7	10000	High-power
LM308	1	10	36	1	2	0.15	0.3	5	Precision, low-bias bipolar
LM312	1	10	40	1	2	0.15	0.3	5	Compensated 308
LM318	1	10	40	200	4	7	15	10	Classic chip
LM343	1	10	68	10	2	2.5	1	10	High-voltage
LM344	1	10	68	19	2	30	10	10	High-voltage
LM833	2	10	36	200	0.3	7	15	10	Bipolar
LM6364	1	5	36	2μA	2	300	160	30	High-speed, bipolar
LT1006A	1	2.7	44	0.5	0.04	0.4	1	20	Bipolar, single-supply, precision
LT1028A	1	8	44	50	0.01	15	75	20	Bipolar precision, low-noise
LT1013C	2	4	44	2	0.06	0.4	0.8	25	Bipolar, single-supply
MC33078	2	10	36	150	0.15	7	16	20	Bipolar
MC34071A	1	3	44	50	0.5	10	4.5	25	Bipolar

TYPE	OP AMPS PER PACKAGE	SUPPLY VOLTAGE		INPUT OFFSET CURRENT, MAX (nA)	INPUT OFFSET VOLTAGE, TYPICAL (μV)	SLEW RATE (V/μS)	FREQUENCY, TYPICAL (MHz)	MAX OUTPUT CURRENT (mA)	COMMENTS
		MIN (V)	MAX (V)						
MC34181	1	3	36	0.05	0.5	10	4	8	JFET, high-speed, low power, low distortion
NE5534	1	6	44	300	0.5	6	10	20	Bipolar
OP-07E	1	6	44	3.8	0.03	0.17	0.6	10	Precision, low noise
OP-37E	1	8	44	35	0.01	17	63	20	Precision, low noise
OP-77E	1	6	44	1.5	0.01	0.3	0.6	12	Improvement over OP-27
OP-90E	1	1.6	36	3	0.05	0.01	0.02	6	Low-power
OP-97E	1	4.5	40	0.1	0.01	0.2	0.9	10	Low-power
TL051C	1	10.5	36	0.1	0.6	24	3	30	JFET, high-speed, low distortion
TL061C	1	4	36	0.2	3	3.5	1	5	JFET, low-power
TLC272A	2	3	18	1 pA	—	4.5	2.3	10	CMOS, low-power
TLC279C	4	3	18	0.1 pA	0.4	4.5	2.3	10	Quad CMOS

Common 4000 Series Logic ICs

4001	Quad 2-input buffered NOR gate	4051	8-channel analog multiplexer/demultiplexer
4002	Dual 4-input NOR gate	4052	Dual 4-channel analog multiplexer/demultiplexer
4006	18-stage static shift register	4053	Triple 2-channel analog multiplexer/demultiplexer
4007	Dual complementary pair and inverter	4063	4-bit magnitude comparator
4009	Hex inverting buffer	4066	Quad bilateral switch
4010	Hex noninverting buffer/converter	4068	8-input NAND gate
4011	Quad 2-input buffered NAND gate	4069	Input protected hex inverter
4012	Dual 4-input NAND gate	4071	Quad 2-input buffered OR gate
4013	Dual D flip-flop	4072	Dual 4-input buffered OR gate
4014	8-bit static shift register	4073	Triple 3-input buffered AND gate
4015	Dual 4-bit static shift register	4075	Triple 3-input buffered OR gate
4016	Quad bilateral switch	4077	Quad 2-input EXCLUSIVE-NOR gate
4017	5-stage decade counter/divider	4078	8-input buffered NOR gate
4018	Presettable divide-by-N counter	4081	Quad 2-input buffered AND gate
4020	14-stage ripple carry binary counter	4082	Dual 4-input buffered AND gate
4021	8-bit static shift register	4093	Quad 2-input NAND Schmitt trigger
4022	Divide-by-8 counter/divider with 8 decimal outputs	4094	8-stage shift-and-store bus register

continued

Common 4000 Series Logic ICs *(Continued)*

4023	Buffered triple 3-input NAND gate	4099	8-bit addressable latch
4024	7-stage ripple carry binary counter	4511	BCD to 7-segment latch/decoder/driver
4025	Triple 3-input NOR gate	4512	8-channel buffered data selector
4027	Dual JK flip-flop	4514	4-bit latched/4-to-16 line decoders
4028	BCD to decimal decoder	4515	4-bit latch/4-to-16 line decoder
4029	Synchronous up/down counter, binary/decade counter	4516	Binary up-own counter
4030	Quad EXCLUSIVE-OR gate	4518	Dual 4-bit decoder counter
4040	12-stage ripple carry binary counter	4520	Dual 4-bit counter
4042	Quad D-clocked latch	4521	24-stage frequency divider and oscillator
4043	Quad cross-couple NOR R/S 3 state latch	4528	Dual monostable multivibrator
4047	Monostable/astable multivibrator	4538	Dual precision monostable multivibrator
4049	Hex inverting buffer	4584	Hex Schmitt-trigger
4050	Hex noninverting buffer	4585	Magnitude comparator

Common 7000 Series Logic ICs

7400	Quad 2-iput NAND gate	74138	3-line to 8-line decoder/demultiplexer
7402	Quad 2-input NOR gate	74139	Dual 2-line to 2-line decoder/demultiplexer
7404	Hex Inverter	74145	BCD-to-decimal decoder/driver
7406	Hex Inverter/buffer diver	74148	8-line to 3-line priority encoder
7408	Quad 2-input positive AND gate	74150	16-bit data selector
7410	Triple 3-input NAND gate	74151	Data selector multiplexer
7411	Triple 3-input AND gate	74153	Dual 4-line to 1-line data selector/multiplexer
7414	Hex Schmitt-trigger inverter	74155	Dual 2-line to 4-line decoder/demultiplexer
7420	Dual 4-input NAND gate	74156	Dual 2-line to 4-line decoder/demultiplexer
7421	Dual 4-input AND gate	74158	Quad 2-line to 1-line multiplexer
7426	Quad 2-input NAND gate	74159	4-line to 16-line decoder/demultiplexer
7427	Triple 3-input NOR gate	74161	Synchronous 4-bit binary counter
7430	8-input NAND gate	74163	Fully synchronous 4-bit binary counter
7432	Quad 2-input OR gate	74164	8-bit parallel-out serial shift register
7433	Quad 2-input NOR buffer	74166	Parallel load 8-bit shift register
7438	Quad 2-input NAND buffer	74169	Synchronous 4-bit up/down counter

7442	4-line to 10-line decoder (1 of 10)	74378	Hex D-type flip-flop
7445	BCD-to-decimal decoder/driver	74390	Dual decade counter
7447	BCD 7-segement decoder/driver	74293	Dual 4-bit decade and binary counter
7451	Dual 2-wide 2-input AND-OR-invert gate	74173	Quad D-type flip-flop 3-state output
7473	Dual JK master slave flip-flop with clear	74174	Hex D-type flip-flop with clear
7474	Dual D-type positive edge-triggered flip-flop	74175	Quad D-type flip-flop with clear
7475	4-bit bistable latch	74191	Synchronous up-down 4-bit counter
7485	4-bit magnitude comparator	74193	Synchronous 4-bit up/down counter
7486	Quad 2-input exclusive OR gate	74194	4-bit bi-directional shift register
7492	Divide-by-12 counter	74195	4-bit parallel access shift register
74107	Dual JK master slave flip-flop	74221	Dual monostable multivibrator with Schmitt-trigger input
74109	Dual JK positive-edge-trig. flip-flop	74240	Octal buffers/drivers with 3-state output
74112	Dual JK negative-edge-triggered flip-flop	74241	Octal bus/line driver
74121	Monostable multivibrator	74243	Quad bus transceiver
74122	Retriggerable monostable multivibrator	74244	Octal buffer, line driver, line receiver
74123	Dual retriggerable monostable multivibrator	74245	Octal bus transceiver with 3-state output
74124	Dual voltage controlled oscillator	74251	Data selector/multiplexer with 3-state output
74125	Quad 3-input buffer with 3-state output	74253	Dual 4-line-to-1 line data selector/multiplexer with 3-state output
74126	Quad 3-state buffer	74257	Quad 2-to-1-line selector/multiplexer with 3-state output
74128	Quad 2-input NOR driver	74258	Quad 2-to-1-line selector/multiplexer with 3-state output
74132	Quad 2-input NAND with Schmitt trigger	74259	8-bit addressable latch
74136	Quad 2-input XOR gate	74521	Octal comparator
74266	Quad 2-input XOR gate	74540	Octal buffer/line driver
74273	Octal D-type flip-flop with clear	74541	Octal buffer/line driver
74279	Quad S-R latch	74573	Octal D-type latch with 3-state output
74280	9-bit odd-even parity generator/checker	74574	Octal D-type flip-flop 3-state output
74283	4-bit binary full adder	74590	8-bit binary counter with 3-state output registers
74298	Quad 2-innput multiplexer with storage	74595	8-bit shift registers with output latches
74348	8-line-to-3-line priority encoder with 3-state output	74640	Octal inverting bus transceiver

continued

Common 7000 Series Logic ICs *(Continued)*

74365	Hex bus driver with 3-state output	74641	Octal non-inverting bus transceiver
74367	Hex buffer/driver, true data	74645	Octal noninverting bus transceiver 3-state output
74373	Octal transparent latch	74646	Octal bus transceiver an register 3-state output
74374	Octal D-type flip-flop	74670	4 × 4 register file with 3-state output
74375	Quad latch	74688	8-bit identity comparator

Semiconductor Codes

JOINT ELECTRON DEVICE ENGINEERING COUNCIL (JEDEC) CODE

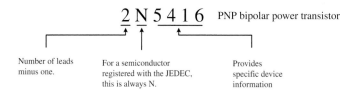

FIGURE G.1

EUROPEAN SEMICONDUCTOR CODE (PRO ELECTRON CODE)

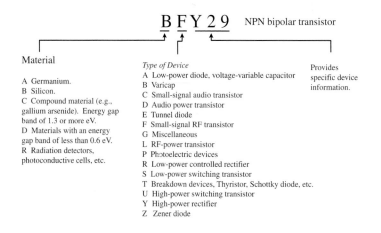

FIGURE G.2

JAPANESE INDUSTRIAL STANDARD (JIS)

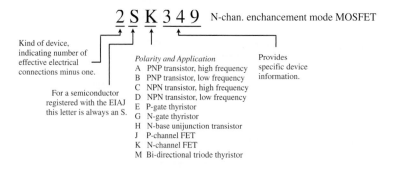

FIGURE G.3

A good semiconductor cross-reference catalog (e.g., NTE) will list both European and Japanese semiconductor devices and will provide you with a list of suitable replacements that are produced in this country.

There are many semiconductor devices that do not one follow of the preceding codes. Instead, they are labeled with a code designated by the device's manufacturer. For example, Motorola uses an MJ*xxxx* code to specify power transistors (metal case), an MPS*xxxx* code to specify low-power transistors (plastic case), and an MRF code to specify HF, VHF, and microwave transistors. Likewise, Texas Instruments uses a TIP*xxxx* code to specify power transistors (plastic case) and a TI*xxxx* code to specify small signal transistors (plastic case).

Analog/Digital Interfacing

There are a number of tricks used to interface analog circuits with digital circuits. In this appendix we'll take a look at two basic levels of interfacing: One level deals with simple on/off triggering; the other level deals with true analog-to-digital and digital-to-analog conversion—converting analog signals into digital numbers and converting digital numbers into analog signal.

H.1 Triggering Simple Logic Responses from Analog Signals

There are times when you need to drive logic from simple on/off signals generated by analog devices. For example, you may want to latch an alarm (via a flip-flop) when an analog voltage, say, one generated from a temperature sensor, reaches a desired threshold level. Or perhaps you simply want to count the number of times a certain analog threshold is reached. For simple on/off applications such as these, it is common to use a comparator or op amp as the interface between the analog output of the transducer and the input of the logic circuit. Often it is possible to simply use a voltage divider network composed of a transducer of variable resistance and a pull-up resistor. Here are some sample networks to illustrate the point.

FIGURE H.1

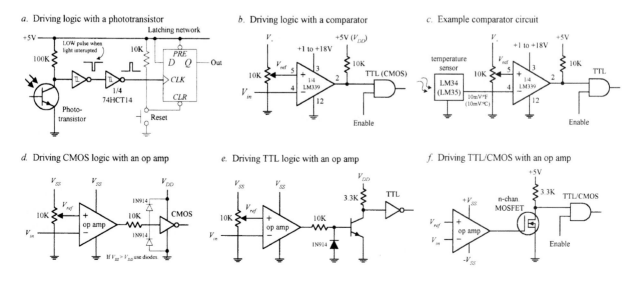

a. Driving logic with a phototransistor

b. Driving logic with a comparator

c. Example comparator circuit

d. Driving CMOS logic with an op amp

e. Driving TTL logic with an op amp

f. Driving TTL/CMOS with an op amp

497

In Fig. H.1*a*, a phototransistor is used to trigger a logic response. Normally, the phototransistor is illuminated, which keeps the input of the first Schmitt inverter low—the output of the second inverter is high. When the light is briefly interrupted, the phototransistor momentarily stops conducting, causing the input to the first inverter to pulse low, while the output of the second inverter pulses high. This high pulse could be used to latch a D flip-flop, which could be used to trigger a LED or buzzer alarm.

In Fig. H.1*b*, a single-supply comparator with open-collector output is used as an analog-to-digital interface. When an analog voltage applied to V_{in} exceeds the reference voltage (V_{ref}) set at the noninverting input (+) via the pot, the output goes low (comparator sinks current through itself to ground). When V_{in} goes below V_{ref}, the output goes high (comparator's output floats, but the pull-up resistor pulls the comparator's output high).

In Fig. H.1*c*, a simple application of the previous comparator interface is shown. The input voltage is generated by an LM34 or LM35 temperature sensor. The LM34 generates 10 mV/°F, while the LM35 generates 10 mV/°C. The resistance of the pot and V_+ determine the reference voltage. If we want to drive the comparator low when 75°C is reached, we set the reference voltage to 750 mV, assuming we're using the LM35.

In Fig. H.1*d*, an op amp set in comparator mode can also be used as an analog-to-digital interface for simple switching applications. CMOS logic can be driven directly through a current limiting resistor, as shown. If the supply voltage of the op amp exceeds the supply voltage of the logic, protection diodes should be used (see figure).

In Fig. H.1*e*, an op amp that is used to drive TTL typically uses a transistor output stage like the one shown here. The diode acts to prevent base-to-emitter reverse breakdown. When V_{in} exceeds V_{ref}, the op amp's output goes low, the transistor turns off, and the logic input receives a high.

In Fig. H.1*f*, an *n*-channel MOSFET transistor is used as an output stage to an op amp.

H.2   Using Logic to Drive External Loads

Driving simple loads such as LEDs, relays, buzzers, or any device that assumes either an on or off state is relatively simple. When driving such loads, it is important to first check the driving logic's current specifications—how much current, say, a gate can sink or source. After that, you determine how much current the device to be driven will require. If the device draws more current that the logic can source or sink, a high-power transistor typically can be used as an output switch. Figure H.2 shows some sample circuits used to drive various loads.

In Fig. H.2*a*, LED's can be driven directly by logic through a current-limiting resistor. Current can be sunk or sourced as shown—though it is better to sink when using TTL. Because a LED's forward voltage drop (V_F) and forward current (I_F) can vary from 1.0 to 2.0 V and 1 to 20 mA, respectively, it is best to select current-limiting resistors that obey the equation shown in the figure. If a LED requires more current than the logic can supply or sink, a transistor output stage like the one shown in Fig. H.2*f* can be used.

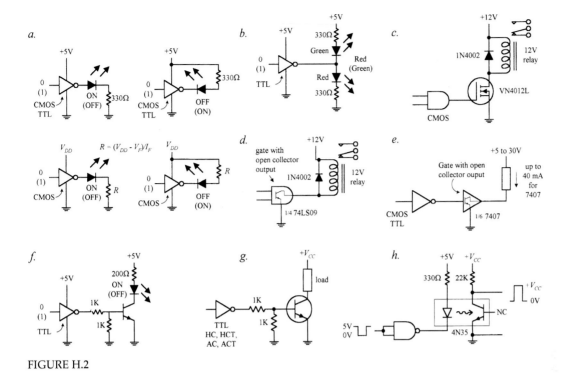

FIGURE H.2

Figure H.2*b* shows a simple way to get dual lighting action from a pair of LEDs. When the gate's output goes low, the upper green LED turns on, while the lower red LED turns off. The LEDs switch states when the output goes high.

Relays will draw considerable current. To avoid damaging the logic device, in Fig. H.2*c*, a power MOSFET transistor is attached to the logic output. The diode is used to protect the circuit from current spikes generated by relay as it switches states.

A handy method for interfacing standard logic with loads is to use a gate with an open-collector output as a go-between. Recall that open collector gates cannot source current—they can only sink current. However, they typically can sink 10 times the current of a standard logic gate. In Fig. H.2*d*, an open-collector gate is used to drive a relay. Check the current ratings of specific open-collector devices before using them to be sure they can handle the load current.

Figure H.2*e* shows another open-collector application. In Fig. H.2*f*, a bipolar transistor is used to increase the output drive current used to drive a high-current LED. Make sure the transistor is of the proper current rating.

Figure H.2*g* is basically the same as last example, but the load can be something other than an LED. In Fig. H.2*h*, an optocoupler is used to drive a load that requires electrical isolation from the logic driving it. Electrical isolation is often used in situations where external loads use a separate ground system. The voltage level at the load side of the optical interface can be set via V_{CC}. There are many different types of optocouplers available (see Chap. 5).

H.3   Analog Switches

Analog switches are ICs designed to switch analog signals via digital control. The internal structure of these devices typically consists of a number of logic control gates interfaced with transistor stages used to control the flow of analog signals.

Figure H.3 shows various types of analog switches. The CMOS 4066B quad bilateral switch uses a single supply voltage from 3 to 15 V. It can switch analog or digital signals within ±7.5 V and has a maximum power dissipation of around 700 mW. Individual switches are controlled by digital inputs A through D. The TTL-compatible AH0014D DPDT analog switch can switch analog signals of ±10 V via the A and B logic control inputs. Note that this device has separate analog and digital supplies: $V+$ and $V-$ are analog; V_{CC} and GND are digital. The DG302A dual-channel CMOS DPST analog switch can switch analog signals within the ±10-V range at switching speeds up to 15 ns.

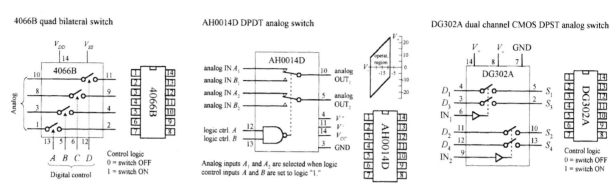

FIGURE H.3

There are a number of circuits that use analog switches. They are found in modulator/demodulator circuits, digitally controlled frequency circuits, analog signal-gain circuits, and analog-to-digital conversion circuits where they often act as sample-hold switches. They can of course be used simply to turn a given analog device on or off.

H.4   Analog Multiplexer/Demultiplexer

Recall from Section 12.3 that a digital multiplexer acts like a data selector, while a digital demultiplexer acts like a data distributor. Analog multiplexers and demultiplexers act the same way but are capable of selecting or distributing analog signals. (They still use digital select inputs to select which pathways are open and which are closed to signal transmission.)

A popular analog multiplexer/demultiplexer IC is the 4051B, shown in Fig. H.4. This device functions as either a multiplexer or demultiplexer, since its inputs and outputs are bidirectional (signals can flow in either direction). When used as a multiplexer, analog signals enter through I/O lines 0 through 7, while the digital code that selects which input gets passed to the analog O/I line (pin 3) is applied to digital inputs A, B, and C. See the truth table in the figure. When used as a demultiplexer, the connections are reversed: The analog input comes in through the analog O/I line (pin 3) and passes out through one of the seven analog I/O lines—the specific output is again selected by the digital inputs A, B, and C. Note that when the inhibit line (INH) is high, none of the addresses are selected.

4051B analog multiplexer/demultiplexer

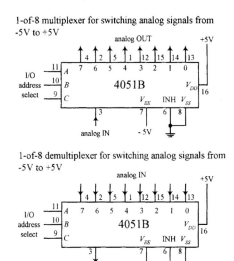

FIGURE H.4

INH	C	B	A	Addr.
L	L	L	L	0
L	L	L	H	1
L	L	H	L	2
L	L	H	H	3
L	H	L	L	4
L	H	L	H	5
L	H	H	L	6
L	H	H	H	7
H	X	X	X	none

The input/output analog voltage levels for the 4051B are limited to a region between the positive supply voltage V_{DD} and the analog negative supply voltage V_{EE}. Note that the V_{SS} supply is grounded. If the analog signals you are planning to use are all positive, V_{EE} and V_{SS} can both be connected to a common ground. However, if you plan to use analog voltages that range from, say, –5 to +5 V, V_{EE} should be set to –5 V, while V_{DD} should be set to +5 V. The 4051B accepts digital signals from 3 to 15 V, while allowing for analog signals from –15 to –15 V.

H.5 Analog-to-Digital and Digital-to-Analog Conversion

In order for analog devices (e.g., speakers, temperature sensors, strain gauges, position sensors, light meters, etc.) to communicate with digital circuits in a manner that goes beyond simple threshold triggering, we use an analog-to-digital converter (ADC). An ADC converts an analog signal into a series of binary numbers, each number being proportional to the analog level measured at a given moment. Typically, the digital words generated by the ADC are fed into a microprocessor or microcontroller, where they can be processed, stored, interpreted, and manipulated. Analog-to-digital conversion is used in data-acquisition systems, digital sound recording, and within simple digital display test instruments (e.g., light meters, thermometers, etc.).

In order for a digital circuit to communicate with the analog world, we use a digital-to-analog converter (DAC). A DAC takes a binary number and converts it to an analog voltage that is proportional to the binary number. By supplying different binary numbers, one after the other, a complete analog waveform is created. DACs are commonly used to control the gain of an op amp, which in turn can be used to create digitally controlled amplifiers and filters. They are also used in waveform generator and modulator circuits and as trimmer replacements and are found in a number of process-control and autocalibration circuits.

H.5.1 ADC and DAC Basics

Figure H.5 shows the basic idea behind analog-to-digital and digital-to-analog conversion. In the analog-to-digital figure, the ADC receives an analog input signal along with a series of digital sampling pulses. Each time a sampling pulse is received, the ADC measures the analog input voltage and outputs a 4-bit binary number that is proportional to the analog voltage measured during the specific sample. With 4 bits, we get 16 binary codes (0000 to 1111) that correspond to 16 possible analog levels (e.g., 0 to 15 V).

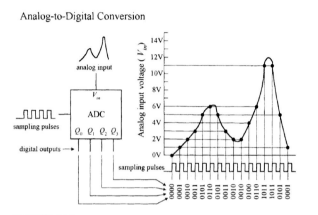

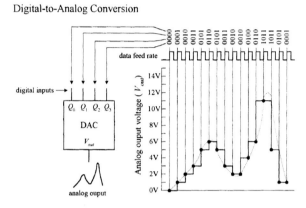

FIGURE H.5

In the digital-to-analog conversion figure, the DAC receives a series of 4-bit binary numbers. The rate at which new binary numbers are fed into the DAC is determined by the logic that generates them. With each new binary number, a new analog voltage is generated. As with the ADC example, we have a total of 16 binary numbers to work with and 16 possible output voltages.

As you can see from the graphs, both these 4-bit converters lack the resolution needed to make the analog signal appear continuous (without steps). To make things appear more continuous, a converter with higher resolution is used. This means that instead of using 4-bit binary numbers, we use larger-bit numbers, such as 6-bit, 8-bit, 10-bit, 12-bit, 16-bit, or even 18-bit numbers. If our converter has a resolution of 8 bits, we have $2^8 = 256$ binary number to work with, along with 256 analog steps. Now, if this 8-bit converter is set up to generate 0 V at binary 00000000 and 15 V at binary 11111111 (full-scale), then each analog step is only 0.058 V high ($\frac{1}{256} \times 15$ V). With an 18-bit converter, the steps get incredibly tiny because we have $2^{18} = 262,144$ binary numbers and steps. With 0 V corresponding to binary 000000000000000000 and 15 V corresponding to 111111111111111111, the 18-bit converter yields steps that are only 0.000058 V high! As you can see in the 18-bit case, the conversion process between digital and analog appears practically continuous.

H.5.2 Simple Binary-Weighted Digital-to-Analog Converter

Figure H.6 shows a simple 4-bit digital-to-analog converter that is constructed from a digitally controlled switch (74HC4066), a set of binary-weighted resistors, and an operational amplifier. The basic idea is to create an inverting amplifier circuit whose gain is controlled by changing the input resistance R_{in}. The 74HC4066 and the resistors together act as a digitally controlled R_{in} that can take on one of 16 possible values. (You can think of the 74HC4066 and resistor combination as a digitally controlled current source. Each new binary code applied to the inputs of the 74HC4066 generates a new discrete current level that is summed by R_F to provide a new discrete output voltage level.) We choose scaled resistor values of R, $R/2$, $R/4$, and $R/8$ to give R_{in} discrete values that are equally spaced. To find all possible values of R_{in}, we use the formula provided in Fig. H.6. This formula looks like the old resistors-in-parallel formula, but we must exclude those resistors which are not selected by the digital input code—that's what the coefficients A through D are for (a coefficient is either 1 or 0, depending on the digital input). Now, to find the analog output voltage, we simply use $V_{out} = -V_{in} (R_F/R_{in})$—the expression used for the inverting amplifier (see Chap. 7). Figure H.6 shows what we get when we set $V_{in} = -5$ V, $R = 100$ kΩ, and $R_F = 20$ kΩ, and take all possible input codes.

Simple binary-weighted digital-to-analog converter

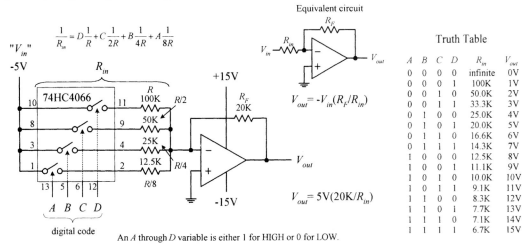

$$\frac{1}{R_{in}} = D\frac{1}{R} + C\frac{1}{2R} + B\frac{1}{4R} + A\frac{1}{8R}$$

$$V_{out} = -V_{in}(R_F/R_{in})$$

$$V_{out} = 5V(20K/R_{in})$$

Truth Table

A	B	C	D	R_{in}	V_{out}
0	0	0	0	infinite	0V
0	0	0	1	100K	1V
0	0	1	0	50.0K	2V
0	0	1	1	33.3K	3V
0	1	0	0	25.0K	4V
0	1	0	1	20.0K	5V
0	1	1	0	16.6K	6V
0	1	1	1	14.3K	7V
1	0	0	0	12.5K	8V
1	0	0	1	11.1K	9V
1	0	1	0	10.0K	10V
1	0	1	1	9.1K	11V
1	1	0	0	8.3K	12V
1	1	0	1	7.7K	13V
1	1	1	0	7.1K	14V
1	1	1	1	6.7K	15V

An *A* through *D* variable is either 1 for HIGH or 0 for LOW.

FIGURE H.6

The binary-weighted DAC shown above is limited in resolution (4-bit, 16 analog levels). To double the resolution (make an 8-bit DAC), you might think to add another 74HC4066 and $R/16$, $R/32$, $R/64$, and $R/128$ resistors. In theory, this works; in reality, it doesn't. The problem with this approach is that when we reach the $R/128$ resistor, we must find a 0.78125-kΩ resistor, assuming $R = 100$ kΩ. Assuming we can find or construct an equivalent resistor network for $R/128$, we're still in trouble because the tolerances of these resistors will screw things up. This scaled-resistor approach becomes impractical when we deal with resolutions of more than a few bits. To increase the resolution, we scrap the scaled-resistor network and replace it with an $R/2R$ ladder network—the manufacturers of DAC ICs do this as well.

H.5.3 R/2R Ladder DAC

An $R/2R$ DAC uses an $R/2R$ resistor ladder network instead of a scaled-resistor network, as was the case in the previous DAC. The benefit of using the $R/2R$ ladder is that we need only two resistor values, R and $2R$. Figure H.7 shows a simple 4-bit $R/2R$ DAC. For now, assume that the switches are digitally controlled (in real DACs they are replaced with transistors).

$R/2R$ ladder 4-bit digital-to-analog converter

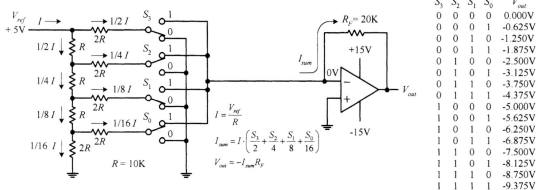

S_3	S_2	S_1	S_0	V_{out}
0	0	0	0	0.000V
0	0	0	1	-0.625V
0	0	1	0	-1.250V
0	0	1	1	-1.875V
0	1	0	0	-2.500V
0	1	0	1	-3.125V
0	1	1	0	-3.750V
0	1	1	1	-4.375V
1	0	0	0	-5.000V
1	0	0	1	-5.625V
1	0	1	0	-6.250V
1	0	1	1	-6.875V
1	1	0	0	-7.500V
1	1	0	1	-8.125V
1	1	1	0	-8.750V
1	1	1	1	-9.375V

FIGURE H.7

The trick to understanding how the $R/2R$ ladder works is realizing that the current drawn through any one switch is always the same, no matter if it is thrown up or thrown down. If a switch is thrown down, current will flow through the switch into ground (0 V). If a switch is thrown up, current will flow toward virtual ground—located at the op amp's inverting input (recall that if the noninverting input of an op amp is set to 0 V, the op amp will make the inverting input 0 V, via negative feedback). Once you realize that the current through any given switch is always constant, you can figure that the total current (I) supplied by V_{ref} will be constant as well. Once you've got that, you figure out what fractions of the total current passes through each of the branches within the $R/2R$ network using simple circuit analysis. Figure H.7 shows that $\frac{1}{2}I$ passes through S_3 (MSB switch), $\frac{1}{4}I$ through S_2, $\frac{1}{8}I$ through S_1, and $\frac{1}{16}I$ through S_0 (LSB switch). If you're interested in how that was figured out, the following circuit reduction should help.

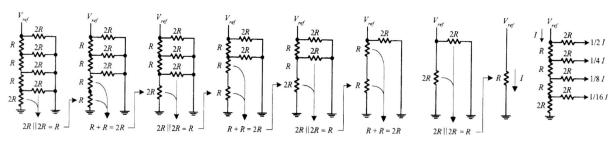

FIGURE H.8

Now that we have a means of consistently generating fractions of $\frac{1}{2}I$, $\frac{1}{4}I$, $\frac{1}{8}I$ and $\frac{1}{16}I$, we can choose, via the digital input switches, which fractions get summed together by the amplifier. For example, if switches $S_3S_2S_1S_0$ are thrown to 0101 (5), $\frac{1}{4}I + \frac{1}{16}I$ combine to form I_{sum}. But what is I? Using Ohm's law, it's just $I = V_{ref}/R = +5 \text{ V}/10 \text{ k}\Omega = 500 \text{ μA}$. This means that $I_{sum} = \frac{1}{4}(500 \text{ μA}) + \frac{1}{16}(500 \text{ μA}) = 156.25 \text{ μA}$. The final output voltage is determined by $V_{\text{out}} = -I_{\text{sum}}R_F = -(156.25\ \mu\text{A})(20\ \text{k}\Omega) = -3.125$ V. The formulas and the table in Fig. H.7 show the other possible binary/analog combinations.

To create an $R/2R$ DAC with higher resolution, we simply add more runs and switches to the ladder.

H.5.4 Integrated Digital-to-Analog Converters

Making DACs from scratch isn't worth the effort. The cost as well as the likelihood for conversion errors is great. The best thing to do is to simply buy a DAC IC. You can buy these devices from a number of different manufacturers (e.g., National Semiconductor, Analog Devices, Texas Instruments, etc.). The typical resolutions for these ICs are 6, 8, 10, 12, 16, and 18 bits. DAC ICs also may come with a serial digital input, as opposed to the parallel input scheme we saw in Figs. H.6 and H.7. Before a serial-input DAC can make a conversion, the entire digital word must be clocked into an internal shift register.

Most often DAC ICs come with an external reference input that is used to set the analog output range. There are some DACs that have fixed references, but these are becoming rare. Often you'll see a manufacturer list one of its DACs as being a *multiplying* DAC. A multiplying DAC can produce an output signal that is proportional to the product of a varying input reference level (voltage or current) times a digital code. As it turns out, most DACs, even those which are specifically designated as multiplying DAC on the data sheets, can be used for multiplying purposes simply by using the reference input as the analog input. However, many such ICs do not provide the same quality multiplying characteristics, such as a wide analog input range and fast conversion times, as those which are called *multiplying* DACs.

Multiplying is most commonly applied in systems that use ratiometeric transducers (e.g., position potentiometers, strain gauges, pressure transducers, etc.). These transducers require an external analog voltage to act as a reference level on which to base analog output responses. If this reference level is altered, say, by an unwanted supply surge, the transducer's output will change in response, and this results in conversion errors at the DAC end. However, if we use a multiplying DAC, we eliminate these errors by feeding the transducer's reference voltage to the DAC's analog input. If any supply voltage/current errors occur, the DAC will alter its output in proportion to the analog error.

DACs are capable of producing unipolar (single-polarity output) or bipolar (positive and negative) output signals. In most cases, when a DAC is used in unipolar mode, the digital code is expressed in standard binary. When used in bipolar mode, the most common code is either offset binary or 2's complement. Offset binary and 2's complement codes make it possible to express both positive and negative values. Figure H.9 shows all three codes and their corresponding analog output levels (referenced from an external voltage source).

Note that in the figure, "FS" stands for full scale, which is the maximum analog level that can be reached when applying the highest binary code. It is important to realize that at full scale, the analog output for an n-bit converter is actually $(2^n - 1)/2^n \times V_{\text{ref}}$, not $2^n/2^n \times V_{\text{ref}}$. For example, for an 8-bit converter, the number of binary numbers is

Common digital codes used by digital-to-analog converters

	Unipolar operation			Bipolar operation				
	Binary	Analog output		Offset binary	Analog output		2's comp.	Analog output
FS	1111 1111	$\text{Vref}\left(\dfrac{255}{256}\right)$	FS	1111 1111	$+\text{Vref}\left(\dfrac{127}{128}\right)$	FS	0111 1111	$+\text{Vref}\left(\dfrac{127}{128}\right)$
FS-1	1111 1110	$\text{Vref}\left(\dfrac{254}{256}\right)$	FS-1	1111 1110	$+\text{Vref}\left(\dfrac{126}{128}\right)$	FS-1	0111 1110	$+\text{Vref}\left(\dfrac{126}{128}\right)$
↓			↓			↓		
$\dfrac{FS}{2}$	1000 0000	$\text{Vref}\left(\dfrac{128}{256}\right)=\dfrac{V_{ref}}{2}$	0 + 1LSB	1000 0001	$+\text{Vref}\left(\dfrac{1}{128}\right)$	0 + 1LSB	0000 0001	$+\text{Vref}\left(\dfrac{1}{128}\right)$
↓			0	1000 0000	$\text{Vref}\left(\dfrac{0}{128}\right)=0$	0	0000 0000	$\text{Vref}\left(\dfrac{0}{128}\right)=0$
LSB	0000 0001	$\text{Vref}\left(\dfrac{1}{256}\right)$	0-1LSB	0111 1111	$-\text{Vref}\left(\dfrac{1}{128}\right)$	0-1LSB	1111 1111	$-\text{Vref}\left(\dfrac{1}{128}\right)$
LSB-1	0000 0000	$\text{Vref}\left(\dfrac{0}{256}\right)=0$	↓			↓		
			−FS+1	0000 0001	$-\text{Vref}\left(\dfrac{127}{128}\right)$	−FS+1	1000 0001	$-\text{Vref}\left(\dfrac{127}{128}\right)$
	(FS = Full Scale)		−FS	0000 0000	$-\text{Vref}\left(\dfrac{128}{128}\right)$	−FS	1000 0000	$-\text{Vref}\left(\dfrac{128}{128}\right)$

FIGURE H.9

$2^8 = 256$, while the maximum analog output level is $255/256V_{ref}$, not $256/256V_{ref}$, since the highest binary number is 255 (1111 1111). The "missing count" is used up by the LSB-1 condition (0 state).

H.5.5 Example DAC ICs

DAC0808 8-Bit DAC (National Semiconductor)

The DAC0808 is a popular 8-bit DAC that requires an input reference current and supplies 1 of 256 analog output current levels. Figure H.10 shows a block diagram of the DAC0808, along with its IC pin configuration and a sample application circuit.

FIGURE H.10

DAC0808 8-bit Digital-to-Analog Converter (resolution = 256 steps)

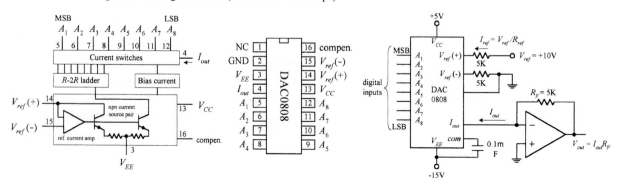

In the application circuit, the analog output range is set by applying a reference current (I_{ref}) to pin 14 ($+V_{ref}$). In this example, I_{ref} is set to 2 mA via an external +10 V/5 kΩ resistor combination. (Note that another 5-kΩ resistor is required between pin 15 ($-V_{ref}$) and ground.) To determine the DAC's analog output current (I_{out}) for all possible binary inputs, we use the following formula:

$$I_{out} = I_{ref}\left(\frac{A_1}{2} + \frac{A_2}{4} + ... + \frac{A_8}{256}\right) = \frac{\text{decimal equivalent of input binary number}}{256}$$

At full scale (all A's high or binary 255), $I_{out} = I_{ref}(255/256) = (2\text{ mA})(0.996) = 1.99$ mA. Considering that the DAC has 256 analog output levels, we can figure that each corresponding level is spaced 1.99 mA/256 = 0.0078 mA apart.

To convert the analog output currents into analog output voltages, we attach the op amp. Using the op amp rules from Chap. 7, we find that the output voltage is $V_{out} = I_{out} \times R_f$. At full scale, $V_{out} = (1.99\text{ mA})(5\text{ kΩ}) = 9.95$ V. Each analog output level is spaced 9.95 V/256 = 0.0389 V apart.

The DAC0808 can be configured as a multiplying DAC by applying the analog input signal to the reference input. In this case, however, the analog input current should be limited to a range from 16 μA to 4 mA to retain reasonable accuracy. See the National Semiconductor's data sheets for more details.

DAC8043 Serial 12-Bit Input Multiplying DAC (Analog Devices)

The DAC8083 is a high-precision 12-bit CMOS multiplying DAC that comes with a serial digital input. Figure H.11 shows a block diagram, pin configuration, and Write cycle timing diagram for this device.

DAC8043 12-bit serial input multiplying D/A converter

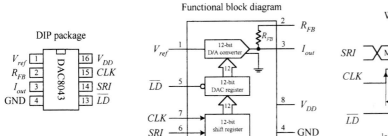
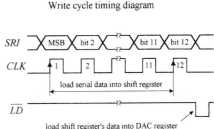

FIGURE H.11

Before the DAC8043 can make a conversion, serial data must first be clocked into the input register by supplying an external clock signal (each positive edge of the clock load one bit). Once loaded, the input register's contents are dumped off to the DAC register by applying a low pulse to the $\overline{LD}$ line. Data in the DAC register are then converted to an output current through the I_{out} terminal. In most applications, this current is then transformed into a voltage by an op amp stage, as is the case within the two circuits shown in Fig. H.12. In the unipolar (2-quadrant) circuit, a standard binary code is used to select from 4096 possible analog output levels. In the bipolar (4-quadrant) circuit, an offset binary code is used again to select from 4096 analog output levels, but now the range is broken up to accommodate both positive and negative polarities. If you're interested in learning more about the DAC8043, go to Analog Device's Web site and check out the data sheet.

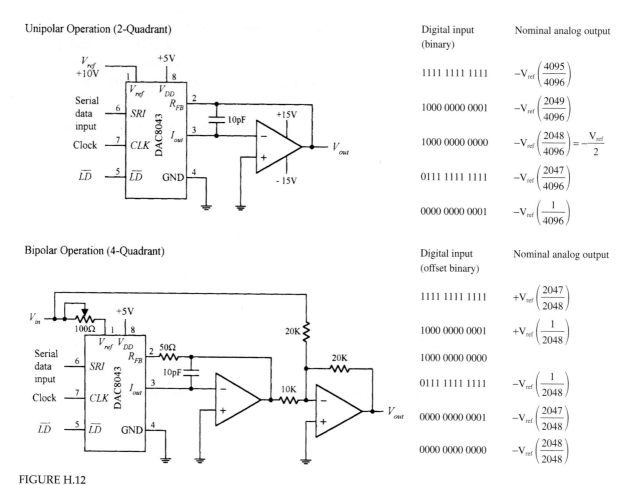

Unipolar Operation (2-Quadrant)

Digital input (binary)	Nominal analog output
1111 1111 1111	$-V_{ref}\left(\dfrac{4095}{4096}\right)$
1000 0000 0001	$-V_{ref}\left(\dfrac{2049}{4096}\right)$
1000 0000 0000	$-V_{ref}\left(\dfrac{2048}{4096}\right) = -\dfrac{V_{ref}}{2}$
0111 1111 1111	$-V_{ref}\left(\dfrac{2047}{4096}\right)$
0000 0000 0001	$-V_{ref}\left(\dfrac{1}{4096}\right)$

Bipolar Operation (4-Quadrant)

Digital input (offset binary)	Nominal analog output
1111 1111 1111	$+V_{ref}\left(\dfrac{2047}{2048}\right)$
1000 0000 0001	$+V_{ref}\left(\dfrac{1}{2048}\right)$
1000 0000 0000	
0111 1111 1111	$-V_{ref}\left(\dfrac{1}{2048}\right)$
0000 0000 0001	$-V_{ref}\left(\dfrac{2047}{2048}\right)$
0000 0000 0000	$-V_{ref}\left(\dfrac{2048}{2048}\right)$

FIGURE H.12

H.6   Analog-to-Digital Converters

There are a number of techniques used to convert analog signals into digital signals. The most popular techniques include successive approximation conversion and parallel-encoded conversion (or flash conversion). Other techniques include half-flash conversion, delta signal processing (DSP), and pulse code modulation (PCM). In this section we'll focus on successive approximation and parallel-encoded conversion techniques.

H.6.1   Successive Approximation

Successive approximation A/D conversion is the most common approach used in integrated ADCs. In this conversion technique, each bit of the binary output is found, one bit at a time—MSB first. This technique yields fairly fast conversion times (from around 10 to 300 μs) with a limited amount of circuitry. Figure H.13 shows a simple 8-bit successive approximation ADC, along with an example analog-to-digital conversion sequence.

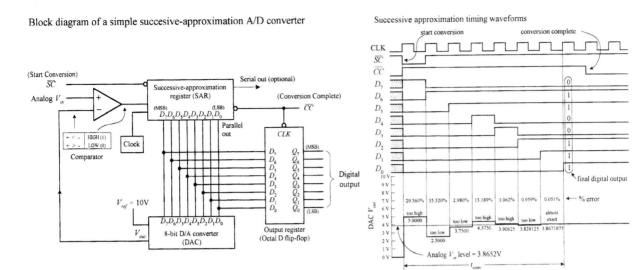

Block diagram of a simple succesive-approximation A/D converter

Successive approximation timing waveforms

FIGURE H.13

To begin a conversion, the $\overline{SC}$ (start conversion) input is pulsed low. This causes the successive approximation register (SAR) to first apply a high on the MSB (D_7) line of the DAC. With only D_7 high, the DAC's output is driven to one-half its full-scale level, which in this case is +5 V because the full-scale output is +10 V. The +5-V output level from the DAC is then compared with the analog input level, via the comparator. If the analog input level is greater than +5 V, the SAR keeps the D_7 line high; otherwise, the SAR returns the D_7 line low. At the next clock pulse, the next bit (D_6) is tried. Again, if the analog input level is larger than the DAC's output level, D_6 is left high; otherwise, it is returned low. During the next 6 clock pulses, the rest of the bits are tried. After the last bit (LSB) is tried, the CC (conversion complete) output of the SAR goes low, indicating that a valid 8-bit conversion is complete, and the binary data are ready to be clocked into the octal flip-flop, where they can be presented to the Q_0–Q_7 outputs. The timing diagram shows a 3.8652-V analog level being converted into an approximate digital equivalent. Note that after the first approximation (the D_7 try), the percentage error between the actual analog level and corresponding digital equivalent is 29.360 percent. However, after the final approximation, the percentage error is reduced to only 0.051 percent.

Until now, we've assumed that the analog input to our ADC was constant during the conversion. But what happens when the analog input changes during conversion time? Errors result. The more rapidly the analog input changes during the conversion time, the more pronounced the errors will become. To prevent such errors, a sample-and-hold circuit is often attached to the analog input. With an external control signal, this circuit can be made to sample the analog input voltage and hold the sample while the ADC makes the conversion. Many ADC ICs contain internal sample-and-hold stages, while others can be interfaced with a sample-and-hold IC, such as the LF198 sample-and-hold amplifier. We'll take a look at the LF198 in greater detail in a second.

H.6.2 Example ADCs

ADC0803 8-Bit A/D Converter (National Semiconductor)

The ADC0803 is an 8-bit successive approximation type ADC that contains an on-chip clock circuit. Figure H.14 shows the pinouts, functional block diagram, and example analog-to-digital conversion circuit.

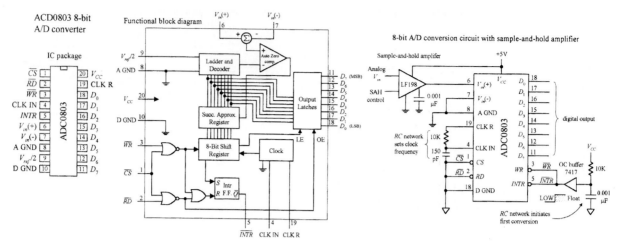

FIGURE H.14

In the example circuit, a conversion is initiated when the $\overline{WR}$ input is set low, while an end-of-conversion is indicated when the $\overline{INTR}$ output goes low. When the $\overline{INTR}$ output is connected directly to the $\overline{WR}$ input, the ADC is set up for continuous conversion. The resistor, capacitor, and buffer act as an automatic reset circuit that is used to ensure that the $\overline{WR}$ input is taken low when power is first applied. The RC circuit connected to the CLK-R and CLK-IN inputs acts to set the converter's clock frequency, where $f = 1/1.1RC$.

The ADC0803 has both a positive analog input (pin 6) and a negative analog input (pin 7). Having two inputs allows for positive, negative, or differential input signals. When working with positive analog input signals, pin 7 is grounded, while pin 6 is used as the analog input. When working with negative analog input signals, pin 6 is grounded, while pin 7 is used as the analog input. Differential signals are based on the voltage difference between pin 6 and pin 7.

The $V_{ref}/2$ input is used to determine which analog input voltage will generate the maximum digital output of 1111 1111 (decimal 255). If this input is left unconnected, the supply voltage (+5 V) sets the analog range from 0 to +5 V. When a voltage is applied to the $V_{ref}/2$ input, a different analog input range can be obtained. For example, to generate a maximum digital output code of 11111111 when an analog input voltage of +4 V is applied, a +2-V level is set at the $V_{ref}/2$ input ($V_{ref}/2 = +4$ V$/2 = +2$ V).

Also included in the example circuit is an LF198 sample-and-hold amplifier. This device is used to prevent any errors due to input signal variations during the conversion process. The LF198 contains two unity-gain noninverting amplifiers, a logic-controlled switch, and an external capacitor. When the LF198's control input is made high, the internal logic-controlled switch is closed, and the external capacitor charges up quickly to equal the analog input voltage. When the logic-controlled switch is made low, the capacitor will retain the analog input voltage level. This stored voltage level can then used by the ADC without worrying about analog input variations. You can find many ADCs out there that have the sample-and-hold section built into the IC. Check the electronics catalogs for these chips.

ADC0808 8-Bit A/D Converter with 8-Chan Multiplexer and Parallel Output (National Semiconductor)

The ADC0808, shown below, is similar to the previous ADC0803; however, it comes with a multiplexer input stage that allows for 8-channel analog input selection. To address a given analog input, a 3-bit binary word is applied to the address select inputs (ADD A, ADD B, ADD C).

ADC0808 successive approximation 8-bit A/D converter with 8-channel multiplexer

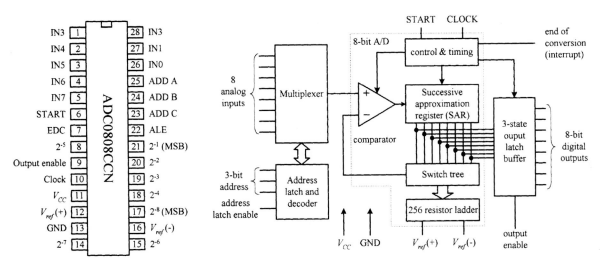

FIGURE H.15

ADCs with built-in multiplexers are frequently used in data-acquisition systems that monitor a number of different input transducers. In these systems, a microprocessor or microcontroller is used to generate the address signals, process the ADC's binary output, and make the appropriate logic decisions based on the data.

AD0831 8-Bit Serial I/O A/D Converter (National Semiconductor)

The AD0831 is an 8-bit A/D converter that comes with a synchronous serial output (D_o) (see Fig. H.16). The serial output comes in handy in microcontroller applications that have limited I/O capability. To illustrate how the AD0831 works, let's take a look at a simple application that uses a BASIC stamp II (microcontroller with PBASIC interpreter). Though we'll discuss the BASIC stamp II (BS2) later, for now, just think of the device as having a number of I/O terminals that can be programmed—via PC host program—to accept or output signals. (The PC-to-BS2 connection via the PC serial port is used when programming the BS2.)

To start a conversion, the AD0831's $\overline{CS}$ input is supplied with a high-to-low signal from the BS2's P_0 output. This signal must stay low for the duration of the conversion. Next, the AD0831's CLK input must receive a single clock pulse (low-to-high-to-low) from the BS2's P_2 output to signify that the conversion should begin at the next clock pulse. After the first clock pulse, 8 more clock pulses are supplied by the BS2 to complete the conversion. During each of these clock pulses, a serial bit is sent out the D_o output where it enters the BS2's P_3 input.

ADC0831 8-bit Serial A/D converter

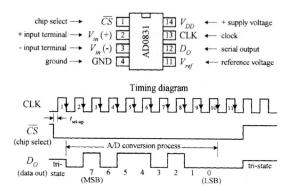

Interfacing the AD0831 with a BASIC Stamp microcontroller

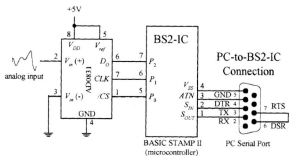

FIGURE H.16

In the example circuit in Fig. H.16, the reference voltage and $V-$ are wired to provide 0 to +5 V, where binary 0000 0000 corresponds to a 0-V analog input, while binary 1111 1111 corresponds to a +5-V analog input. If we had 3.53-V analog input, the binary serial output would be 1011 0100.

AD0838 8-Bit Serial I/O A/D Converter with Multiplexer (National Semiconductor)

The AD0838 resembles the previous AD0831 but comes with an 8-channel analog multiplexer stage. In order to use the multiplexer to select a given analog input, additional multiplexer address bits are needed during the priming phase before the actual conversion takes place. Check out the AD0838's data sheet for more details.

ADC0838 8-bit serial I/O A/D converter
with multiplexer options

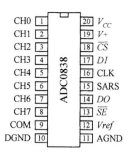

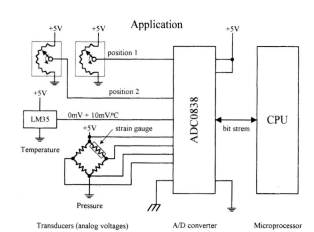

FIGURE H.17

H.6.3   Parallel-Encoded A/D Conversion (Flash Conversion)

Parallel-encoded A/D conversion, or flash conversion, is perhaps the easiest conversion process to understand. To illustrate the basics behind parallel encoding (also referred to as *simultaneous multiple comparator* or *flash converting*), let's take a look at the simple 3-bit converter below.

Simple 3-bit parallel-encoded A/D converter

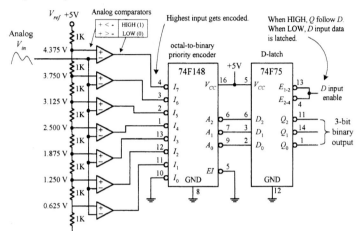

Truth Table

Analog V_{in} (V)	I_7	I_6	I_5	I_4	I_3	I_2	I_1	I_0	$\overline{A}_2$	$\overline{A}_1$	$\overline{A}_0$	$\overline{Q}_2$	$\overline{Q}_1$	$\overline{Q}_0$
0.000 - 0.625	0	0	0	0	0	0	0	0	1	1	1	0	0	0
0.625 - 1.250	0	0	0	0	0	0	1	0	1	1	0	0	0	1
1.250 - 1.875	0	0	0	0	0	1	1	0	1	0	1	0	1	0
1.875 - 2.500	0	0	0	0	1	1	1	0	1	0	0	0	1	1
2.500 - 3.125	0	0	0	1	1	1	1	0	0	1	1	1	0	0
3.125 - 3.750	0	0	1	1	1	1	1	0	0	1	0	1	0	1
3.750 - 4.375	0	1	1	1	1	1	1	0	0	0	1	1	1	0
4.375 or greater	1	1	1	1	1	1	1	0	0	0	0	1	1	1

Table assumes that the D-latch's enable input is set HIGH, making the latch appear transparent (D inputs follow Q outputs).

FIGURE H.18

The set of comparators is the key feature to note in this circuit. Each comparator is supplied with a different reference voltage from the 1-kΩ voltage divider network. Since we've set up a +5-V reference voltage, the voltage drop across each resistor within the voltage divider network is 0.625-V. From this you can determine the specific reference voltages given to each comparator (see figure). To convert an analog signal into a digital number, the analog signal is applied to all the comparators at the same time, via the common line attached to the inverting inputs of all the comparators. If the analog voltage is between, say, 2.5000 and 3.125 V, only those comparators with reference voltage below 3.125 V will output a high. To create a 3-bit binary output, the eight comparator outputs are feed into an octal-to-binary priority encoder. A D-latch also can be incorporated into the circuit to provide enable control of the binary output. The truth table should fill in the rest.

TLC5501 6-Bit Flash ADC IC (Texas Instruments)

TLC5501 IC incorporates similar circuitry to the "homemade" flash ADC shown in Fig. H.18. However, unlike the previous circuit, it has a 6-bit output. This device comes with separate analog and digital supplies—both of which have a maximum operating range from –0.5 to +7 V. The analog input range is limited to the range of –0.5 to V_{CC} + 0.5 V. Two reference leads ($V_{ref,T}$ and $V_{ref,B}$) are used to set the analog operating range, while a clock enable input is used as an analog sampling control (see timing diagram).

The function table shown in Fig. H.19 shows all possible digital outputs when $V_{ref,T}$ and $V_{ref,B}$ have been adjusted so that the voltage at the transition from digital 0 to 1 is 4.000 V and the transition to full scale is 4.993 V. By changing the reference voltages, this range can be altered. See Texas Instrument's data sheet for more details.

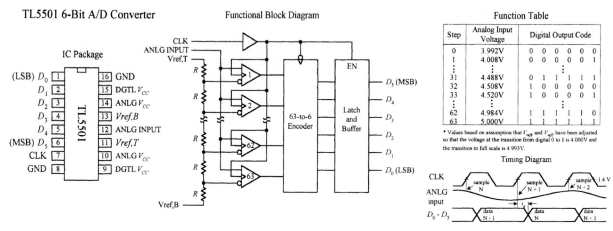

FIGURE H.19

Flash converters are the fastest ADCs available. With a single clock cycle, you can tell exactly what the input voltage is. However, a disadvantage of flash converters is that to achieve 8-bit accuracy requires 256 comparators, while a 10-bit device requires 1024 comparators. To make these comparators operate at high speeds, they must draw considerable amounts of current, and beyond 10 bits, the number of comparators required becomes unmanageable. Flash ADCs have conversion times from 10 to 50 ns and therefore are frequently used in high-speed data acquisition systems such as TV video digitizing (encoding), radar analysis, high-speed digital oscilloscopes, robot vision applications, etc.

Displays

A number of displays can be interfaced with control logic to display numbers, letters, special characters, and even graphics. Two popular displays that we'll consider here include the light-emitting diode (LED) display and the liquid-crystal display (LCD).

I.1   LED Displays

LED displays come in three basic configurations, numeric (numbers), alphanumeric (numbers and letters), and dot matrix forms (see Fig. I.1). Numeric displays consist of 7 LED segments. Each LED segment is given a letter designation, as shown in the figure. 7-segment LED displays are most frequently used to generate numbers (0–9), but they also can be used to display hexadecimal (0–9, A, B, C, D, E, F). The 14-segment, 16-segment, and special 4×7 dot matrix displays are alphanumeric, while the 5×7 dot matrix display is both alphanumeric and graphic—you can display unique characters and simple graphics.

Various types of displays

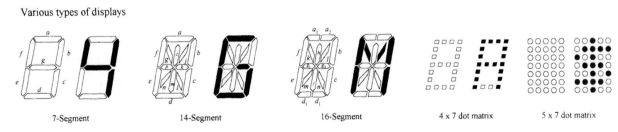

7-Segment   14-Segment   16-Segment   4 x 7 dot matrix   5 x 7 dot matrix

FIGURE I.1

I.1.1   Direct Drive of Numeric LED Displays

Seven-segment LED displays come in two varieties, common anode and common cathode. Figure I.2 shows single digital 8-segment (7 digit segments + decimal point) displays of both varieties.

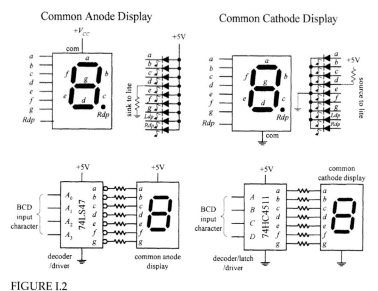

FIGURE I.2

To drive a given segment of a common anode display, current must be sunk out through the corresponding segment's terminal. With the common cathode display, current must be sunk into the corresponding segment's terminal. A simple way to drive these displays is to use BCD to 7-segment display decoder/drivers, like the ones show in the figure. Applying a BCD input character results in a decimal digit being displayed (e.g., 1001 applied to A_0–A_3, or A–D displays a "5"). The 74LS47 active-low open-collector outputs are suited for a common anode display, while the 74HC4511's active-high outputs are suited for a common cathode display. Both ICs also come with extra terminals used for lamp testing and ripple blanking, as well as leading zero suppression (controlling the decimal point). We discussed how to use these terminals in Section 12.6.

When driving a multidigit display, say, one with eight digits, the previous technique becomes awkward. It requires eight discrete decoder/driver ICs. One way to avoid this problem is to use a special direct-drive LED display driver IC. For example, National Semiconductor's MM5450, shown in Fig. I.3, is designed to drive 4- or 5-digit alphanumeric common anode LED displays. It comes with 34 TTL-compatible outputs that are used to drive desired LED segments within a display. Each of these outputs can sink up to 15 mA. In order to specify which output lines are driven high or low, serial input data are clocked into the driver's serial input. The serial data chain that is entered is 36 bits long; the first bit is a start bit (set to 1), while the remaining 35 bits are data bits. Each data bit corresponds to a given output data line that is used to drive a given LED segment within the display. At the thirty-sixth positive clock signal, a LOAD signal is generated that loads the 35 data bits into the latches—see block diagram in Fig. I.3. At the low state of the clock, a $\overline{\text{RESET}}$ signal is generated that clears the shift register for the next set of data. You can learn more about the MM5450 from National's Web site (*www.national.com*).

I.1.2   Multiplexed LED Displays

Another technique used to drive multidigit LED displays involves multiplexing. Multiplexing can drastically reduce the number of connections needed between display and control logic. In a multiplexed display, digits share common segment lines. Also, only one digit within the display gets lighted at a time. To make it appear that a complete readout is displayed, all the digits must be flashed very rapidly in sequence, over and over again. The simple example in Fig. I.4 shows multiplexing in action.

MM5450 (National Semiconductor) LED Dispaly Driver

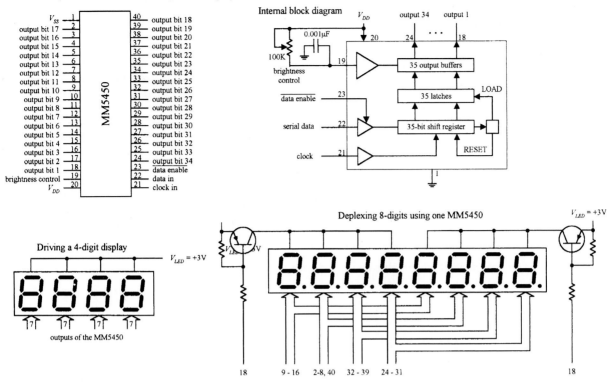

FIGURE I.3

Simple multiplexing scheme

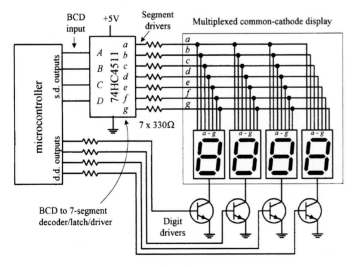

FIGURE I.4

Here we have a multiplexed common-cathode display—all digits share common segment lines (a–g). To supply a full one-digit readout, digits must be flashed rapidly, one at a time. To enable a given digit, the digit's common line is grounded via one of the digital drivers (transistors)—all other digits' common lines are left floating. In this example, the drivers are controlled by a microcontroller. Now, to light the segments of a given digit, the microcontroller supplies the appropriate 4-bit BCD code to the 7-segment decoder/driver (74HC4511). As an example, if we wanted to display 1234, we'd have to program the microcontroller (using software) to turn off all digits except the MSD (leftmost digit) and then supply the decoder/driver with the BCD code for 1. Then the next significant digit (2) would be driven, and then the next significant digit (3), and then the LSD (4). After that, the process would recycle for as long as we wanted our program to display 1234.

I.2   Alphanumeric LED Displays

I.2.1   Simple Alphanumeric Display

Alphanumeric dual display (internally wired for mutliplexing)

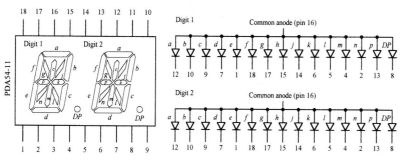

FIGURE I.5

Figure I.5 shows a common anode, 2-character, 14-segment (+ decimal) alphanumeric display. Notice that the segments of the two characters are internally wired together. This means that the display is designed for multiplexing. Though it is possible to use a microcontroller along with transistor drivers to control this display, the number of lines required is fairly large. Another option is to use a special driver IC, like Intersil's ICM7423B 14-segment 6-bit ASCII driver. Another alternative is simply to avoid using this kind of display and use a "smart" alphanumeric display that contains all the necessary control logic (drivers, code converters, etc.).

I.2.2   "Smart" Alphanumeric Display

HPDL-1414 (Hewlett Packard) four character smart alphanumeric display

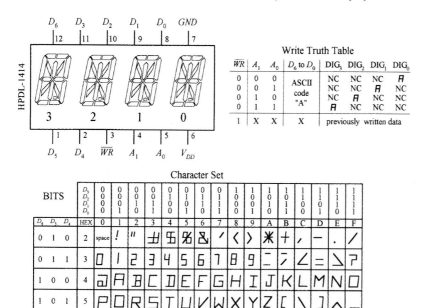

FIGURE I.6

The HPDL-1414 is a "smart," 4-character, 16-segment display. This device is complete with LEDs, onboard four-word ASCII memory, a 64-word character generator, 17-segment drivers, 4-digit drivers, and scanning circuitry necessary to multiplex the four LED characters. It is TTL compatible and is relatively easy to use. The seven data inputs D_0 to D_6 accept a 7-bit ASCII code, while the digital select inputs A_0 and A_1 accept a 2-bit binary code that is used to specify which of the four digits is to be lighted. The WRITE ($\overline{WR}$) input is used to load new data into memory. After a character has been written to memory, the IC decodes the ASCII data, drives the display, and refreshes it without the need for external hardware or software.

I.2.3   "Smart" Hexadecimal and Numberic Dot Matrix Displays

Figure I.7 shows Hewlett Packard's HDSP-076x series of 4×7 dot matrix displays: −0760 (numeric, right-hand decimal point), −0761 (numeric, left-hand decimal point), −0762 (hexadecimal), and the −0763 (over range ±1). These displays are "smart" solid-state devices that contain onboard data memory, decoder, and display drivers. For both the numeric and hex displays, a 4-bit positive BCD code, applied to pins 8, 1, 2, and 3, is used to select a number or character—see truth table within the figure. The numeric displays also come with decimal and enable inputs. When an enable input is high, any changes in BCD input levels have no effect on display memory, displayed character, or decimal point. The hexadecimal displays do not have decimal points but come with blanking control inputs. The blanking input has no effect on the display memory. The over-range display uses pins 1, 2, 3, 4, and 8 to select either "+" or "−," a "1," a decimal point, or blanking condition—see truth table in figure. Check out Hewlett Packard's data sheets for more information about these displays.

HDSP-076x series (Hewlett Packard) of hexadecimal and numeric displays

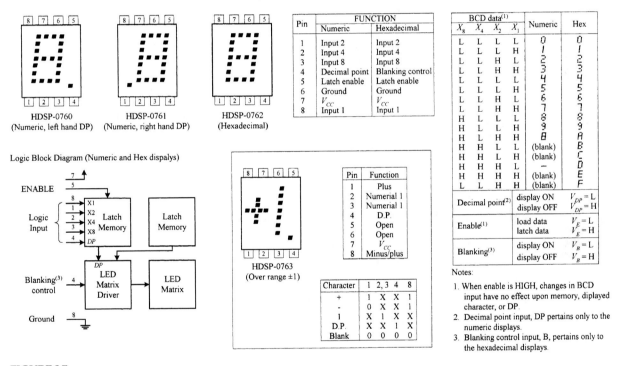

FIGURE I.7

I.2.4 5 × 7 Dot Matrix Displays

Figure I.8 shows Infineon's DL07135 intelligent 5 × 7 dot matrix display. This device has a 96 ASCII character set (uppercase and lowercase characters included), built-in memory, built-in character generator, built-in multiplex and LED drive circuitry, built-in lamp test, and intensity control (4 levels). The tables and timing diagrams should provide you with a general understanding of how this device works. For specifics, check out Infineon's data sheet.

DL07135 (Infineon Technologies) 5 x 7 Dot Matrix Intelligent Display

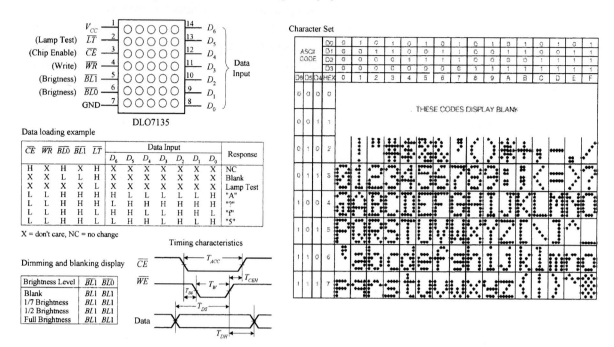

FIGURE I.8

Multidigit 5 × 7 Dot Matrix Displays by Siemens

SLR2016: 4 DIGIT 5 × 7 DOT MATRIX DISPLAY
PDSP1880: 8-CHARACTER 5 × 7 DOT MATRIX ALPHANUMERIC PROGRAMMABLE DISPLAY

The SLR2016 is a 4-digit 5 × 7 dot matrix display module with a built-in CMOS circuit. The integrated circuit contains memory, a 128 ASCII ROM decoder, multiplexing circuitry, and drivers. The SLR2016 has two address bits (A_0, A_1), seven parallel inputs (D_0–D_6) for ASCII code, a Write input ($\overline{WR}$), Blanking input ($\overline{BL}$), and Clear function input ($\overline{CLR}$) used to clear the ASCII character RAM. The character set consists of 128 special ASCII characters for English, German, Italian, Swedish, Danish, and Norwegian. The desired data code (D_0–D_6) and digit address (A_0, A_1) must be held stable during the write cycle for storing new data. Data entry may be asynchronous. Digit 0 is defined as a right-hand digit with $A_1 = A_2 = 0$. Clearing the entire internal 4-digit memory can be accomplished by holding the Clear input ($\overline{CLR}$) low for at least 1 ms. A multicharacter display system can be built using a number of SLR2016s because each digit can be addressed independently and will continue to display the character last stored until replaced with another.

The PDSP1880 is similar to the preceding display module, but it also has a programmable ROM that can be programmed by the manufacturer to provide specified custom characters. It also has 8 digits. This device is a bit technical to understand, but if you're interested, check out manufacturer's data sheet.

SLR2016 (Siemens) 4-Digit 5 x 7 Dot Matrix
Alphanumeric Intelligent Display

SLR2016 Character Set

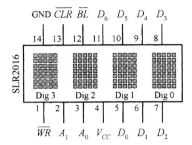

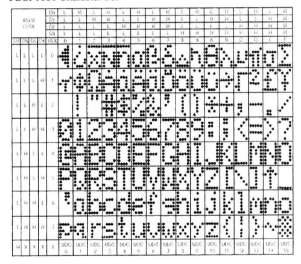

PDSP1880 (Siemens) 8-Character 5 x 7 Dot
Matrix Alphanumeric Programmable Display

PDSP1880 Character Set

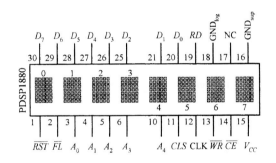

FIGURE I.9

I.3   Liquid-Crystal Displays

In low-power CMOS digital systems (e.g., battery- or solar-powered electronic devices), the dissipation of an LED display can consume most of a system's power requirements, something you want to avoid, especially since you are looking to save power when using CMOSs anyway. Liquid-crystal displays (LCDs), on the other hand, are ideal for low-power applications. Unlike a LED display, an LCD is a passive device. This means that instead of using electric current to generate light, it uses light that is already externally present (e.g., sunlight, room lighting). For the LCD's optical effects to occur, the external light source need only supply a minute amount of power (within the mW/cm^2 range).

One disadvantage with LCDs is their slow switching speeds (time it takes for a new digit/character to appear). Typical switching speeds for LCDs range from around 40 to 100 ms. At low temperature, the switching speeds get even worse. Another problem with LCDs is the requirement that external light be present. Though there are LCD displays that come with backlighting (e.g., LED behind the display), the backlighting alone tends to defeat attempts at keeping power consumption to a minimum.

I.3.1   Basic Explanation of How an LCD Display Works

An LCD display consists of a number layers that include a polarizer, a set of transparent electrodes, a liquid-crystal element, a transparent back electrode, a second polarizer, and a mirror. See leftmost figure below.

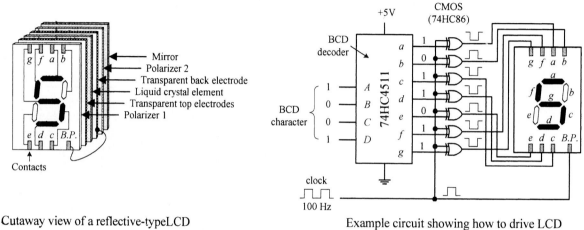

Cutaway view of a reflective-typeLCD

FIGURE I.10

Example circuit showing how to drive LCD

The transparent top electrodes are used to generate the individual segments of a digit, character, etc., while the transparent back electrode forms a common plane, often referred to as the *back plane* (BP). The top electrode segments and the back electrode are wired to external contacts. With no potential difference between a given top electrode and the back electrode, the region where the top electrode is located appears silver in color against a silver background. However, when a potential is applied between a given top electrode and back electrode, the region where the top electrode is located appears dark against a silver background.

The circuit in Fig. I.10 shows a simple way to drive a 7-segment LCD. It uses a 74HC4511 BCD decoder and XOR gates to generate the prior drive signals for the LCD. A very important thing to note in this circuit is the clock. As it turns out, an LCD actually requires ac drive signals (e.g., squarewaves) instead of dc drive signals. If dc were used, the primary component of the display, namely, the liquid crystal, would undergo electrochemical degradation (more on the liquid crystal in a moment). The optimal frequency of the applied ac drive signal is typically from around 25 Hz to a couple hundred hertz. Now that we understand that, it is easy to see why we need the XOR gates. As the clock delivers squarewaves to the back electrode (back plane, or BP), the XOR gates act as enable gates that pass and invert a signal and apply it to a given top electrode segment. For example, if a

BCD code of 1001 (5) is applied to the decoder, the decoder's outputs a, c, d, f, and g go high, while outputs b and e go low. When a positive clock pulse arrives, XOR gates attached to the outputs that are high invert the high levels. XOR gate attached to outputs that are low pass on the low levels. During the same pulse duration, the back plane is set high. Potentials now are present between a, c, d, f, and g segments and the back plane, and therefore these segments appear dark. Segments b and e, along with the background, appear silverish because no potential exists between them and the back plane. Now, when the clock pulse goes low, the display remains the same (provided the BCD input hasn't changed), since all that has occurred is a reverse in polarity. This has no effect on the optical properties of the display.

I.3.2 Detailed Explanation of How an LCD Works (The Physics)

Figure I.11 shows how an LCD generates a clear (silverish) segment. When control signals sent to the transparent top and back electrodes are in phase, no potential exists between the two electrodes. With no potential present, the cigar-shaped organic liquid crystals (nematic crystals) arrange themselves in spiral state, as shown in the figure. The upper crystal aligns itself horizontal to the page, while the lowest crystal aligns itself perpendicular to the page. The upper crystal and the lower crystal are held in place by tiny groves that are edged into the inner surfaces of the glass surfaces of the cell. Crystals in between the upper crystal and the lower crystal progressively spiral 90° due to electrostatic forces that exist between neighboring crystals. When polarized light passes through a region of the display that contains these spirals, the polarization angle of the light is rotated 90°. Now, looking at the display as a whole, when incident unpolarized light passes through polarizer 1 (as shown in the figure), the light becomes polarized in the same direction of the plane of polarization of the first polarizer. The polarized light then passes through the transparent top electrode and enters the liquid-crystal cell. As it passes through the cell, its polarization angle is rotated 90°. The polarized light that exits the cell then passes through the transparent back electrode and the second polarizer without problems. (If we were to remove the liquid-crystal cell, all polarized light that passed through the first polarizer would be absorbed, since we'd have crossed polarizers.) The light that passes through the second polarizer then reflects off the mirror, passes through the second polarizer, on through the liquid-crystal cell (getting rotated 90°), through the first polarizer, and finally reaches the observer's eye. This reflected light appears silver in color. Note that the background of LCD displays constantly appears silver because no potential exists across the liquid-crystal cell in the background region.

Figure I.12 shows how an LCD generates a dark segment. When control signals sent to top and back electrodes are out of phase, a potential difference exists between the two electrodes. This causes the crystals to align themselves in a parallel manner, as shown in the figure. When the polarized light from the first polarizer passes through the cell region containing these parallel crystals, nothing happens—the polarization angle stays the same. However, when the light comes in contact with the second polarizer, it is absorbed because the angle of polarization of the light and the plane of polarization of the second polarizer are perpendicular to each other. Since

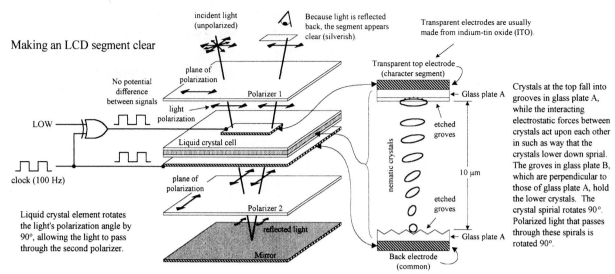

Making an LCD segment clear

No potential difference between signals

LOW

clock (100 Hz)

Liquid crystal element rotates the light's polarization angle by 90°, allowing the light to pass through the second polarizer.

incident light (unpolarized)

Because light is reflected back, the segment appears clear (silverish).

Transparent electrodes are usually made from indium-tin oxide (ITO).

plane of polarization

Polarizer 1

light polarization

Liquid crystal cell

plane of polarization

Polarizer 2

reflected light

Mirror

Transparent top electrode (character segment)

Glass plate A

nematic crystals

etched groves

10 μm

etched groves

Glass plate A

Back electrode (common)

Crystals at the top fall into grooves in glass plate A, while the interacting electrostatic forces between crystals act upon each other in such as way that the crystals lower down sprial. The groves in glass plate B, which are perpendicular to those of glass plate A, hold the lower crystals. The crystal spirial rotates 90°. Polarized light that passes through these spirals is rotated 90°.

Note: LCD's require ac waveforms. Using dc would result in electrochemical degradation of the cell material.

FIGURE I.11

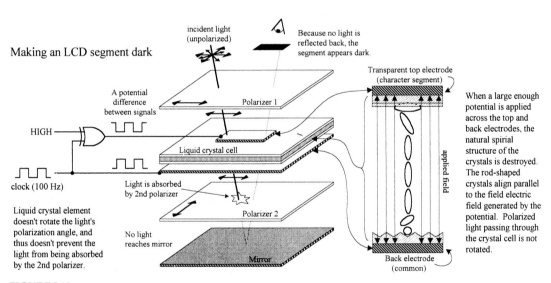

Making an LCD segment dark

A potential difference between signals

HIGH

clock (100 Hz)

Liquid crystal element doesn't rotate the light's polarization angle, and thus doesn't prevent the light from being absorbed by the 2nd polarizer.

incident light (unpolarized)

Because no light is reflected back, the segment appears dark.

Polarizer 1

Liquid crystal cell

Light is absorbed by 2nd polarizer

Polarizer 2

No light reaches mirror

Mirror

Transparent top electrode (character segment)

applied field

Back electrode (common)

When a large enough potential is applied across the top and back electrodes, the natural spirial structure of the crystals is destroyed. The rod-shaped crystals align parallel to the field electric field generated by the potential. Polarized light passing through the crystal cell is not rotated.

FIGURE I.12

light reaches the mirror, no light is reflected back to the observer's eye, and hence the segment appears dark.

The LCD display shown in Fig. I.11 represents what is referred to as a *standard twisted nematic display*. Another common LCD is the supertwist nematic display. Unlike the standard twisted display, this display's nematic crystals rotate 270° from top to bottom. The extra 180° twist improves the contrast and viewing angle.

I.3.3 Driving Liquid-Crystal Displays

CD4543B CMOS BCD-to-7-Segment Latch/Decoder/Driver (Texas Instruments)

The CD4543B, shown below, is a BCD-to-7-segment latch/decoder/driver that is designed for LCDs, as well as for LED displays. When used to drive LCDs, a squarewave must be applied simultaneously to the CD4543B's Phase (*Ph*) input and to the LCD's backplane. When used to drive LED displays, a high is required at the Phase input for common cathode displays, while a low is required for common anode displays. To blank the display (set outputs *a–g* low), the *BL* input is set high. The CD4543B also comes with a Latch Disable input (*LD*) which can be used to latch onto input data, preventing new input data from altering the display.

CMOS BCD-to-Seven-Segment Latch/Decoder/Driver for LCDs

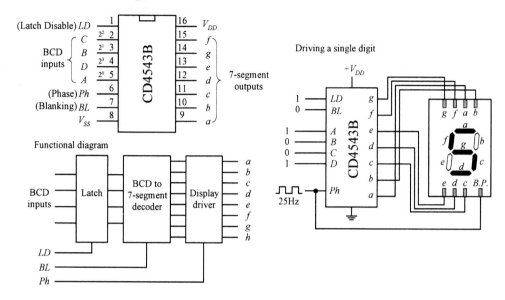

FIGURE I.13

MM5453 Liquid-Crystal Display Driver (National Semiconductor)

The MM5453 is a 40-pin IC that can drive up to 33 segments of LCD, which can be used to drive 4½-digit 7-segment displays. It houses an internal oscillator section (requiring an external *RC* circuit) that generates the necessary squarewaves used to drive the LCD. To activate given segments within the display, a serial code is applied to the data input. The code first starts out with a start bit (high) followed by data bits that specify which outputs should be driven high or low. Figure I.14 shows an example display circuit, along with corresponding data format required to drive a 4½-digit display.

VI-322-DP LCD Display and ICL7106 3½-Digit LCD, A/D Converter Driver

There are a number of specialized LCD displays that can be found in the electronics catalogs. An example is Varitronix's VI-322-DP 3½-digit (plus ~, +, BAT, Δ) LCD, shown below. This display is configured in a static drive arrangement (each segment has a separate lead) and is found in many test instruments. To drive this display, you first check to see what kind of driver the manufacturer suggests. In this case, the manufacturer suggest using Intersil's ICL7106. This IC is a 3½-digit LCD/LED display driver as well as an A/D converter. This dual-purpose feature makes it easy to interface transducers directly to the same IC that driving the display. To learn how to use the ICL7106, check out Intersil's data sheet at *www.intersil.com*.

MM5453 (National Semiconductor) Liquid Crystal Display Driver

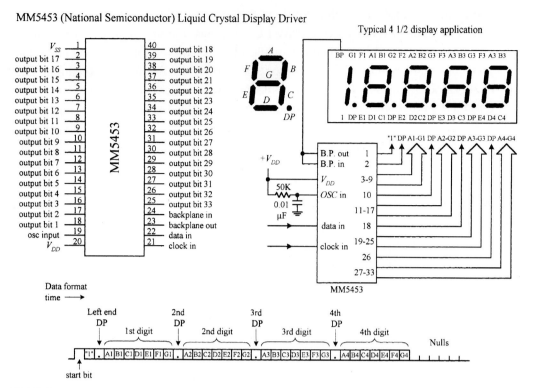

FIGURE I.14

VI-322-DP (Varitronix) static-drive LCD

ICL7106PL 3 1/2 digit driver

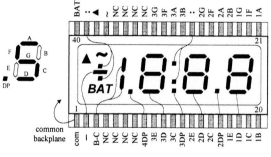

FIGURE I.15

I.3.4 Multiplexed LCD Displays

We have just seen examples of static-drive-type LCDs, where each segment (to electrode) had its own lead, and a single common plane was used as the back electrode. Another type of LCD is designed with multiplexing in mind and is referred to as *dynamic drive* or *multiplexed display*. As with the multiplexed LED display, multiplexed LCD displays can greatly reduce the number of external connections required between display and driver. However, they require increased complexity in drive circuitry (or software) to drive. In a multiplexed LCD, appropriate segments are connected together to form groups which are sequentially addressed by means of multiple backplane electrodes.

VIM-1101-2 Multiplexed LCD and ICM7231 LCD Driver

Figure I.16 shows Vartronix's VIM-1101-2 multiplexed LCD. This display is actually referred to as a *triplexed display* (⅓ multiplexed) because the actual segments and decimals are grouped in threes. Notice that the pin count for this display is quite small when compared with that of the previous static-drive VIM-322-DP LCD. The reduction in external connections is made possible by joining top electrode segments together, as shown in the figure. Also notice how backplane electrodes all share three common lines. To light the various segments of the display is very difficult. It requires exacting address sequencing and specialized waveforms. Again, to see if there is an easy way out, we check to see if Vartronix specifies an appropriate driver IC. One such driver Vartronix specifies is Intersil's ICM7232. This device is designed to generate the voltage levels and switching waveforms required to drive triplexed displays with ten 7-segment digits with two independent annunciators—like the VIM-1101-2 display. To write to the display, 6 bits of data and 4 bits of digit address are clocked serially into a shift register and then decoded and written to the display. Check out Intersil's data sheet for the specifics.

VIM-1101-2 (Varitronix) multiplexed LCD and ICM7231 (Intersil) LCD diver

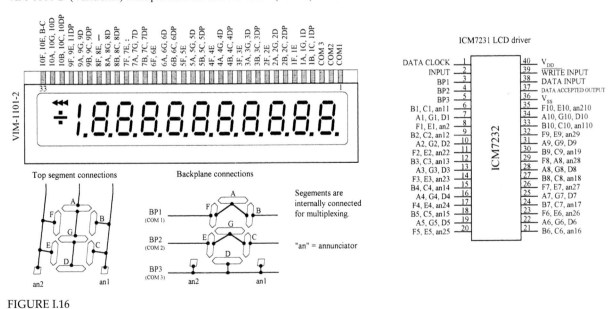

FIGURE I.16

I.3.5 "Intelligent" Dot Matrix LCD Modules

Dot matrix LCD displays are used to display alphanumeric characters and other symbols. These displays are used in cell phones, calculators, vending machines, and many other devices that provide the user with simple textual information. Dot matrix LCDs are also used in laptop computer screens; however, these displays incorporate special filters, multicolor back lighting, etc. For practical purposes, we'll concentrate on the simple alphanumeric LCD displays.

An alphanumeric LCD screen is usually divided into a number of 5×8 pixel blocks, with vertical and horizontal spaces separating each block. Figure I.17 shows a display with 20 columns and 4 rows of 5×8 pixel blocks. Other standard configurations come with 8, 16, 20, 24, 32, or 40 columns and 1, 2, or 4 rows. To generate a character within a given block requires that each pixel within the block be turned on or off. Now, as you can imagine, to control so many different pixels (electrode segments) requires a great deal of sophistication. For this reason, an intelligent driver IC is required.

Alphanumeric LCD module

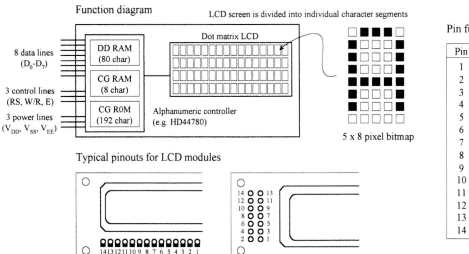

FIGURE I.17

Almost all alphanumeric LCD modules are controlled by Hitachi's HD44780 (or equivalent) driver IC. This driver contains a permanent memory (CG ROM) that stores 192 alphanumeric characters, a random access memory (DD RAM) used to store the display's contents, a second random access memory (CG RAM) used to hold custom symbols, input lines for data and instruction control signals, multiplexed outputs for driving LCD pixels, and additional outputs for communicating with expansion chips to drive more LCD pixels. This driver is built right into the LCD module. (You could attempt to construct your own module be interfacing the driver with an LCD, but it would not be worth the effort—the numerous tiny connections would drive you nuts.) From now on, all modules described in this section are assumed to be HD44780-driven.

Basic Overview of the Pins

The standard LCD module comes with a 14-pin interface: 8 data lines (D_0–D_7), 3 control lines (RS, W/R, E), and three power lines (V_{DD}, V_{SS}, V_{EE}). V_{DD} (pin 2) and V_{SS} (pin 1) are the module's positive and negative power supply leads. Usually V_{DD} is set to +5 V, while V_{SS} is grounded. V_{EE} (pin 3) is the display's contrast control. By changing the voltage applied to this lead, the contrast of the display increases or decreases. A potentiometer placed between supply voltages, with its wiper connected to V_{EE}, allows for manual adjustment. D_0–D_7 (pins 7–14) are the data bus lines. Data can be transferred to and from the display either as a single 8-bit byte or as two 4-bit nibbles. In the latter case, only the upper four data lines (D_4–D_7) are used. RS (pin 4) is the Register Select line. When this line is low, data bytes transferred to the display module are interpreted as commands, and data bytes read from the display module indicate its status. When the RS line is set high, character data can be transferred to and from the display module. R/W (pin 5) is the Read/Write control line. To write commands or character data to the module, R/W is set low. To read character data or status information from the module R/W is set high. E (pin 6) is the Enable control input, which is used to initiate the actual transfer of command or character data to and from the module. When writing to the display, data on the D_0–D_7 lines is transferred to the display when the enable input receives a high-to-low transition. When reading from the display, data become available to the D_0–D_7 lines shortly after a low-to-high transition occurs at the enable input and will remain available until the signal goes low again.

Figure I.18 shows the instruction set and standard set of characters for a LCD module. In a moment we'll go through some examples illustrating how to use the instructions and how to write characters to the display.

Test Circuit Used to Demonstrate How to Control the LCD Module

Figure I.19 shows a simple test circuit that is quite useful for learning how to send commands and character data to the LCD module. (In reality, the LCD module is connected to a microprocessor or microcontroller, as shown to the left in the figure. We'll discuss these interfaces later in this section.) In this circuit, switches connected to data inputs use pullup resistors in order to supply a high (1) when the switch is open or supply a low (0) when the switch is closed. The enable input receives its high and low levels from a debounced toggle switch. Debouncing the enable switch prevents the likelihood of multiple enable signals being generated. Multiple enable signals tend to create unwanted effects, such as generating the same character over and over again across the display. The 5-kΩ pot is used for contrast. Note that in this circuit we've grounded the R/W line, which means we'll only deal with writing to the display.

When Power Is First Applied

When power is first applied to the display, the display module resets itself to its initial settings. Initial settings are indicated in the LCD instruction set with an asterisk. As indicated, the display is actually turned off during the initial setting condition. If we attempt to write character data to the display now, nothing will show up. In order to show something, we must issue a command to the module telling it to turn on its display. According to the instruction set, the Display & Cursor On/Off instruction can be used to turn on the display. At the same time, this instruction also selects the cursor style. For example, if we apply the command code 0000 1111 to D_7–D_0, making sure to keep RS low so the module will interpret data as a command, a blinking cursor with underline should appear at the top leftmost position on the display. But before this command can take effect, it must be sent to the module by momentarily setting the Enable line (E) low.

Another important instruction that should be implemented after power-up is the Function Set command. When a 2-line display is used, this command tells the module to turn on the second line. It also tells the module what kind of data transfer is

LCD Instruction Set

INSTRUCTION	R/S	R/W	D_7	D_6	D_5	D_4	D_3	D_2	D_1	D_0
Clear Display	0	0	0	0	0	0	0	0	0	1
Display & Cursor Home	0	0	0	0	0	0	0	0	1	X
Character Entry Mode	0	0	0	0	0	0	0	1	I/D	S
Display & Cursor On/Off	0	0	0	0	0	0	1	D	C	B
Display/Cursor Shift	0	0	0	0	0	1	D/C	R/L	X	X
Function Set	0	0	0	0	1	DL	N	F	X	X
Set CGRAM Address	0	0	0	1	A	A	A	A	A	A
Set Display Address	0	0	1	A	A	A	A	A	A	A
Poll the "Busy Flag"	0	0	BF	X	X	X	X	X	X	X
Write Character to Display [a]	1	0	D	D	D	D	D	D	D	D
Read Character on Dsplay [b]	1	1	D	D	D	D	D	D	D	D

I/D = Increment (I/D = 1)*/Decrement (I/D = 0) each byte written to display
S = Display shift on (S = 1), Display shift off (S = 0)*
D = Turn display on (D = 1), Turn display off (D = 0)*
C = Show cursor (C = 1), Hide cursor (C = 0)
B = Underline cursor (B = 0, C = 1), Blink cursor (B = 1, C = 1)
D/C = Move display (D/C = 1), Move cursor (D/C = 0)
R/L = Direction of shift: Shift right (R/L = 1), Shift left (R/L = 0)
DL = Set data interface length: 8-bit interface (DL = 1)*, 4-bit interface (DL = 0)
N = Number of display lines: 2 line mode (N = 1), 1 line mode (N = 0)*
F = Character font format: 5 x 10 dot (F = 1), 5 x 7 dot (F = 0)*
BF = Poll the Busy Flag: controller not busy (BF = 0), controller busy (BF = 1)
A = CGRAM or display address bit
D = Character data bit
a = Write character to display at the current cursor position
b = Read character on display at the current cursor position
X = Don't care
* = Initialization settings

Standard LCD Character Table

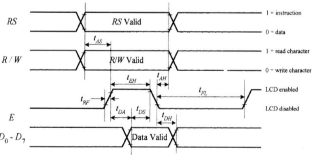

Steps used to Read and Write data to and from LCD module (HD44780 controlled)

HD44780 Timing Diagram

Steps for displaying a character
Set to Write Character to Display mode: R/W = 0, RS = 1.
Apply data bits (character code) to D_7 - D_0.
Breifly set E = 1, then set E = 0

Steps for reading data from display
Set to Read Character on Display mode: R/W = 1, RS = 1.
Set E = 1.
Read data from D_7 - D_0.
Set E = 0.

1 = instruction
0 = data
1 = read character
0 = write character
LCD enabled
LCD disabled
Data Valid

t_{AS} (Address setup time) - For data inputs to be interpreted correctly, they must be held for a minimum time of t_{AS} (~140ns) prior to the Enable signal.
t_{EH} (Enable high time) - E must be held HIGH for a minimum of t_{EH} (~450ns) for proper operation.
t_{DS} (Data setup time) - Data inputs must be held stable for a time t_{DS} (~200ns) prior to the E signal for proper operation.
t_{AH} (Address hold time) - Control lines RS and R/W lines must not change for a duration of t_{AH} (~10ns) after the E line goes LOW for proper operation.
t_{DH} (Data hold time) - Data lines D_0 - D, must not change for a duration of t_{DH} (~20ns) after the E line goes LOW for prior operation.
t_{EL} (enable low time) - E line must not be set HIGH again (for the next command), for at least t_{EL} (~500ns) for proper operation.
t_{RF} (rise and fall time) - Rise and fall times are each ~25ns each.

FIGURE I.18

going to be used (8-bit or 4-bit; more on this later), and whether a 5 × 10 or 5 × 7 pixel format will be used (5 × 10 is found in some 1-line displays). Assuming that the display used in our example circuit is a 2-line display, we can send the command 0011 1000 telling the display to turn on both lines, use an 8-bit transfer, and provide a 5 × 7 pixel character format. Again, to send this command, we set RS low, then supply the command data to D_7–D_0, and finally pulse E low.

Simple experimental setup that uses switches to write to LCD module

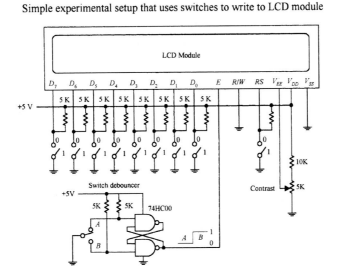

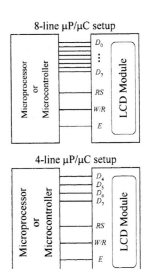

FIGURE I.19

Now that the module knows what format to use, we can try writing a character to the display. To do this, we set the module to character mode by setting *RS* high. Next, we apply one of the 8-bit codes listed in the Standard LCD Character Set table to the data inputs D_7–D_0. For example, if we wanted to display the letter *Q*, we'd apply 01010001 (hex 51 or 51_H). To send the character data to the LCD module, we pulse *E* low. A *Q* should then appear on the display. To clear the screen, we use the Clear Display command 0000 0001, remembering to keep *RS* low and then pulsing *E* low.

Addressing

After power-up, the module's cursor is positioned to the far-left corner of the first line of the display. This display location is assigned a hexadecimal address of 00_H. As new characters are entered, the cursor automatically moves to the right to a new address of 01_H, then 02_H, etc. Though this automatic incrementing feature makes life easy when entering characters, there are times when it is necessary to set the cursor position to a location other than the first address location.

To set the cursor to another address location, a new starting address must be entered as a command. There are 128 different addresses to choose from, although not all these addresses have their own display location. In fact, there are only 80 display locations laid out on a single line in one-line mode or 40 display locations laid out on each line in two-line mode. Now, as it turns out, not all display locations are necessarily visible on the screen at one time. This will be made more apparent in a moment. Let's first try a simple address example with the LCD module set to two-line mode (provided that two lines are actually available).

To position the cursor to a desired location, we use the Set Address command. This command is specified with the binary code 1000 0000 + (binary value of desired hex address). For example, to send a command telling the cursor to jump to the 07_H address location, we would apply (1000 0000 + 0000 0111) = 1000 0111 to the D_7–D_0 inputs, remembering to hold *RS* low and then pulsing *E* low. The cursor should now be located at the eighth position over from the left.

It is important to realize that the relationship between addresses and display locations varies from module to module. Most displays are configured with two lines of characters, the first line starting at address 00_H and the second line at address 40_H. Figure I.20 shows the relationship between address and display locations for various LCD modules. Note that the four-line module is really a two-line type with the two lines split, as shown in the figure.

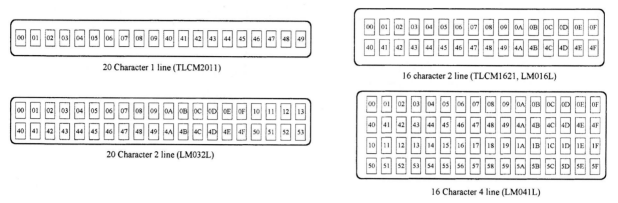

FIGURE I.20

Shifting the Display

As mentioned a second ago, regardless of size, LCD modules have 80 display locations that can be written to. With smaller displays, not all 80 locations can be displayed at once on the screen. For example, if we were to enter all the letters of the alphabet onto the first line of a 20-character display, only letters *A* through *T* would appear on the screen. Letters *S* through *Z*, along with the cursor, would be "pushed off" to the right of the screen, hidden from view. To bring these hidden characters into view, we can apply the Cursor/Display Shift command to shift all display locations to the left. The command for shifting to the left is 0001 1000. Every time this command is issued, the characters shift one step to the left. In our example, it would take 7 of these commands to bring *T* through *Z* and the cursor into view. To shift things to the right, we apply the command 0001 1100. To bring the cursor back to address 00_H and shift the display address 00_H back to the left-hand side of the display, a Cursor Home command (0000 0010) can be issued. Another alternative is to use the Clear Display command 0000 0001. However, this command also clears all display locations.

Character Entry Mode

If you do not want to enter characters from left to right, you can use the Character Entry Mode to enter characters from right to left. To do this, the cursor must first be sent to the rightmost display location on the screen. After that, the Character Entry Mode command 0000 0111 is entered into the module. This sets the entry mode to autoincrement/display shift left. Now, when characters are entered, they appear on the right-hand side, while the display shifts left for each character entered.

User-Defined Graphics

Commands 0100 0000 to 0111 1111 are used to program user-defined graphics. To program these graphics on-screen, the display is cleared, and the module is sent a Set Display Address command to position the cursor at address 00_H. At this point, the contents of the eight user character locations can be viewed by entering binary data 0000 0000 to 0000 0111 in sequence. These characters will appear initially as garbage.

To start defining the user-defined graphics, a Set CGRAM command is sent to the module. Any value between 0100 0000 (40H) and 0111 1111 (7F) will work. Data entered from now on will be used to construct the user-defined graph, row by row. For example, to create a light bulb, the following data entries are made: 0000 1110, 0001 0001, 0001 0001, 0001 0001, 0000 1110, 0000 1010, 0000 1110, 0000 0100. Notice that the first three most significant bits are always 0 because there are only 5 pixels per row. Other user-defined graphics can be defined by entering 8 byte sequence, and so on. Figure I.21 shows how the CGRAM address corresponds to the individual pixels of the user-defined graphic.

There are up to eight user-defined graphics that can be programmed. These then become part of the character set and can be displayed by using codes 0000 0000 to 0000 1111 or 0000 1000 to 0000 1111, both of which produce the same result.

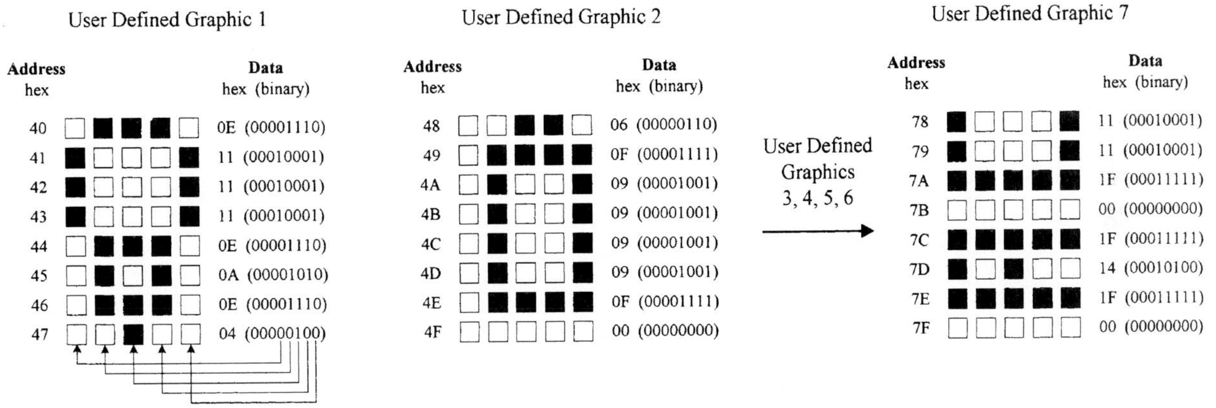

FIGURE I.21

One problem when creating user-defined graphics is they will be lost when power is removed from the module—a result of the volatile CGRAM. Typically, the user-defined graphic data are actually stored in an external nonvolatile EPROM or EEPROM, where they are copied by a microprocessor and loaded into the display module sometime after power-up.

4-Bit Data Transfer

As indicated in the Function Set command, the LCD module is capable both 8-bit and 4-bit data transfer. In 4-bit mode, only data lines D_4–D_7 are used. The other four lines, D_0–D_3, are left either floating or tied the power supply. To send data to the display requires sending two 4-bit chunks instead of one 8-bit word.

When power is first applied, the module is set up for 8-bit transfer. To set up 4-bit transfer, the Function Set command with binary value 0010 0000 is sent to the display. Note that the since there are only 4 data line in use, all 8 bits cannot be sent. However, this is not a problem, since the 8-bit/4-bit selection is on data bit D_4. From now on, 8-bit character and command bits must be sent in two halves, the first 4 most significant bits and then the remaining 4 bits. For example, to write character data 0100 1110 to the display would require setting RS high, applying 0100 to the data lines, pulsing E low, then applying 1110 to the data lines, and pulsing E low again.

The 4-bit transfer is frequently used when the LCD module is interfaced with a microcontroller that has limited I/O lines. See Fig. I.19.

Memory Devices

Memory devices provide a means of storing data on a temporary or permanent basis for future recall. The storage medium used in a memory device may be a semiconductor-based IC (primary memory), a magnetic tape, magnetic disk, or optical disk (secondary memories). In most cases, the secondary memories are capable of storing more data than primary memories because their surface areas on which data can be stored are larger. However, at the same time, secondary memories take much longer to access (read or write) data because memory locations on a disk or tape must be physically positioned to the point where they can be read or written to by the read/write mechanism. Within a primary memory device, memory locations are arranged in tiny regions within a large matrix, where each memory location can be accessed quickly (matter of nanoseconds) by applying the proper address signals to the rows within the matrix.

Figure J.1 shows an overview of primary and secondary memories. In this section we'll discuss only the primary memories, since these devices are used more frequently in designing gadgets than secondary memories. Secondary memories are almost exclusively used for storing large amounts of computer data, audio data, or video data.

Today, the technology used in the construction of primary memory devices is almost exclusively based on MOSFET transistors. Bipolar transistors are also used within memory ICs. However, these devices are less popular because the amount of data they can store is significantly smaller than that of a memory IC built with MOSFET transistors. At one time, bipolar memories had a significant edge in speed over MOSFET memories, but today the speed gap has almost disappeared.

The primary family of memory devices consists of two basic subfamilies: *read-only memories* (ROMs) and *read/write memories* (RWM)—more commonly referred to as *random access memories* (RAM). Within each of these subfamilies there exist more subfamilies, as shown in Fig. J.1. Let's start out by discussing the ROM devices first.

J.1 Read-Only Memories (ROMs)

A read-only memory, or ROM, is used to store data on a permanent basis. These devices are capable of random access, like RAMs, but unlike RAMs, they do not lose stored data when power is removed from the IC. ROMs are used in nearly all com-

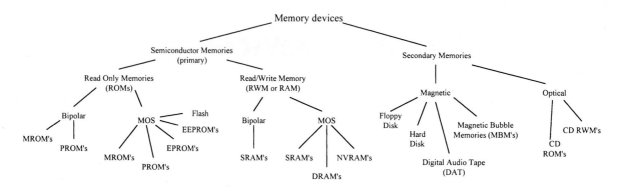

FIGURE J.1

puters to store boot-up instructions (stack allocation, port and interrupt initializations, instructions for retrieving the operating system from disk storage, etc.) that get enacted when the computer is first turned on. ROMs are used in microcontroller applications (simple-function gadgets: appliances, toys, etc.) where the entire stand-alone program resides in ROM. The microcontroller's CPU (central processing unit) retrieves the program instructions and uses volatile RAM for temporary data storage as it runs through the ROM's stored instructions. In some instances, you find ROMs within discrete digital hardware where they are used to store look-up tables or special code-conversion routines. For example, digital data from an analog-to-digital converter could be used to address stored words that represent, say, a binary equivalent to a temperature reading in Celsius or Fahrenheit. They also can be used to replace a complex logic circuit, where, instead of using a large number of discrete gates to get the desired function table, you simply program the ROM to provide the designed output response when input data are applied. The last few applications mentioned, however, are becoming a bit obsolete—the microcontroller seems to be taking over everything.

ROMs are generally used for read-only operations and are not written to after they are initially programmed. However, there are some ROM-like devices, such as EPROM, EEPROMs, and flash memories, that are capable of erasing stored data and rewriting data to memory. Before we take a look at these erasable ROM-like devices, as well as the nonerasable ROM devices, let's first cover some memory basics.

J.2   Simple ROM Made Using Diodes

To get a general idea of how read-only memory works, let's consider the simple circuit in Fig. J.2.

In reality, today's ROMs rarely use diode memory cells. Instead, they typically use transistorlike memory cells formed unto silicon wafers. Also, a more realistic ROM device comes with three-state output buffers that can be enabled or disabled (placed in a high Z state) by applying a control signal. The three-state buffers make it possible to effectively disconnect the memory from a data bus to which it is attached. (In our simple diode memory circuit, the data are always present on the output lines.) The basic layout, with address decoder and memory cells, is pretty much the same

Basic Diode ROM

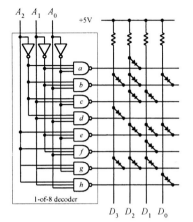

$A_2\ A_1\ A_0$	$D_3\ D_2\ D_1\ D_0$	Active gate (sink)
0 0 0	1 0 1 1	a
0 0 1	0 0 1 0	b
0 1 0	1 0 0 1	c
0 1 1	0 1 1 1	d
1 0 0	1 0 0 0	e
1 0 1	1 1 0 1	f
1 1 0	0 0 0 1	g
1 1 1	1 1 1 0	h

FIGURE J.2

Here is a simple ROM device that uses an address decoder IC to access eight different 4-bit words stored in a diode matrix. Data to be read are output via the D_3–D_0 lines. The diode matrix is broken up into rows and columns. The intersection of a row and column represents a bit location. When a given row and column are linked together with a diode, the corresponding data output line goes low (0) when the corresponding column is selected by the address decoder via the A_2–A_0 inputs. (When a specific row is addressed, the NAND gate sinks current, so the current from the supply passes through the diode and into the NAND gate's output. This makes the corresponding data line low.) When no diode is placed between a given column and row, the corresponding data line goes high (0) when the corresponding row is selected by the address decoder. (There is no path to ground in this case.) In this particular example, we have what's called an 8×4 ROM (8 different 4-bit words). Now, by increasing the width of the matrix (adding more columns), it is possible to increase the word size. By increasing the height of the matrix (adding more rows—more addresses), it is possible to store more words. In other words, we could make an $m \times n$ ROM.

for all memory devices. There are additional features, however, and we'll discuss these in a minute. First, let's cover some memory nomenclature.

J.3 Terms Used to Describe Memory Size and Organization

A ROM that is organized in an $n \times m$ matrix can store n different m-bit words or, in other words, can store $n \times m$ bits of information. To access n different words requires $\log_2 n$ address lines. For example, our simple ROM in Fig. J.2 requires $\log_2 8 = 3$ address inputs (this may look more familiar: $2^3 = 8$). (Note that within multiplexed memories and memories that come with serial inputs, the actual physical number of address inputs is ether reduced or the address information is entered serially—along with data and other protocol information. More on this later.)

In terms of real memory ICs, the number of address inputs is typically 8 or higher (for parallel input devices at any rate). Common memory sizes are indicated in Table J.1. Note that in the table, 1K is used to represent 1024 bits, not 1000 bits, as the k (kilo) would lead you to believe. By digital convention, we say that $2^1 = 2$, $2^2 = 4$, $2^3 = 8$, ... $2^8 = 256$, $2^9 = 512$, $2^{10} = 1,024$ (or 1 K), $2^{11} = 2,048$ (or 2 K), ... $2^{18} = 262,144$ (256K), $2^{19} = 524,288$ (540 K), $2^{20} = 1,048,576$ (or 1 M—mega), $2^{21} = 2,097,152$ (2 M), ... $2^{30} = 1,073,741,824$ (or 1 G—giga), ... etc. If this convention confuses you, it should, it is not exactly obvious, and leaves you scratching your head. Also, when a data sheet says 64 K, you have to read further to figure out what the actual organization is, say, 2048×32 (2 K $\times$ 32), or 4096×16 (4 K $\times$ 16), or 8192×8 (8 K $\times$ 8), $16,384 \times 4$ (16 K $\times$ 4), etc.

Now, watch out for terms such as kB, MB, and GB. These terms do not refer to 1024 (1 K), 1,048,576 (1 M), and 1,073,741,824 (1 G) bits of data, respectively. Here, the B signifies 1 byte, or 8 bits. This means that a memory that stores 1 kB actually stores 1 K $\times$ 8 (8 K) bits of data. Likewise, memories that store 1 MB and 1 GB actually store 1 M $\times$ 8 (8 M) and 1 G $\times$ 8 (8 G) bits of data, respectively.

TABLE J.I   **Common Memory Sizes**

NO. OF ADDRESS LINES	NO. OF MEMORY LOCATIONS	NO. OF ADDRESS LINES	NO. OF MEMORY LOCATIONS	NO. OF ADDRESS LINES	NO. OF MEMORY LOCATIONS
8	256	14	16,384 (16 K)	20	1,048,576 (1 M)
9	512	15	32,768 (32 K)	21	2,097,152 (2 M)
10	1,024 (1 K)	16	65,536 (64 K)	22	4,194,304 (4 M)
11	2,048 (2 K)	17	131,072 (128 K)	23	8,388,608 (8 M)
12	4,096 (4 K)	18	262,144 (256 K)	24	16,777,216 (16 M)
13	8,192 (8 K)	19	524,288 (540 K)	25	33,554,432 (32 M)

J.4   Simple Programmable ROM

Figure J.3 shows a more accurate representation of a ROM-type memory. Unlike the diode ROM, each memory cell contains a transistor and fusible link. Initially, the ROM has all programmable links in place. With every programmable link in place, every transistor is biased on, causing high voltage levels (logic 1's) to be stored throughout the array. When a programmable link is broken, the corresponding memory cell's transistor turns off, and the cell stores a low voltage level (logic 0). Note that this ROM contains three-state output buffers that keep the output floating until a low is applied to the Chip Enable ($\overline{CE}$) input. This feature allows the ROM to be interfaced with a data bus.

FIGURE J.3

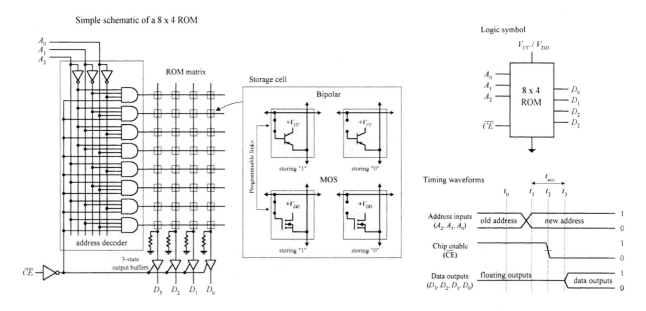

A basic ROM circuit schematic is shown in Fig. J.3, along with the appropriate address and chip-enable waveforms needed to enact a Read operation. To read data stored at a given address location, the Chip Enable input is set high to disable the chip (remove old data from data outputs)—see time t_0. At time t_2, a new address is placed on the 3-bit address bus (A_2, A_1, A_0). At time t_2, the Chip Enable input is set low, which allows addressed data stored in memory to be output via D_3, D_2, D_1, D_0. In reality, the stored data are not output immediately but are delayed for a very short time (from t_2 to t_3) due to the propagation delay that exists between the initial chip enable signal and the signal that reaches the enable leads of the output buffers. In memory lingo, the time from t_1 to t_4 is referred to as the *access time,* which is between around 10 ns and a couple hundred nanoseconds, depending on the specific technology used.

Now there are two important questions that need addressing. First, how does one "break" a programmable link? In other words, how do we program the ROM? Second, is it possible to restore a "broken" programmable link back to its "unbroken" state? In other words, is it possible to reprogram the ROM? This leads to the next topic.

J.5 Various Kinds of ROM Devices

There are basically two kinds of ROMs: those which can only be programmed once and those which can be reprogrammed any number of times. One-time programmable memories include the mask ROMs (MROMs) and the programmable ROM (PROM). ROMs that can be reprogrammed include the erasable programmable ROM (EPROM), electrically erasable programmable ROM (EEPROM), and flash memory.

J.5.1 Mask ROMs (MROMs)

A mask ROM or MROM is a custom memory device that is permanently programmed by the manufacturer simply by adding or leaving out diodes or transistors within a memory matrix. In order to create a desired memory configuration, you must supply the manufacturer with a truth table stating what data configuration is desired. Using the truth table, the manufacturer then generates a *mask* that is used to create the interconnections within the memory matrix during the fabrication process. As you can imagine, producing a custom MROM is not exactly cheap—in fact, it is rather costly (upwards of $1000). It is only worthwhile using a MROM if you plan to mass-produce some device that requires the same data instructions (e.g., program instructions) over and over again—no upgrades to memory needed in the future. In this case, the cost for each IC—after the initial mask is made—is relatively cheap, assuming you need more than a couple thousand chips. MROMs are commonly found in computers, where they are used to store system-operating instructions and are used to store data that is used to decode keyboard instructions into data instructions that drive the cathode-ray tube or the monitor.

J.5.2 Programmable ROMs (PROMs)

PROMs are fusible-link programmable ROMs. Unlike the MROM, data are not etched in stone. Instead, the manufacturers provide you with a memory IC whose matrix is clean (full of 1's). The number of bits and the configuration ($n \times m$) of the matrix vary depending on specific ROM. To program the memory, each fusible link must be blown with a high-voltage pulse (e.g., 21 V). The actual process of blowing individual fuses requires a PROM programming unit. This PROM programmer typically includes a hardware unit (where the actual PROM IC is attached) along with programming cable that is linked to a computer (e.g., via a serial or parallel port). Using software provided by the manufacturer, you enter the desired memory configuration in the program running on the computer and then press a key, which causes the software program to instruct the external programming unit to blow the appropriate links within the IC. PROMs are relatively easy to program once you have figured out how to use the software, but as with MROMs, once the device is programmed, the memory cannot be altered. In other words, if you screw things up, you must begin afresh with a new chip. These devices were popular some years ago, but today they are considered obsolete.

J.5.3   EPROMs, EEPROMs, and Flash Memories

The most popular ROM-type devices used today are the EPROM, EEPROM, and flash memory. These devices, unlike the previous MROM and PROM, can be erased and reprogrammed—a very useful feature when prototyping or designing a gadget that requires future memory alterations.

EPROMs

An EPROM (erasable programmable ROM) is a device whose memory matrix consists of a number of specialized MOSFET transistors. Unlike a conventional MOSFET transistor, the EPROM transistor has an additional floating gate that is buried beneath the control gate—insulated from both control gate and drain-to-source channel by an oxide layer (see Fig. J.4). In its erased state (unprogrammed state), the floating gate is uncharged and does not affect the normal operation of the control gate (which when addressed results in a high voltage or logic 1 being passed through to the data lines). To program an individual transistor, a high-voltage pulse (around 12 V) is applied between the control gate and the drain terminal. This pulse, in turn, forces energetic electrons through the insulating layer and onto the floating gate (this is referred to as *hot electron injection*). After the high voltage is removed, a negative charge remains on the floating gate and will stay there for decades under normal operating conditions. With the negative charge in place, the normal operation of the control gate is inhibited; when the control gate is addressed, the charge on the floating gate prevents a high voltage from reaching the data line—the addressed data appears as a low or logic 0.

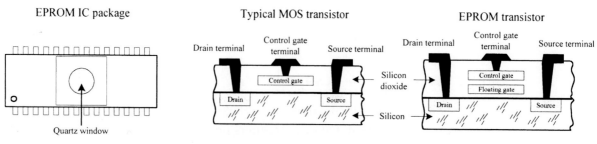

FIGURE J.4

In order to reprogram (erase) an EPROM, you must first remove the device from the circuit and then remove a sticker covering its quartz window. After that, you remove all stored charges on the floating gates by shining ultraviolet (UV) light through the window unto the interior transistor matrix. The UV light liberates the stored electrons within the floating gate region by supplying them with enough energy to force them through the insulation. It usually takes 20 minutes of UV exposure for the whole memory matrix to be erased. The number of times an EPROM can be reprogrammed is typically limited to a couple hundred cycles. After that, the chip degrades considerably.

EPROMs are often used as nonvolatile memories within microprocessor-based devices that require the provision for future reprogramming. They are frequently used in prototyping and then substituted with MROMs during the mass-production phase. EPROMs are also integrated within microcontroller chips where their soul purpose is to store the microcontroller's main program (more on this in Appendix K).

EEPROMs

An EEPROM (electrically erasable programmable ROM) is a technology somewhat related to the EPROM, but it does not require out-of-circuit programming or UV erasing. Instead, an EEPROM is capable of selective memory cell erasure by means of controlled electrical pulses. In terms architecture, an EEPROM memory cell consists of two transistors. One transistor resembles the EPROM transistor and is used to store data, while the other transistor is used to clear charge from the first transistor's floating gate. By supplying the appropriate voltage level to the second transistor, it is possible to selectively erase individual memory cells instead of having to erase the entire memory matrix, as was the case with the EPROM. The only major disadvantage with an EEPROM over an EPROM is size—due to the two transistors. However, today, with new fabrication processes, size is becoming less of an issue.

In terms of applications, EEPROMs are ideal for remembering configuration and calibration settings of a device when the power is turned off. For example, EEPROMs are found within TV tuners, where they are used to remember the channel, volume setting of the audio amplifier, etc. when the set is turned off. EEPROMs are also found on board microcontrollers, where they are used to store the main program.

Flash Memory

A flash memory is generally regarded as the next evolutionary step in ROM technology that combines the best features of EPROM and EEPROM. These devices have the advantage of both in-circuit programming (like EEPROM) and high storage density (like EPROM). Some variants of flash are electrically erasable, like EEPROMs, but must be erased and reprogrammed on a device-wide basis similar to an EPROM. Other devices are based on a dual transistor cell and can be erased and reprogrammed on a word-by-word basis. Flash devices are noted for their fast write and erase times, which exceed those of the EEPROM.

Flash memories are becoming very popular as mass-storage devices. They are found in digital cameras, where a high-capacity flash memory card is inserted directly into a digital camera and can store hundreds of high-resolution images. They are also used in digital music players, cellular phones, palmtops, etc.

J.5.4   Sample EPROM and EEPROM ICs from Microchip

27LV64 Low-Voltage CMOS EPROM (Microchip)

Microchip's 27LV64 is an 8 K × 8 low-voltage (3 V) CMOS EPROM designed for battery-powered applications (see Fig. J.5). This device is capable of accessing individual bytes within memory at an access speed greater than 200 ns at 3 V. To access all 8192 (8 K) 8-bit words requires 13 address lines (A_0–A_{12}). Data to be read are output via data outputs O_0–O_7. Data that are to be written to memory also use these lines. Other important control lines include a Chip enable ($\overline{CE}$), an Output Enable ($\overline{OE}$), and a Program Enable ($\overline{PGM}$). V_{PP} is reserved for the programming voltage, V_{CC} is the positive power supply lead (+5 or +3 V), V_{ss} is the ground lead, NC represents no internal connections, and NU represents not used (no external connections allowed).

27LV64 (Microchip) 64K (8K x8) CMOS EPROM

DIP/SOIC Package

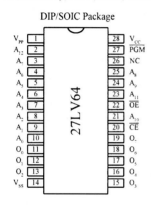

Modes of Operation

Operation Mode	$\overline{CE}$	$\overline{OE}$	$\overline{PGM}$	V_{PP}	A_9	O_0 - O_7
Read	V_{IL}	V_{IL}	V_{IL}	V_{CC}	X	D_{OUT}
Program	V_{IL}	V_{IL}	V_{IL}	V_H	X	D_{IN}
Program Verify	V_{IL}	V_{IL}	V_{IL}	V_H	X	D_{OUT}
Program inhibit	V_{IL}	X	X	V_H	X	High Z
Standby	V_{IL}	X	X	V_{CC}	X	High Z
Output Disable	V_{IL}	V_{IH}	V_{IH}	V_{CC}	X	High Z
Identify	V_{IL}	V_{IL}	V_{IH}	V_{CC}	V_H	Identity Code

V_{IH} = Logic "1" input voltage (2.0V to V_{CC} + 1V)
V_{IL} = Logic "0" input voltage (-0.5V to 0.8V)
X = Don't care

Programming Waveforms

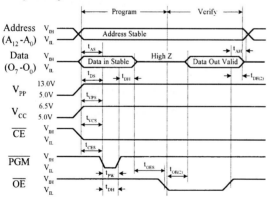

Read Waveforms

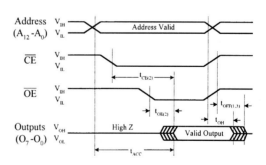

FIGURE J.5

Figure J.5 presents a table indicating the various modes that the 27LV64 can be placed in, as well as the timing waveforms required for writing (programming) and reading data. Before data can be written to memory, the device must be placed in programming mode: V_{CC} must be brought to the proper voltage, V_{PP} is set to the proper V_H level (high), the $\overline{CE}$ pin is set low, the $\overline{OE}$ pin is set high, and the $\overline{PGM}$ pin is set low. Initially in its erased state, the EPROM's array of memory cells is set high (logic 1). To program the device requires changing 1's to 0's. To do this, the address location to be programmed is applied to A_0–A_{12}, while the data to be programmed are applied to pins O_0–O_7. When address and data are stable, the location is programmed by setting $\overline{OE}$ high, $\overline{CE}$ low, and then applying a low-going pulse to the $\overline{PGM}$ line.

After the memory has been programmed, it must be verified to see if everything checks out. For this purpose, the verify mode is used. For this mode to take affect, V_{CC} must be at the proper level, V_{PP} must be at the proper V_H level, $\overline{CE}$ and $\overline{OE}$ must be low, and $\overline{PGM}$ must be high.

To read data from memory, the read mode must be accessed by setting $\overline{CE}$ low to enable the chip and setting $\overline{OE}$ low to gate the data to the output pins. For Read operations to be accurate, the address lines must remain stable for an address access time t_{ACC} of at least 200 ns (max.). There are many other timing parameters used within the timing waveforms. You can learn about what these parameters mean by checking out microchip's data sheet.

Other modes that can be enacted include an inhibit mode, identity mode, a standby mode, and an output disable mode, as shown in the truth table in Fig. J.5. The inhibit mode is used when programming multiple devices in parallel with different data. The identity mode is used to identify both the manufacturer of the EPROM and identify the device type. With A_0 low, an 8-bit binary number is generated that identifies the manufacturer; with A_0 high, another 8-bit binary number identifies the device type. Standby mode is used to place the IC in rest condition. During standby, the memory's supply current drops from 20 mA to 100 μA. Output disable mode places the outputs of the EPROM in a high-impedance state. This feature eliminates bus contention in microprocessor-based systems in which multiple devices share the same bus.

To erase the memory (set all cells to 1), the EPROM's window is exposed to ultraviolet light. To ensure complete erasure, the manufacturer specifies exposing the device with a dose of 15 W-s/cm² UV light, within 1 inch and directly underneath an ultraviolet lamp with a wavelength of 2537 Å, intensity of 12,000 μW/cm² for approximately 20 minutes.

28LV64A 64 K (8 × 8) Low-Voltage CMOS EEPROM (Microchip)

This 64 K EEPROM is organized into 8 K by 8-bit words. This device has four basic modes of operations that include read, standby, write inhibit, and byte write, as shown in the table in the following figure.

Here's a rundown on the various modes for the 27LV64A EEPROM.

READ MODE There are two control functions that must be logically satisfied in order to obtain data at the outputs (I/O pins) of the EEPROM. The chip enable ($\overline{CE}$) is the power control and is used to select the device. The output enable ($\overline{OE}$) is the output control and is used to gate data to the output pins independent of device selection. Assuming that data on the address lines are stable, address access time (t_{ACC}) is equal to the delay from $\overline{CE}$ to output (t_{CE}). Data are available at the output at a time t_{OE} after the falling edge of $\overline{OE}$, assuming that $\overline{CE}$ has been low and addresses have been stable for at least $t_{ACC} - t_{OE}$. See the read timing waveforms shown in Fig. J.6.

WRITE MODE A write cycle is initiated by applying a low going pulse to the $\overline{WE}$ pin. On the falling edge of $\overline{WE}$, the address information is latched. On rising edge of $\overline{WE}$, the data and the control pins ($\overline{CE}$ and $\overline{OE}$) are latch. The READ/$\overline{BUSY}$ pin goes to a logic low level, indicating that the device is in a write cycle, which signals the micro-

27LV64A (Microchip) 64K (8K x8) CMOS EEPROM

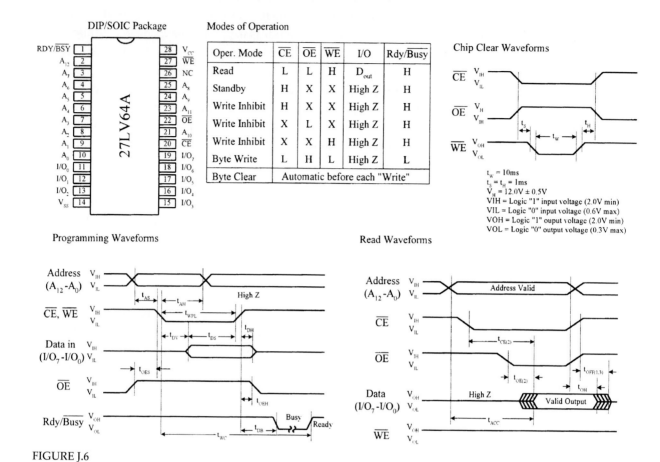

FIGURE J.6

processor host that the system bus is free for other activity. When READ/$\overline{\text{BUSY}}$ goes back to a high level, the device has completed writing and is ready to accept another cycle. See the programming waveforms shown in Fig. J.6.

STANDBY MODE The EEPROM can be placed in standby mode by applying a high to the $\overline{\text{CE}}$ input. During standby mode, outputs are placed in a high impedance state, independent of the $\overline{\text{OE}}$ input.

CHIP CLEAR All data may be cleared to 1's in a chip clear cycle by raising $\overline{\text{OE}}$ to 12 V and bringing the $\overline{\text{WE}}$ and $\overline{\text{CE}}$ pins low.

WRITE INHIBIT This mode is used to ensure data integrity, especially during critical power-up and power-down transitions. Holding $\overline{\text{WE}}$ or $\overline{\text{CE}}$ high or $\overline{\text{OE}}$ low inhibits a write cycle during power-on and power-off.

Serial Access Memories

So far we have only seen memories that incorporated parallel access. These devices sit directly on the address and data buses, making it easy for processors to quickly access the memory. These devices are easy to use in principle; however, since all their address lines are typically tied to an address bus within a microprocessor-based system, it is not uncommon for the data to be inadvertently destroyed when the processor runs amuck (issues an undesired write).

Another type of memory that can "hide" the memory from the processor, as well as reduce the total number of pins, uses a serial access format. To move data to/from memory and processor, a serial link is used. This serial link imposes a strict protocol on data transfers that practically eliminates the possibility that the processor can destroy data accidentally.

Figure J.7 shows a few serial EPROM and EEPROMs from Microchip. The SDA pin found in the EEPROM devices acts as a bidirection data lead used to transfer address and data information into the memory IC, as well as transfer data out to the processor. The SCL pin is the serial clock input used to synchronize the data transfer from and to the device. The 24xx64 and 24LC01B/02B EEPROMs also come with special device address inputs A_0, A_1, and A_2 that are used for multiple device operation. WP is used to enable normal memory operation (read/write entire memory) or inhibit write operations.

Sample serial EPROMs and EEPROMs from Microchip

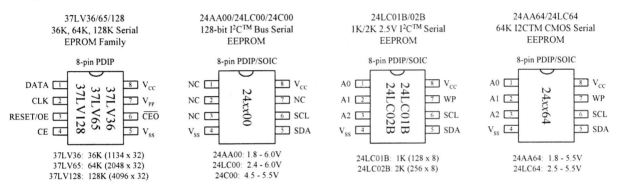

FIGURE J.7

The serial data entered are comprised of a string of information, starting with a start bit, address bits, data bits, and other specialized control information. Controlling a serial memory is a bit complex, due to the serial protocol and variations in protocol from IC to IC. If you want to learn more about these serial memories (and you should—they are very handy in microcontroller applications for logging data and storing programs, etc.), check out the various manufacturers' Web sites and read through their data sheets.

J.6 Random-Access Memories (RAMs)

For applications that require constant and quick read and write cycles, it is necessary to use a random-access memory, or RAM. (The erasable programmable ROM devices, like the EEPROM, have limited read/write endurance—around 100,000 cycles—and take considerable time to write to memory.) A random-access memory (RAM) is used for temporary storage of data and program instructions in microprocessor-based applications. Unlike ROMs, however, RAMs are volatile devices, which means they lose their data if power to the IC is interrupted.

J.6.1   Static and Dynamic RAM

There are two basic types of RAM: static RAM (SRAM) and dynamic RAM (DRAM). In an SRAM, data are stored in memory cells that consist of flip-flops, whereas in a DRAM, data are stored as charges within capacitors etched into the semiconductor integrated circuit. A bit that is written into an SRAM memory cell stays there until overwritten or until power is turned off. In a DRAM, a bit written to the memory cell will disappear within milliseconds if not *refreshed*—supplied with periodic clocking to replenish capacitor charge lost to leakage.

In general, the major practical differences between SRAM and DRAM include overall size, power consumption, speed, and ease of use. In terms of size, DRAMs can hold more data per unit area than SRAMs, since a DRAM's capacitor takes up less space than a SRAM's flip-flop. In terms of power consumption, SRAMs are more energy efficient because they do not require constant refreshing. In terms of speed and ease of use, SRAMs are superior because they do not require refresh circuitry.

In terms of applications, SRAMs are used when relatively small amounts of read/write memory are needed and are typically found within application-specific integrated chips that require extremely low standby power. For example, they are frequently used within portable equipment such as pocket calculators. SRAMs are also integrated into all modern microprocessors, where they act as on-chip cache memories that provide a high-speed link between processor and memory. Dynamic RAMs, on the other hand, are used in applications where a large amount of read/write memory (within the megabyte range) is needed, such as within computer memory modules.

In most situations, you do not have to worry about dealing with discrete RAM memory ICs. Most of the time RAM is already build into a microcontroller or conventionally housed on printed circuit board memory modules that simply plug into a computer's memory banks. In both these cases, you really do not have to know how to use the memory—you let the existing hardware and software take care of the addressing, refreshing, etc. For this reason, we will not discuss the finer details of the various discrete SRAM and DRAM ICs out there. Instead, we will simply take a look at some SRAM and DRAM block diagrams that illustrate the basics, and then we will take a look at some memory packages, such as SIMMs and DIMMs, that are used within computers.

J.6.2   Very Simple SRAM

Figure J.8 shows a very elementary SRAM that is set up with a 4096 (4 K) $\times$ 1-bit matrix. It uses 12 address lines to address 4096 different memory locations—each location contains a flip-flop. The memory matrix is set up as a 64×64 array, with A_0 to A_5 identifying the row and A_6 to A_{11} identifying the column to pinpoint the specific location to be used. The box labeled *Row Select* is a 6-to-64 decoder for identifying the appropriate 1-of-64 row. The box labeled *Column Select* is also a 6-to-64 decoder for identifying the appropriate 1-of-64 column.

To write a new bit of data to memory, the bit is applied to D_{IN}, the address lines are set, the Chip Select input ($\overline{CS}$) is set low (to enable the chip), and the Write Enable input ($\overline{WE}$) is set low (to enable the D_{IN} buffer). To read a bit of data from memory, the address lines are set, $\overline{CS}$ is set low, and $\overline{WE}$ is set high (to enable the D_{OUT} buffer). See timing waveforms in Fig. J.8.

By combining eight 4 K × 1 SRAM ICs together, as shown in the lower circuit in Fig. J.8, the memory can be expanded to form a 4 K × 8 configuration—useful in simple 8-bit microprocessor systems. When an address is applied to the address bus, the same address locations within each memory IC are accessed at the same time. Therefore, each data bit of an 8-bit word applied to the data bus is stored in the same corresponding address locations within the memory ICs.

There are other SRAM ICs that come with configurations larger than $n × 1$. For example, they may come in, say, an $n × 4$ or $n × 8$ configuration. As with the $n × 1$ devices, these SRAMs can be expanded (e.g., two $n × 8$ devices could be combined to form an $n × 16$ expanded memory; four $n × 8$ devices could be combined to form an $n × 32$ expanded memory, etc.).

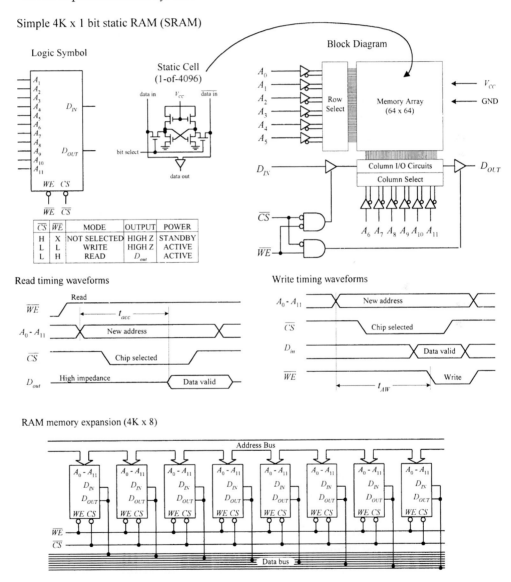

Simple 4K x 1 bit static RAM (SRAM)

FIGURE J.8

J.6.3   Note on Nonvolatile SRAMs

In many applications it would be ideal to have a memory device that combines both the speed and cycle endurance of a SRAM with the nonvolatile characteristics of ROM devices. To solve this problem, manufacturers have created what are called *nonvolatile SRAMs*. One such device incorporates a low-power CMOS SRAM together with a lithium battery and power-sensory circuitry. When the power is removed from the chip, the battery kicks in, providing the flip-flops with sufficient voltage to keep them set (or reset). SRAMs with battery backup, however, have limited lifetimes due to the life expectancies of the lithium batteries—around 10 years.

Another nonvolatile SRAM that requires no battery backup is referred to as a *NOVRAM* (*nonvolatile RAM*). Instead, these chips incorporate a backup EEPROM memory array in parallel with an ordinary SRAM array. During normal operation, the SRAM array is written to and read from just like an ordinary SRAM. When the power supply voltage drops, an onboard circuit automatically senses the drop in supply voltage and performs a store operation that causes all data within the volatile SRAM array to be copied to the nonvolatile EEPROM array. When power to the chip is turned on, the NOVRAM automatically performs a recall operation that copies all the data from the EEPROM array back into the SRAM array. A NOVRAM has essentially unlimited read/write endurance, like a conventional SRAM, but has a limited number of store-to-EEPROM cycles—around 10,000.

J.6.4 Dynamic Random-Access Memory (DRAM)

Figure J.9 shows a very basic 16 K × 1 dynamic RAM. Normally, to access all 16,384 memory locations (capacitors) would require 14 address lines. However, in this DRAM (as within most large-scale DRAMs), the number of address lines is cut in half by multiplexing. To address a given memory location is a two-step process. First, a 7-bit row address is applied to A_0–A_6, and then Row Address Strobe ($\overline{RAS}$) is sent low. Second, a 7-bit column address is applied to A_0–A_6, and then the Column Address Strobe ($\overline{CAS}$) is sent low. At this point the memory location is latched and can now be read or written to by using the $\overline{WE}$ input. When $\overline{WE}$ is low, data are written to the RAM via D_{in}. When $\overline{WE}$ is high data are read from the RAM via D_{out}. See timing waveforms in Fig. J.9.

Simple DRAMs like this must be refreshed every 2 ms or sooner to replenish the charge on the internal capacitors. For our simple device, there are three ways to refresh the cells. One way is to use a Read cycle, another way is to use a Write cycle, and the last way is to use an $\overline{RAS}$-only cycle. Unless you are reading or writing to/from all 128 row every 2 ms, the $\overline{RAS}$-only cycle is the preferred technique. To perform this cycle, $\overline{CAS}$ is set high, A_0–A_6 are set up with the row address 000 0000, $\overline{RAS}$ is pulsed low, the row address is then incremented by 1, and the last two steps are repeated until all 128 rows have been accessed.

As you can see, having to come up with the timing waveforms needed to refresh the memory is a real pain. For this reason, manufacturers produce dynamic RAM controllers or actually incorporate automatic refreshing circuitry within the DRAM IC. In other words, today's DRAMs have all the "housekeeping" functions built in. Practically speaking, this makes the DRAM appear static to the user.

DRAM technology is changing very rapidly. Today there are a number of DRAM-like devices that go by such names as ECC DRAM, EDO DRAM, SDRAM, SDRAMII, RDRAM, SLDRAM, etc. We will discuss these technologies in a moment.

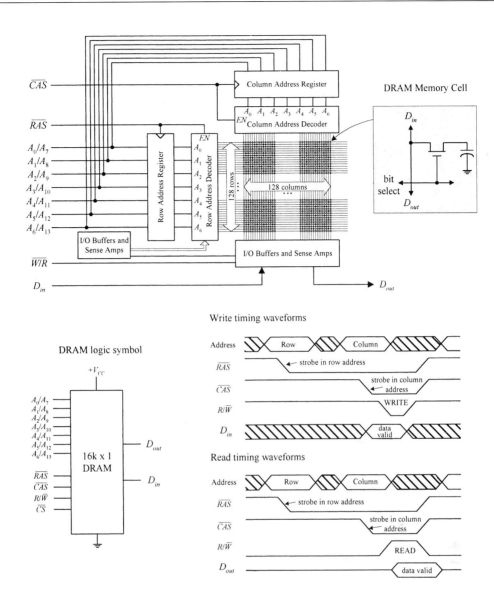

FIGURE J.9

J.6.5 Computer Memory

As mentioned earlier, you typically do not have to worry about RAM. (The only real exception would be nonvolatile RAM that is used in many EEPROM-like applications.) RAM is usually either already integrated into a chip, such as a microcontroller, or is placed in reduced pin devices like SIMMs (single in-line memory modules) or DIMMs (dual in-line memory modules) that slide (snap) into a computer's memory bank sockets. In both cases, not much thought is needed—assuming you are not trying to design a microcontroller or computer from scratch. The main concern nowadays is figuring out what kind of RAM module to buy for your computer.

Within computers, RAM is used to hold temporary instructions and data needed to complete tasks. This enables the computer's central processing unit (CPU) to access instructions and stored data in memory very quickly. For example, when the CPU loads an application, such as a word processor or page layout program into memory, the CPU can quickly find what it needs, instead of having to search for bits and pieces from, say, the hard drive or external drive. For RAM to be quick, it must be in direct communication with the computer's CPU. Early on, memory was soldered directly onto the computer's system board (motherboard). However, over time, as memory requirements increased, having fixed memory onboard became impractical. Today, computers house expansion slots arranged in memory banks. The number of memory banks and the specific configuration vary, depending of the computer's CPU and how the CPU receives information.

Most desktop computers today use either SIMM or DIMM memory modules. Both types of modules use dynamic RAM ICs as the core element. The actual SIMM or DIMM module resembles a printed circuit board and houses a number RAM ICs that are expanded onboard to provide the necessary bit width required by the CPU using the module. To install a SIMM or DIMM module, simply insert the module into one of the computer memory banks sockets found on the motherboard. Current computer systems typically use 168-pin DIMMs. Older Pentium and later 486 PCs commonly use 72-pin SIMMs, while still older 486 PCs commonly use 30-pin SIMMs.

30-Pin SIMMs

Figure J.10 shows an example of how 30-pin SIMMs are used in conjunction with a 32-bit processor. Each SIMM provides 8 bits, and therefore, four SIMMs are required to support the CPU's 32-bit format. The memory configuration on such a system is typically divided into two memory banks—bank zero and bank one. Each memory bank consists of four 30-pin SIMM sockets. The CPU addresses one memory bank at a time. Typical 30-pin SIMM memory formats (and capacities) are 256 K $\times$ 8 (256 kB), 1 M $\times$ 8 (1 MB), and 4 M $\times$ 8 (4 MB). Another type of SIMM, called a *parity SIMM*, uses an $n \times 9$. These SIMMs add a single parity bit to every 8 bits of data. Parity bits are used for error detection—we'll talk about parity in more detail in a moment.

30-pin SIMMs: Each supports 8 data bits; four 30-pin SIMMs are needed to supply 32-bit processors.

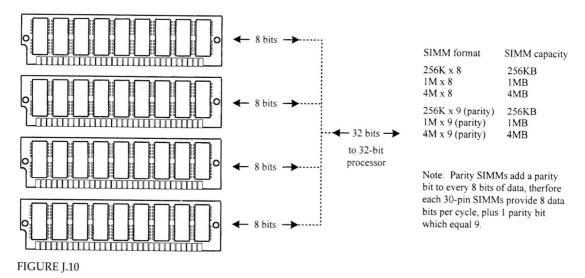

FIGURE J.10

With most computer models, mixing of different-capacity SIMMs within the same bank is to be avoided. If mixing occurs, either the computer will not boot up, or the computer will not recognize or use some of the memory bank. For example, if a bank had three 1-MB SIMMs and one 4-MB SIMM, it would recognize them all as 1-MB SIMMs.

72-Pin SIMMs

The 72-pin SIMM is an improvement over the 30-pin SIM. One 72-pin SIMM supports 32 data bits, which is four times the number of data bits supported by a single 30-pin SIMM. If you have a 32-bit CPU—such as a 486 from Intel or 68040 from Motorola—you need only one 72-pin SIMM per bank. Figure J.11 shows the standard SIMM memory formats and capacities. Note that the parity SIMMs use an $n \times 36$ format. The additional 4 bits are parity bits—1 for every 8 bits within 32 bits.

72-pin SIMMs: Supports 32-bit processors (e.g. Intel's 486 or Motorola's 68040)

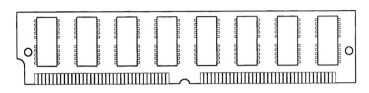

SIMM format	SIMM capacity
256K x 32	1MB (or 8M)
1M x 32	4MB
2M x 32	8MB
4M x 32	16MB
8M x 32	32MB
256K x 36 (parity)	1MB
1M x 36 (parity)	4MB
2M x 36 (parity)	8MB
4M x 36 (parity)	16MB
8M x 36 (parity)	32MB

Note: Parity SIMMs add a parity bit to every 8 bits of data, therfore 72-pin SIMMs provide 32 data bits per cycle, plus 4 parity bits, which equal 36 bits.

FIGURE J.11

168-Pin DIMMs

Dual in-line memory modules, or DIMMs, closely resemble SIMMs. The principal difference between the two is that on a SIMM, opposing pins on either side of the board are tied together to form one electrical contact; on a DIMM, opposing pins remain electrically isolated to form two separate contacts. DIMMs are often used in computer configurations that support a 64-bit or wider memory bus. In many cases, these computer configurations are based on powerful 64-bit processors like Intel's Pentium or IBM's PowerPC processor. Figure J.12 shows a sample 16 M × 64-bit synchronous DRAM that comes in the standard 168-pin DIMM package.

16M x 64 bit synchronous DRAM module
Includes 16 DRAMs on a printed circuit board (pipeline architecture) with 168-pin DIMM package.
Supports Intel's Pentium or IBM's PowerPC processors.

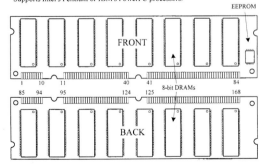

A0 to A11	Address Inputs
BA0, 1	Bank Select
DQ0 to DQ63	Data Inputs/Outputs
/CS0 to 3	Chip Select
/RAS	Row Address Strobe
/CAS	Column Address Strobe
/WE	Write Enable
DQMB0 to 7	Output Disable/Write Mask
CLK0 to 3	Clock input
CKE0, 1	Clock enable
SDA	Serial Data/Address for PD
SCL	Clock for PD
SA0 to 2	Address for PD
VDD	Power (+3.3V)
VSS	Ground
NC	No Connection

FIGURE J.12

J.6.6   Memory Data Integrity Checking

There are two primary methods used to ensure the integrity of data stored in memory. One is parity checking; the other is error correction code (ECC) checking. Parity has been the most common method used to date. The process adds 1 additional bit to every 8 bits (1 byte) of data—as we saw in the SIMM × 9 and × 36 parity memory modules. ECC is a more comprehensive method of data integrity checking that can detect and correct single-bit errors. This technique, however, is a bit more expensive and is often avoided merely to keep computer prices down. Most computers designed for use as high-end servers are designed to support ECC. Most desktop computers designed for use in business are designed to support parity, while most low-cost computer designed for use in homes and small businesses use nonparity memory.

J.6.7   DRAM Technology Used in Computer Memories

There are a number of DRAM technologies that are incorporated into computer memory modules these days. EDO memory, or extended data out, is a technology that allows the CPU (ones that support EDO) to access memory 10 to 20 percent faster than standard DRAM chips. Another variation of DRAM is the synchronous DRAM (SDRAM), which uses a clock to synchronize signals input and output on a memory chip. The clock is coordinated with the CPU clock so the timing of the memory chips and the timing of the CPU are in synch. Synchronous DRAMs save time in executing commands and transmitting data, thereby increasing the overall performance of the computer. SDRAM allows the CPU to access memory approximately 25 percent faster than EDO memory. DDR or SDRAMM II (double-data-rate SRAM) is a faster version of SDRAM that is able to read data on both the rising and falling edges of the system clock, thus doubling the data rate of the memory chip. RDRAM (Rambus DRAM) is an extremely fast DRAM technology that uses a high-bandwidth "channel" to transmit data at speeds about ten times faster than a standard DRAM.

Microprocessors and Microcontrollers

Within almost every mildly complex electronic gadget you find these days, there is a microprocessor or microcontroller running the show. Also, within these gadgets you will find practically none of the discrete logic ICs (e.g., logic gates, flip-flops, counters, shift registers, etc.) that we covered earlier in this book. Now, before doing anything else, let's stop and address the second statement above, since it seems to go against everything we learned earlier.

When writing a book on electronics geared toward beginners, it is often necessary to introduce somewhat obsolete devices and techniques in order to illustrate some important principle. A perfect example is the logic gates and the flip-flop. These discrete devices are hardly ever used, but they provide the beginner with a basic understanding of logic states, logical operations, memory, etc. To illustrate more complex principles, such as counting, the earlier principles (and the devices that go with them) are used. This rather historical approach of doing electronics is important foundation building; however, it should not be dwelt on if you aim is to design complex electrical gadgets that mesh with today's technological standards.

What should be dwelt on is the *microcontroller*. This device is truly one of the great achievements of modern electronics. With a single microcontroller it possible to replace entire logic circuits comprised of discrete devices. The microcontroller houses an onboard central processing unit (CPU) that performs the same basic function as a computer's microprocessor (performs logical operations, I/O control operations, etc.). It also houses extra onchip goodies such as a ROM, a RAM, serial communication ports, often A/D converters, and more. In essence, microcontrollers are mini-computers without the keyboard and monitor. With a single microcontroller you can build a robot that performs various functions. For example, you can use the micro-controller to control servo motors, generate sound via a speaker, monitor an infrared sensor to avoid objects, record input data generated by analog transducers, etc. Microcontrollers are also used within microwaves, TVs, VCRs, computer peripherals (laser printers, disk drives), automobiles control systems, security systems, toys, environmental data logging instruments, cellphones, and any device that requires programlike control.

Now *microprocessors,* such as the Intel's Pentium processors, are similar to micro-controllers but are designed primarily for fast-paced "number crunching," which is important for running today's sophisticated multimedia programs and games. Microprocessors also require mass amounts of additional support devices, such as RAM, ROM, I/O controllers, etc.—something the microcontroller has built in. For this reason, we will only briefly introduce microprocessors in this appendix and instead focus mainly microcontrollers. Microcontrollers are definitely the "in thing" these days, mainly because they are so easy to use and can be used as the "brains" within so many different battery-powered gadgets. Microprocessors, on the other hand, are mostly used within computers and tend not to have much direct practical use to the inventor who likes to tinker with actual IC's.

K.1 Introduction to Microprocessors

Microprocessors are integrated circuits which through address lines, data lines, and control lines have the ability to read and execute instructions from external read-only memory (ROM), hold temporary data and programs in external random-access memory (RAM) by writing and reading, and receive and supply input and output signals to and from external support circuitry. Figure K.1 shows a simple microprocessor-based system with support circuitry.

Simple microprocessor-based system

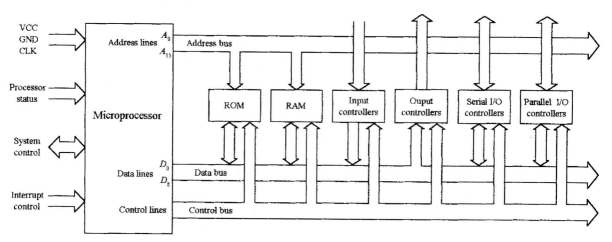

FIGURE K.1

K.1.1   The Microprocessor

At the heart of a microprocessor-based system is the microprocessor or central processing unit (CPU). The microprocessor is responsible for performing computations on chunks of data that are organized into words. In this particular microprocessor-based system, the CPU consists of an 8-bit processor—one that works with 8-bit words. In more sophisticated microprocessor-based systems, the processors work with larger bit words, such as 16-bit, 32-bit, or higher (e.g., 64-bit Pentium processor). The processor's role is to read program instructions from memory and then execute those instructions by supplying the three external buses with the proper levels and timing to make the connected devices (memory, I/O controllers, etc.) perform their specific operations. Within the processor itself, there are a number of integrated subsections designed to perform specific tasks. One of these subsections is the instruction decoder, which accepts and interprets instructions fetched from memory and then figures out what steps to take based on the fetched instructions. The processor also contains an arithmetic logic unit (ALU) that can perform the instructed operations, such as add, complement, compare, shift, move, etc., on quantities stored on chip registers or in external memory. A program counter keeps track of the current location in the executing program. There are many other subsections within a processor—we'll take a closer look at an actual microprocessor in a second.

K.1.2 Address Bus, Data Bus, and Control Bus

There are basically three buses that are used to relay address, data, and control information between the microprocessor and external devices, such as RAM, ROM, and various I/O controllers. These include the address bus, the data bus, and the control bus. (Recall that a *bus* is a group of conductors shared by various devices.)

The address bus is used by the microprocessor to select a particular address location within an external device, such as memory. In our processor-based system, the address bus is 16 bits wide, which means that the processor can access 2^{16} (65,536) different addresses.

The data bus is used to transfer data between the processor and external devices (memory, peripherals). The data bus for our microprocessor-based system is 8 bits wide. With more sophisticated processors, such as the early 486 and the later Pentium processors, the number of data lines is larger—32 and 64 bits, respectively.

The control bus is of varying width, depending on the microprocessor being used. It carries control signals that are tapped into by other ICs to specify what operation is being performed. From these signals, the ICs can tell if the operation is a read, write, interrupt, I/O, memory access, or some other operation.

K.1.3 Memory

Computers typically use three types of memory: ROM, RAM, and mass memory (e.g., hard drives, floppies, CD-ROMs, ZIP drives, etc.) ROM acts as a nonvolatile memory used to store the boot-up sequence of instructions, which include port allocations, interrupt initializations, and codes needed to get the operating system read from the hard drive. RAM is used to hold temporary data and programs used by the microprocessor. Mass memory is used for long-term data storage.

K.1.4 Input and Output Controllers

In order for the microprocessor to receive data from or send data to various input/output devices, such as keyboards, displays, printers, etc., special I/O controllers are needed as an interface. The controller is linked to the bus system and is controlled by the microprocessor. In order for the microprocessor to keep tabs on the various I/O devices, the microprocessor can either directly address the controller by means of program instructions or have the controller issue an interrupt signal when the external device generates a "read me" type of signal. Computers come with all sorts of controllers, from sound controllers, SCSI controllers, universal serial bus controllers, game controllers, floppy-disk controllers, hard-disk controllers, etc. All of these you purchase from the manufacturer, or they are already included in the computer. Different hardware requires different control signals. For this reason, software drivers (special programs) designed to create a uniform programming interface to each particular piece of hardware are used.

K.1.5 Sample Microprocessor

To get a basic idea of what the interior of a simple processor looks like, let's examine Intel's 8085A processor. Though this chip is obsolete by today's standards, it shares many core features with the modern microprocessors and is much easier device to understand.

Intel's 8085A microprocessor : functional block diagram, IC, timing

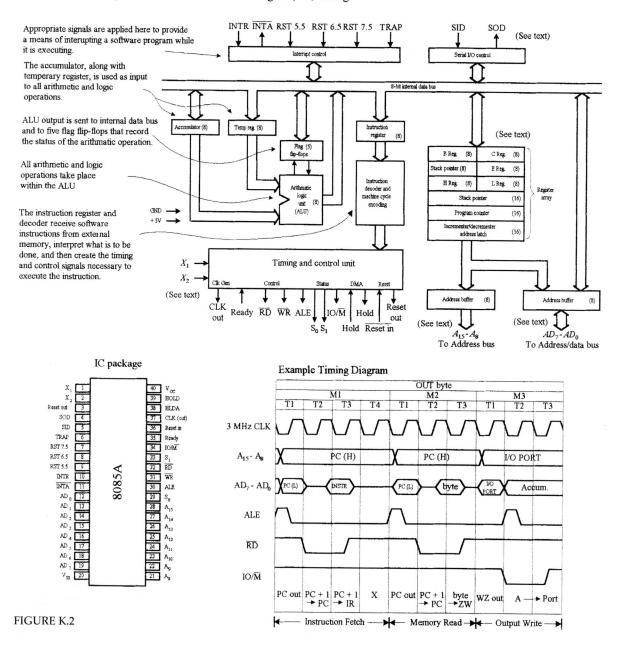

FIGURE K.2

ALU

The 8085A, as with all processors, contains an arithmetic logic unit (ALU) comprised of gates that perform fundamental arithmetic and logical operations (+, −, ×, /, OR, AND, NOT, etc.), including comparisons (=, <, <=, >, <=), as well as shifting data values sent to it from memory. Support for these operations is supplied by one or more ALU registers, called *accumulators,* which can receive initial values from memory, hold the cumulative results of the arithmetic and logic operations, and transmit the final result back to memory. A group of binary indicators (or flags) associated with the ALU provides control feedback information in the form of a summary of the ALU's status after each operation. This might typically include whether the result produced was positive or negative and zero or nonzero, respectively.

Control Unit

The control unit is the directing element within the system, having the responsibility of executing the stored program sequences. It performs this role by repeatedly following a basic instruction execution cycle for each program instruction. It first fetches the instruction from the main store memory into a special control unit register called the *instruction register.* The instruction is then decoded (separated) into parts indicating what is to be done (the operation part) and what information is involved (the operand parts). After that, it executes the instruction by sending the appropriate control signals to the ALU, I/O, and memory. This is referred to as a *fetch-decode-execute cycle.*

Interrupts

Another important feature in microprocessor systems is interrupt control, which provides a way for external digital signals to interrupt a software program while it is running. There are five interrupt inputs for the 8085A: INTR, RST, RST 5.5, RST 6.5, RST 7.5, and TRAP. The interrupts are arranged in a fixed priority that determines which interrupt is to be recognized if more than one is pending. The order of priority is as follows: TRAP (highest priority), RST 7.5 (second highest), RST 6.5 (third highest), RST 5.5 (four highest), and INTR (lowest priority). The TRAP interrupt is useful for catastrophic events such as power failure or bus error.

Address and Data Signals

HIGH ADDRESS (A_{15}–A_8) The higher-order 8 bits of a 16-bit address.

ADDRESS/DATA (AD_7–AD_0) The lower-order 8 bits of 16-bit address or 8 bits of data. Note that this address scheme requires multiplexing—a feature used to reduce pin count.

SERIAL OUTPUT DATA (SID) A single-bit input used to accommodate devices that transmit serially (one bit at a time).

SERIAL OUTPUT DATA (SOD) A single-bit output to accommodate devices that receive serially.

Timing and Control Signals

CLK (OUT) The system clock output that is sent to peripheral chips and is used to synchronize their timing.

X_1, X_2 Signals applied here come from an external crystal or other device used to drive the internal clock generator.

ADDRESS LATCH ENABLE (ALE) This signal occurs during the first clock state of a machine cycle and enables the address to get latched into onchip latch peripherals. This allows the address module (e.g., memory, I/O) to recognize that it is being addressed.

STATUS (S_0, S_1) Control signals used to indicate whether a read or write operation is taking place.

I/O/M Used to enable either I/O or memory modules for read and write operations.

READ CONTROL (RD) Indicates that the selected memory or I/O module is to be read and that data bus is available for data transfer.

WRITE CONTROL (WR) Indicates that data on the data bus are to be written into the selected memory or I/O location.

Memory and I/O Symbols

HOLD Used to request that CPU relinquish control and use of the external system bus. The CPU will complete execution of the instruction presently in the instruction register and then enter a hold state, during which no signals are inserted by the CPU to the control, address, or the data buses.

HOLD ACKNOWLEDGE (HLDA) Output signal from control unit that acknowledges the hold signal and indicates that the bus is now available.

READY Used to synchronize the CPU with slower memory or I/O devices. When an address device asserts ready, the CPU may proceed with an input (DB_{in}) or (WR) operation; otherwise, the CPU enters a wait state until the device is ready.

CPU Initialization

RESET IN Causes the contents of the program counter to be set to zero. The CPU resumes execution at location zero.

RESET OUT Acknowledges that the CPU has been reset. The signal can be used to reset the system.

General-Purpose Registers

The 8085A contains a set of six general-purpose 8-bit registers: B, C, D, E, H, and L. These registers are called *general-purpose* because they can be used in any manner deemed by the individual programming of the microprocessor. The general-purpose registers can hold numeric data, BCD data, ASCII data, or any other type of information required. They can be used as six 8-bit registers or as three 16-bit register pairs (BC, DE, HL). Register pairs hold 16-bit numeric data or any other 16-bit coded information. Besides holding 16 bits of data, register pairs also address memory data. A memory address, placed into a register pair, allows the contents of the location's address by the register pair to be manipulated. Register pair memory addressing is called *indirect addressing*.

Special-Purpose Registers

Special-purpose registers accumulate results from arithmetic and logic operations and provide "housekeeping" for the microprocessor. Housekeeping registers are not normally programmed with an instruction but instead are used by the microprocessor. Special-purpose registers for the 8085A include an accumulator, flag register, program counter, and stack pointer.

All microprocessors contain an accumulator register that accumulates the answers of most arithmetic and logic operations carried out by the ALU.

Flag registers used in the 8085A contain 5 bits used as flags or indicators for the ALU. The flags change whenever the 8085A executes most arithmetic operations, and they are used to indicate the status of the arithmetic operation. The 5 flag bits are broken down as follows: (1) one sign flag bit that is used to tell whether a result of an arithmetic or logic operation is positive or negative, (2) one zero flag bit that is used to indicate whether the result of an ALU operation is zero or nonzero, (3) one auxiliary carry flag bit that holds any carry that occurs between the least significant and most significant half-byte result from the ALU, (4) one parity flag bit that shows the parity of the result from the ALU, (5) one carry flag bit that holds any carry that occurs out of the most significant bit of the accumulator after an addition and also holds a borrow after a subtraction.

The program counter does not counter programs but rather locates the next software instruction to be executed by the processor. It counts through the memory locations beginning at lower addresses and progressing toward higher addresses.

The stack pointer stores the address of the last entry on the stack. The stack is a data storage area in RAM used by certain processor operations.

Timing Diagram for the 8085A

Figure J.2 shows an example of 8085A timing. In essence, three machine cycles (M1, M2, M3) are needed. During the first cycle, an OUT instruction is fetched, while during the second cycle, the second half of the instruction is fetched, which contains the number of I/O devices selected for output. During the third machine cycle, the contents of the accumulator are written out to the selected device via the data bus. At the beginning of each machine cycle, an address-latch-enable (ALE) pulse is supplied by the control unit that alerts external circuits. During timing state T1 of machine cycle M1, the control unit sets the IO/M signal to indicate that a memory operation is to occur. Also during this cycle, the control unit instructs the program counter to place its contents on the address bus (A_{15}–A_8) and address/data bus (AD_7–AD_0). During timing state T2, the address memory module places the contents of the addressed memory location on the address/data bus. The control unit sets the read control (RD) signal to indicate a read, but it waits until T3 to copy the data from the bus. This gives the memory time to put the data on the bus and for the signal levels to stabilize. The final state, T4, is a bus idle state during which the CPU decodes the instructions. The remaining machine cycles proceed in a similar manner.

K.1.6   *Programming a Microprocessor*

Each microprocessor has its own unique set of instructions for performing tasks such as reading from memory, adding numbers, and manipulating data. For example, the Intel's Pentium processor found in IBM compatibles use a completely different set of instructions from the Motorola's PowerPC found in MacIntosh computers.

The actual language used by any given microprocessor is machine code, which consists of 1's and 0's. Directly programming a microprocessor in machine language, however, is never done—it is too hard. A simple task such as multiplying two numbers may take hundreds of machine instructions to accomplish. Keeping track of all the columns of binary numbers and making sure that each column is bit-perfect would be a nightmare.

A simple alternative to machine code is to first write out a program in assembly language, which uses mnemonic abbreviations and symbolic names for memory locations and variables. The conversion from assembly language to machine language involves translating each mnemonic into the appropriate hexadecimal machine code (called *operation code,* or *opcode* for short) and storing the codes in specific memory locations, which are also given hexadecimal address numbers. This can be done by a software package called an *assembler* provided by the microprocessor manufacturer or can be done by the programmer by looking up the codes (also given by manufacturer) and memory addresses, which is called *hand assembly.* When using hand assembly, you must determine which memory locations (within ROM) are to be reserved for the program. Assembly language allows the programmer to write the most streamlined, memory-efficient programs, which lead to very fast execution times. However, as with machine code, programming in assembly language is also tedious.

To make life easier on the programmer, a compiled or interpreted high-level language such as BASIC, PASCAL, or C can be used. Using one of these languages, you write out you program within algebraic commands and user-friendly control commands such as if . . . then, for . . . next, etc. When using these languages, you do not have to worry about addressing memory locations or figuring out which bits get shifted into which registers, etc. You simply declare variables and write out arithmetic and logic statements. This is referred to as *source code.* After the source code is created, there are two possible routes that must be taken before the microprocessor can run the program. One route involves compiling the source code, while the other route involves interpreting the source code. Languages such as C, PASCAL, and FORTRAN require *compiling,* a process in which a language compiler converts the source-code statements into assembly code. Once the assembly code is created, an assembler program converts the assembly code into machine code that the microprocessor can use. A language such as BASIC, on the other hand, is an interpreted language. Instead of compiling an assembly-language program from the source program, an interpreter program looks at the statements within the program and then executes the appropriate computer instructions on the spot. In general, interpreted languages run much more slowly than complied languages, but there is no need for compiling and no delay after entering a program before it is run. We'll talk more about compilers and interpreters later on.

Table K.1 provides codes in BASIC, assembly (8085A), and machine (8085A) languages that count from 5 down to 0. Note that the assembly and machine codes are specific to the 8085A. In BASIC, the code is pretty much self-explanatory. The variable COUNT holds the counter value, and line 30 checks the COUNT to see if it is equal to zero. If so, then the program jumps back to the beginning; otherwise, it jumps back to line 20 (subtract 1 from COUNT) and checks the COUNT again.

The 8085A's assembly-language mnemonics for this program include MVI, DCR, and START. A complete listing of the 8085A's mnemonics is given in Table K.2. The first mnemonic, MVI, means "Move Immediate." The instruction MVI A, 05H will

TABLE K.1   **Counting from 5 to 0 Using BASIC, Assembly Language, and Machine Language**

BASIC LANGUAGE		8085A ASSEMBLY LANGUAGE		8085A MACHINE LANGUAGE	
LINE	INSTRUCTION	LABEL	INSTRUCTION	ADDRESS (HEX)	CONTENTS
10	COUNT = 5	START:	MVI A,05H	4000	3E (opcode)
20	COUNT = COUNT−1			4001	05 (data)
30	IF COUNT = 0	LOOP:	DCR A	4002	3D (opcode)
	THEN GOTO 10		JZ START	4003	CA (opcode)
40	GOTO 20			4004 4005	00 40 }(address)
			JUMP LOOP	4006	C3 (opcode)
				4007 4008	02 40 }(address)

move the data value of 05 (the H stands for hexadecimal) into register A (accumulator). The next instruction, DCR A, decrements register A by 1. The third instruction, JZ START, is a conditional jump. The condition that this statement is looking for is the zero condition. As the register is decremented, if A equals 0, then a flag bit, called a *zero flag*, gets set (set to logic 0). The instruction JX START is interpreted as jump to statement label START if the zero flag is set. If this condition is not met, then the program jumps to the next instruction, JMP LOOP, which is an unconditional jump. This instruction is interpreted as jump to label LOOP regardless of any condition flags.

In terms of machine code, the assembly-language instruction MVI A,05H requires 2 bytes to complete. The first byte is the opcode, 3E (binary 0011 1110), which identifies the instruction for the microprocessor. They second byte (referred to as the *operand*) is the data value 05. The opcode for JZ is CA (binary 1100 1010), while the following 16-bit (2-byte) address to jump to if the condition (zero) is not met. This makes it a 3-byte instruction; byte 2 of the instruction (location 4004 hex) is the lower-order byte of the address, and byte 3 is the higher-order byte of the address to jump to. The opcode for JMP is C3 (binary 1100 0011) and must be followed by a 16-bit (2-byte) address specifying the location to jump to.

K.2   Microcontrollers

As you can see from what we have covered until now, programming a microprocessor (even a fairly simple one), as well as linking all the external support chips, such as the RAM, ROM, I/O controllers, etc., is difficult work. It requires burning startup instructions into ROM, connecting various I/O devices to the buses, writing programs in order to communicate with the I/O devices, understanding interrupt protocols, etc. In reality, microprocessor systems, such as those found within personal computers, are usually too difficult to construct or are simply not worth the effort; you can buy fully assembled units or at least buy the motherboard and then attach memory modules (DIMMs, SIMMs), sound cards, disk drives, etc. In the latter case,

TABLE K.2   Instruction Set Summary for the Intel 8085A Microprocessor

MNEMONIC	OPCODE	SZAPC	~S	DESCRIPTION	NOTES
ACI n	CE	*****	7	Add with Carry Immediate	A = A + n + CY
ADC r	8F	*****	4	Add with Carry	A = A + r + CY (21X)
ADC M	8E	*****	7	Add with Carry to Memory	A = A + [HL] + CY
ADD r	87	*****	4	Add	A = A + r (20X)
ADD M	86	*****	7	Add to Memory	A = A + [HL]
ADI n	C6	*****	7	Add Immediate	A = A + n
ANA r	A7	****0	4	AND Accumulator	A = A&r (24X)
ANA M	A6	****0	7	AND Accumulator and Memory	A = A&[HL]
ANI n	E6	**0*0	7	AND Immediate	A = A&n
CALL a	CD	-----	18	Call unconditional	−[SP] = PC, PC = a
CC a	DC	-----	9	Call on Carry	If CY = 1 (18~s)
CM a	FC	-----	9	Call on Minus	If S = 1 (18~s)
CMA	2F	-----	4	Complement Accumulator	A = ~A
CMC	3F	----*	4	Complement Carry	CY = ~CY
CMP r	BF	*****	4	Compare	A − r (27X)
CMP M	BF	*****	7	Compare with Memory	A − [HL]
CNC a	D4	-----	9	Call on No Carry	If CY = 0 (18~s)
CNZ a	C4	-----	9	Call on No Zero	If Z = 0 (18~s)
CP a	F4	-----	9	Call on Plus	If S = 0 (18~s)
CPE a	EC	-----	9	Call on Parity Even	If P = 1 (18~s)
CPI n	FE	*****	7	Compare Immediate	A − n
CPO a	E4	-----	9	Call on Parity Odd	If P = 0 (18~s)
CZ a	CC	-----	9	Call on Zero	If Z = 1 (18~s)
DAA	27	*****	4	Decimal Adjust Accumulator	A = BCD format
DAD B	09	----*	10	Double Add BC to HL	HL = HL + BC
DAD D	19	----*	10	Double Add DE to HL	HL = HL + DE
DAD H	29	----*	10	Double Add HL to HL	HL = HL + HL
DAD SP	39	----*	10	Double Add SP to HL	|HL = HL + SP
DCR r	3D	****_	4	Decrement	r = r −1 (0X5)
DCR M	35	****_	10	Decrement Memory	[HL] = [HL] − 1
DCX B	0B	-----	6	Decrement BC	BC = BC − 1
DCX D	1B	-----	6	Decrement DE	DE = DE − 1
DCX H	2B	-----	6	Decrement HL	HL = HL − 1
DCX SP	3B	-----	6	Decrement Stack Pointer	SP = SP − 1

MNEMONIC	OPCODE	SZAPC	~S	DESCRIPTION	NOTES
DI	F3	-----	4	Disable Interrupts	
EI	FB	-----	4	Enable Interrupts	
HLT	76	-----	5	Halt	
IN p	DB	-----	10	Input	A = [p]
INR r	3C	****-	4	Increment	r = r + 1 (0X4)|
INR M	3C	****-	10	Increment Memory	[HL] = [HL] + 1
INX B	03	-----	6	Increment BC	BC = BC + 1
INX D	13	-----	6	Increment DE	DE = DE + 1
INX H	23	-----	6	Increment HL	HL = HL + 1
INX SP	33	-----	6	Increment Stack Pointer	SP = SP + 1
JMP a	C3	-----	7	Jump unconditional	PC = a
JC a	DA	-----	7	Jump on Carry	If CY = 1 (10~s)
JM a	FA	-----	7	Jump on Minus	If S = 1 (10~s)
JNC a	D2	-----	7	Jump on No Carry	If CY = 0 (10~s)
JNZ a	C2	-----	7	Jump on No Zero	If Z = 0 (10~s)
JP a	F2	-----	7	Jump on Plus	If S = 0 (10~s)
JPE a	EA	-----	7	Jump on Parity Even	If P = 1 (10~s)
JPO a	E2	-----	7	Jump on Parity Odd	If P = 0 (10~s)
JZ a	CA	-----	7	Jump on Zero	If Z = 1 (10~s)
LDA a	3A	-----	13	Load Accumulator direct	A = [a]
LDAX B	0A	-----	7	Load Accumulator indirect	A = [BC]
LDAX D	1A	-----	7	Load Accumulator indirect	A = [DE]
LHLD a	2A	-----	16	Load HL Direct	HL = [a]
LXI B,nn	01	-----	10	Load Immediate BC	BC = nn
LXI D,nn	11	-----	10	Load Immediate DE	DE = nn
LXI H,nn	21	-----	10	Load Immediate HL	HL = nn
LXI SP,nn	31	-----	10	Load Immediate Stack Ptr	SP = nn
MOV r1,r2	7F	-----	4	Move register to register	r1 = r2 (1XX)
MOV M,r	77	-----	7	Move register to Memory	[HL] = r (16X)
MOV r,M	7E	-----	7	Move Memory to register	r = [HL] (1X6)
MVI r,n	3E	-----	7	Move Immediate	r = n (0X6)
MVI M,n	36	-----	10	Move Immediate to Memory	[HL] = n
NOP	00	-----	4	No Operation	
ORA r	B7	**0*0	4	Inclusive OR Accumulator	A = Avr (26X)|
ORA M	B6	**0*0	7	Inclusive OR Accumulator	A = Av [HL]

continued

TABLE K.2   **Instruction Set Summary for the Intel 8085A Microprocessor** *(Continued)*

MNEMONIC	OPCODE	SZAPC	~S	DESCRIPTION	NOTES
|ORI n	F6	**0*0	7	Inclusive OR Immediate	A = Avn
|OUT p	D3	-----	10	Output	[p] = A
PCHL	E9	-----	6	Jump HL indirect	PC = [HL]
POP B	C1	-----	10	Pop BC	BC = [SP] +
POP D	D1	-----	10	Pop DE	DE = [SP] +
POP H	E1	-----	10	Pop HL	HL = [SP] +
POP PSW	F1	-----	10	Pop Processor Status Word	{PSW,A} = [SP] +
PUSH B	C5	-----	12	Push BC	− [SP] = BC
PUSH D	D5	-----	12	Push DE	− [SP] = DE
PUSH H	E5	-----	12	Push HL	− [SP] = HL
PUSH PSW	F5	-----	12	Push Processor Status Word	− [SP] = {PSW,A}
RAL	17	----*	4	Rotate Accumulator Left	A = {CY,A} < −
RAR	1F	----*	4	Rotate Accumulator Right	A = −> {CY,A}
RET	C9	-----	10	Return	PC = [SP] +
RC	D8	-----	6	Return on Carry	If CY = 1 (12~s)
RIM	20	-----	4	Read Interrupt Mask	A = mask
RM	F8	-----	6	Return on Minus	If S = 1 (12~s)
RNC	D0	-----	6	Return on No Carry	If CY = 0 (12~s)
RNZ	C0	-----	6	Return on No Zero	If Z = 0 (12~s)
RP	F0	-----	6	Return on Plus	If S = 0 (12~s)
RPE	E8	-----	6	Return on Parity Even	If P = 1 (12~s)
RPO	E0	-----	6	Return on Parity Odd	If P = 0 (12~s)
RZ	C8	-----	6	Return on Zero	If Z = 1 (12~s)
RLC	07	----*	4	Rotate Left Circular	A = A <−
RRC	0F	----*	4	Rotate Right Circular	A = −> A
RST z	C7	-----	12	Restart	(3X7) − [SP] = PC,PC = z
SBB r	9F	*****	4	Subtract with Borrow	A = A − r − CY
SBB M	9E	*****	7	Subtract with Borrow	A = A − [HL] − CY
SBI n	DE	*****	7	Subtract with Borrow Immed	A = A − n − CY
SHLD a	22	-----	16	Store HL Direct	[a] = HL
SIM	30	-----	4	Set Interrupt Mask	mask = A
SPHL	F9	-----	6	Move HL to SP	SP = HL
STA a	32	-----	13	Store Accumulator	[a] = A
STAX B	02	-----	7	Store Accumulator indirect	[BC] = A

MNEMONIC	OPCODE	SZAPC	~S	DESCRIPTION	NOTES
STAX D	12	- - - - -	7	Store Accumulator indirect	[DE] = A
STC	37	- - - - I	4	Set Carry	CY = I
SUB r	97	*****	4	Subtract	A = A − r (22X)
SUB M	96	*****	7	Subtract Memory	A = A − [HL]
SUI n	D6	*****	7	Subtract Immediate	A = A − n
XCHG	EB	- - - - -	4	Exchange HL with DE	HL <–> DE
XRA r	AF	**0*0	4	Exclusive OR Accumulator	A = Axr (25X)
XRA M	AE	**0*0	7	Exclusive OR Accumulator	A = Ax [HL]
XRI n	EE	**0*0	7	Exclusive OR Immediate	A = Axn
XTHL	E3	- - - - -	16	Exchange stack Top with HL	[SP] <–> HL

Notes:

	S Z A P C	
PSW	0 * 0 1	Flag unaffected/affected/reset/set
S	S	Sign (Bit 7)
Z	Z	Zero (Bit 6)
AC	A	Auxilary Carry (Bit 4)
P	P	Parity (Bit 2)
CY	C	Carry (Bit 0)

a p	Direct addressing
M z	Register indirect addressing
n nn	Immediate addressing
r	Register addressing

DB n (,n)	Define Byte(s)
DB 'string'	Define Byte ASCII character string
DS nn	Define Storage Block
DW nn (,nn)	Define Word(s)

A B C D E F L	Registers (8-bit)
BC DE HL	Register pairs (16-bit)
PC	Program Counter register (16-bit)
PSW	Processor Status Word (8-bit)
SP	Stack Pointer register (16-bit)

a nn	16-bit address/data (0 to 65535)
n p	8-bit data/port (0 to 255)
r	Register (X = B, C, D, E, H, L, M, A)
z	Vector (X = 0H, 8H, 10H, 18H, 20H, 28H, 30H, 38H)

+ −	Arithmetic addition/subtraction
& ~	Logical AND/NOT
v x	Logical inclusive/exclusive OR
< – – >	Rotate left/right
< – >	Exchange
[ ]	Indirect addressing
[ ] + − [ ]	Indirect address auto-inc/decrement
{ }	Combination operands
(X)	Octal op code where X is a 3-bit code
If (~s)	Number of cycles if condition is true

Source: From J. P. Bowen, 1985 Programming Research Group, Oxford University Computing Laboratory.

not much thought is needed, since the microprocessor and all the major hardwiring is already taken care on the motherboard. In all, the microprocessor is simply too difficult and impractical for most gadget-oriented applications.

When designing programmable gadgets, it is best to avoid microprocessors entirely. Instead, you should use a microcontroller. The microcontroller is a specialized microprocessor that houses much of the support circuitry onboard, such as ROM, RAM, serial communications ports, A/D converters, etc. In essence, a microcontroller is a minicomputer, but without the monitor, keyboard, and mouse. They are called *microcontrollers* because they are small (micro) and because they control machines, gadgets, etc. With one of these devices, you can build an "intelligent" machine, write a program on a host computer, download the program into the microcontroller via the parallel or serial port of the PC, and then disconnect the programming cable and let the program run the machine. For example, in the microwave oven, a single microcontroller has all the essential ingredients to read from a keypad, write information to the display, control the heating element, and store data such as cooking time.

There are literally thousands of different kinds of microcontrollers available. Some are one-time-programmable (OTP), meaning that once a program is written into its ROM (OTP-ROM), no changes can be made to the program thereafter. OTP microcontrollers are used in devices such as microwaves, dishwashers, automobile sensor systems, and many application-specific devices that do not require changing the core program. Others microcontrollers are reprogrammable, meaning that the microcontroller's program stored in ROM (which may either be an EPROM, EEPROM, or flash) can be changed if desired—a useful feature when prototyping or designing test instruments that may require future I/O devices.

Microcontrollers are found in pagers, bicycle light flashers, data loggers, toys (e.g., model airplanes and cars), antilock breaking systems, VCRs, microwave ovens, alarm systems, fuel injectors, exercise equipment, etc. They also can be used to construct robots, where the microcontroller acts as the robot's brain, controlling and monitoring various input and output devices, such as light sensors, stepper and servo motors, temperature sensors, speakers, etc. With a bit of programming, you can make the robot avoid objects, sweep the floor, and generate various sounds to indicate that it has encountered difficulties (e.g., low power, tipped over, etc.) or has finished sweeping. The list of applications for microcontrollers is endless.

K.2.1 Basic Structure of a Microcontroller

Figure K.3 shows the basic ingredients found within many microcontrollers. These include a CPU, ROM (OTP-ROM, EPROM, EEPROM, FLASH), RAM, I/O ports, timing circuitry/leads, interrupt control, a serial port adapter (e.g., UART, USART), and a analog-digital (A/D, D/A) converter.

The CPU is equivalent to a microprocessor (often referred to as an *imbedded processor*)—it is the "thinking" element within the microcontroller. The CPU retrieves program instructions that the user programs into ROM, while using RAM to store temporary data needed during program execution. The I/O ports are used to connect external devices that send or receive instructions to or from the CPU.

A very simplist view of the basic components of a microcontroller

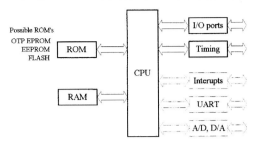

Possible ROM's
OTP EPROM
EEPROM
FLASH

ROM

RAM

CPU

I/O ports

Timing

Interupts

UART

A/D, D/A

FIGURE K.3

Possible internal architectures: RISC, SISC, CISC, Harvard, Von-Neuman

The serial port adapter is used to provide serial communications between the microcontroller and a PC or between two microcontrollers. It is responsible for controlling the different rates of data flow common between devices. Example serial port adapters found within microcontrollers are the UART (universal asynchronous receiver transmitter) or the USART (universal synchronous/asynchronous receiver transmitter). The UART can handle asynchronous serial communications, while the USART can handle either asynchronous or synchronous serial communications.

An interrupt system is used to interrupt a running program in order to process a special routine called the *interrupt service routine*. This boils down to the ability of a microcontroller to sample external data that requires immediate attention, such as data conveyed by an external sensor indicating important shutdown information, say, when things get too hot, objects get too close, etc. A timer/counter is used to "clock" the device—provide the driving force needed to move bits around. There are a number of microcontrollers that come with built-in A/D and D/A converters that can be used to interface with analog transducers, such as temperature sensors, strain gauges, position sensors, etc.

Example Microcontrollers: PIC16C56 and PIC16C57

Figure K.4 shows Microchip's PIC16C56 and PIC16C57 microcontrollers. As you can see in the internal architecture diagram, both microcontrollers house onchip CPU, EPROM, RAM, and I/O circuitry. The architecture is based on a register file concept that uses separate buses and memories for programs and data (Harvard architecture). This allows execution to occur in parallel. As an instruction is being "prefetched," the current instruction is executing on the data bus.

The PIC16C56's program memory (EPROM) has space for 1024 words, while the PIC16C57 has space for 2048 words. An 8-bit-wide ALU contains one temporary working register and performs arithmetic and Boolean functions between data held in the working register and any file register. The ALU and register file are composed of up to 80 addressable 8-bit registers, and the I/O ports are connected via the 8-bit-wide data bus. Thirty-two bytes of RAM are directly addressable, while the access to the remaining bytes work through bank switching.

In order for bit movement to occur (clock generation), the PIC controllers require a crystal or ceramic resonator connected to pins OSC1 and OSC2. The PIC microcontrollers reach a performance of 5 million instructions per second (5 MIPS) at a clock

Microchip's PIC16C56 and PIC16C57 microcontrollers

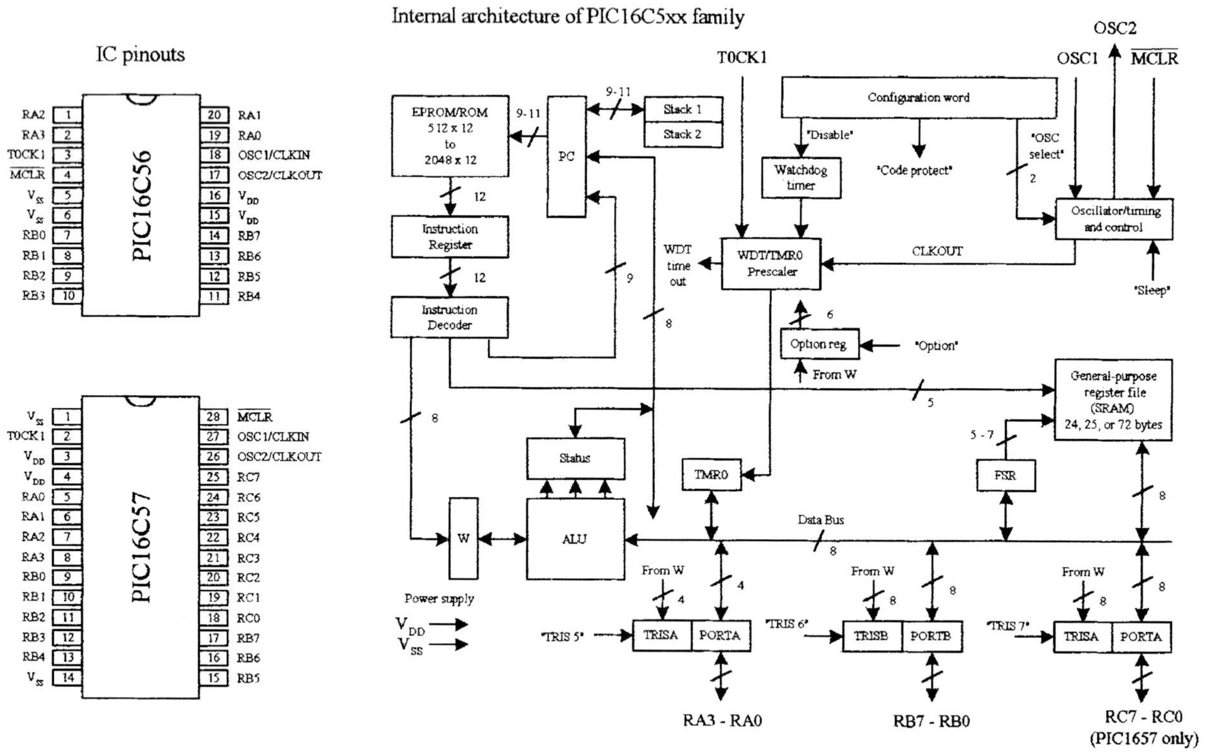

Internal architecture of PIC16C5xx family

IC pinouts

PIC16C56

RA2	1		20	RA1
RA3	2		19	RA0
T0CK1	3		18	OSC1/CLKIN
$\overline{MCLR}$	4		17	OSC2/CLKOUT
V_{SS}	5		16	V_{DD}
V_{SS}	6		15	V_{DD}
RB0	7		14	RB7
RB1	8		13	RB6
RB2	9		12	RB5
RB3	10		11	RB4

PIC16C57

V_{SS}	1		28	MCLR
T0CK1	2		27	OSC1/CLKIN
V_{DD}	3		26	OSC2/CLKOUT
V_{DD}	4		25	RC7
RA0	5		24	RC6
RA1	6		23	RC5
RA2	7		22	RC4
RA3	8		21	RC3
RB0	9		20	RC2
RB1	10		19	RC1
RB2	11		18	RC0
RB3	12		17	RB7
RB4	13		16	RB6
V_{SS}	14		15	RB5

[Internal architecture block diagram with labels: T0CK1, OSC2, OSC1, MCLR, EPROM/ROM 512 x 12 to 2048 x 12, PC, Stack 1, Stack 2, Configuration word, "Disable", "Code protect", "OSC select", Watchdog timer, Oscillator/timing and control, Instruction Register, 12, WDT time out, WDT/TMR0 Prescaler, CLKOUT, "Sleep", Instruction Decoder, 12, 9, 8, Option reg, 6, "Option", From W, General-purpose register file (SRAM) 24, 25, or 72 bytes, 5, Status, TMR0, 5-7, FSR, 8, W, ALU, Data Bus, 8, Power supply V_{DD} V_{SS}, From W, 4, "TRIS 5", TRISA, PORTA, From W, 8, "TRIS 6", TRISB, PORTB, From W, 8, "TRIS 7", TRISA, PORTA, RA3 - RA0, RB7 - RB0, RC7 - RC0 (PIC1657 only)]

FIGURE K.4

frequency of 20 MHz. A watchdog timer is also included, which is a free-running onchip *RC* oscillator that works without external components. It continues running when the clock has stopped, making it possible to generate a reset independent of whether the controller is working or sleeping.

These chips also come with a number of I/O pins that can be linked to external devices such as light sensors, speakers, LEDs, or other logic circuits. The PIC16C56 comes with 12 I/O pins that are divided into three ports, port A (RA3–RA0), port B (RB7–RB0), and port C (RC7–RC0); the PIC16C57 comes with eight more I/O pins than the PIC16C56.

K.2.2 Programming the Microcontroller

Like microprocessors, microcontrollers use a set of machine-code instructions (1's and 0's) to perform various tasks such as adding, comparing, sampling, and outputting data via I/O ports. These machine-code instructions are typically programmed into onboard ROM (EPROM, EEPROM, flash) via a programming unit linked to a personal computer (PC). The actual programming, however, isn't written out in machine code but rather is written out in high-level language within an editor program running on the PC. The high-level language used may be a popular language such as C or a specially tailored language that the manufacturer has created to optimize all the features present within its microcontrollers.

Using a manual and software you get from the manufacturer, you learn to write out humanlike statements that tell the microcontroller what to do. You type the statements in an editor program, run the program, and check for syntax errors. Once you think the program is OK, you save it and run a compiler program to translate it into machine language. If there is an error in your program, the compiler may refuse to perform the conversion. In this case, you must return to the text editor and fix the bugs before moving on. Once the bugs are eliminated and the program is compiled successfully, a third piece of software is used to load the program into the microcontroller. This may require physically removing the microcontroller from the circuit and placing into a special programmer unit linked to the host PC via a serial or parallel port.

Now there is another way to do things, which involves using an interpreter instead of a compiler. An interpreter is a high-level language translator that does not reside in the host PC but resides within the microcontroller's ROM. This often means that an external ROM (EPROM, EEPROM, flash) is needed to store the actual program. The interpreter receives the high-level language code from the PC and, on the spot, interprets the code and places the translated code (machine code) into the external ROM, where it can be used by the microcontroller. Now this may seem like a waste of memory, since the interpreter consumes valuable onchip memory space. Also, using an interpreter significantly slows things down—a result of having to retrieve program instructions from external memory. However, when using an interpreter, a very important advantage arises. By having the interpreter onboard to translate on the spot, an immediate, interactive relationship between host program and microcontroller is created. This allows you to build your program, immediately try out small pieces of code, test the code by downloading the chunks into the microcontroller, and then see if the specific chunks of code work. The host programs used to create the source code often come with debugging features that let you test to see where possible programming or hardwiring errors may result by displaying the results (e.g., logic state at a given I/O pin) on the computer screen while the program is executing within the microcontroller. This allows you to perfect specific tasks within the program, such as perfecting a sound-generation routine, a stepper motor control routine, and so forth.

K.2.3   BASIC Stamps (Microcontroller with Interpreter and Extra Goodies)

The BASIC Stamp is, at the heart, a microcontroller with interpreter software built in. These devices also come with additional support circuitry, such as an EEPROM, voltage regulator, ceramic oscillator, etc. BASIC Stamps are ideal for beginners because they are easy to program, quite powerful, and relatively cheap—a whole startup package costs around $150 dollars or so. These devices are also very popular among inventors and hobbyist, and you'll find a lot of helpful literature, application notes, and fully tested projects on the Internet.

The original stamp was introduced in 1993 by Parallax, Inc. It got its name from the fact that it resembled a postage stamp. The early version of the BASIC Stamp was the REV D, while later improvements lead to the BASIC Stamp I (BSI) and to the BASIC Stamp II (BSII). Here we'll focus mainly on the BSI and the BSII. Both the BSI and BSII have a specially tailored BASIC interpreter firmware built into the microcontroller's EPROM. For both stamps, a PIC microcontroller is used. The actual program that is to be run is stored in an onboard EEPROM. When the battery is connected, stamps run the BASIC program in memory. Stamps can be reprogrammed at any time by temporarily connecting them to a PC running a simple host program. The new program is typed in, a key is hit, and the program is loaded into the stamp. Input/output pins can be connected with other digital devices such as sense switches, LED, LCD displays, servos, stepper motors, etc.

BASIC Stamp II (BSII-IC)

The BSII is a module that comes in a 28-pin DIL package (see Fig. K.5). The brain of the BSII is the PIC16C57 microcontroller that is permanently programmed with a PBASIC2 instruction set within its internal OTP-EPROM (one-time program ROM). When programming the BSII, you tell the PIC16C57 to store symbols, called *tokens*, in external EEPROM memory. When the program runs, the PIC16C57 retrieves tokens from memory, interprets them as PBASIC2 instructions, and carries out those instructions. The PIC16C57 can execute its internal program at a rate of 5 million machine instruction per second. However, each PBASIC2 instruction takes up many machine instructions, so the PBASIC2 executes more slowly, around 3000 to 4000 instructions per second.

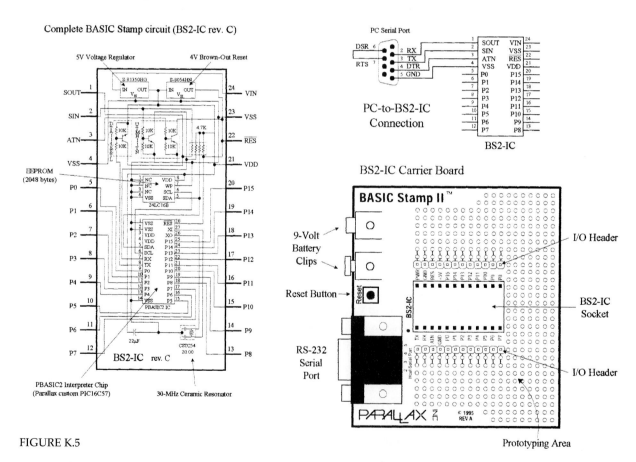

FIGURE K.5

The BSII comes with 16 I/O pins (P0–P15) that are available for general use by your programs. These pins can be interfaced with all modern 5-V logic, from TTL through CMOS (technically, they have characteristics like the 74HCT logic series). The direction of a pin—either input or output—is set during the programming phase. When a pin is set as an output pin, the BSII can send signal to other devices, like

LEDs, servos, etc. When a pin is set as an input pin, it can receive signals from external devices, such as switches, photosensors, etc. Each I/O pin can source 20 mA and sink 25 mA. Pins P0–P7 and pins P8–P15, as groups, can each source a total of 40 and sink 50 mA.

2048-Byte EEPROM

The BSII's PIC's internal OTP-EPROM (one-time programmable read-only memory) is permanently programmed at the factory with Parallax's firmware which turns this memory into a PBASIC2 interpreter chip. Because they are interpreters, the Stamp PICs have the entire PBASIC language permanently programmed into their internal program memory. This memory cannot be used to store your PBASIC2 program. Instead, the main program must be stored in the EEPROM (electrically erasable, programmable read-only memory). This memory retains data without power and can be reprogrammed easily. At run time, the PBASIC2 program created on the host computer is loaded into the BSII's EEPROM starting at the highest address (2047) and working downward. Most programs do not use the entire EEPROM, which means that PBASIC2 lets you store data in the unused lower portion of the EEPROM. Since programs are stored from the top of the memory downward, data are stored in the bottom of the memory working upward. If there is an overlap, the Stamp host software will detect this problem and display an error message.

Reset Circuit

The BSII comes with a reset circuit. When power is first connected to the Stamp, or if it falters due to a weak battery, the power supply voltage can fall below the required 5 V. During such brownouts, the PIC is in a voltage-deprived state and will have the tendency to behave erratically. For this reason, a reset chip is incorporated into the design, forcing the PIC to rest to the beginning of the program and hold until the supply voltage is within acceptable limits.

Power Supply

To avoid supplying the BSII with unregulated supply power, a 5-V regulator is incorporated into the BSII. This regulator accepts a voltage range from slightly over 5 V up to 15 V and regulates it to a steady 5 V. It provides up to 50 mA. The regulated 5 V is available at output V_{DD}, where it can be used to power other parts of your circuits so long as no more than 50 mA is required.

Connecting BSII to a Host PC

To program a Stamp requires connecting it to a PC that runs host software to allow you to write, edit, download, and debug PBASIC2 programs. The PC communicates with the BSII through an RS-232 (COM port) interface consisting of pins S_{IN}, S_{OUT}, and *ATM* (serial in, serial out, and attention, respectively). During programming, the BSII host program pulses *ATM* high to reset the PIC and then transmits a signal to the PIC through S_{IN} indicating that it wants to download a new program. PC-to-BSII connector hookup is shown in Fig. K.5. This connection allows the PC to reset the BSII for programming, download programs, and receive debug data from the BSII. The additional pair of connections, pin 6 and 7 of the DB9 socket, lets the BSII host software identify the port to which the BSII is connected. Usually, when programming a BSII, you use a special BSII carrier board, which comes with a prototyping area, I/O header, BSII-IC socket, 9-V battery clips, and an RS-232 serial port connector, as shown in Fig. K.5. These boards, along with programming cable and software, can be purchased as startup packages.

PBASIC Language

Even though the BASIC Stamp has "BASIC" in its name, it cannot be programmed in Visual BASIC or QBASIC. It does not have a graphic user interface, a hard drive, or a lot of RAM. The BASIC Stamp must only be programmed with Parallel's BASIC (PBASIC), which has been specifically designed to exploit all the BASIC Stamp's capabilities. PBASIC is a hybrid form of BASIC programming language with which many people are familiar. Currently, there are two version of PBASIC: PBASIC1 for the BASIC Stamp I and PBASIC2 for the BASIC Stamp II. Each version is specifically tailored to take advantage of the inherent features of the hardware it runs on. PBASIC is called a *hybrid* because, while it contains some simplified forms of normal BASIC control constructs, it also has special commands to efficiently control I/O pins. PBASIC is an easy language to master and includes familiar instructions such as GOTO, FOR . . . NEXT, and IF . . . THEN. It also contains Stamp-specific instructions, such as PULSOUT, DEBUG, BUTTON, etc., which will be discussed shortly.

The actual program to be downloaded into the Stamp is first written using the BASIC Stamp I (for BSI) or BASIC Stamp II (for BSII) editor software running on an IBM-compatible PC or MacIntosh that is running SoftPC or SoftWindows 2.0. After you write the code for your application, you simply connect the Stamp to the computer's parallel port (for BSI) or serial port (for BSII), provide power to the Stamp, and press ALT-R within the editor program to download the code into the Stamp. As soon as the program has been downloaded successfully, it begins executing its new program from the first line of code.

The size of the program that can be stored in a Stamp is limited. For the BSI, there are 256 bytes of program storage, enough for around 80 to 100 lines of PBASIC code. For the BSII, there are 2048 bytes worth of program space, enough for around 500 to 600 lines of PBASIC code. The amount of program memory for the Stamps cannot be expanded, since the interpreter chip (PIC) expects the memory to be specific and fixed in size. However, in terms of data memory, expansion is possible. It is possible to interface EEPROMs or other memory devices to the Stamp's I/O pins to gain more data storage area. This requires that you supply the appropriate code within your PBASIC program to make communication between the Stamp and external memory device you choose possible. Addition data memory is often available with Stamp-powered applications that monitor and record data (e.g., environmental field instrument).

Debugging

To debug PBASIC programs, the BASIC Stamp editor comes with two handy features—syntax checking and a DEBUG command. Syntax checking alerts you to any syntactical error and is automatically performed on your code the moment you try to download to the BASIC Stamp. Any syntax errors will cause the download process to abort and will cause the editor to display an error message, pointing out the error in the source code. The DEBUG command, unlike syntax checking, is an instruction that is written into the program to find logical errors—ones that the Stamp does not find, but ones that the designer had not intended. DEBUG operates similar to the PRINT command in the BASIC language and can be used to print the current status of specific variable within your PBASIC program as it is executed within the BASIC stamp. If your PBASIC code includes a DEBUG command, the editor opens up a special window at the end of the download process to display the result for you.

Overview of PBASIC II Programming Language

The PBASIC II language, like other high-level computer languages, involves defining variables and constants and using address labels, mathematical and binary operators, and various instructions (e.g., branching, looping, numerics, digital I/O, serial I/O, analog I/O, sound I/O, EEPROM access, time, power control, etc.) Here's a quick rundown on the elements of the PBASIC II language.

COMMENTS Comments can be added within the program to describe what you're doing. They begin with an apostrophe (') and continue to the end of the line.

VARIABLES These are locations in memory that your program can use to store and recall numbers. These variables have limited range. Before a variable can be used in a PBASIC2 program, it must be declared. The common way used to declare variables is to use a directive VAR:

```
symbol   var   size
```

where the symbol can be any name that starts with a letter; can contain a mixture of letters, numbers, and underscore; and must not be the same as PBASIC keywords or labels used in the program. The size establishes the number of bits of storage the variable is to contain. PBASIC2 provides four sizes: bit (1 bit), nib (4 bits), byte (8 bits), and word (16 bits). Here are some examples of variable declarations:

```
'Declare variables.
sense_in     var bit      'Value can be 0 or 1.
speed        var nib      'Value in range 0 to 15.
length       var byte     'Value in range 0 to 255.
n            var word     'Value in range 0 to 65535.
```

CONSTANTS Constant are unchanging values that are assigned at the beginning of the program and may be used in place of the numbers they represent within the program. Defining constants can be accomplished by using the CON directive:

```
beeps        con   5        'number of beeps
```

By default, PBASIC2 assumes that numbers are in decimal (base 10). However, it is possible to use binary and hexadecimal numbers by defining them with prefixes. For example, when the prefix "%" in placed in front of a binary number (e.g., %0111 0111), the number is treated as a binary number, not a decimal number. To define a hexadecimal number, the prefix "$" is used (e.g., $EF). Also, PBASIC2 will automatically convert quoted text into the corresponding ASCII codes(s). For example, defining a constant as "A" will be interpreted as the ASCII code for A (65).

ADDRESS LABELS The editor uses address labels to refer to addresses (locations) within the program. This is different from some versions of BASIC, which use line numbers. In general, an address label name can be any combination of letters, numbers, and underscores. However, the first character in the label name cannot be a number, and the label name must not be the same as a reserved word, such as a PBASIC instruction or variable. The program can be told to go to the address label and follow whatever instructions are listed after. Address labels are indicated with a terminating colon (e.g., loop:).

MATHEMATICAL OPERATORS   BPASIC2 uses two types of operators: unary and binary. Unary operators take precedence over binary operator. Also, unary operations are always performed first. For example, in the expression 10 − SQR 16, the BSII first takes the square root of 16 and then subtracts it from 10.

Unary Operators

ABS   Returns absolute value

SQR   Returns square root of value

DCD 2^n-power decoder

NCD   Priority encoder of a 16-bit value

SIN   Returns two's compliment sine

COS   Returns two's compliment cosine

Binary Operators

+   Addition

−   Subtraction

/   Division

//   Remainder of division

*   Multiplication

**   High 16-bits of multiplication

*/   Multiply by 8-bit whole and 8-bit part

MIN   Limits a value to specified low

MAX   Limits a value to specified high

DIG   Returns specified digit of number

<<   Shift bits left by specified amount

>>   Shift bits right by specified amount

REV   Reverse specified number of bits

&   Bitwise AND of two values

| Bitwise OR of two values

&   Bitwise XOR of two values

PBASIC Instructions Used by BASIC Stamp II

BRANCHING

IF *condition* THEN *addressLabel*

Evaluate condition and, if true, go to the point in the program marked by *address-Label*. (Conditions: =, <> not equal, >, <1, > =, < =).

BRANCH *offset*, [*address0, address1, . . . addressN*]

Go to the address specified by offset (if in range).

GOTO *addressLabel*

Go to the point in the program specified by *addressLabel*.

GOSUB *addressLabel*

Store the address of the next instruction after GOSUB, then go to the point in the program specified by *addressLabel*.

RETURN

Return from subroutine.

LOOPING

FOR *variable* = *start* to *end* {*STEP stepVal*} . . . NEXT

Create a repeating loop that executes the program lines between For and Next, incrementing or decrementing *variable* according to *stepVal* until the value of the variable passes the *end* value.

NUMERICS

LOOKUP *index*, [*value0, value1, . . . valueN*], *resultVariable*

Look up the value specified by the index and store it in a variable. If the index exceeds the highest index value of the items in the list, variable is unaffected. A maximum of 256 values can be included in the list.

LOOKDOWN *value, {comparisonOp,}* [*value0, value1, . . . valueN*], *resultVariable*

Compare a value to a list of values according to the relationship specified by the comparison operator. Store the index number of the first value that makes the comparison true into *resultVariable*. If no value in the list makes the comparison true, *resultVariable* is unaffected.

RANDOM *variable*

Generates a pseudo-random number using byte or word variable that's bits are scrambled to produce a random number.

DIGITAL I/O

INPUT *pin*	Make the specified pin an input.
OUTPUT *pin*	Make the specified pin an output.
REVERSE *pin*	If pin is an output, make it an input. If pin is an input, make it an output.
LOW *pin*	Make specified pin's output low.
HIGH *pin*	Make specified pin's output high.
TOGGLE *pin*	Invert the state of a pin.
PULSIN *pin, state, resultVariable*	Measure the width of a pulse in 2-μs units.
PULSOUT *pin, time*	Outputs a timed pulse to by inverting a pin for some time ($\times$ 2 μs).
BUTTON *pin, downstate, delay,rate,bytevariable, targetstate, address*	Debounce button input, perform auto-repeat, and branch to address if button is in target state. Button circuits may be active-low or active-high.
SHIFTIN *dpin,cpin,mode, [result{\bits}{,result{\bits} . . . }]*	Shift data in from a synchronous-serial device.
SHIFTOUT *dpin,cpin,mode, [data{\bits}{,data{\bits} . . . }]*	Shift data out to a synchronous-serial device.
COUNT *pin, period, variable*	Count the number of cycles (0-1-0 or 1-0-1) on the specified pin during *period* number of milliseconds and store that number in *variable*.
XOUT *mpin,zpin, [house\keyORCommand{\cycles} {,house\keyOrCommand{\cycles} . . . }]*	Generate X-10 powerline control codes.

SERIAL I/O

SERIN *rpin{\fpin},baudmode, {plabe}{timeout,tlabe,}[input Data]*	Receive asynchronous serial transmission.
SEROUT *tpin,baudmode,{pace,} [outputData]*	Send data serially with optional byte pacing and flow control.

ANALOG I/O

PWM *pin, duty, cycles*	Output fast pulse-width modulation, then return pin to input. This can be used to output an analog voltage (0–5 V) using a capacitor and resistor.
RCTIME *pin, state, resultVariable*	Measure an *RC* charge/discharge time. Can be used to measurepotentiometers.

SOUND

FREQOUT *pin, duration, freq1{,freq2}*

Generate one or two sine-wave tones for a specified duration.

DTMFOUT *pin,{ontime,offtime,}{,tone . . . }*

Generate dual-tone, multifrequency tones (DTMF, i.e., telephone "touch" tones).

EEPROM ACCESS

DATA

Store data in EEPROM before downloading PBASIC program.

READ *location, variable*

Read EEPROM location and store value in variable.

WRITE *address,byte*

Write a byte of data to the EEPROM at appropriate address.

TIME

PAUSE *milliseconds*

Pause the program (do nothing) for the specified number of milliseconds. Pause execution for 0–65,535 ms.

POWER CONTROL

NAP *period*

Enter sleep mode for a short period. Power consumption is reduced to about 50 μA assuming no loads are being driven. The duration is $(2^{period}) \times 18$ ms.

SLEEP *seconds*

Sleep from 1–65,535 seconds to reduce power consumption by ~50 μA.

END

Sleep until the power cycles or the PC connects ~50 μA.

PROGRAM DEBUGGING

DEBUG *outputData{,outputData . . . }*

Display variables and messages on the PC screen within the BSII host program; *outputData* consists of one or more of the following: text strings, variables, constants, expressions, formatting modifiers, and control characters

Making a Robot Using the BASIC Stamp II

To illustrate how easy it is to make interesting gadgets using the BASIC Stamp II (BSII), let's take a look at a robot application. In this application, the main objective is to prevent the robot from running into objects. The robot aimlessly moves around, and when it comes close to an object, say, to the left, the robot is to stop and then back up and move off in another direction. In this example, a BSII acts as the robot's brain, two servos connected to wheels acts as its legs, a pair of infrared transmitters and sensors acts as its eyes, and a piezoelectric speaker acts as it voice. Figure K.6 shows the completed robot, along with the various individual components.

Components and connections used to create object-avoiding robot

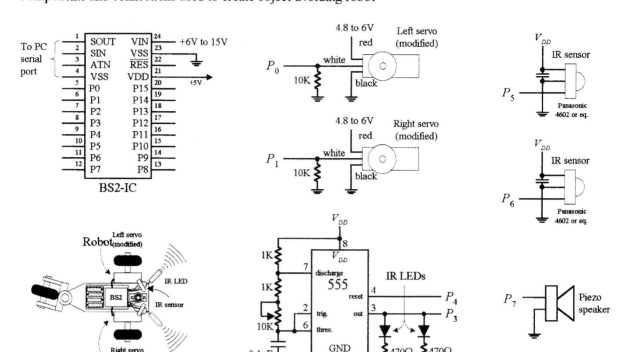

FIGURE K.6

THE SERVOS

The directional movement of the robot is controlled by right and left servo motors that have been modified so as to provide a full 360° worth of rotation—modifying a servo is discussed in Chap. 13. To control a servo requires generating pulses ranging from 1000 to 2000 μs in width at intervals of approximately 10 to 20 ms. With one of the servos used in our example, when the pulse width sent to the servo's control line is set to 1500 μs, the servo is centered—it doesn't move. However, if the pulse width is shortened to, say, 1300 μs, the modified servo rotates clockwise. Conversely, if the pulse width is lengthened, say, to 1700 μs, the modified servo rotates counterclockwise.

The actual control pulses used to drive one of the servos in the robot are generated by the BSII using the PULSOUT *pin, time1* and the PAUSE *time2* instructions. The *pin* represents the specific BSII pin that is linked to a servo's control line, while *time1* represents how long the pin will be pulsed high. Note that for the PULSOUT instruction, the decimal placed in the *time1* slot actually represents half the time, in microseconds (μs), that the pin is pulsed high. For example, PULSOUT 1, 1000 means that the BSII will pulse pin 1 high for 2000 μs, or 2ms. For the PAUSE instruction, the decimal placed in the *time2* slot represents a pause in milliseconds (ms). For example, PAUSE 20, represents a 20-ms pause. Figure K.7 shows sample BSII code used to generate

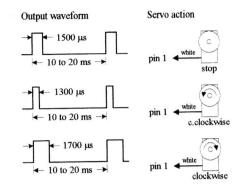

FIGURE K.7

```
BS2 code              Comments

pulsout 1, 750       'pulse width of 1500us on pin 1
pause 20             'pause for 20 ms


pulsout 1, 650       'pulse width of 1300us on pin 1
pause 20             'pause for 20 ms


pulsout 1, 850       'pulse width of 1700us on pin 1
pause 20             'pause for 20 ms
```

IR TRANSMITTERS AND RECEIVERS

The robot's object-detection system consists of a right and left set of infrared (IR) LED transmitters and IR detector modules. The IR LEDs are flashed via a 555 timer at a very high frequency, which in this example happens to be 38 kHz, 50 percent duty cycle. The choice of this frequency is used to avoid interference from other household sources of infrared, primarily incandescent lights. (Many types of IR LED transmitters and sensors could be used in this robot, and they may work best using a different frequency.) The IR photons emitted by the LED rebound off objects in the path of the robot and reflect back to the IR detector module. When a detector module receives photons, the I/O pin of the BSII connected to the module goes low. (Note that the BSII can only execute around 4000 instruction per second, while the number of pulses generated by the detector module is 38,000. In this case the actual number of pulses received by the BSII will be less, around 10 or 20.)

PIEZOELECTRIC SPEAKER

A piezoelectric speaker is linked to one of the BSII I/O terminals and is used generate different sounds when the robot is moving forward or backing up. To provide the piezoelectric speaker with a sinusoidal waveform needed to generate sound, the FREQOUT *pin, time, frequency* instruction is used. The instruction FREQOUT 7, 1000, 440 creates a 440-Hz sinusoidal frequency on pin 7 that lasts for 1000 ms.

THE PROGRAM

The following is a program used to control the robot. It is first created using the PBA-SIC2 host software and then downloaded into the BSII during runtime.

```
'Program for object-avoiding robot

'Define variables and constants
'--------------------------------------------------------------------------------
n             var word        'n acts as a variable that changes.
right_IR      var in5         'Sets pin 5 as an input for right IR detector.
left-IR       var in6         'Sets pin 6 as an input for left IR detector.
right_servo   con 0           'Assigns 0 which will be used to identify right servo.
left_servo    con 1           'Assigns 1 to identify left servo.
IR_out        con 3           'Assigns 3 to identify IR output.
delay         con 10          'A constant that we be used in program.
speed         con 100         'Used to set servo speed.
turn_speed    con 50          'Used to set turn speed of robot.

'Main program
'--------------------------------------------------------------------------------
high IR_out                               'Sets pin 6 "high"
pause 50                                  'Pauses for 50 milliseconds
sense:                                    'Label used to specify IR-sense routine.
if left_IR = 0 and right_IR = 0 then backup   'Object in path, jump to back_up routine.
if left_IR = 0 then turn_right            'Object on left side, jump to turn_right routine.
if right_IR = 0 then turn_left            'Object on right, jump to "turn_left" routine.

'Sound Routines
'--------------------------------------------------------------------------------
forward_sound:                            'Label
freqout 7,1000, 440                       'Generate 1000ms, 440 Hz tone on pin 7

back_sound:                               'Label
freqout 7,1000,880                        'Generate 1000ms, 880 Hz tone on pin 7

'Motion routines
'--------------------------------------------------------------------------------
forward:                                  'Label used to specify forward routine.
gosub forward_sound                       'Tells program to jump to forward sound subroutine.
debug "forward"                           'Tells stamp to display the word "forward" on debug window.
pause 50                                  'Pause for 50ms.
for n = 1 to delay*2                       'For…Next loop that starts x = 1 and repeats until x = 20.
pulsout left_servo, 750-speed             'Make left servo spin to make robot move forward.
pulsout right_servo, 750+speed            'Make right servo spin to make robot move forward.
pause 20                                  'Pauses for 20ms, path of servo control.
next                                      'End of For…Next loop.
goto sense                                'Once forward routine is finished go back to sense routine.

backup:                                   'Label used to specify back-up routine.
gosub backup_sound                        'Tells program to jump back-up sound subroutine.
debug "backward"                          'Displays "backward" on the debug window.
pause 50                                  'Pause for 50ms to ensure
for n = 1 to delay*3                       'For…Next loop that starts x = 1 and repeats until x = 60
```

```
pulsout left_servo, 750+speed        'Makes left servo spin to make robot move backward.
pulsout right_servo, 700-speed       'Makes right servo spin to make robot move backward.
pause 20                             'Pauses for 20ms, part of servo control.
next                                 'End of For…Next loop.

turn_left:                           'Label used to specify turn-left routine.
debug "left"                         'Displays "left" on the debug window.
pause 50                             'Pause for 50ms.
for x = 1 to delay*1                 'For…Next loop that starts x = 1 and repeats until x = 10.
pulsout left_servo, 750-turn_speed   'Makes left servo spin to make robot turn left.
pulsout right_servo, 700-turn_speed  'Makes right servo spin to make servo spin left.
pause 20                             'Pause for 20ms, part of servo control.
next                                 'End of For…Next loop.
goto sense                           'Once left-turn routine is finished, jump back to sense.

turn_right:                          'Label used to specify turn-right routine.
debug "right"                        'Displays "right" on debug window.
pause 50                             'Pause for 50 ms.
for x = 1 to delay*1                 'For…Next loop.
pulsout left_servo, 750+turn_speed   'Makes left servo spin to make robot turn right.
pulsout right_servo, 750+turn_speed  'Makes right servo spin to make robot turn right.
pause 20                             'Pause for 20 ms, part of servo control.
next                                 'End of For…Next loop
goto sense                           'Once right-turn is finished, jump back to sense.
```

BASIC Stamp I (BSI-IC)

Figure K.8 shows the BASIC Stamp I (BSI), which was mentioned briefly earlier. This device is actually the predecessor of the BSII, but it is still used frequently enough that it is worth mentioning here. The BSI contains most of the same features as the BSII, but not all. It has only 8 I/O pins instead of 16, and it uses a PIC16C56 instead of a PIC16C57. It also uses a PBASIC1 programming language instead of PBASIC2, and its link with the host computer is via the parallel port instead of the serial port. It has a smaller instruction set, is a bit slower, and doesn't have as many variables for RAM.

Things Needed to Get Started with the BASIC Stamp

These include programming software, programming cable, manual, the BASIC Stamp module, and an appropriate carrier board (optional). If you are interested in using a particular Stamp, either the BASIC Stamp I or II, purchase either startup kit. These kits include all five items listed, at a lower cost than purchasing each part separately. However, if you intend to use both BASIC Stamp I and II, it is best to purchase the BASIC Stamp Programming Package (which includes manual, software, and cables for both versions of the Stamp) and then purchase the BASIC Stamp modules and optionally the carrier boards separately.

Learning More about BASIC Stamps

To fully understand all the finer details needed to program BASIC Stamps, it is necessary to read through the user's manual. However, reading the user's manual alone

Complete BASIC Stamp circuit (BS1-IC rev. A)

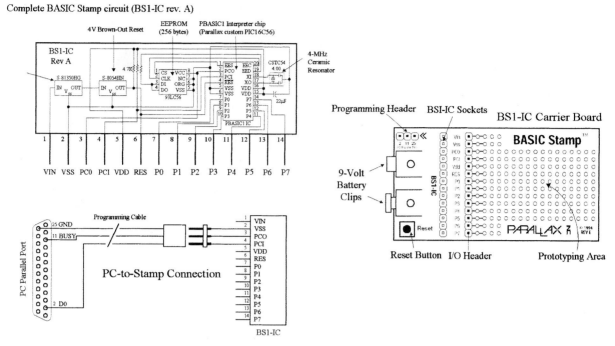

FIGURE K.8

tends not to be the best learning strategy—it is easy to lose your place within all the technical terms, especially if you are a beginner. If you are a beginner, I recommend visiting *www.stampsinclass.com*. At this site you'll find a series of tutorials, along with the Stamp user manual, that you can download. The tutorials are written is plain English and assume no prior knowledge of microcontrollers (or electronics for that matter). Another good source to learn more about BASIC Stamps is the book *Programming and Customizing the Basic Stamp Computer,* written by Scott Edwards (McGraw-Hill). This book is also geared toward beginners and is easy reading.

Thinking about Mass Production

Recall that the major components of the BASIC Stamp circuit are the PIC (houses CPU and ROM for storing PBASIC interpreter), external EEPROM (stores program), and the resonator. In large-scale runs it would be nice to get rid of the external memory and remove the interpreter program and simply download a compiled PBASIC code directly into the PIC—this saves space and money. As it turns out, the BASIC Stamp editor software includes a feature to program PBASIC code directly into a PIC microcontroller using Parallax's PIC16Cxx programmer. (The major benefit for starting out with the Stamp is that you can easily fine-tune your code, test out chunks, and immediately see if it works—an important feature when creating prototypes. When prototyping with a PIC, however, checking for errors is much harder because you must compile everything at once—you can't test out chunks of code.)

K.2.4 Other Microcontrollers Worth Considering

There are thousands of different microcontrollers on the market made by a number of different manufacturers. Each manufacturer has a number of different devices, each designed with different features, making them suitable for different applications.

Perhaps the most popular microcontrollers used by hobbyists are Microchip's PICs. These microcontrollers are relatively simple to program, inexpensive, and have loads of development software available. There is a very complete line of CMOS PIC microcontrollers with varying features.

Microchip supplies a PICSTART Plus programmer that is an entry-level development kit that will support PIC12C5XX, PIC12CX, PIC16C5XX, PIC12C6X, PIC16C, PIC17C, PIC14000, PIC17C4X, and PIC17C75X. The PICSTART Plus programming kit includes RS232 cable, power supply, PIC16C84 EEPROM microcontroller sample, software, and manual. Check out Microchip's Web site for more details.

Other major manufacturers of microcontrollers include Motorola, Hitachi, Intel, NEC, Philips, Toshiba, Texas Instruments, National Semiconductor, Mitsubishi, Zilog, etc. Here are some sample microcontrollers worth mentioning.

8051 (Intel and Others)

The 8051 is a very popular microcontroller that comes with modified Harvard architecture (separate address spaces for program, memory and data memory) and has comes with 64 K worth of program memory. This chip is very powerful and easy to program, and there is loads of development software, both commercial and freely available. It is often featured in construction projects in the popular hobbyist magazines.

8052AH-BASIC

This is popular with hobbyists, and like the BASIC Stamp, it is easy to work with.

68HC11 (Motorola)

This is a popular 8-bit controller which, depending on the variety, has built-in EEPROM/OTPROM, RAM, digital I/O, timers, A/D converter, PWM generator, pulse accumulator, and synchronous and asynchronous communications channels.

COP800 Family (National Semiconductor)

Basic family is fully static 8-bit microcontroller, which contains system timing, interrupt logic, ROM, RAM, and I/O. Depending on the device, features include 8-bit memory-mapped architecture, serial I/O, UART, memory-mapped I/O, many 16-bit timer/counters, a multisourced vectored interrupt, comparator, watchdog timer and clock monitor, modulator/timer, 8-channel A/D converter, brownout protection, halt mode, idle mode, and high-current I/O pins. Most within the family operate over a 2.5- to 6.0-V range.

DS5000/DS2250 (Dallas Semiconductor)

All you need to add is a crystal and two capacitors to end up with a working system. These chips come complete with nonvolatile RAM.

TMS370 (Texas Instruments)

Similar to the 8051 in having 256 registers, A and B accumulators, stack in the register page, etc. The peripherals include RAM ROM (mask, OTP, or EEPROM), two timers, SCI (synchronous serial port), SPI (asynchronous serial port), A/D (8-bit, 8-chan), and interrupts.

K.2.5   Evaluation Kits/Board

Many manufacturers offer assembled evaluation kits or board which usually allow you to use a PC as a host development system, as we saw with Parallax BASIC Stamp. Among some of the other popular evaluation kits/boards are the following.

Motorola's EVBU, EVB, EVM, EVS

A series of very popular evaluation/development systems based on the 68HC11. Comes complete with BUFFALO monitor and varying types of development software. Commonly used in university courses.

Motorola 68705 Starter Kit

Motorola supplies a complete development system—software, hardware, simulator, emulator, manuals, etc.—for about 100 dollars.

National Semiconductor's EPU

The COP8780 evaluation/programming unit (EPU) offers you a low-cost ($125) tool for an introduction to National's COP BASIC family of 8-bit microcontrollers. The system includes the EPU board, assembler and debugger software, sample code, C compiler, wall power supply, documentation, etc.

INDEX

Supplementary Index of ICs

7400 Series ICs

4000 Series ICs

Other ICs

ABOUT THE AUTHOR

Paul Scherz is a physicist/mechanical engineer who received his B.S. in physics from the University of Wisconsin. His area of interest in physics currently focuses on elementary particle interactions, or high-energy physics, and he is working on a new theory on the photon problems with Nikolus Kauer (Ph.D. in high-energy physics, Munich, Germany). Paul is an inventor/hobbyist in electronics, an area he grew to appreciate through his experience at the University's Department of Nuclear Engineering and Engineering Physics and the Department of Plasma Physics.